Modern Materials and Manufacturing (MMM 2021)

IOP Conference Series: Materials Science and Engineering
Volume 1140

Tallinn, Estonia
27 - 29 April 2021

Editors:

Kristo Karjust
Tauno Otto
Jakob Kubarsepp
Irina Hussainova

ISBN: 978-1-7138-4617-8
ISSN: 1757-8981

Printed from e-media with permission by:

Curran Associates, Inc.
57 Morehouse Lane
Red Hook, NY 12571

Some format issues inherent in the e-media version may also appear in this print version.

Institute of Physics
Dirac House, Temple Back
Bristol BS1 6BE UK

Phone: 44 1 17 929 7481
Fax: 44 1 17 920 0979

techtracking@iop.org

Additional copies of this publication are available from:

Curran Associates, Inc.
57 Morehouse Lane
Red Hook, NY 12571 USA
Phone: 845-758-0400
Fax: 845-758-2633
Email: curran@proceedings.com
Web: www.proceedings.com

TABLE OF CONTENTS

IOP Conf. Series: Materials Science and Engineering 1140 (2021) 011001 doi:10.1088/1757-899X/1140/1/011001

Preface

In this volume of IOP Conference Series: Materials Science and Engineering contains papers contributed to the conference of "Modern Materials and Manufacturing" – MMM2021 held on April 27 – 29, 2021 in Tallinn, Estonia. The conference was organized by Tallinn University of Technology (TalTech). Selected papers of conference are published in special issue of Proceedings of the Estonian Academy of Sciences and journal Mechanics of Composite Materials, prepared with participation of PhD students.

The conference on Modern Materials and Manufacturing is an international forum focused on Industry 4.0 and Manufacturing Engineering, Materials Engineering and Tribology, and brings together materials and manufacturing experts from the universities and industry. It encourages new collaborations between different partners not only from academia, but also from national competence centers, industry and funding bodies.

Due to the high quality of presentations and subsequent networking opportunities, this event, hosting combined 13th International DAAAM Baltic Conference and 29th International Baltic Conference BALTMATTRIB 2021, has achieved worldwide interest and active participation from researchers from national and global technical communities.

The technical program contained presentations results of the scientific and engineering researches in important areas of materials engineering as well as materials processing technologies with particular emphasis on the current trends in the development and application of advanced materials.

The organizing committee believe this issue will be a valuable resource, not only for the participants of MMM2021 Conference, but also for the whole research community in the field of materials and manufacturing engineering.

Scientific Committee of DAAAM Baltic 2021

Prof. Tauno Otto, Chair (Tallinn University of Technology, Estonia)
Prof. Dago Antov (Tallinn University of Technology, Estonia)
Prof. Metin Aydogdu (Trakya University, Turkey)
Dr. Rūta Banelienė (Vilnius Gediminas Technical University, Lithuania)
Prof. Algirdas Bargelis (Kaunas University of Technology, Lithuania)
Prof. Andris Chate (Riga Technical University, Latvia)
Prof. Martin Eerme (Tallinn University of Technology, Estonia)

IOP Conf. Series: Materials Science and Engineering 1140 (2021) 011001 doi:10.1088/1757-899X/1140/1/011001

Dr. Tatjana Karaulova (Tallinn University of Technology, Estonia)
Prof. Kristo Karjust (Tallinn University of Technology, Estonia)
Prof. Branko Katalinic (Vienna University of Technology, Austria)
Prof. Petri Kuosmanen (Aalto University, Finland)
Prof. Minna Lanz (Tampere University, Finland)
Dr. Tõnu Leemet (Estonian University of Life Sciences, Estonia)
Prof. Andrei Lobov (Norwegian University of Science and Technology, Norway)
Prof. Jüri Majak (Tallinn University of Technology, Estonia)
Prof. Jüri Riives (Tallinn University of Technology, Estonia)
Prof. Anna Zawada Tomkiewitcz (Koszalin University of Technology, Poland)
Prof. Toms Torims (Riga Technical University, Latvia)
Prof Raine Viitala (Aalto University, Finland)
Dr Andrius Vilkauskas (Kaunas University of Technology, Lithuania)

Scientific Committee of BALTMATTRIB 2021

Prof. Jakob Kübarsepp, Chair (Tallinn University of Technology, Estonia)
Dr. Andrei Surzenkov, Scientific Secretary (Tallinn University of Technology, Estonia)
Dr. Ewald Badisch (AC2T Research GmbH, Excellence Centre of Tribology, Austria)
Ass. Prof. Regita Bendikiene (Kaunas University of Technology, Lithuania)
Prof. Michael Gasik (Aalto University, Finland)
Prof. Janis Grabis (Riga Technical University, Latvia)
Prof. Irina Hussainova (Tallinn University of Technology, Estonia)
Dr. Pavol Hvizdoš (Institute of Materials Research, Slovak Academy of Sciences, Slovakia)
Prof. Vytenis Jankauskas (Aleksandras Stulginskis University, Lithuania)
Prof. Kaupo Kukli (Institute of Physics, University of Tartu, Estonia)
Prof. Emeritus Priit Kulu (Tallinn University of Technology, Estonia)
Dr. Janis Locs (Riga Technical University, Latvia)
Prof. Remigiusz Michalczewski (Institute of Sustainable Technologies – National Research Institute, Poland)
Dr. Vitali Podgurski (Tallinn University of Technology, Estonia)
Dr. Miguel A. Rodriguez (Instituto de Ceramica y Vidrio, Spain Academy of Sciences, Spain)
Prof. Renno Veinthal (Tallinn University of Technology, Estonia)

Editors

Prof. Kristo Karjust (Tallinn University of Technology, Estonia)
Prof. Tauno Otto, Chair (Tallinn University of Technology, Estonia)
Prof. Jakob Kübarsepp, Chair (Tallinn University of Technology, Estonia)
Prof. Irina Hussainova (Tallinn University of Technology, Estonia)

IOP Conf. Series: Materials Science and Engineering 1140 (2021) 011002 doi:10.1088/1757-899X/1140/1/011002

Peer review declaration

All papers published in this volume of *IOP Conference Series: Materials Science and Engineering* have been peer reviewed through processes administered by the Editors. Reviews were conducted by expert referees to the professional and scientific standards expected of a proceedings journal published by IOP Publishing.

Detailed information about peer review process:

- Type of peer review: Single-blind
- Reviewers rated the following criteria in a three-point system: Quality of Content, Significance for theory or practice, Originality and level of innovativeness, Thematic Relevance and Overall recommendation.
  Reviewers also gave comments for the authors, pointed out strengths and weaknesses of the submitted contribution and provided suggestions for improvements. Authors must resubmit articles after the revision and then final decision about acceptance/rejections was made.
- Conference submission management system: Conference Software - ConfTool Pro
- Number of submissions received: 65
- Number of submissions sent for review: 65
- Number of submissions accepted: 58
- Acceptance Rate (Number of Submissions Accepted / Number of Submissions Received X 100): 89,2
- Average number of reviews per paper: 2
- Total number of reviewers involved: 65
- Contact persons for queries:
  - Aigar Hermaste aigar.hermaste@taltech.ee
  - Andrei Surženkov andrei.surzenkov@taltech.ee

Acoustic emission monitoring of composite specimens with impact damage during static compression test

Margarita Urbaha[1], **Ivans Agafonovs**[1], **Vladislav Turko**[1], **Jurijs Feščuks**[1], **Juris Korhs**[1]

1 Department of Transport systems and logistics, Institute of Aeronautics, Riga Technical University, Latvia,

[1]urbaha.m@gmail.com

Abstract. The standard specimens have been made by anisotropic carbon fiber plastic with an epoxy matrix. The specimens were subject to static load after its shock impact. During static compression loading, the parameters of acoustic emission (AE) signals were synchronously recorded: AE impulse amplitudes, its energy and the number of oscillations. According to the analysis of the acoustic emission signals the following events were observed: matrix cracking, fiber delamination and fiber destruction. The character of the AE signals change made it possible to determine the energy of the acoustic emission signals as a diagnostic witness of the onset of two moments of the specimen destruction. The first = near 70% of ultimate load's (the destruction of matrix) and total destruction under ultimate static load.

Keywords: acoustic emission, composite materials, impact damage, compression test

1. Introduction

This test method covers compression residual strength properties of multidirectional polymer matrix composite laminated plates, which have been subjected to drop-weight impact per test method ASTM D7136 [1] prior to application of compressive force. After impact damage specimens were tested on compression test ASTM D7137 with acoustic emission (AE) signals synchronously recorded. The method utilizes a flat, rectangular composite plate, previously subjected to a damaging event, which is tested under compressive loading using a stabilization fixture [2].

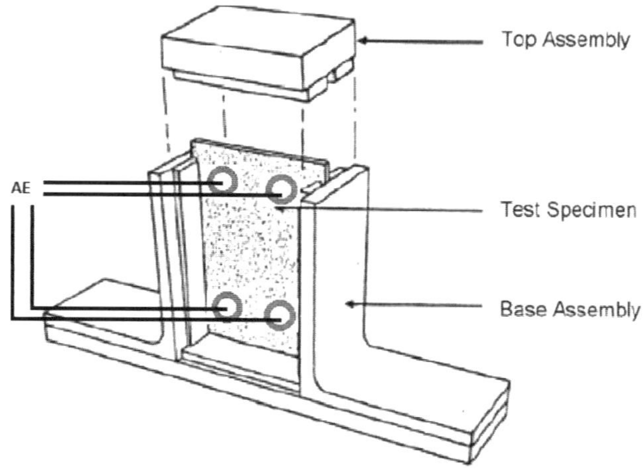

Figure 1. Schematic of compressive residual strength support fixture with specimen in place and AE sensors

2. Experimental

Specimens were made of carbon fiber HexPly M21 it is a high performance, very tough epoxy matrix for use in primary aerospace structures [3]. It exhibits excellent resistance to damage, especially at high energy impacts. HexPly M21 is a toughened epoxy resin system supplied with unidirectional woven carbon. In total were tested 5 specimens. General view of specimen and testing rig are shown on Fig.2. The compressive test fixture, shown in Fig. 1 and Fig. 2, utilizes adjustable retention plates to support the specimen edges and inhibit buckling when the specimen is end-loaded. The fixture consists of one base plate, two base slide plates, two angles, four side plates, one top plate, and two top slide plates.

Figure 2. General view of testing rig and specimen with AE sensors support glued

Specimens have dimensions of 150 mm length x 100 mm width and a thickness of 3 mm.In total four AE sensors used. Vallen system AMSY-6 is a digital multi-channel acoustic emission (AE) acquisition system to obtain AE energy and amplitude signals during loading. A measurement channel consists of an AE sensor, preamplifier and one channel of an ASIP-2 (dual channel acoustic signal processor). Each channel combines an analogue measurement section and a digital signal processing unit. AE features, such as time of the first threshold crossing (arrival time), rise time, duration, peak amplitude, energy and counts, are extracted by the ASIP-2 [4]. According to the procedure, the calibration was performed using a standardized source of Hsu-Nielsen (Hsu-Nielsen source). In this case, the threshold value was 30.1 dB [5].

The AMSY-6 provides 4 types of measurement data:

- Hit data: data that is generated when the AE signal exceeds the threshold.
- Status data: data that is generated in regular time intervals
- Parametric data: data of external parametric sensors which are measured at regular time intervals. Parametric data is stored with every hit and at user-specified time intervals.
- Waveform data: sampled AE sensor signal

The Instron 8801 machine have been used for specimen compressive loading . The Instron 8801 100 kN is a compact servo hydraulic fatigue testing system that meets the challenging demands of various static and dynamic testing requirements. 8801 systems provide complete testing solutions to satisfy the needs of advanced materials and component testing, and are ideally suited for fatigue testing and fracture mechanics [6].

Instron 9400 series is a drop tower test system that performs mono-multi axial impact tests on products and speciments. The system allows to set impact energies ranging from 0,3 to 405J, allowing to test a wide range of products and materials, also thanks to the large volume of work available for the positioning of the sample. Specimens of carbon fiber HexPly M21 were damaged with 25J impact. After damage specimen is installed in a multi-piece support fixture (Fig.2), that has been aligned to minimize loading eccentricities and induced specimen bending. The specimen assembly is placed between flat platens and end-loaded under compressive force until failure. Applied force, crosshead displacement, and AE energy and amplitude data are recorded while loading.

3. Result and discussion

Analysis of the results of changes in the energy of acoustic emission signals for the specimens and their location showed that AE location practically does not affect the general nature of the change in the AE signal energy. Nevertheless, the location of the AE sensor at position 4 most characteristically showed a three-stage process of sample destruction. Considering the general structure of the composite material (Fig. 3), it can be noted:

1. Protection cover destruction,
2. Boundary sheetmatrices destruction,
3. Structural significant items (SSI) destruction.

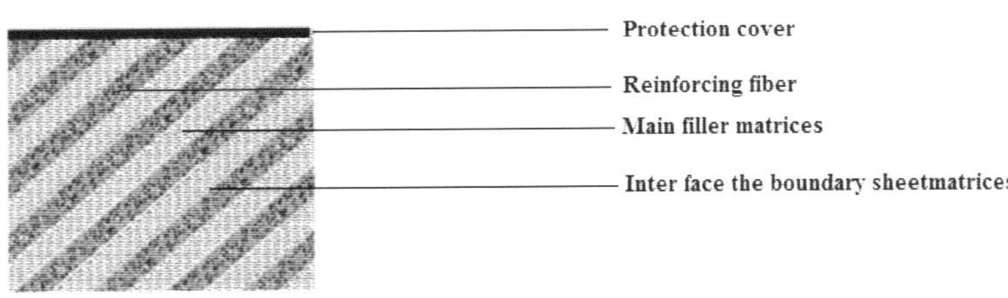

Figure 3. View of specimen structure

IOP Conf. Series: Materials Science and Engineering 1140 (2021) 012001 doi:10.1088/1757-899X/1140/1/012001

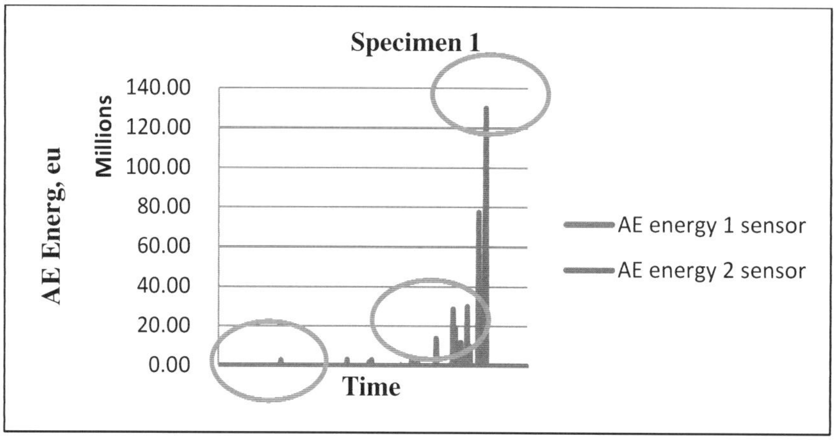

Figure 4. Acoustic energy signals of 1-2 sensors measured on specimen 1

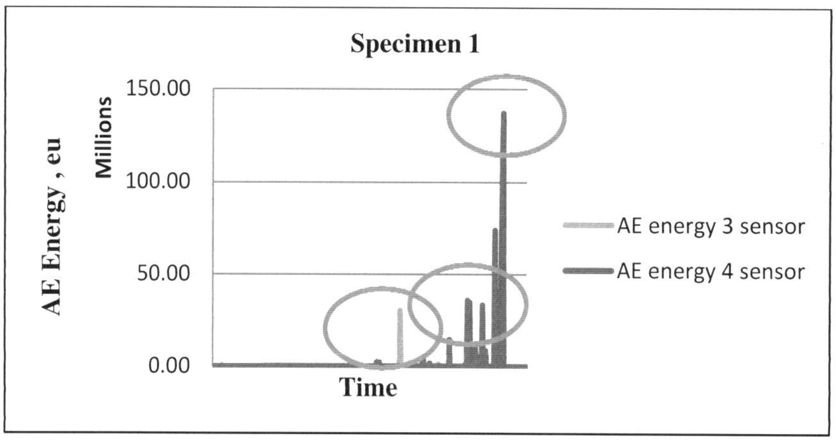

Figure 5. Acoustic energy signals of 3-4 sensors measured on specimen 1

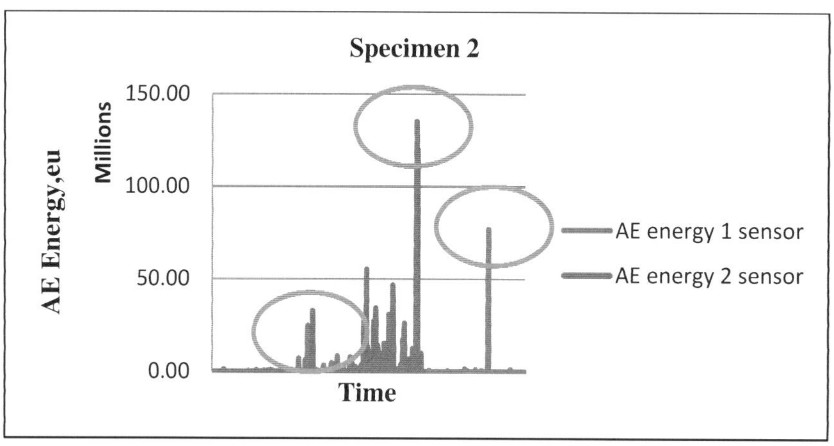

Figure 6. Acoustic energy signals of 1-2 sensors measured on specimen 2

IOP Conf. Series: Materials Science and Engineering 1140 (2021) 012001 doi:10.1088/1757-899X/1140/1/012001

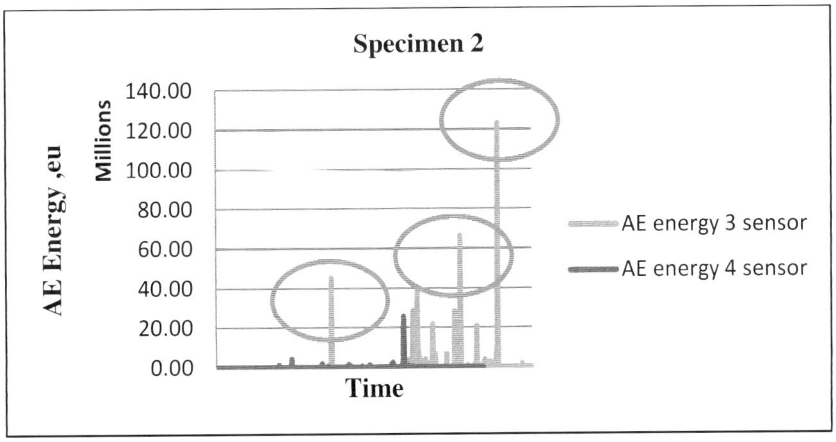

Figure 7. Acoustic energy signals of 3-4 sensors measured on specimen 2

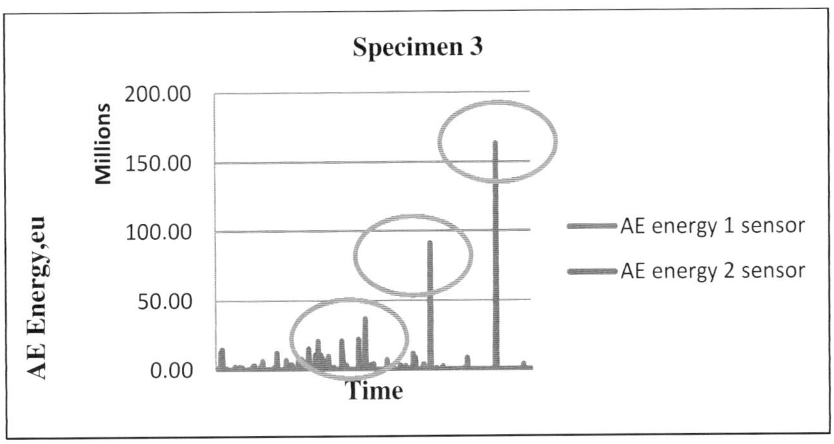

Figure 8. Acoustic energy signals of 1-2 sensors measured on specimen 3

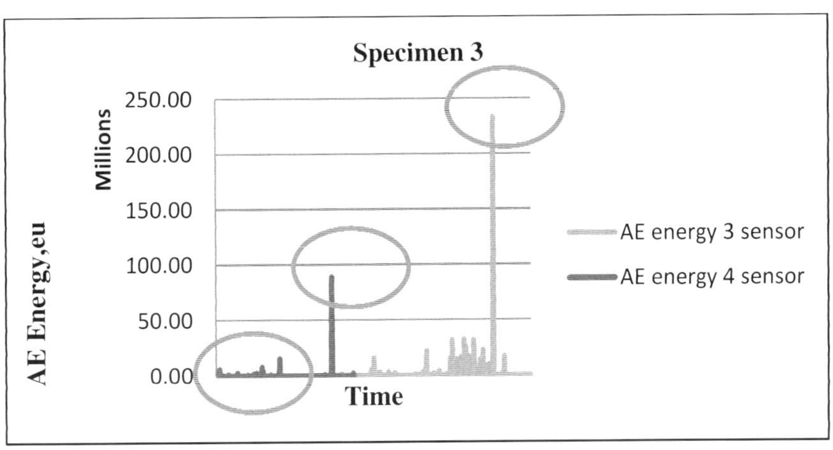

Figure 9. Acoustic energy signals of 3-4 sensors measured on specimen 3

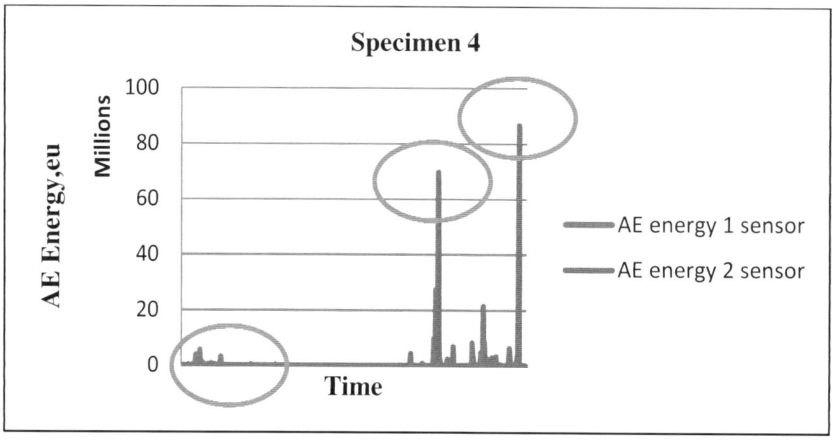

Figure 10. Acoustic energy signals of 1-2 sensors measured on specimen 4

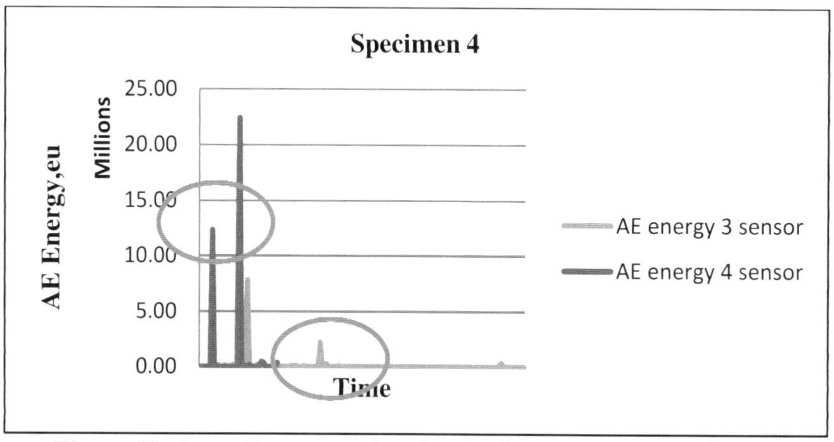

Figure 11. Acoustic energy signals of 3-4 sensors measured on specimen 4

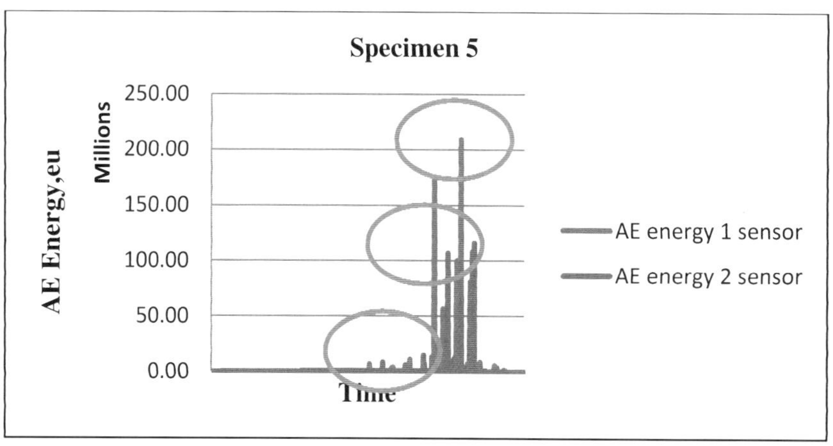

Figure 12. Acoustic energy signals of 1-2 sensors measured on specimen 5

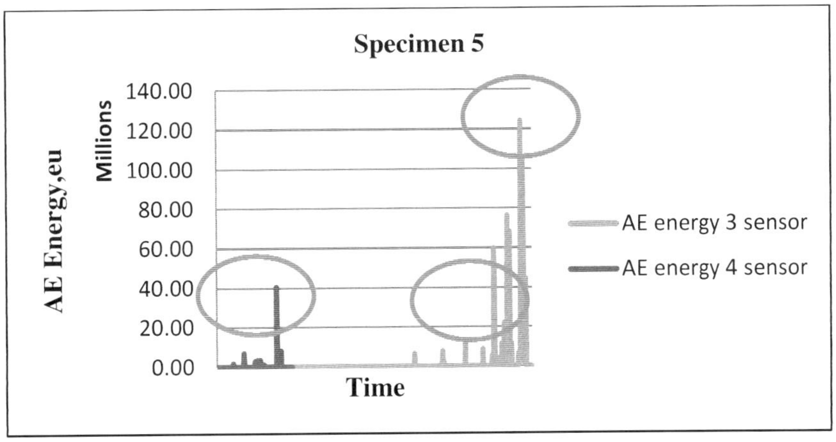

Figure 13. Acoustic energy signals of 3-4 sensors measured on specimen 5

Determining the destruction of the composite material with AE signals it is obvious that the onset of destruction will begin with auxiliary components the initial, first stage, recorded by low-power AE signals. Further, the process of destruction affects SSI by about 70% of the load and the third, final stage, a final loss of the bearing capacity is observed, accompanied by the final destruction of the sample occurs by the release of energy recorded by the AE sensors.

Thus, it becomes possible to predict the destruction process of the specimen at the moment of the onset of the second stage of its destruction. A similar picture is observed in the analysis of changes in the acoustic emission energy signals for 5 specimens. The most typical view of three zones destructions we can see on specimen №3 (Fig.8 and Fig.9).

The energy is given in energy units (eu) and the software takes the selected gain into account, when calculating and displaying the energy. Energy is calculated by squaring the digitized AE signal and integrating the results during the hit. 1 eu corresponds to 10-14V2 sec at the sensor output. The Energy is processed in either a so-called true energy mode, or a signal strength mode. In both modes, the energy is given in energy units (eu) and the software takes the selected gain into account, when calculating and displaying the energy. The energy is stored in a logarithm format with a resolution of 0.1%.Application of AE signals as a diagnostic feature for on-line monitoring for working with Helicopter health and usage monitoring (HUMS) and Structural Health Monitoring (SHM) systems.[7-9]

Conclusion

Analysis of the results of the acoustic emission signals recorded by the energy sensors under static loading before the destruction of the composite samples allows us to do the following conclusions:

1) The process of destruction of composite specimens includes three stages:

- Stage initial, accompanied by the destruction of non-force elements of the composite material approximately up to 20% of the destruction load;
- Stage of the beginning of the destruction of auxiliary SSI, approximately up to 70 % breaking load without loss of bearing bearing capacity;
- Stage of the final destruction of the specimen, with a breaking load (100%), accompanied by a complete loss of bearing capacity.

2) The presence of several stages of destruction of the composite material allows using the second stage of destruction of SSI as a diagnostic sign of the onset of destruction failure of SSI and its use in monitoring systems for the technical condition of SSI, such as Structural Health Monitoring (SHM) and Health and Usage Monitoring System (HUMS)

3) The stages of destruction of composite materials and the ability to control their destruction by acoustic emission methods allows designers to design SSI on the principles of safe destruction of structures (Fail-Safe Concept)

Acknowledgements

This work has been supported by the European Regional Development Fund within 1.1.1. Measure 1.1.1.2 “Aid to Post-Doctoral Research” of the Specific Aid Objective “To increase the research and innovative bearing capacity of scientific institutions of Latvia and the ability to attract external financing, investing in human resources and infrastructure” of the Operational Programmed “Growth and Employment” (No. 1.1.1.2/VIAA/1/16/104 Structural Health Monitoring system for Inflight Monitoring (FLY SAFE System).

PostDoc Latvia
NACIONĀLAIS ATTĪSTĪBAS PLĀNS 2020
EIROPAS SAVIENĪBA
Eiropas Reģionālās attīstības fonds
IEGULDĪJUMS TAVĀ NĀKOTNĒ

References

1. Standard Test Method for Measuring the Damage Resistance of a Fiber-Reinforced Polymer Matrix Composite to a Drop-Weight Impact Event. Available at https://www.astm.org/Standards/D7136.htm
2. Standard Test Method for Compressive Residual Strength Properties of Damaged Polymer Matrix Composite Plates. Available at https://www.astm.org/Standards/D7137.htm
3. HexPly® 21 epoxy matrix. Available at https://www.alquilujointernacional.eu/wp-content/uploads/2019/01/Caracter%C3%ADsticas-fibra-de-carbono-unidireccional.pdf
4. Multi-Channel Systems. Available at https://www.vallen.de/products/multi-channel-systems/
5. Hsu-Nielsen source https://www.ndt.net/article/az/ae/hsunielsensource.htm
6. 8801 (100kN) Fatigue Testing Systems. Available at https://www.instron.us/products/testing-systems/dynamic-and-fatigue-systems/servohydraulic-fatigue/8801-floor-model
7. Urbaha, M., Turko, V., Agafonovs, I., Sorokins, A. (2020). Experimental evaluation of static and fatigue strength of aluminium-based structural metallic alloys. Engineering for Rural Development, 19, pp. 487-493.
8. Urbahs, A., Banovs, M., Turko, V., Urbaha, M., Nedelko, D., Lebedevs, I. (2019). Research into the features of service damage in the composite material of helicopter rotor blades. Transport Means - Proceedings of the International Conference, 2019-October, pp. 466-469.
9. Nedelko, D., Urbahs, A., Turko, V., Urbaha, M., Carjova, K., Nagaraj, P.(2019). Assessment of the limits of signs of health and usage monitoring system for helicopter transmission. Procedia Computer Science, 149, pp. 252-257.

Acoustic performance of an additive manufactured lattice structure

Fabio Auriemma[1] and Le Liu[2]

1 Senior researcher.
2 Early stage researcher.
Department of Mechanical and Industrial Engineering, TalTech – Tallinn University of Technology, Tallinn, Estonia.

fabio.auriemma@taltech.ee

Abstract. In this paper, a new type of fibreless, bulk, sound absorber is studied. The acoustic material is an additive manufactured lattice structure constituted by a mesh of pins generated by uniform spatial translation of a regular hexagonal prismatic open cell. The material presents remarkable advantages of metal foams: it is "green", since no deteriorating, air polluting fibres are used; it is thermally and mechanically resistant, thus fire retardant, self-supporting and impact resistant. Compared to fibrous acoustic materials, it has low moisture absorption although the weight is higher. In addition to these properties, this additive manufactured material is provided with regular open cells whose geometry can be precisely controlled according to the desired performance. Moreover, the acoustical behaviour can be well modelled by simulating the air-solid thermo-viscous interaction within one single cell. The high, broad band sound absorption is a major result of the large viscous losses occurring in the flow which oscillates within the small hexagonal apertures. This lattice, referred to as honeycomb skeleton absorber (HAS), is a multifunctional material which can be used in a variety of structural applications where the sound levels must be controlled and/or a certain thermal resistance/insulation is required (for instance car floors).

1. Introduction

The recent developments in AM technologies has been utilized in the field of applied acoustics mainly to produce 3D printed acoustic metamaterials, in a variety of forms and for diverse purposes [1] and, very recently, the Selective Laser Melting has been used to 3D print metal cores for thermo-acoustic applications [2-4].

At the current state, technical solutions for broad band medium-high frequency ranges still rely on the use of *traditional* porous materials, such as fibrous elements (e.g. glass- or rock-wools), foams (typically of polyurethane) and textiles [5]. In these materials the dissipation of acoustic energy occurs by means of viscous losses within the small air saturated apertures. The main advantage of these solutions is represented by the wide frequency range of sound absorption, although the absorption at low frequencies is poor. Several drawbacks are associated to these materials, especially to the fibrous ones and to the plastic foams, which are also the most widely spread: they are often non-renewable; their performance can deteriorate over time; small particles can separate and pollute the surrounding media; the oil particles/moisture absorption makes these materials unsuitable in applications like turbine liners and compressor silencers of supercharged engines. For similar reasons, a recent trend also in HVAC systems has seen gradual reduction in the use of these solutions.

Recently, micro-perforated and micro-grooved panels have been designed to be used in modern HVAC systems, turbine liners and silencers [6,7]. These panels are constituted by one of more thin metal layers provided with sub-millimeter apertures [8-10].

Special mention is deserved by metal foams, because these materials are not affected by major deterioration, they preserve air cleanliness while being thermally and mechanically resistant, thus fire retardant, self-supporting and impact resistant [11-13]. Compared to fibrous materials, these foams have larger weight. However, by providing also lower moisture absorption than fibrous materials, metal foams can be used in several applications where fibrous and plastic absorbers are unsuitable. Traditional metal foams are usually manufactured by means of high-pressure infiltration process, which provides closed cells specimens, resulting in low absorption coefficient. Often, such operations as compressive deformation or hole-drilling are needed to increase the open porosity in order to improve the absorption performance.

In this paper the metal AM technology is used to produce a new type of open-cell lattice structure, referred to as honeycomb skeleton absorber (HSA), where sound absorption can be combined together with mechanical stiffness and thermal resistant properties. The HSA is provided with a mesh of pins generated by uniform spatial translation of a regular hexagonal prismatic unit open cell. As a major advantage, the additive manufacturing process gives the possibility to precisely control the geometry of the unit cell in order to meet the acoustical/mechanical/thermal requirements. Moreover, due to the regular geometry, the performance of the HSA can be modelled accurately. In this paper, the acoustic behavior of several HSA provided with different cell sizes has been studied by means of FEM accounting for thermo-viscous interaction between air and solid structure. The samples have been additive manufactured and tested in an impedance tube.

This work has been inspired by the contribution provided by Fotsing et al [14] and Deshmukh et al [15] where the AM has been first utilized for designing acoustic absorbers.

2. Theory

The propagation of sound waves in porous materials and, generally, in geometries with small dimensions, is primarily affected by viscous and thermal losses. In fact, nearby the solid walls, viscous and thermal boundary layers emerge, due to the viscosity and the thermal conductivity of the oscillating fluid medium. A measure of the length scale for the oscillatory thermal and viscous diffusion is give by the thickness of the viscous boundary layer, $\delta_v = \sqrt{2\mu/\omega}$, and the thickness of the viscous thermal layer, $\delta_\kappa = \sqrt{2\kappa/\omega}$. Here μ is the dynamic viscosity of the fluid, κ is the thermal conductivity and ω is the angular frequency (linked to the frequency f by means of $\omega=2\pi f$). The two layers are linked by the dimensionless Prandtl number, $Pr = v/\kappa = (\delta_v/\delta_\kappa)^2$ (it is ~0.7 for air at 20°C). At longer distances or higher frequencies, the fluid responds adiabatically, and with no major viscous losses, to the acoustic perturbation, whilst in the very proximity of the walls the response is isothermal and lossy. Within small (e.g. submillimeter sized) apertures and audible/low frequencies, the boundary layers cover large part of the volumes. This condition represents the principal mechanism on which porous absorbers work.

In order to model the propagation of acoustic waves in porous media it is, therefore, necessary to include thermal conduction effects and viscous losses explicitly in the governing equations. In this sense, a suitable set of equations is represented by the linearized Navier-Stokes equations in quiescent background conditions which account for mass continuity, momentum, and energy balance. In absence of thermal gradient, these equations simplify as follow:

$$i\omega\rho_m u_1 = -\frac{dp_1}{dx} + \mu\nabla^2 u\ , \quad i\omega\rho_m c_p T_1 = \kappa\nabla^2 T + i\omega p_1 \tag{1}$$

In Eq.s 1 the subscript 1 indicates a complex term, i.e. the oscillating part of a certain physical quantity, the subscript m indicates the mean value, u is the particle velocity, p is the pressure, ρ is the density and T is the temperature (thus, for instance, $p = p_m + Re[p_1 e^{i\omega t}]$).

3. Additive manufactured HSE

The honeycomb skeleton shape of acoustic material has been chosen to provide quasi-circular geometries within the sound absorptive cross sections. In fact, although the circularly shaped apertures are well known to be the most effective for the energy dissipation process, the hexagonal shape of the apertures allows minimizing the solid region between adjacent apertures in order to keep the porosity of the material as high as possible.
Several cylindrical samples, provided with 27mm diameter and 50mm height, have been manufactured by means of Selective Laser Melting (SLM). Magnified images of a HSE sample are reported in Fig.1.

a. 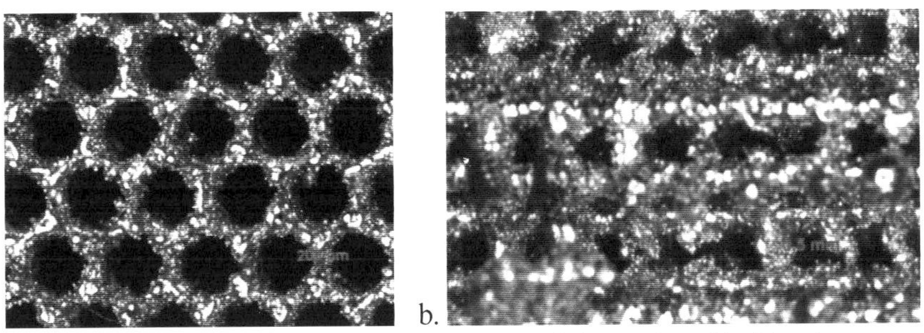 b.

Figure 1. 15x-magnified images of the cells within the HSE. a) front view, b) side view.

The fabrication of the parts is based on solid CAD 3D models (see Figure 2a). has it has been performed using SLM280 HL from SLM Solutions, which uses stainless steel AISI 316L powder as feedstock. The AM process is thoroughly described in reference [2].

4. Modelling and experiments

The Eq.s 2 and 3 can be modelled for complex geometries by means of finite element method. For the regularly shaped unit cell in Fig.2a, only the fluid inside of it is modelled by means of FEM. The symmetry of the geometry can be used to limit the discretized domain and an isothermal condition is applied on the surfaces where the solid walls are located. This is shown in Fig.2b.

Both the numerical model and the experimental approaches are based on *transfer-matrix*. Named S_i and S_o the surfaces which limit the domain modelled, the acoustic pressure and particle velocity at these surfaced are by means of the transfer matrix **T** so that:

$$\begin{bmatrix} p_1 \\ u_1 \end{bmatrix}_{S_i} = \begin{bmatrix} T_{11} & T_{12} \\ T_{21} & T_{22} \end{bmatrix} \begin{bmatrix} p_1 \\ u_1 \end{bmatrix}_{S_o} \tag{2}$$

As demonstrated in [16], the wavenumber of the fluid within the acoustic material, k_p , and its characteristic impedance $Z = \rho_p c_p$ can be obtained as:

$$k_p = \frac{1}{d} \cos^{-1} T_{11} \quad \text{and} \quad Z = \sqrt{\frac{T_{12}}{T_{21}}} \tag{3}$$

By utilizing the definition of transfer matrix and its matricial properties, it is possible to show that, if a rigid wall ($u=0$) condition is imposed at the surface S_o and an acoustic pressure is imposed at S_i, the terms T_{11}, T_{12} and T_{21} reduce to:

$$T_{11} = \frac{p_1|_{S_i}}{p_1|_{S_d}},\ T_{12} = \frac{\left(p_1|_{S_i}\right)^2-\left(p_1|_{S_o}\right)^2}{p_1|_{S_o}u_1|_{S_i}},\ T_{21} = \frac{\left(u_1|_{S_i}\right)}{p_1|_{S_o}} \tag{4}$$

As a consequence, k_p and Z can be obtained by using Eq.3. When FEM is utilized, the averaged values of pressure and particle velocity at S_i and S_o can be directly computed. In case of experiments, those values are carried out by using a *3-microphone impedance tube* [16]. The surface impedance Z_s, the reflection coefficient R_x and, finally, the absorption coefficient α of the acoustic element of length L can be computed as follows:

$$Z_s = -1iZ\cot\left(k_pL\right) \quad R_x = \frac{Z_s-\rho c}{Z_s+\rho c} \quad \alpha = 1-|R_x|^2 \tag{5}$$

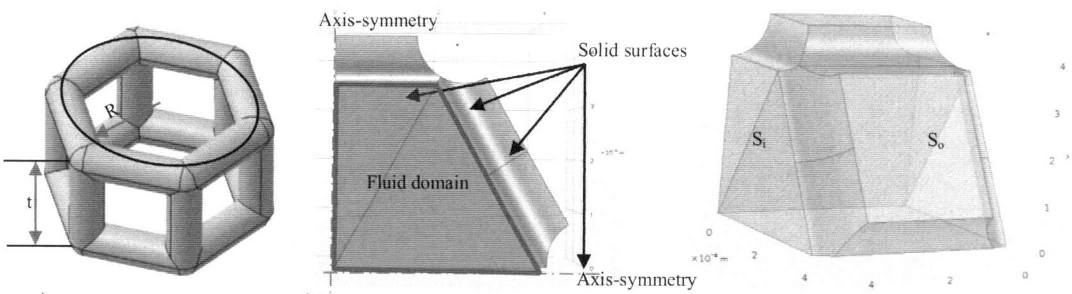

a) b) c)

Figure 2. HSA modelling. a) CAD, b) axis-symmetry of the cell, c) side view of modelled cell.

5. Results

The results are expressed, in Figure 3a, in terms of absorption coefficient as a function of the sound frequency for three different geometries (S01, S02, S03). The figure also includes the geometrical data of the samples, which have different values of the radius R of the circle inscribed in the hexagon and of the height t of the unit cell (see Figure 2a). For all samples, the diameter of the struts of the constitutive mesh is 0.18mm, which is considered to be the minimum value for an acceptable printing process. Only the sample S01 has been additive manufactured and tested in impedance tube. However, this is enough to validate the FEM approach, thus to provide reliability to the simulation of the other geometries. The graphs of the absorption coefficient show that the best performance, with total sound absorption at certain frequencies, is reached for R=0.5mm, and t=0.25÷0.5mm.

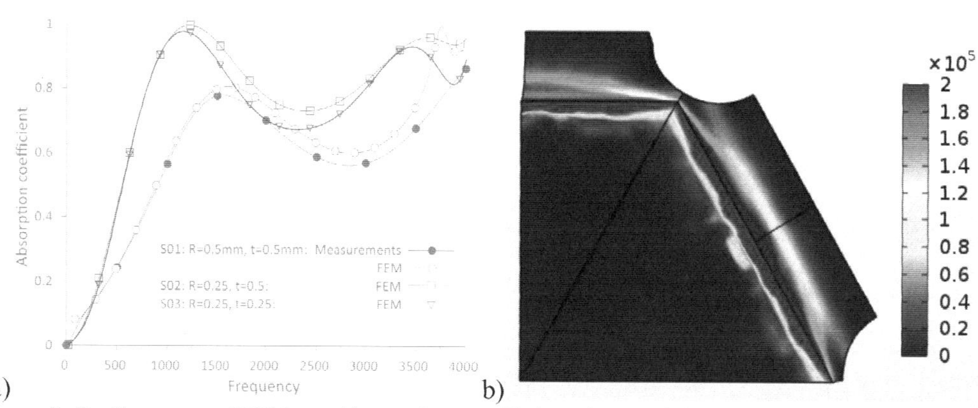

a) b)

Figure 3. Performance of HSA: a. Absorption coefficient, b. Total thermo-viscous power dissipation density (W/m^3)

The Figure 3b confirms that major thermo-viscous losses take place in proximity of the struts, i.e. within the previously mentioned viscous and thermal boundary layers. It must be pointed out that 3D printing geometries provided with very small apertures might be critical because of possible clogging due to the powder bed used in the process. As a consequence, the actual porosity might appear to be fairly different from the desired one and modelling of sound propagation within the geometry provided by the CAD drawing might be misleading.

6. Conclusions

The acoustic performance of the additive manufactured HSA open cell metal lattice structure has been studied. The HSA provides good sound absorption and its multifunctional potentialities will be explored in future investigations by testing mechanical strength and thermal resistance/insulation. Differently from metal foams, the HSA is provided with a regular inner geometry and open cells, which result in good sound absorption. Several configurations of HSA have been studied, their acoustic behavior has been modelled by means of FEM and the results, in terms of absorption coefficient, have been confirmed experimentally. The absorption coefficient reaches unitary values for the samples with smallest apertures, but the printing process in this case still requires further optimization.

References

[1] 2013 *Acoustic Metamaterials and Phononic Crystals* ed Deymier P A (Springer).
[2] Auriemma F and Holovenko Y 2019 Use of Selective Laser Melting for Manufacturing the Porous Stack of a Thermoacoustic Engine *KEM* **799** doi:10.4028/www.scientific.net/kem.799.246.
[3] Auriemma F and Holovenko Y 2019 Performance of additive manufactured stacks in a small scale thermoacoustic heat engine *SAE Technical paper 2019-01-1534,* doi:10.4271/2019-01-1534.
[4] Auriemma F, Di Giulio E, Napolitano M, Dragonetti R 2020 Porous cores in small thermoacoustic devices for building applications *Energies* **13**, 2941. doi:10.3390/en13112941.
[5] Allard J F, Atalla N 1993, *Modelling Sound Absorbing Materials: Propagation of Sound in Porous Media* (New York: Elsevier Applied Science).
[6] Kabral R et al 2014 A new type of compact silencer for high frequency noise *Proc. 9th Int. DAAAM Baltic Conf.* pp 29–35 ISSN 9789949236206.
[7] Kabral R et al 2012 A Novel Design for Cruiser Type Motorcycle Silencer Based on Micro-Perforated Elements *SAE Techincal paper 2012-32-0109* doi:10.4271/2012-32-0109.
[8] Auriemma F, Rämmal H, Lavrentjev J 2013 Micro-grooved elements – a novel solution for noise control. *SAE Int J Mater Manuf* **6**(3):599-610 doi: 10.4271/2013-01-1941.
[9] Auriemma F, Tiikoja H 2015 On the Acoustic Impedance of a Fibreless Sound Absorptive Element *SAE Int. J. Engines* **8**(5):2268-2275 doi:10.4271/2015-24-2462.
[10] Auriemma F, Rämmal H, Lavrentjev J 2013 Application of novel micro-grooved elements to small engine silencer. *SAE technical paper JSAE 2013-10-15* doi:10.4271/2013-32-9001.
[11] Lu T J, Hess A, Ashby M F 1999 Sound absorption in metal foams *J. App Phys.* **85**(11) pp 7528-7539.
[12] Han F et al 2003 Acoustic absorption behaviour of an open-celled aluminuim foam *J.Phys. D: Appl. Phys.* **36** (249) pp 294-301.
[13] Perrot C, Panneton R, Olny X 2007 Computation of the dynamic thermal dissipation properties of porous media by Brownian motion simulation: application to an open-cell aluminum foam *J. App. Phys.* **102** 074917 pp 1-13.
[14] Forsing E R et al. 2019 Acoustic properties of periodic micro-structures obtained by additive manufacturing *App. Acou.* **148** pp 322-331 doi:10.1016/j.apacoust.2018.12.030.
[15] Deshmukh S, Harshavardhan R, Ramamoorthy S 2019 Design of periodic structures for acoustic applications: concept, parametric study and experimental validation, *Materials and Design* **175** 107830 pp 1-14, doi:10.1016/j.matdes.2019.107830.
[16] Song B H and Bolton J S 1999 A transfer-matrix approach for estimating the characteristic impedance and wavenumbers of limp and rigid porous materials *J. Acoust. Soc. Am.* **107**(3).

Isothermal Oxidation of SLM fabricated $Mo(Si_{1-x}Al_x)_2$-based composite

Tatevik Minasyan[1,2*], Le Liu[1], Ehsan Toyserkani[2], Irina Hussainova[1*]

[1]Department of Mechanical and Industrial Engineering, Tallinn University of Technology, Ehitajate 5, 19086 Tallinn, Estonia

[2]Multi-scale Additive Manufacturing Laboratory, University of Waterloo, 200 University Ave., West Waterloo, Ontario N2L 3G1, Canada

*tatevik.minasyan@taltech.ee

Abstract. Near full dense C40 hexagonal structured $Mo(Si_{1-x}Al_x)_2$-based composite coupons with 7 mm diameter by 7 mm height were prepared by selective laser melting (SLM) of $MoSi_2$-30 wt.% AlSi10Mg powder mixture and subjected to isothermal oxidation. As a reference, $MoSi_2$ bulk sample was prepared by spark plasma sintering (SPS) technique. The quantitative oxidation tests were conducted in dry air atmosphere at 500 °C for 72 and 120 hours with three specimens for each dwell time. The estimated weight change after 120 h oxidation accompanied with the SEM analysis revealed the slow oxidation of the SPS prepared $MoSi_2$ bulk sample without evidence of pesting, however, the formed oxide layer was not protective and contained large cracks and multiple microcracks. The SLM produced sample showed remarkably slower oxidation rate as well as the formation of only a nanoscale oxide layer, highlighting the significant influence on Al additive to enhance $MoSi_2$ oxidation resistance, thus making the developed material promising to be used as heating element.

1. Introduction

The focus on $MoSi_2$-based materials over the last decades is mainly being pursued by heavy industrial complexes using materials to meet application requirements in the oil industry and aviation field. $MoSi_2$ possesses high melting point, hardness, thermal and electrical conductivities, and stable chemistry under oxidizing and reducing environments at elevated temperatures [1]. As such, these combined properties provide good prospects for $MoSi_2$ to qualify as material for high temperature structural applications. During elevated temperature oxidation, $MoSi_2$ innately generates a continuous layer of self-healing silica on the surface; moreover, when heated above 1700°C, the molten silica actively fills in pores and surface defects, inhibiting further oxidation of the material [2]. On the contrary, SiO_2 instead decomposes into unstable silicon monoxide (SiO) under reducing environment.
Regardless of the promising characteristics of $MoSi_2$, the inferiority in fracture toughness at room temperature and low strength at high temperatures, intensive oxidation and pest disintegration at intermediate temperatures (around 500°C) demands that improvement be made to the as-is $MoSi_2$ material [3,4]. The accelerated oxidation at intermediate temperatures is a result of intense MoO_3 formation, a volatile compound that causes pores and defects within the silica scale promoting oxide permeation [5]. Numerous approaches have been investigated to enhance the mechanical properties of

$MoSi_2$ by the addition of compatible compounds or additive elements, including ZrO_2, [6], SiC [7], Si_3N_4 [8], TiC [9] etc. This is all in the effort to strengthen the oxidation resistance of innate $MoSi_2$ by alloying with aluminum resulting in formation of $Mo(Si,Al)_2$ [5, 10, 11]. Molybdenum aluminosilicide, when oxidized forms alumina or mullite layer, which primarily acts to prevent pesting and to promote the high temperature oxidation resistance of the materials in both oxidizing and reducing atmospheres. It is important to note that numerous researchers have found that the pesting at low temperatures (400-600°C) does not occur in dense $MoSi_2$ samples (>95% relative density) and that the intensive oxidation of $MoSi_2$ is affected by various factors, including the composition, impurities, surface defects (pores and cracks), oxidation temperature and oxidizing environment [12-14].

In the present research, the oxidation behavior of $Mo(Si,Al)_2$-based composite bulks, prepared by laser powder-bed fusion (LPBF)/selective laser melting (SLM) technique, was examined at 500 °C in dry air for 72 and 120 hours. Detailed analysis of the SLM sample preparation is presented in [15,16]. The novelty of the work relies on the study of oxidation resistance of SLM prepared $Mo(Si,Al)_2$ samples, and the main goal rests on finding the applicability of the material as heating element.

As a reference, $MoSi_2$ bulk specimen prepared by spark plasma sintering technique was used. The weight change and the growth of oxide scale was examined along with microstructural and phase composition analysis, and the oxidation rate was evaluated.

2. Experimental

$MoSi_2$ powder was prepared by combustion synthesis of commercial Mo (<5 μm, 99% purity, Pacific Particulate Materials (PPM)) and Si (<20 μm, 99% purity, Silgrain-Elkem) powders mixture with Mo:Si molar ratio of 1:2.

For the preparation of $Mo(Si,Al)_2$-based materials cylindrical samples with 7 mm of height and 7 mm of diameter were prepared by selective laser melting of combustion synthesized $MoSi_2$ powder mixed with 30 wt.% AlSi10Mg alloy (15-63 μm, 99% purity, SLM Solutions). The Renishaw AM400 customized machine (Renishaw, UK) equipped with ytterbium laser (wavelength of 1070 nm,) was used for powder consolidation. The samples were prepared at 200 W laser power and 500 $mm \cdot s^{-1}$ scanning speed. The layer thickness and hatching space were set 35 μm and 85 μm, respectively [15].

As a reference bulk $MoSi_2$ sample was prepared by spark plasma sintering technique. The sintering was conducted at 100° min^{-1} heating rate to 1400°C, applying 50 MPa pressure for 10 min. With reference to the consolidation technique, the samples were labelled as SLM-sample and SPS-sample, respectively.

The top and bottom surfaces of SLM-sample were polished to 0.5 μm using standard metallurgical methods, while the side surfaces of the coupons were not polished. The reference SPS sample was polished to 0.5 μm finish. Three samples were prepared under each condition and subjected to oxidation and further investigation.

The quantitative oxidation tests were conducted in Nabertherm L9/13 furnace (with P330 PID controller) in dry air atmosphere at 500°C for 72 and 120 hours. Prior to oxidation the specimens were cleansed by ultrasonic cleaner with deionized water for 10 min, afterwards weighed at room temperature with analytical balance.

Zeiss EVO MA15 and TESCAN VEGA3 scanning electron microscopes (SEM) equipped with INCA Energy and Bruker QUANTAX 200 (USA) EDS detectors were used to study the microstructure and elemental composition of the starting powders, compacted and oxidized samples. The phase composition analysis was performed by Rigaku SmartLab SE X-ray diffractometer (Tokyo, Japan) (with CuKα radiation, λ = 1.54056 Å), with 1D detector D/teX Ultra 250. The density of the fabricated specimens was measured by Archimedes' method.

3. Results and discussion

As shown in Fig. 1a, the combustion synthesized $MoSi_2$ powder is composed of agglomerates of fine particulates with submicron to few microns in size. The SPS-sample (Fig. 1 b, c) is primarily composed of $MoSi_2$ phase (dark gray), while small and infrequent inclusions of well dispersed Mo_5Si_3 were disclosed (coloured with light gray shade). Several apparent pores in few hundred nanometers to 2-micron size range are detected.

The calculated relative density of spark plasma sintered $MoSi_2$ sample is 97.8%. $MoSi_2$-30 wt. % AlSi10Mg powder is composed of fine $MoSi_2$ particles and relatively coarser and round shaped AlSi10Mg alloy particles with satellite structure. SLM produced sample (labelled SLM-sample) is composed of cellular grains, cellular and columnar dendritic structures. The relative density of SLM prepared sample is 99.3%. Detailed information about SLM-sample preparation is described in [15]. According to compositional analysis results, the SLM-sample is composed of 70.9 wt.% $Mo(Si_{0.67},Al_{0.33})_2$, 0.7 wt. % $Mo(Si_{0.7},Al_{0.3})_2$, 1.1 wt.% $MoSi_2$, 17.5 % $Al_{0.85}Si_{0.15}$ and 9.7 wt.% Si [15].

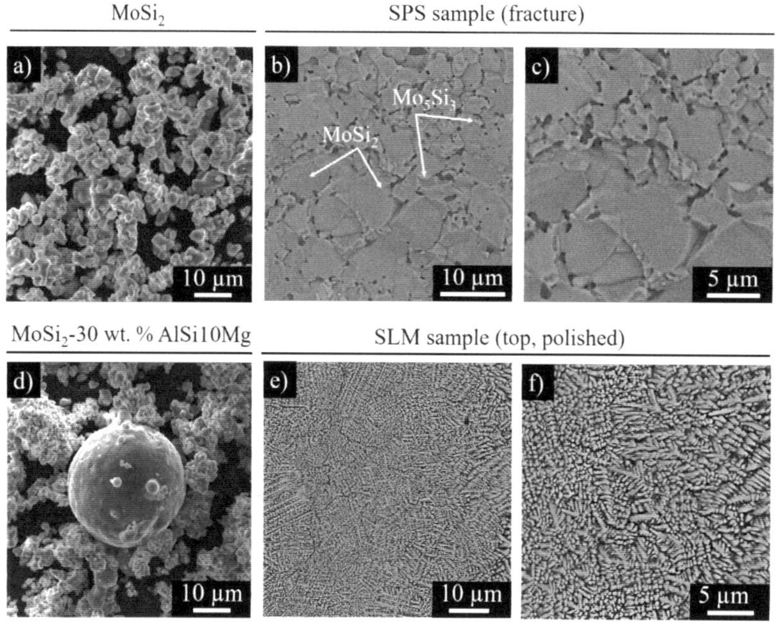

Figure 1. SEM images of combustion synthesized $MoSi_2$ powder (a), spark plasma sintered $MoSi_2$ (SPS-sample) (b, c), $MoSi_2$-30 wt.% AlSi10Mg powder (d) and SLM-sample (e, f)

Fig. 2 shows the SEM images of the top and vertical fracture surfaces of SPS-sample after 72 and 120 hours of oxidation. After 72 h, the sample was coloured with white and after 120 h with light green layer (because of the intensive yellowish-green MoO_3 formation).

From top surface secondary electron-detected (SE) images of the oxidized SPS-samples (Fig. 2 a, b and e, f), the whisker and platelet-like crystals of MoO_3 and lumps of amorphous SiO_2 were disclosed, as confirmed by EDS (Fig. 3) and XRD analyses (Fig. 4). The oxide layers are clearly visible at backscattered electron-detected (BSE) images of vertical fractures of the oxidized samples (Fig. 2 d and h), and the approximate oxide layer thickness is around 16-19 µm after 72 h oxidation (Fig. 2d). After 120 h of oxidation the grown oxide layer thickness was about 25-28 µm (Fig. 2h).

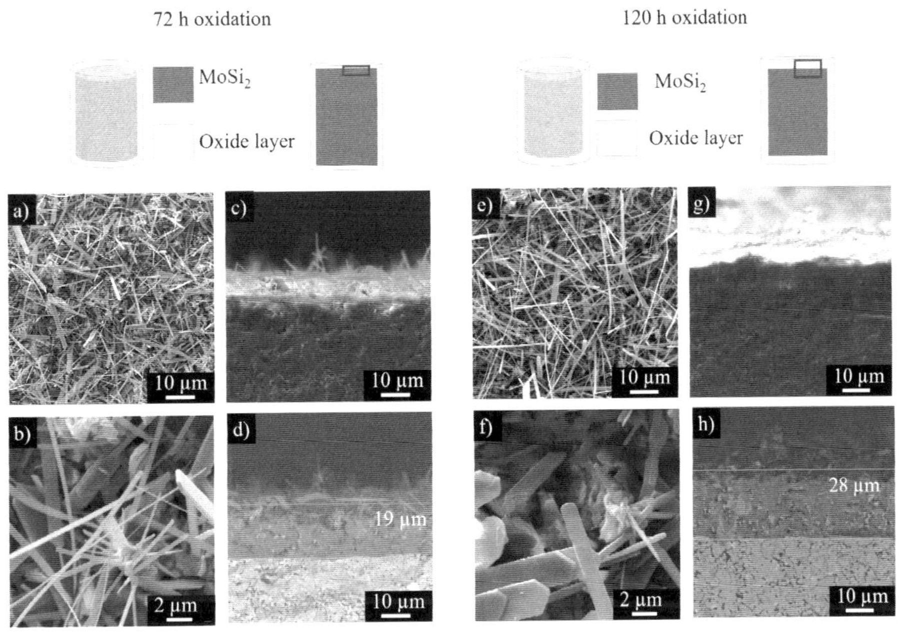

Figure 2. Top surface SE-detected images of SPS-sample after 72 (a, b) and 120 h (e, f) of oxidation. Vertical fracture SE (c) and BSE (d) images after 72 h; and SE (g) and BSE (h) images after 120 h oxidation

According to performed EDS analysis of 120 h oxidized SPS-sample, the platelet-like crystals are mostly composed of MoO_3 (Fig. 3, Spectrum 1), the overall composition of the surface represents the mixture of MoO_3 and SiO_2 oxides, with 1:2.2 molar ratio (approximate calculation, based on EDS analysis results).

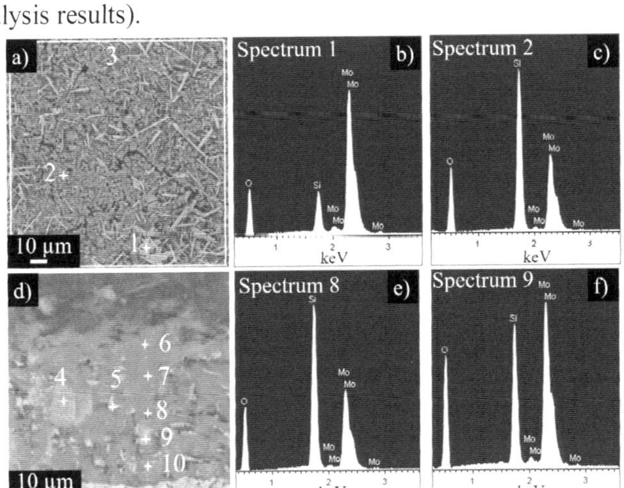

g)

Spectrum number	MoO_3:SiO_2 Molar ratio (approx.)	MoO_3:SiO_2 Weight ratio (approx.)
1	2.4 : 1	5.7 : 1
2	1 : 3.9	1 : 1.6
3	1 : 2.2	1 : 1
4	1.5 : 1	3.7 : 1
5	1.15 : 1	2.75 : 1
6	1 : 2.4	1 : 1
7	1 : 2.3	1 : 1
8	1 : 2.1	1.1 : 1
9	1 : 1	2.5 : 1
10	1 : 2.3	1 : 1

Figure 3. BSE-detected images of top surface (a) and vertical fracture (d) of the 120 h oxidized SPS-sample, respective EDS spectrums (b, c, e, f), the weight and molar ratio of MoO_3 and SiO_2 oxides based on the elemental composition analysis (g)

Spectra 6, 7, 8 and 10 at vertical fracture of the oxidized sample show almost the same molar ratio of MoO_3 and SiO_2 (Fig. 3g) coinciding the top surface EDS results. This signifies that neither serious enrichment nor depletion of elements within the oxide layer occurred; the distribution of elements throughout the layer remained relatively constant. However, looking at BSE image of the fracture (Fig. 3d), there are certain brighter areas and spots, which according to EDS analysis are proven to be MoO_3 rich inclusions (Fig. 3g, Spectrum 4 and 5). The oxides scales generated on $MoSi_2$ after 72 and 120 h of oxidation are porous, and massive cracks along with small microcracks are observed on the surface (Fig. 3a). The formation of cracks is explained by the volume expansion because of the oxide formation and partial evaporation of MoO_3, which prevents the generation of protective oxide layer [4]. The diffractograms of the 72 and 120 h oxidized samples show the weakening of $MoSi_2$ peaks and strengthening of respective peaks of MoO_3. SiO_2 was in amorphous form (the halo of amorphous silica is detected at 2θ=15-32°), as the crystalline state is formed at higher temperatures.

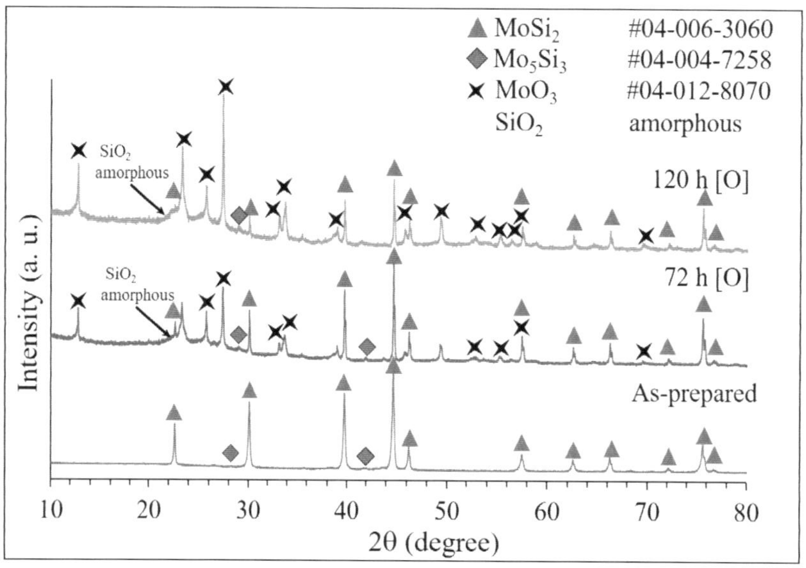

Figure 4. The XRD patterns of the SPS-sample: as prepared, after 72 h and 120 h of oxidation

The weight change per surface area of 72 h and 120 h oxidized SPS-sample was 1.272 $mg{\cdot}cm^{-2}$ and 1.919 $mg{\cdot}cm^{-2}$, respectively (Table 1). The oxidation rate for 72 h and 120 h oxidized SPS-sample were 6.172 - $7.330{\cdot}10^{-11}$ $m{\cdot}s^{-1}$ and 5.787 - $6.481{\cdot}10^{-11}$ $m{\cdot}s^{-1}$ respectively, proving that after the formation of denser layer, the oxidation rate is decreased (Table 1).

Table 1. The weight change, oxide layer thickness and corresponding oxidation rate of reference SPS-sample and SLM-sample after 72 and 120 hours of oxidation

Sample	Oxidation time	Weight change	Oxide layer thickness	Oxidation rate
Label	(hours)	($mg{\cdot}cm^{-2}$)	(μm)	($m{\cdot}s^{-1}$)
SPS-sample	72	1.2720 ± 0.0059	16 -19	6.172 - $7.330{\cdot}10^{-11}$
SPS-sample	120	1.9195 ± 0.0065	25 - 28	5.787 - $6.481{\cdot}10^{-11}$
SLM-sample	120	0.0488 ± 0.0003	0.05 - 0.08	1.157 - $1.851{\cdot}10^{-13}$

These results are in good agreement with literature data [3, 4, 17, 18]. No pesting was observed after even 120 h of oxidation, which was also noted in [12]. It is important to mention, that relatively dense samples with only few internal defects (cracks and pores) are not experiencing pesting, however the formation of non-protective oxide scales takes place.

Coming to SLM-sample, from the general observation, the appearance of the oxidized SLM-sample is similar to the as-prepared non-oxidized one. Upon SEM examination, the BSE-detected images of the top surface (Fig. 5a) showed similar morphology seen between the top oxidized surface and non-oxidized one (Fig. 1e, f) affirming that no obvious oxide formation took place at this temperature during 120 h of oxidation. The thickness of the oxide layer was in 50-80 nm range (Fig. 5b). No MoO_3 whiskers were found on the surface.

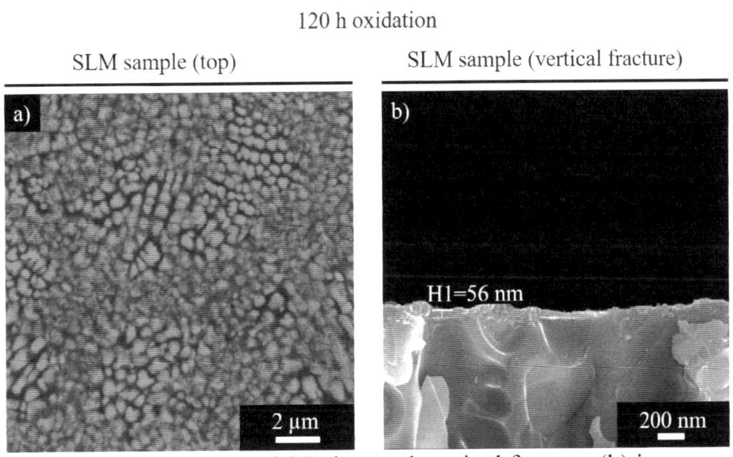

Figure 5. BSE-detected top surface (a) and SE-detected vertical fracture (b) images of SLM-sample after 120 h of oxidation

Therefore, the SLM prepared $Mo(Si_{1-x},Al_x)_2$ -based material experiences only negligible oxidation, and the weight change per surface area was calculated as 0.0488 ± 0.0003 mg·cm^{-2}, being significantly less than the reference SPS-sample (Table 1). It can be concluded, that the addition of Al to $MoSi_2$ and the formation of C40 hexagonal aluminosilicide dramatically improves the oxidation resistance of the material.

4. Conclusions

The SLM prepared $Mo(Si,Al)_2$-based material was subjected to oxidation for 72 and 120 hours at a dry air atmosphere and temperature of 500°C. As a reference, $MoSi_2$ bulk sample was prepared by spark plasma sintering technique and oxidized under the same conditions. It was revealed that after 72 and 120 h of oxidation, the oxide scale of $MoSi_2$ SPS-sample was 16-19 and 25-28 µm, respectively, and no pesting was observed; however, the generated oxide layer was non-protective, as it contained a network of large cracks and fine microcracks. The SLM prepared sample after 120 h of oxidation showed no obvious oxide layer formation (around 50-80 nm) and the weight change was calculated as 0.0488 ± 0.0003 mg·cm^{-2}. This signifies, that the presence of $Mo(Si,Al)_2$ remarkably improves the oxidation resistance of the material and suppresses the potential pesting. Therefore, the developed material can be successfully exploited as a heating element in many applications as this material solves the critical drawback of the widely used $MoSi_2$ heaters.

Acknowledgments

This work was supported by the Estonian Research Council grant PRG643 (I. Hussainova). The authors would like to thank PhD O. Volobujeva and M. Viljus for SEM imaging of the materials.

References

[1] Pan, Y. and Wang, S. Insight into the oxidation mechanism of MoSi2: Ab-initio calculations. *Ceramics International*, 2018, **44(16)**, 19583-19589.

[2] Zhu, L., Ren, X., Wang, X., Kang, X., Zheng, R. and Feng, P. (2020). Microstructure and high-temperature oxidation resistance of MoSi2-ZrO2 composite coatings for Niobium substrate. *Journal of the European Ceramic Society*. 2020

[3] Hansson, K., Halvarsson, M., Tang, J. E., Pompe, R., Sundberg, M. and Svensson, J. E. Oxidation behaviour of a MoSi2-based composite in different atmospheres in the low temperature range (400–550° C). *Journal of the European Ceramic Society*, 2004, *24(13)*, 3559-3573.

[4] Zhang, F., Zhang, L., Shan, A. and Wu, J. Oxidation of stoichiometric poly-and single-crystalline MoSi2 at 773 K. *Intermetallics,* 2006, **14(4)**, 406-411.

[5] Ingemarsson, L., Halvarsson, M., Engkvist, J., Jonsson, T., Hellström, K., Johansson, L. G. and Svensson, J. E. Oxidation behavior of a Mo (Si, Al) 2-based composite at 300–1000° C. *Intermetallics*, 2010, **18(4)**, 633-640.

[6] Ma, Q., Yang, Y., Kang, M. and Xue, Q. Microstructures and mechanical properties of hot-pressed MoSi2–matrix composites reinforced with SiC and ZrO2 particles. *Composites science and technology*, 2001, **61(7)**, 963-969.

[7] Han, X. X., Wang, Y. L., Xiong, X., Heng, L. I., Chen, Z. K. and Wei, S. U. N. Microstructure, sintering behavior and mechanical properties of SiC/MoSi2 composites by spark plasma sintering. *Transactions of Nonferrous Metals Society of China*, 2018, **28(5)**, 957-965.

[8] Suryanarayana, C. Structure and properties of ultrafine-grained MoSi2+Si3N4 composites synthesized by mechanical alloying. *Materials Science and Engineering: A*, 2008, **479(1-2)**, 23-30.

[9] Meng, J., Lu, J., Wang, J. and Yang, S. Preparation and properties of MoSi2 composites reinforced by TiC, TiCN, and TiB2. *Materials Science and Engineering: A*, 2005, **396(1-2)**, 277-284.

[10] Hellström, K., Persson, P. and Ström, E. Oxidation behaviors and microstructural alterations of a Mo (Si, Al) 2-based composite after heating at 1580° C either in a furnace (ex-situ) or via alternating current (in-situ). *Journal of the European Ceramic Society*, 2015, **35(2)**, 513-523.

[11] Minasyan, T., Aydinyan, S., Liu, L., Volubujeva, O., Toyserkani, E. and Hussainova, I. (2020). Mo (Si1-x,Alx)2-based composite by reactive laser powder-bed fusion. *Materials Letters*, **281**, 128776.

[12] Kuchino, J., Kurokawa, K., Shibayama, T. and Takahashi, H. Effect of microstructure on oxidation resistance of MoSi2 fabricated by spark plasma sintering. *Vacuum*, 2004, **73(3-4)**, 623-628.

[13] Samadzadeh, M., Oprea, C., Sharif, H. K. and Troczynski, T. Comparative studies of the oxidation of MoSi2 based materials: low-temperature oxidation (300–900° C). *International Journal of Refractory Metals and Hard Materials*, 2017, **66**, 11-20.

[14] Knittel, S., Mathieu, S. and Vilasi, M. The oxidation behaviour of uniaxial hot pressed MoSi2 in air from 400 to 1400° C. *Intermetallics*, 2011, **19(8)**, 1207-1215.

[15] Minasyan, T., Aydinyan, S., Toyserkani, E. and Hussainova, I. In Situ Mo(Si,Al)2-Based Composite through Selective Laser Melting of a MoSi2-30 wt.% AlSi10Mg Mixture. *Materials*, 2020, **13(17)**, 3720.

[16] Minasyan, T., Aydinyan, S., Toyserkani, E. and Hussainova, I. Parametric Study on In Situ Laser Powder Bed Fusion of Mo (Si1− x, Alx) 2. *Materials*, 2020, **13(21)**, 4849.

[17] Liu, Y. Q., Shao, G. and Tsakiropoulos, P. On the oxidation behaviour of MoSi2. *Intermetallics*, 2001, **9(2)**, 125-136.

[18] Kurokawa, K., Houzumi, H., Saeki, I. and Takahashi, H. Low temperature oxidation of fully dense and porous MoSi2. *Materials Science and Engineering: A*, 1999, **261(1-2)**, 292-299.

Functionally Gradient Ti6Al4V-TiB Composite Produced by Spark Plasma Sintering

Le Liu[1*], Roman Ivanov[1], Rahul Kumar[1], Tatevik Minasyan[1], Maksim Antonov[1], Irina Hussainova[1*]

[1]Tallinn University of Technology, Ehitajate 5, 19180 Tallinn, Estonia

*le.liu@taltech.ee

Abstract. Functionally gradient materials (FGM) are progressive materials, where the final characteristics are sequentially changing with dimensions. Direct fabrication of functionally gradient Ti6Al4V-TiB composite was performed by spark plasma sintering (SPS) technique. Powder mixtures of six different ratios of Ti-alloy and TiB_2 were subjected to sintering at a temperature of 1350 °C and a pressure of 50 MPa with a dwelling time of 15 min to obtain a six-layer specimen. Scanning electron microscope images demonstrate smooth transition between layers. Vikers hardness (HV5) is 350 on the top side and 1038 on the bottom side, and a corresponding coefficient of friction (CoF) is 0.99 and 0.76, respectively, which demonstrates their different performance under sliding wear.

1. Introduction

The functionally gradient materials with heterogeneous (FGM) phase distribution throughout the volume possess the comprehensive properties of the constituent compounds gradually changing with dimensions. The coherent change in the content of an FGM component allows to overcome the interface disparity of different phases, provide the coherent coexistence and broaden the applicability of the material [1-4].

Titanium alloys (specifically Ti6Al4V) possessing light weight, high chemical stability, high strength and elastic modulus are capturing favored positions in aerospace, automotive and biomedical applications; however, their inferior tribological properties, mainly low wear resistance and a high coefficient of friction are suppressing the further conquer of new applications [5]. Thus, numerous compounds in powder or fiber form, have been reported to reinforce Ti, including carbides (WC, B_4C, TiC, SiC), borides (TiB_2, TiB), etc [6-8]. Adding TiB into Ti matrix is an advantageous approach to increase strength and the stiffness of the titanium matrix material (TMC), considering matching density and thermal expansion coefficients [8].

Spark plasma sintering (SPS) is a leading powder consolidation technology adopting the advantages of relatively low compaction temperatures, achieving full density of the materials, wide range of applied pressure and temperature, suppression of undesired grain growth and negligible material loss in comparison with conventional powder metallurgy techniques [9].

In this work, a set of functionally gradient Ti6Al4V-TiB composites was manufactured by spark plasma sintering using Ti6Al4V-TiB_2 powders mixtures. Hardness and sliding volume loss values were measured to confirm the different side performance of gradient sample; morphology were studied to show the gradient structure and component were analyzed to show the physico-chemical development during the consolidation process.

2. Experimental

2.1. Samples preparation

Commercially available Ti6Al4V (SLM Solutions, <63 μm) and TiB_2 (Alfa Aesar, Thermo Fisher Scientific, <44 μm) powders were chosen as starting materials. Considering the large size of Ti6Al4V particles, the powder was initially sieved and the fraction with less than 20 μm was used for powder mixture preparation.

Six powders mixtures of Ti6Al4V-TiB_2 with different weight ratio starting from pure Ti6Al4V to Ti6Al4V-TiB_2 50-50 wt% with step size of 10 wt% TiB_2 addition were prepared, corresponding to 100-0, 90-10, 80-20, 70-30, 60-40 and 50-50 wt% mixtures. For powder mixing the disintegrator DSL-175 with a combined inertial-centrifugal classifier-was used in protective atmosphere (argon). The velocity of collision was set as 156 $m{\cdot}s^{-1}$, thus, the specific energy of treatment was calculated as 18.4 $kJ{\cdot}kg^{-1}$.

For gradient sample, powder mixtures with six different weight ratio of Ti6Al4V and TiB_2 were layer by layer added to the graphite mold of 25.4 mm diameter, then consolidated using SPS device (FCT Systeme GmbH, Germany) in vacuum at temperature of 1350°C with simultaneous application of 50 MPa pressure for a dwell time of 15 min. The powder mixtures were heated up to sintering temperature with a heating rate of 100 $°C{\cdot}min^{-1}$.

As a reference, individual powder mixtures with six different Ti6Al4V and TiB_2 ratios were also prepared as independent samples using the same sintering parameters, as mentioned above. The preparation scheme of the independent and gradient samples are illustrated in Figure 1.

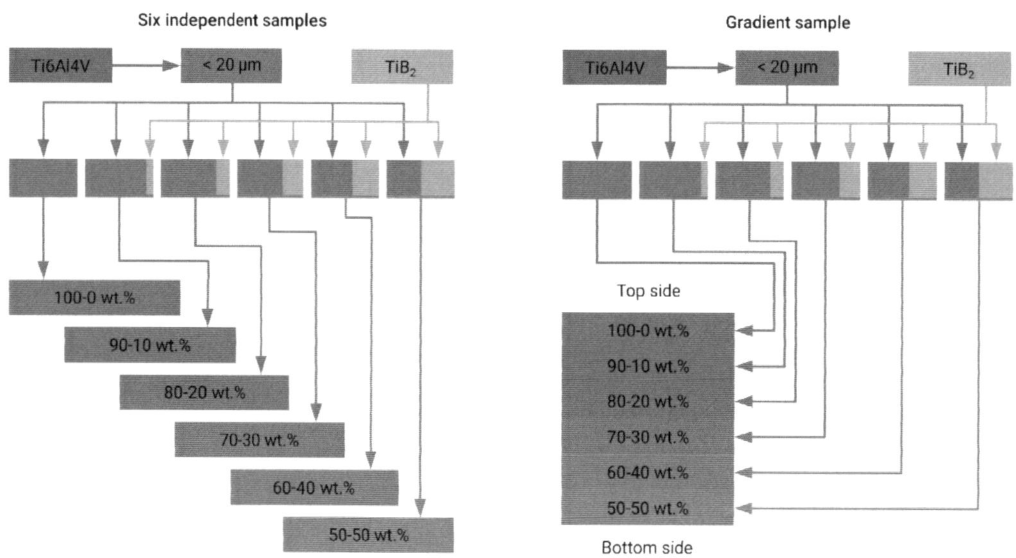

Figure 1. Schematics of Ti6Al4V- (0-50 wt%) TiB_2 independent samples and gradient sample preparation process

2.2. Characterization

Sample section was polished and microstructural examination of the specimens was conducted using a Hitachi TM1000 scanning electron microscope (SEM). The Vickers hardness was estimated on the polished surface using Indentec 5030 SKV hardness tester applying a load of 49 N for 10 s. X-ray diffraction (XRD) data were collected using a Rigaku SmartLab SE diffractometer with a D/teX Ultra 250 1D detector. The measurements were performed at room temperature from 18 to 108° (2θ) with 0.01° step with 5 seconds per step.

2.3. Sliding wear test

A universal materials test device (UMT-2) supplied by CETR (Bruker) was engaged for materials testing under dry reciprocating sliding condition in ball-on-plate configuration. A ball of alumina with Ø3 mm

(Redhill Precision, Czech Republic), hardness HV10=1450 and roughness Ra=0.02 µm, was used as a counterbody. The frequency and amplitude of sliding was 5 Hz and 1 mm, respectively. The sliding load was 1N. The test consisted of 10 steps of 1000 sec each. The surface of the wear tracks evaluated by means of the 3D profilometer Bruker Contour GT-K0+ to determine the volume of material loss (net missing volume). Experiments were conducted in air at room temperature (20±2 °C) with a relative humidity (RH) of 45±5%.

3. Results and discussion

Figure 2 shows the SEM images of developed Ti6Al4V- (0-50 wt%) TiB_2 powders mixtures and respective produced samples. The sieved Ti6Al4V spherical shape powders with particle size <20 µm and TiB_2 powder of plate-like particles with mean size of 20 µm were homogeneously mixed by the disintegrator treatment. For the sintered gradient sample, each layer shows similar image as corresponding independent samples, and there was smooth transformation between each layer boundaries.

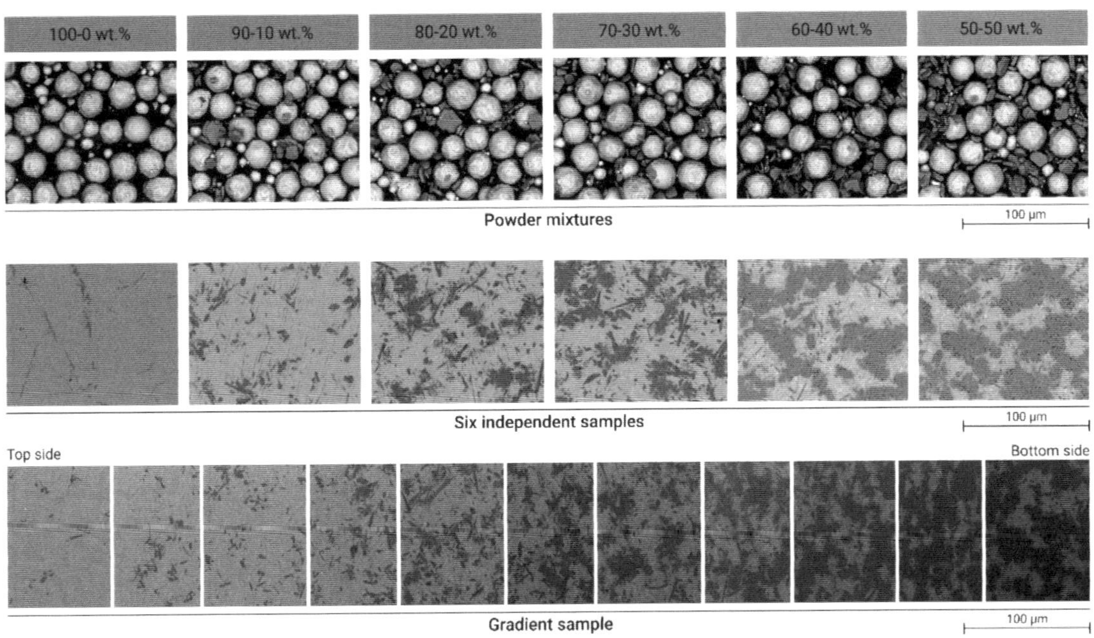

Figure 2. SEM images of six disintegrator mixed Ti6Al4V-(0-50wt%) TiB_2 powders (first row), the sintered individual samples from each of these powder mixtures (second row), and different layers of the gradient sample (third row)

Figure 3 shows the diffractograms of sintered independent samples and different regions of the gradient sample. The TiB phase formation occurs due to the reaction between Ti and TiB_2 following $Ti+TiB_2 \rightarrow 2TiB$. The XRD patterns of independent samples demonstrate the gradual increase in TiB peak, when TiB_2 content in the initial mixture was increased, highlighting the intensive reaction between Ti and TiB_2 under the given conditions. Because of thin quasi-continuous layers of gradient sample, it was not possible to accurately divide all six regions, hence 4 regions were studied, as shown in Figure 3. Stepping from Region 1 to 4, a weakening of Ti peaks is visible in contrast to strengthened TiB peaks, as the Ti is consumed for TiB during reaction.

Figure 4 demonstrates the average volume loss (wear rate) and coefficient of friction (CoF) values of the materials tested. All the composites show low wear rates and CoF values in comparison to pure Ti6Al4V specimen, and a significant drop in wear rate with the increase in TiB_2 addition is noted. It is also revealed that the tribological behaviour of the bottom side of the gradient sample (produced from

Ti6Al4V-50%TiB_2) is closer to the independent sample produced from Ti6Al4V-40%TiB_2 mixture. The volume loss and CoF of the top surface of gradient sample (produced from pure Ti6Al4V) is more close to independent sample prepared from Ti6Al4V-10%TiB_2 mixture, than to the sample sintered from pure Ti6Al4V. Hence, during sintering the components of gradient sample diffused and tend to homogenize.

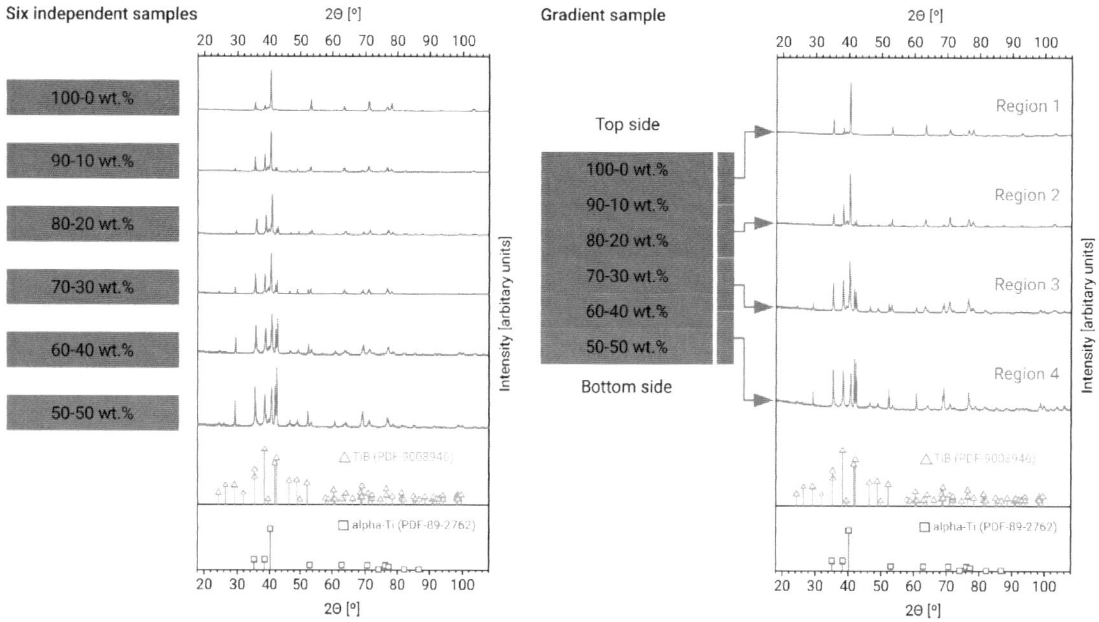

Figure 3. XRD patterns of six independent samples (sintered from Ti6Al4V- (0-50 wt%) TiB_2 mixtures) (left side) and XRD patterns of functionally gradient sample composing of altering quasi-continuous Ti-TiB composite layers (right side) along with reference Ti and TiB diffractograms

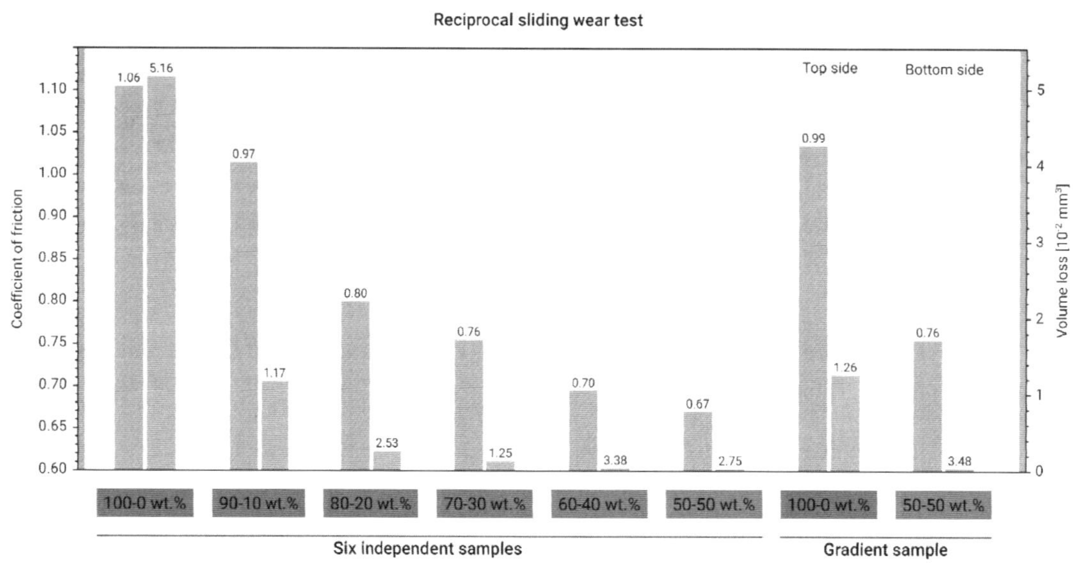

Figure 4. Average Volume loss and coefficient of friction for independent and gradient samples after sliding test

A higher wear rate for pure Ti6Al4Vspecimen is owed to its lower hardness and higher ductility. Severe ploughing, heavily deformed surface and abrasive cutting marks on the pure Ti6Al4V bulks during sliding is commonly reported as the reason behind poor tribological performance [10, 11, 12]. The intensification of wear of pure Ti6Al4V specimen is accompanied by its thermal softening due to frictional heating during sliding passes. A similar mechanism of wear in Ti6Al4V alloy was reported by others [11]. Development of surface cracks due to fatigue (cyclic) during reciprocating sliding is a practical phenomenon. Magaziner et. al. reported a similar fatigue induced cracks on Ti6Al4V alloy in reciprocation sliding and termed it to be 'fretting wear' which in turn results in early crack initiation and its subsequent growth [12, 13].

A limited range of CoF fluctuation is noted for the composites prepared from mixtures containing 20-50 wt% of TiB_2 addition. This is due to the possibility that the worn debris (or worn TiB particles) during sliding acted as third body and intensifies wear and CoF fluctuation. Generated third bodies are stated to induce localized stresses and accompanies the removal of soft matrix phase [13, 14]. Considering the measured hardness values (Vickers) of gradient sample are 350 HV5 for top surface and 1038 HV5 for bottom surface, which are similar to the corresponding individual samples (331 HV5 for sample produced from pure Ti6Al4V and 1024 HV5 for sample prepared from Ti6Al4V-50%TiB_2 mixture), the significant improvement in wear resistance of the composites are attributed to the reduced amount of plastic deformation resulting from improved bulk hardness of composites upon TiB_2 addition.

The different performance of hardness and tribological property from opposite side makes the produced FGM good candidates in many favoured regions. For example, in the machine where good tribological property is needed, the composite side can be the working face, while the metal side can be better connected (such as by welding) with the moving part of the machine.

4. Conclusions

Functionally gradient Ti6Al4V-TiB composite with smooth layer transition was produced from powder mixtures with different ratio of Ti6Al4V and TiB_2 by spark plasma sintering. SEM and XRD analyses showed the in-situ TiB formation, which plays a key role on microstructure, hardness and tribological behavior of the composite.

A significant reduction in wear and CoF values was noted due to an increase in TiB phase in the sintered samples (as a result of increase in TiB_2 content in powder mixtures). The increase of hardness of the composites upon TiB inclusion is held responsible for it.

Acknowledgements

This work was supported by the Estonian Research Council grants PRG643 (I. Hussainova), M-ERA.Net project "HOTselflub" MOBERA18 N.20097582-CA and "DuraCer" MOBERA9 ETAG18012.

References

[1] Santhosh V, Prakash DA, Murugan K, Babu N. 2020 Thermo-mechanical analysis of Tailor-made functionally graded materials through Friction stir processing. *Mater. Today. Proceedings.* **Sep 30**.

[2] Rajasekhar K, Babu VS, Davidson MJ. 2020 Microstructural and mechanical properties of Al-Cu functionally graded materials fabricated by powder metallurgy method *Mater. Today. Proceedings.* **Feb 20**.

[3] Zhong Z, Zhang B, Jin Y, Zhang H, Wang Y, Ye J, Liu Q, Hou Z, Zhang Z, Ye F. 2020 Design and anti-penetration performance of TiB/Ti system functionally graded material armor fabricated by SPS combined with tape casting. *Ceram. Int.* **Jul 31**.

[4] Saleh B, Jiang J, Fathi R, Al-hababi T, Xu Q, Wang L, Song D, Ma A. 2020 30 Years of functionally graded materials: An overview of manufacturing methods, Applications and Future Challenges. *Compos. B. Eng.* **Aug 29**:108376.

[5] Yuan S, Lin N, Zou J, Lin X, Liu Z, Yu Y, Wang Z, Zeng Q, Chen W, Tian L, Qin L. 2020 In-situ fabrication of gradient titanium oxide ceramic coating on laser surface textured Ti6Al4V alloy with improved mechanical property and wear performance. *Vacuum.* **Mar 16**:109327.
[6] Hirai H, Kurita H, Gourdet S, Fujita K, Nakazawa K, Kikuchi S. 2020 Effect of TiB orientation on four-point bending fatigue properties of TiB-reinforced Ti-3Al-2.5V alloy treated with heat extrusion. *Eng. Fract. Mech.* **Oct 15**; 238:107284.
[7] Nasr MM, Anwar S, Al-Samhan AM, Abdo HS, Dabwan A. 2020 On the machining analysis of graphene nanoplatelets reinforced Ti6Al4V matrix nanocomposites. *J. Manuf. Process.* **Nov 10**.
[8] Liu L, Minasyan T, Ivanov R, Aydinyan S, Hussainova I. 2020 Selective laser melting of TiB2-Ti composite with high content of ceramic phase. *Ceram. Int.* **May 27**.
[9] Yung DL, Antonov M, Hussainova I. 2016 Spark plasma sintered ZrC-Mo cermets: Influence of temperature and compaction pressure. *Ceram. Int.* **Aug 15**;42(11):12907-13.
[10] Mao YS, Wang L, Chen KM, Wang SQ, Cui XH. 2013 Tribo-layer and its role in dry sliding wear of Ti–6Al–4V alloy. *Wear.* **Jan 15**;297(1-2):1032-9.
[11] Dalili N, Edrisy A, Farokhzadeh K, Li J, Lo J, Riahi AR. 2010 Improving the wear resistance of Ti–6Al–4V/TiC composites through thermal oxidation (TO). *Wear.* **Aug 19**;269(7-8):590-601.
[12] Magaziner RS, Jain VK, Mall S. 2009 Investigation into wear of Ti–6Al–4V under reciprocating sliding conditions. *Wear.* **Jun 15**;267(1-4):368-73.
[13] Zikin A, Badisch E, Hussainova I, Tomastik C, Danninger H. 2013 Characterization of TiC-NiMo reinforced Ni-based hardfacing. Surface and Coatings Technology. *Surf. Coat. Technol.* **Dec 15**;236:36-44.
[14] Umanskii AP, Storozhenko MS, Hussainova IV, Terentiev AE, Kovalchenko AM, Antonov MM. 2015 Structure, phase composition, and wear mechanisms of plasma-sprayed NiCrSiB-20wt.% TiB2 coatings. Powder Metallurgy and Metal Ceramics. *Powder Metall. Met. Ceram.* **Mar 1**;53(11-12):663-71.

IOP Conf. Series: Materials Science and Engineering 1140 (2021) 012005 doi:10.1088/1757-899X/1140/1/012005

Towards Individualized Production with Flexible Shopfloor IT based on Distributed OPC UA Services

Tobias Wolff, Axel Vick

Fraunhofer Institute for Production Systems and Design Technology, Pascalstraße 8-9, Berlin, GER

E-mail: axel.vick@ipk.fraunhofer.de

Abstract. Nowadays, manufacturers face the problems to adjust their production according to the fast-changing global competition. Car manufacturers, for example, are in front of a new era of individual mobility. They need to switch from combustion engines to electric drives. In general, manufacturers have to develop production strategies to lower set up time and meet requirements faster. The classical production line is most effective if the product variety is low and the lot size is high but facing the rapid growing demand for individualized products and fast-changing requirements, manufacturers must discover new ways of manufacturing. One possibility is the flexible shopfloor. In this work, we will present our concept of a flexible shopfloor IT for controlling a robotic demonstrator. We use OPC UA over TSN as a platform for communication and CODESYS as SoftSPS for the control of sensors and actuators. In order to create valid data for ongoing discussions we use function point analysis and counting lines of sourcecode to measure software size. Our expectation is that the initial effort for implementing a flexible shopfloor IT is higher compared to the classic approach, but for all the following implementation it will take less time, because of the possibility of reuse already developed systems.

1. Introduction

The topic of flexiblity within a company is research topic since the 1980s. In [1], Grob discussed flexiblity with a broad view and named different variation of flexibility like flexiblity of employees, of process and organisation structure and flexiblity of production systems. Within the manufacturing flexibility can be splitted into more appropiate definitions like batch size, variations of products and production concepts. In this paper we focus on the flexible production systems which is one concept of production. In [1] Grob named reasons why companies are forced to increase their flexiblity within their organisation.
The key factors are:

- Rapid changing Global Market,
- Changing Technology,
- Laws and International Regulations,
- Mindset of Society regarding Labor.

For each key factor the recent history provides examples like climate change; rethinking individual mobility; news are distributed faster and trends arise faster and stronger; and young people with different ideals. Companies face those different challenges and much more in the nearest future. They have to find new ways to increase their flexiblity.
Concluding the chapter we can formulate the research question: Does a flexible production reduce the time needed, when reconfiguration is required?

2. Related Work

Industry 4.0 and digitization are intended to find solutions to the problems mentioned. With OPC UA, MQTT, CoAP technologies and standards have been developed that enable stronger networking. OPC UA is seen here as the de facto standard for Industry 4.0, since it enables vertical and horizontal networking of systems. With Time-Sensitive Networking (TSN), OPC UA also becomes real-time capable and thus time-critical applications can be controlled with OPC UA. One example is the control of robots. Communication in OPC UA is based on a client/server principle. This can be a disadvantage in complex systems, since every connection must be actively maintained from both sides. The OPC Foundation extends the standard by the Publish/Subscribe principle. Subscribers create a subscription to a topic and then receive a notification in case of a change or a defined time interval. With [10] Pfrommer et al. present a flexibly structured manufacturing IT. They use OPC UA to network their developed systems - which are on different levels in the automation pyramid - with each other. Pfrommer et al. use AutomationML to describe the capabilities offered by the system. The system sends this description to the central controller when it logs into the network. In [3] Jaekel et al. present their concept of flexible manufacturing, the modular shop-floor IT. The core elements here are the process definition using Integrated Enterprise Modeling (IUM), the Execution Engine [4] and the Cyper Physical System Adapter (CPS Adapter). With IUM, processes and the required machines, resources and products can be defined. The Execution Engine controls the machines and is the link between the IUM and the machines. The CPS Adapter is responsible for connecting the machines to the Execution Engine and thus to the higher-level IT. It forms an intermediate layer and enables machines without OPC UA to be integrated into an IT infrastructure based on OPC UA. Older machines can thus continue to exist in Industry 4.0. With TSN a real time communication based on Ethernet is developed. In [7], [9], [10], and [6] it is presented in combination with OPC UA to enable real-time capable and distributed control.

3. Methods and Material

The work presented in chapter 2 is extended by some points. In figure 1 you can see our architecture and our idea of a flexible production system. Instead of the idea of the CPS Adapter we use CODESYS as a link between machine and IT infrastructure. This has the advantage that a programmable logic controller (PLC) is redundant and the control of the sensors and actuators is done directly from a software. The CODESYS instance is therefore hardware independent and in combination with TSN can run physically far away from the machine in a container - in our case, a Linux server. The robot in Figure 1 is controlled this way. We also follow the edge computing approach. In complex systems the centralization of computing capacities is critical. The idea of edge computing is to distribute the computing resources and bring them closer to

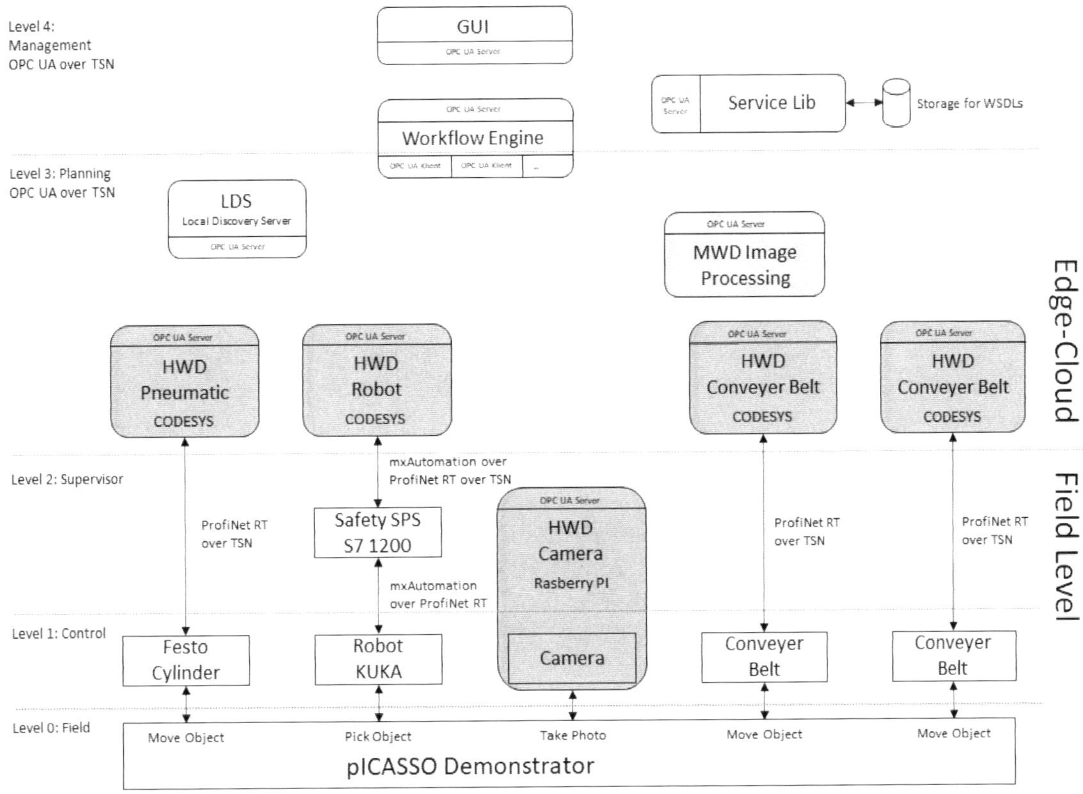

Figure 1. IT Architecture of Components within the Flexible Production System

the source of the data, so that processing of the data is done at the source and only the derived data is sent to the central controller.

On the left side of Figure 1 the levels of the automation pyramid are illustrated so that the system can be categorized by their purpose. With OPC UA this hierachy is redundant because systems can be directly connected to each other. The core system is the Workflow-Engine as centralized control system. We outsourced some functionalities into new systems to meet the Edge-Computing paradigm. The ServiceLib and the Local Discovery Server (LDS) are two of them. The Workflow-Engine is responsible for executing processes and creating service descriptions of newly added services. A web service description language file is created for every service. After creation the Workflow-Engine calls an OPC UA method of the serviceLib which saves the file. Those files are needed when a user defines a new process in the graphical user interface (GUI). To be as flexible as possible we used a LDS [11]. Every newly added OPC UA server registers itself at the discovery server. Every client can use the OPC UA service *FindServers* to get a list of available OPC UA server within the system.

To create data on which we can discuss our approach and find a answer on our research question we need quantization. Our development is software only and that is why we use software estimation methods to analyze it. In [13] the first method of software sizing was introduced and this approach is still used. It is based on counting lines of source code. This counting is good for software at the development phase or for finished products. Another well known approach is the Function Point Analysis [14]. It is standardized by the International Function Point User Group (IFPUG) and was recently recommend by the European Parliament for price calculations of software based on its size [15]. Function Point Analysis is used at the early stages of projects where only the architecture exists. In our architecture we use different programming languages like C++ for the Workflow Engine, Java for the GUI and structured text for the PLCs. The function point analysis is best suitable for comparing systems programmed in different languages. Indeed it is not applicable for PLC programming. A programm of a PLC is executed in cycles whereas in C++ or Java works sequential. In one cycle the PLC reads all inputs, executes all subprograms and writes the outputs. The function point analysis relies on functions, which do not exist in PLC programming. Furthermore function point analysis focus on the written code and not on the compiled program. Reading and writing inputs and outputs in a cycle is an analog digital conversion and not manipulating data. That is why we chose counting lines of source code.

4. Experiments

To prove or disprove our statement we did three experiments. At first we definied a process and applied it to our three use cases. Those three scenarios are our three experiments. Our first case is the classic approach where we develop a PLC programm. We need this reference value in order to compare our other results. The second experiment is our initial implementation of our concept. In our third experiment we replace the services with C++ programs. Those programs simulate a cyber physical system. Replacing services in our architecture is our kind of reconfiguration.

5. Results and Discussion

Our exception was, that a flexible organized production reduces the time needed if reconfiguration is needed. We cannot confirm that. The amount of lines of the classic approach is nearly the same compared to flexible production. The reason for this is the increased amount of functionalities within our architecture. The distributed system based on OPC UA and Cloud Computing results also in a vertical integration within the automation pyramid. The implementation of OPC UA needs a lot development effort which leads to an increased amount of source code. Critical is that the amount of source code over time is not linear. We gained a lot experience of OPC UA an open62541 as C++ library so that we can develop the same application faster than at the beginning of this project. A better method für quantization is needed for our use case. We welcome the further development of the function point analysis. In the current version it offers the best solution to compare software programmed in different languages, but it lacks by analysing a PLC program.

6. Conclusion

The work presented in this paper has two outcomes. The first one is a combination of different technologies to create a flexible production system. Here, the edge-computing paradgima took a place and we showed with CODESYS, OPC UA and TSN how we achieved this. CODESYS makes the hardware programmable logic controller redundant and can run on different systems like servers or computers with different operating systems. CODESYS implements the Micro Embedded Device Server which summarizes functionalities and is therefore not able to implement methods or Publish/Subscribe [12]. A future implementation of both will make CODESYS even more suitable. OPC UA as promising standard is implemented and supports the edge-computing concept with the LDS and the platform for connecting several devices from different vendors from different levels of the automation pyramid.
TSN makes real-time critical applications possible. Especially in Level 1 and Level 2 of the classic automation pyramid the time delay has to be within a few milliseconds with a few microseconds of jitter to meet PLC taks cycle times requirements.

We neglected the definition of processes in our layout. Jaekel et al. [3] and Torka [4] contributed a lot of research here. The implementation of a process definition tool into our architecture would increase the flexibility of our proposed production system. Furthermore, the integration of an enterprise ressource planning system would overcome the last remaining layer of the automation pyramide. Another part which is out of focus of this paper is the time needed for reassembling the hardware, organization and supply chain management.
The second outcome is the analysis of developing effort of such a system. With the lines of code and Function Points we used two different methods for software size measurement. Our initial implementation takes more time than the classical approach by directly programming PLCs. We could not show that our system reduces the effort needed for configuration, because our proposed architecture has more functionalities than implementing a process directly in a PLC. Besides this analysis we discovered the problem of appropriate methods for quantization.
We showed that the combination of edge-computing with OPC UA over TSN and CODESYS as software PLC is one concept of many flexible production systems concepts which was one of our goals. The second contribution of this paper is our analysis, where we tried to show that a flexbile production saves development time. Here we showed that measuring software size is still a challenge. More research has to be done on both: flexible production systems and appropriate methods for quantization.

7. References

[1] Grob R 1986 *Flexibilität in der Fertigung* (Berlin, Heidelberg: Springer) p 2 ff.
[2] Wintrich N, Knothe T, Eggers A, Riedel O, Stölzle S, Margraf T 2014 *Modellbasierte modulare Shopfloor IT - Integration in die Werkzeuge der Digitalen Fabrik* (Germany Berlin: Fraunhofer IPK)
[3] Frank-Walter J, Torka J, Eppelein M, Schliephack W, Knothe T 2017 *Model Based, Modular Configuration of Cyber Physical Systems for the Information Management on Shop-Floor* (Greece Rhodos: On the Move to Meaningful Internet Systems (OTM))
[4] Torka J 2017 *Modellbasierte Execution Engine für die Ablauflogik der modularen Shopfloor IT in der Produktionstechnik* (Germany Berlin: Beuth Hochschule für Technik Berlin)
[5] Saracco R 2017 *Carrying 10 Exabyte of cars generated data*
[6] Zhou Z, Shou G 2019 *An Efficient Configuration Scheme of OPC UA TSN in Industrial Internet* (China Hangzhou: Chinese Automation Congress (CAC))
[7] Gogolev A, Mendoza F, Braun R 2018 *TSN-Enabled OPC UA in Field Devices* (Italy Turin: 2018 IEEE 23rd International Conference on Emerging Technologies and Factory Automation (ETFA))
[8] Eckhardt A, Müller 2019 *Analysis of the Round Trip Time of OPC UA and TSN based Peer-to-Peer Communication* (Spain Zaragoza: 2019 24th IEEE International Conference on Emerging Technologies and Factory Automation (ETFA))
[9] Eymüller C, Hanke J, Hoffmann A, Kugelmann M, Wolfgang R 2020 *Real-time capable OPC-UA Programs over TSN for distributed industrial control* (Austria Vienna: 2020 25th IEEE International Conference on Emerging Technologies and Factory Automation (ETFA))
[10] Pfrommer J, Stogl D, Kiril A, Schubert V, Hein B 2014 *Real-time capable OPC-UA Programs over TSN for distributed industrial control* (Spain Barcelona: 2014 IEEE International Conference on Emerging Technology and Factory Automation (ETFA))
[11] OPC Foundation 2018 *OPC Unified Architecture Part 12: Discovery and Global Services Release 1.04*
[12] OPC Foundation 2018 *OPC Unified Architecture Part 7: Profiles Release 1.04*
[13] Boehm B, Clark B, Horowitz E, Westland C, Madachy R, Selby R 1995 *Cost models for future software life cycle processes: COCOMO 2.0* (Annals of Software Engineering volume 1) p 57 ff.
[14] Albrecht A, Gaffney J 1983 *Software Function, Source Lines of Code, and Development Effort Prediction: A Software Science Validation* (IEEE Transactions on Software Engineering Volume SE 9) p 639 ff.
[15] European Parliament 2020 *European Union Agency for the Operational Management of Large Scale IT Systems in the Area of Freedom, Security and Justice (eu LISA)* (Belgium Brussel)

Range Sensor Overview and Blind-Zone Reduction of Autonomous Vehicle Shuttles

Junyi Gu[1], **Tek Raj Chhetri**[2]

[1]Department of Mechanical and Industrial Engineering, Tallinn University of Technology
[2]Department of Computer Science, University of Innsbruck

E-mail: junygu@taltech.ee

Abstract. In recent years, with the advancement in sensor technologies, computing technologies and artificial intelligence, the long-sought autonomous vehicles (AVs) have become a reality. Many AVs today are already driving on the roads. Still, we have not reached full autonomy. Sensors which allow AVs to perceive the surroundings are keys to the success of AVs to reach full autonomy. However, this requires an understanding of sensor configurations, performance and sensor placements. In this paper, we present our experience on sensors obtained from AV shuttle iseAuto. An AV shuttle iseAuto designed and developed in Tallinn University of Technology is used as an experimental platform for sensor configuration and set-up.

1. Introduction

Recently, there has been growing interest in autonomous vehicles (AVs), which are regarded as a potential trend of transportation in the future. A reliable AV perceives the environment consistently by different sensors, then transfers the sensory data to a computer for post-processing. Sensors in AVs produce information with different characteristics, Global Navigation Satellite System (GNSS) provides the approximated location of vehicles with a general reference; Inertial Measurement Unit (IMU) measures angular rates, linear velocities and orientation of the vehicles base body; range sensors include cameras, LiDARs, radars and sonars detect the objects that are around vehicles in different scales and properties. Sensor fusion algorithms combine sensory data to create more coherent and certain results than using the data individually. Path planning module uses real-time perception of the surrounding environment to update paths of the vehicle in short and long ranges. Fully AVs are supposed to be able to control the self-motions, as well as auxiliary functions in practical situations, for example, the closing and opening of the door. The motion control module, at last, controls the movement of the vehicle to follow the paths and execute the motion commands that are computed by path and motion planners. Security measurements like emergency braking and obstacle avoidance are invoked to the control system directly to improve the safety and reduce the accidents. Fig.1 summarizes the general workflow of AV modules.

Range sensors provide 3D geometry information of vehicles surrounding environment and reflect the properties (speed and acceleration) of objects that are expensive to compute from vision-based perception systems. Examples of range sensors for AVs include radar, LiDAR, sonar and infrared sensors. In particular, radar sensors make a crucial contribution to Advanced Driver Assistance System (ADAS) in the aspects of emergency braking/brake assist, collision

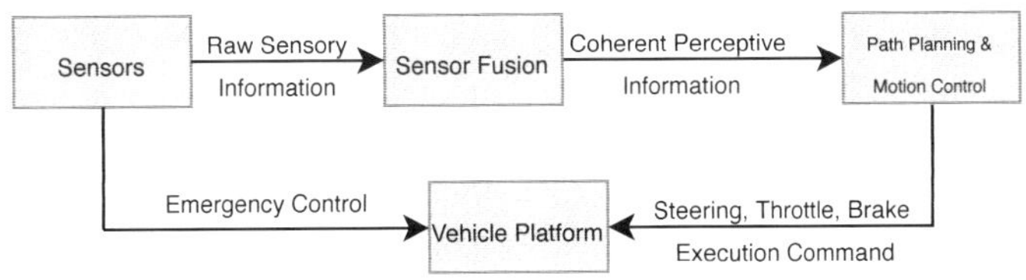

Figure 1. Modules workflow of autonomous vehicles

warning/avoidance, park assist, distance control and so on. LiDAR sensors are a relatively new technology in the AVs field and have attracted much attention in recent years. LiDARs are widely used to measure the distance and describe the environment in three dimensions. However, because of the natural characteristics of laser-based systems, LiDARs are limited by the low level of target reflectivity, resolution and refresh rate. Contrarily, the energy source of sonar sensors is acoustic/ultrasonic waves in a specific frequency, which is less affected by the reflectivity level of targets. Compared with LiDARs, sonar sensors have the advantage of low-cost and are widely used in underwater applications. Typical infrared (IR) sensors are a relatively well-developed technology, which has the advantages of cost, size and reliability. Active IR sensors share the similar principle of sonar but rely on infrared waves (wavelength usually bigger than 780nm that above the visible red light). Passive IR sensors only have receivers to detect infrared radiation and are irreplaceable in the scenarios of human/human-motion detection. This paper analyzes range sensor configurations and blind zones reductions for particular AV shuttle iseAuto.

2. Primary Perceptive Range Sensor Set-ups

In recent years, autonomous vehicles commonly use laser-based range sensors as the primary approach to perceive the environment and measure distances. One of the most popular laser-based sensors is LiDAR, a light-based detection and ranging remote sensing tool that has contributed significantly to AV technology due to the high accuracy and precision.

Currently, decreasing cost and power consumption of LiDARs promote their usage in applications that are sensitive to the vehicle's size and weight, such as Unmanned Aerial Vehicles for mapping and navigation purposes. Other related research and experimental platforms are:

- On the roof of Stanley, the vehicle that won the 2005 DARPA Grand Challenge, there are five lasers measuring cross-sections of the approaching front terrain in different distances out to 25 meters [1];
- VIAC vehicles were equipped with four laser scanners (two lateral laser scanners, one off-road laser scanner and one central laser scanner) which have different characteristics [2];
- More recently, Gao *et al.* [3] set four laser sensors (two single-line lasers, one four-line laser and one 64-line laser) in their Mengshi autonomous vehicle;
- Other experimental vehicles that have laser-sensors installed [4] [5] [6];

AV shuttles are the most common low-speed vehicles using LiDARs, and not only for ranging but also for localization and object classification. AV shuttles that were deployed on the real traffic pilot cases around the world are in limited numbers. Most known brands are Navya and Easymile, followed by iseAuto and GACHA. All these vehicles are relying mostly on LiDARs as the main localization and ranging sensor. Fig.2 presents the main sensor locations on these vehicles.

In this paper we are focusing on the TalTech iseAuto that was designed and developed in the Autonomous Vehicles lab in TalTech, Estonia [7] [8]. The initial design and sensor configuration

Figure 2. Real traffic AV shuttles: Navya Evo, Easymile EZ10, TalTech iseAuto and GACHA

development of iseAuto was supported by mechatronic modeling methodology [9] [10], which emphasizes the importance of early design stage. The output of the initial conceptual design stage proposed to use two Velodyne VLP-16 Puck LiDARs on the front top corners, as shown in Fig.3(a). To detect as many blind zones that are in front of the shuttle as possible, the sensor plane inclined toward the front and side (in practical, 8.3° toward the front and 6.9° toward the side), as shown in Fig.3(b) and 3(c).

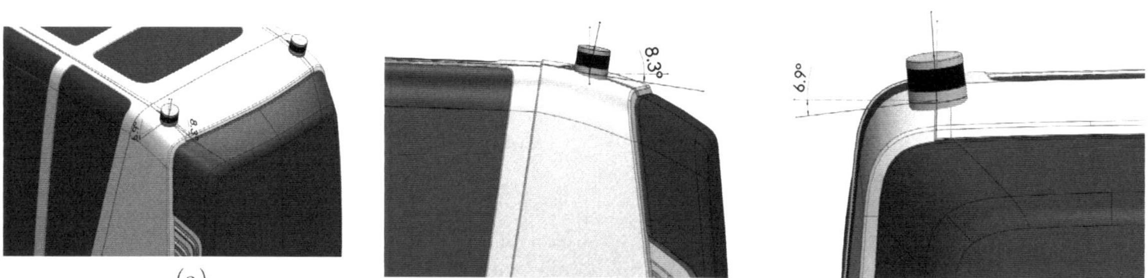

(a) (b) (c)

Figure 3. (a) Initial design of LiDARs location, (b) Front-tilted angle, (c) Side-tilted angle

However, in practical situations, our initial Velodyne VLP-16 sensor location (Fig.3(a)) has no vision of the shuttle's backside because laser beams shooting toward the back were heavily blocked by the shuttle itself. The points cloud of initial location configuration were showed in Fig.4(c). Additionally, the vision of left and right sides is too limited to cover the blind zones of the automatic door, which directly affects the safety of the shuttle. Therefore, our latest configuration of two VLP-16 sensors is locating them in the middle of the front and back sides with some inclines, as shown in Fig.4(a). Moreover, an adjustable mount base allows us to change the front-tilted-angle of the VLP-16 sensor, and a bigger angle helps to detect more blind zones in the front/back of the shuttle but reduce the maximum detection range correspondingly. The points cloud based on the latest configuration were presented in Fig.4(b). Compared with the previous configuration (Fig.4(c)), the coverage of the left and right sides is reduced, but the full view of the backside is available. On the other hand, current configuration helps to reduce the occasional interference patterns and shadowed azimuth ranges that may appear in data when using multiple Velodyne sensors close to one another (especially on top of the vehicle).

3. Blind Zones Reduction

Blind zones detection is an essential task for AVs because it has straight affections to safety. LiDAR-based sensors generally are installed on top of the vehicles to have wider horizontal Field of View (FoV) and further detection range. However, a top-placement configuration of LiDARs results in bigger blind zones around the vehicles, which raise problems in many post-practical

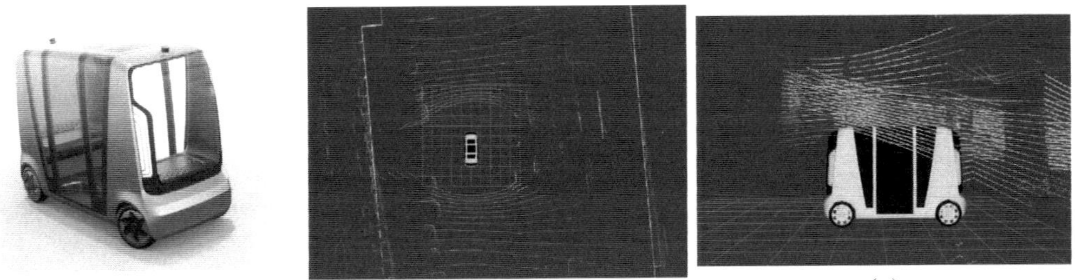

(a) (b) (c)

Figure 4. (a) Latest design of LiDARs location, (b) Points cloud of latest design, (c) Points cloud of initial design

processes such as motion planning in multi-interaction environments and lane change alert in ADAS.

The common solution to reduce the blind zones is installing specific sensors in corresponding positions. Variety types of sensors can be used to detect the objects in blind zones.

- Jamaluddin *et al.* [11] installed an ultrasonic sensor above the rear tire to measure the distance of approaching vehicles. The selection of the ultrasonic sensor maximally prioritizes the cost of the total sensor setup but compromises the performance and accuracy in some real-life scenarios [12].
- Using LiDARs to cover the blind zones is a popular topic in recent because they create detailed 3D points cloud. Researchers can carry out complex post-processes that are based on points cloud data to pursue the best performance. The work in [13] formulated the blind zone problems by occupancy grid and proposed a generic algorithm to optimize the configuration of LiDAR placements. Meadows *et al.* [14] introduced a system that has three LiDARs and used neural work to evaluate the effectiveness of various LiDAR poses.
- Other sensor choices include cameras and radars. Image-based information is usually processed alongside other sensory data. Rangesh *et al.* [15] described a multi-object tracking approach which is capable of working with varying camera FoVs and LiDARs. Dey *et al.* [16] put the camera and radar together and proposed a framework that can optimize the location and orientation for a heterogeneous set of sensors on a given target vehicle.

Installing sensors in corresponding areas provides direct sensory information of the objects in blind zones. However, in the scenarios that the objects' detailed detections are not vital, mathematical processes can be used to calculate the states of the objects when they are in blind zones. Zhou *et al.* [17] proposed to use Kalman Filter to estimate the movement of the approaching vehicles in blind zones for traffic intersection motion planning. Correspondingly, substitute sensors with mathematical algorithms help to reduce power consumption and hardware maintenance work.

In our case, because of the structure of the iseAuto shuttle, the front-top and back-top Velodyne LiDARs cannot detect the blind zones on two sides. Accurate and detailed detection of the objects in the right blind zone, especially the area that is close to the shuttle, is vital for us because the control of the automatic door should be strictly based on it. Our solution is installing two RS-Bpearl LiDARs on the left and right sides of the shuttle.

RS-Bpearl is a short-range LiDAR specifically designed for the detection of the blind zones. Compared with VLP-16 Puck, RS-Bpearl has a shorter range of detection but a much wider 90° vertical FoV and 32 channels. For iseAuto shuttle, the unique FoV design of RS-Bpearl helps to cover more areas on two sides, and the dense points cloud data provides more details of the objects in blind zones. Fig.5 presents the points cloud data that was produced by an

RS-Bpearl LiDAR that was installed on the right side of the shuttle. The scenario in Fig.5(a) is the outdoor environment that has buildings and parking cars. Fig.5(b) shows the ability of the right RS-Bpearl LiDAR to detect the object details (human and ladder) that are close to the automatic door.

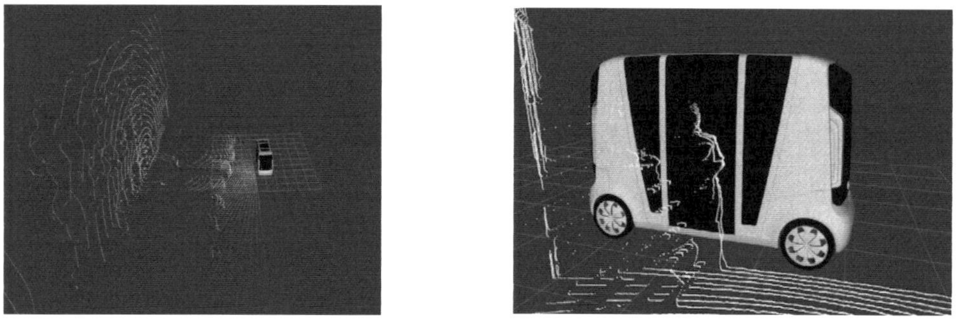

(a) (b)

Figure 5. (a) Points cloud in outdoor environment, (b) Details around automatic door area

Another key blind zone for iseAuto shuttle is the close front area, which is not able to be detected by either front-top VLP-16 or side RS-Bpearl LiDARs, as shown in Fig.6(a). The perception of small objects (kids, pets, etc.) in this area is important for the shuttle's safety system. Available sensor choices to detect this blind zone such as IR and ultrasonic sensors have the economic advantages but compromise in the accuracy. In terms of cost and performance, solid-state LiDARs are believed to be more suitable for large-scale deployment, because solid-state LiDARs are relatively cheaper and do not have inside complex mechanical mirror systems.

(a) (b)

Figure 6. (a) Front blind zone, (b) Points cloud from bottom-front and top-front LiDARs

We deployed a Benewake CE30-C LiDAR on the front-bottom of the shuttle to detect the blind zones. Benewake CE30-C is a typical solid-state LiDAR that is based on the Time of Flight (ToF) ranging principle. The measurement is performed based on the received emitted modulated near-infrared light, which is reflected by the objects. Fig.6(b) shows the merged points cloud data from front-top Velodyne and front-bottom Benewake LiDARs. Benewake solid-state LiDAR can detect the down part of traffic signs and human legs (white points) that cannot be seen by the front-top Velodyne LiDAR.

4. Summary and Future Work

This paper provided an overview of the most common range sensors that are used for AVs and more specifically for AV shuttle, iseAuto. We evaluated the configuration and location of all range sensors that were deployed on iseAuto shuttle for primary perception and blind zones detection. As a result of the analysis, we managed to get a full view of the shuttle surroundings and cover most of the vital blind zones by five LiDARs that have different characteristics.

The future work will focus on the sensor fusion and integration of the long and short range radars into the range-sensor set of the iseAuto as well as implementing AI-based situation awareness defined in the research [18]. The second target is to create a digital twin, which is compliant to our other research results [19] of the vehicle in order to simulate all critical traffic situations and increase the total safety of the deployed system.

Acknowledgments
The research is supported by the EU H2020 project Finest Twins (grant No. 856602).

References

[1] Thrun S *et al.* 2007 Stanley: The Robot That Won the DARPA Grand Challenge *Springer Tracts in Advanced Robotics* vol 36 (Springer, Berlin, Heidelberg)
[2] Bertozzi M *et al.* 2011 VIAC: An out of ordinary experiment *IEEE Intelligent Vehicles Symposium (IV)* pp 175-180
[3] Gao H, Cheng B, Wang J, Li K, Zhao J and Li D 2018 Object Classification Using CNN-Based Fusion of Vision and LIDAR in Autonomous Vehicle Environment *IEEE Transactions on Industrial Informatics* vol 14 no 9 pp 4224-31
[4] Wei J, Snider J M, Kim J, Dolan J M, Rajkumar R and Litkouhi B 2013 Towards a viable autonomous driving research platform *IEEE Intelligent Vehicles Symposium (IV)* pp 763-770
[5] Broggi A *et al.* 2013 Extensive Tests of Autonomous Driving Technologies *IEEE Transactions on Intelligent Transportation Systems* vol 14 no 3 pp 1403-15
[6] Zhang J and Singh S 2018 Laser-visual-inertial odometry and mapping with high robustness and low drift *J. Field Robotics* pp 1242-64.
[7] Sell R, Leier M, Rassõlkin A, and Ernits J 2018 Self-driving car ISEAUTO for research and education *Proc. of the 19th International Conference on Research and Education* (Mechatronics, Delft, Netherlands)
[8] Rassõlkin A, Sell R and Leier M 2018 Development case study of the first estonian self-driving car, iseauto. *Electrical, Control and Communication Engineering* 14 pp 81–88
[9] Sell R, Coatanea E and Christophe F 2008 Important Aspects of Early Design in Mechatronic *Proc. of the 6th international conference of DAAAM Baltic industrial engineering* (Tallinn) pp 177-182
[10] Christophe F, Sell R, Bernard A and Coatanéa E 2009 OPAS: Ontology processing for assisted synthesis of conceptual design solutions *Proc. of the ASME International Design Engineering Technical Conferences and Computers and Information in Engineering Conference* pp 249-260
[11] Jamaluddin M H, Shukor A, Miskon M F and Redzuan M 2016 An Analysis of Sensor Placement for Vehicle's Blind Spot Detection and Warning System *J. of Telecommunication, Electronic and Computer Engineering 8* pp 101-106
[12] Lim B S, Keoh S L and Thing V L L 2018 Autonomous vehicle ultrasonic sensor vulnerability and impact assessment *IEEE 4th World Forum on Internet of Things (WF-IoT)* (Singapore) pp 231-236
[13] Kim T and Park T 2020 Placement Optimization of Multiple Lidar Sensors for Autonomous Vehicles *IEEE Transactions on Intelligent Transportation Systems* vol 21 no 5 pp 2139-45
[14] Meadows W *et al.* 2019 Multi-LiDAR placement, calibration, co-registration, and processing on a Subaru Forester for off-road autonomous vehicles operations *Autonomous Systems: Sensors, Processing, and Security for Vehicles and Infrastructure 2019.*
[15] Rangesh A and Trivedi M M 2019 No Blind Spots: Full-Surround Multi-Object Tracking for Autonomous Vehicles Using Cameras and LiDARs *IEEE Transactions on Intelligent Vehicles* 4 pp 588–599
[16] Dey J, Taylor W and Pasricha S 2020 VESPA: A Framework for Optimizing Heterogeneous Sensor Placement and Orientation for Autonomous Vehicles *IEEE Consumer Electronics Magazine* p 1
[17] Zhou D, Ma Z and Sun J 2020 Autonomous Vehicles' Turning Motion Planning for Conflict Areas at Mixed-Flow Intersections *IEEE Transactions on Intelligent Vehicles* vol 5 no 2 pp 204-216
[18] Wang R, Sell R, Rassõlkin A, Otto T and Malayjerdi E 2020 Intelligent functions development on autonomous electric vehicle platform *J. of Machine Engineering* vol 20 no 2 pp 114-125
[19] Kuts V, Otto T, Bondarenko Y, and Yu F 2020 Digital Twin: Collaborative Virtual Reality Environment for Multi-Purpose Industrial Applications. *ASME 2020 International Mechanical Engineering Congress and Exposition*, (ASME, November 16-19, 2020, Portland, OR, USA.)

Constructive compensation of the thermal behaviour for industrial robots

Eckart Uhlmann[1,2], **Julian Polte**[1], **Christian Mohnke**[1]

[1]Fraunhofer Institute for Production Systems and Design Technology IPK, Pascalstraße 8-9, 10587 Berlin, Germany
[2]Institute of Machine Tools and Factory Management, Pascalstraße 8-9, 10587 Berlin, Germany

E-Mail address: christian.mohnke@ipk.fraunhofer.de

Abstract. Industrial robot systems offer a flexible, adaptable basis due to their kinematics and their mobility An influencing variable, which is particularly relevant for processes with long process times, is the thermal heating and the associated thermal drift of the tool center point. The maximum deviation from the actual nominal position can reach up to $\Delta AP_t = 1.5$ mm. Currently, there are no procedures and methods established in practice which compensate the effects of thermal drift without expensive calibration measures and system downtime. In these investigations a system was developed which allows the reduction of thermally induced displacement by using controlled heating elements. The aim is to keep the entire robot system at a permanent, balanced temperature level. The heating elements are adapted to the geometry of the respective axis and heat the material to a temperature ϑ close to the steady state. A comparison of the drift through the heating system with the error occurring in normal operation shows that the drift of the heating system is comparable with the drift of normal heating. With the heating mats, a thermally induced error of $\Delta AP_t = 0.234$ mm can be generated within $t = 20$ min. While normal heating requires up to $t = 600$ min. The achieved error deviation of the drift of the self-heating to the heated variants is with $\Delta AP = 0.04$ mm only minimal. The results can help to reduce the influence of thermal heating and the associated thermal drift of the TCP without using cost-intensive measures with additional hardware and software on external computers for compensating the errors.

1. Introduction

In the last decade, robotic processing has evolved from a basic research topic to a production technology for industrial use [1]. The main reasons are technological progress and ongoing research [2], which have led to an increase in accuracy AP. Robot-guided machining is an innovative production technology that combines the advantages of a highly accurate machine tool with the flexibility of a 6-axis jointed arm robot. Due to their relatively low investment costs K_I compared to machine tools, milling robot systems represent a cost-effective alternative.

Continuous improvement of the industrial robots with regard to stiffness s and path accuracy AT results permanently in new industrial applications [3]. Another advantage of inexpensive industrial robots is the relatively large, usable work space A_R. Currently, the use of robot-guided machining systems in industrial serial production is mainly limited to the deburring and fettling of components with relatively rough accuracy requirements [4].

IOP Conf. Series: Materials Science and Engineering 1140 (2021) 012007 doi:10.1088/1757-899X/1140/1/012007

Current research concentrates, for example, on the thermally induced influence on the processing accuracy [3] or tries to increase the accuracy by compensating the deviation of the TCP with an improved robot control [5]. The thermally induced machining drift, the so-called thermal drift ΔAP_t, is a significant obstacle to the development of new applications for machining robots in series production.

2. Motivation

Various disturbances occur during robot-guided machining. These disturbances have a considerable influence on the machining accuracy. In order to implement industrial robots in production, it is of great importance to increase the machining accuracy. Thermal deformation Δl in continuous operation has the greatest influence on accuracy, see Figure 1 [6]. In addition, this influence is relevant for handling and assembly without dynamic loads as well as for high-precision machining processes such as milling or grinding, regardless of the type of use. Various studies have shown that the thermal influence has a great impact on the accuracy of six-axis industrial robots. In this context, drift from $\Delta AP_t = 0.10$ mm to $\Delta AP_t = 1.78$ mm was measured [7, 8, 9].

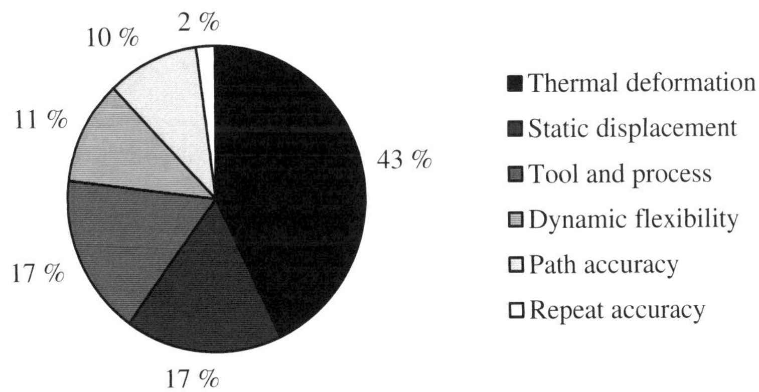

Figure 1. Factors influencing the machining deviation ΔAP [6]

The majority of robotic systems continue to be used in handling and processes with low accuracy requirements. Due to the current low number of robot systems in manufacturing, robot manufacturers have no need for retrofit solutions to compensate for thermal drift ΔAP_t. Currently, there are no established procedures and methods in practice that compensate for the effects of thermal drift ΔAP_t without costly calibration measures with corresponding system downtime t_S. In the field of university and institutional research, there are still a few approaches to compensatory and constructive measures, but there is no development approach with direct practical relevance [5, 10, 11].

3. Measures to reduce thermal influences

3.1. Preliminary investigations

The thermal drift ΔAP_t is time-variant and is constantly changing during the manufacturing process through different starting and processing speeds c_P, operating loads m_O and process times t_P [3]. This is particularly relevant in continuous operation, however, also in the production of individual parts, since a large part of the temperature-induced deviation already occurs in the first 30 minutes of heating. Due to the knowledge of the robot system and the process sequence, the thermally induced deviations can only be assessed to a limited extent or not at all.

All investigations for the experimental system analysis as well as the technological investigations are carried out on a six-axis vertical articulated robot of the type KR 60 HA from the company KUKA ROBOTER GMBH, Augsburg. A high-frequency spindle of the type ES350 from HSD SPA, Pesaro, Italy, is mounted on the end effector of the robot. The drive power is P = 8 kW. The maximum speed is specified as n_s = 36.000 1/min Tools can be accommodated with the standardized form HSK-E

40. The position of the tool center point is recorded with the optical measuring system Leica Absolute Laser Tracker AT960-MR from HEXAGON METROLOGY GMBH, Wetzlar. The reflector required for the laser tracker can be mounted directly in the holder of the milling spindle and thus in the tool axis via an adapter.

On the given robot system, the KUKA KR 60 HA, a maximum thermal drift of the tool center point of $\Delta AP_t = 1.78$ mm could be determined. It must be taken into account that all axes reach a quasi-stationary state after t = 300 min and do not continue to heat up, i. e, the displacement does not increase any further. Compared to the robot-specific repeatability of RP = 0.05 mm, the deviation is significant and has a detectable effect on the accuracy achievable with a processing robot system.

During a machining process, all axes are rarely in operation simultaneously and the axis speeds c are much lower. This results in a significantly lower structure heating and thus also in a low drift ΔAP_t. Figure 2 shows the comparison of the directions of thermal drift ΔAP_t of the sample machining with the maximum heating of the system. Due to the reduced range of motion and the low axis speeds c during sample machining, there is less heating of the servo motors and gears and thus less heat input Q into the robot structure. The result can be explained by the sample machining and the related axis movement. Machining is carried out within a small working space on the machining table. The manual axes, i.e. axes 3, 4 and 5, are mainly used for machining. These axes mainly cause heating of the arm and consequently a thermally induced drift ΔAPt in the X-direction in world coordinates. The difference in the Z-direction can be explained by the stronger heating and the associated expansion of the swingarm. Due to the expansion, the TCP lifts in the Z-direction and leads to a positive drift ΔAP_t. For sample machining, the heating of the swingarm is low. This results in a significantly lower thermal drift ΔAP_t. The stationary state is reached for sample machining after a time of t = 600 min and thus after about twice the time t as for maximum heating.

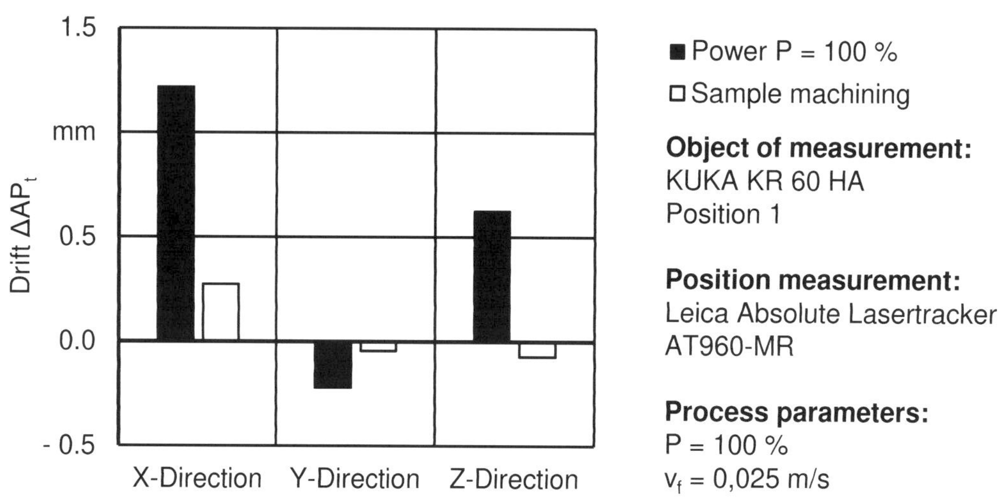

Figure 2. Comparison influence of direction on drift ΔAP_t on TCP between sample machining and maximum power P in world coordinates

The IPK and the WINKLER AG, Heidelberg have developed a system that allows the reduction of thermal caused displacement ΔAP_t or different temperature gradients $\Delta\vartheta$ by controlled heating elements. From thermographic photographs of an industrial robot it can be seen that temperatures of 40 °C $\leq \vartheta \geq$ 90 °C are present at gear and motor components, whereas the axes are in a temperature range of 30 °C $\leq \vartheta \leq$ 50 °C.

According to the current state of knowledge on thermal drift ΔAP_t, the heating of affected axles is preferable to cooling [8, 11, 12]. The aim is to keep the entire robot system at a permanent, balanced temperature level, the steady state. In this case, no additional heating and thus no additional increase in thermal drift ΔAP_t is to be expected.

3.2. Heating system

The heating elements are adapted to the geometry of the axis and keep the material at a temperature ϑ close to the highest expected heat level ϑ_{max}. The heating elements are switched off when the temperature ϑ on the axes of the industrial robot no longer changes or the heat input of the motors and gears automatically keeps the system at a constant level. The robot system must then be calibrated once. High-performance surface heating systems are used, which have a new type of material and heating wire base, resistant cover materials suitable for use in a production environment and a geometry adapted to six-axis industrial robots.

With a total weight of m = 680 kg of a KUKA KR 60 HA, heating up the entire mass in the time t = 30 min resulted in a heating power of approximately P_H = 10 kW. For this approximate calculation, a temperature of ϑ = 50 °C is assumed as steady-state temperature. Not the entire surface of the robot system is available for the heating mats. For usable surface area A_H, the maximum heating power for the robot system is P_H = 2.6 kW. The total heating surface of the robot results to A_H = 1.05 m^2 and the surface capacities of the individual heating elements are between P_A = 0.16 W/m^2 and P_A = 0.30 W/m^2. The heating surface is divided into 14 individual mats and the mats are operated with a voltage of 12 V $\leq$ U $\leq$ 60 V. Figures 3 and 4 show the heating mats attached to axes 1, 2 and 3.

Figure 3. Heating mats attached to axes 1 and 2.

Figure 4. Heating mats attached to axes 2 and 3.

In addition, a solution for the measurement and control of the new surface heating mat systems was implemented. The systems have integrated regulators to control the heating power P_H and individual settings of the individual heating mats. These can be parameterized by an infrared interface. The controllers use a Fast-Adaptive-Tuning-Algorithm (FAT), which automatically and continuously adapts the control behaviour to the controlled system. Heating elements of the company WINKLER AG, Heidelberg, are used for targeted heating. The required heating systems are individually manufactured and adapted for this investigation.

4. Results

For a real consideration of the heating and the associated drift ΔAP_t a simplified milling process is considered. The working area is limited to an area near the working table. The machining was performed in a continuous loop and the heating behaviour and the resulting drift ΔAP_t was recorded. The loop simulates a series production of a milling component and the realistic temperature-related error development associated with it. The temporal error development is shown in Figure 5.

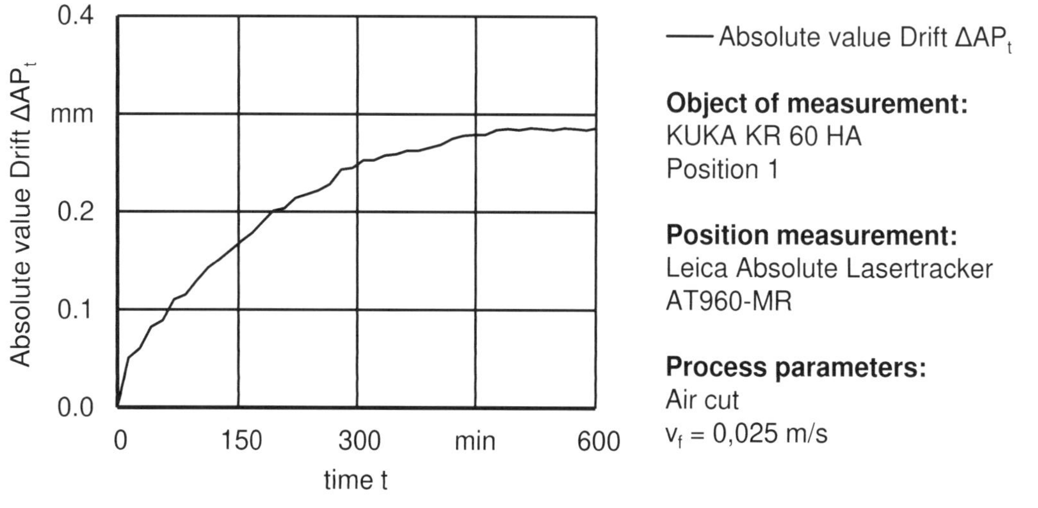

Figure 5. Experimentally determined drift ΔAP_t as a function of time t for sample machining using the of measuring position 1 of the KUKA KR 60 HA

The maximum drift for sample machining is $\Delta AP_{t,max}$ = 0,28 mm. The value is reached after a time of t = 600 min. This results in a significantly lower drift ΔAP_t for real machining, but a longer heating phase until the material is stationary.

For a real machining process, the heating system can be used to reach the stationary state and the maximum drift after a time of t = 20 min. For this purpose, the heating mats on the robot system were used in addition to the movement loop of the processing. The result is shown in Figure 6. In comparison to the results of Figure 5, the time to reach the steady state can be reduced from t = 600 min to t = 20 min. The heating system was used for only t = 11 min and then switched off. Subsequently, overshooting occurred due to the residual heat in the system. After a short cooling down period, the system is now in a stationary state. For real processes and motion sequences, a thermally stable state can be achieved with the help of the heating system after less than t = 30 min.

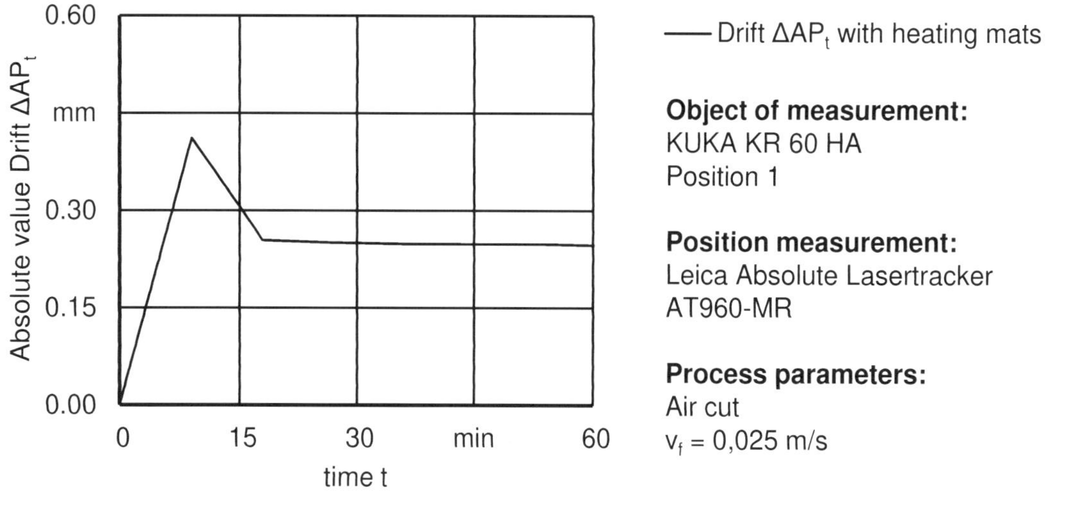

Figure 6. Experimentally determined drift ΔAP_t for sample processing including heating mats

In a first series of tests a machining process without heating system is simulated. Obomodulan® is chosen as the material for the test components. These are sheets or blocks of polyurethane foam. The material was chosen because it is cost-effective and very easy to machine. Compared to metallic materials, considerably lower cutting forces F_S and thus a reduced displacement ΔAP can be expected. In the present investigations, Garant milling tools from the company HOFFMANN SE, Munich, are used. With its rake angle, the milling cutter is particularly suitable for soft materials such as aluminum or plastic. Furthermore, the milling cutter with its face cutting geometry enables plunging, which is necessary for the drilling operation in the sample component. The robot system processes a first component in a cold state and then heats up in the programmed loop for a time of t = 600 min. In the stationary state a further component is machined. Afterwards the two components are measured and the drift ΔAP_t is observed. The visual evaluation of the scanned components shows how the thermal drift ΔAP_t affects the milling process. The geometric shapes and the hole in Figure 7 show that the robot shifts the entire milling image in the drift direction as the heat is generated. A distortion of the shapes cannot be detected or measured. It can therefore be concluded that the mutual influence of the individual axis drifts ΔAP_t and also the overall drift ΔAP_t in the area of the component remain almost constant.

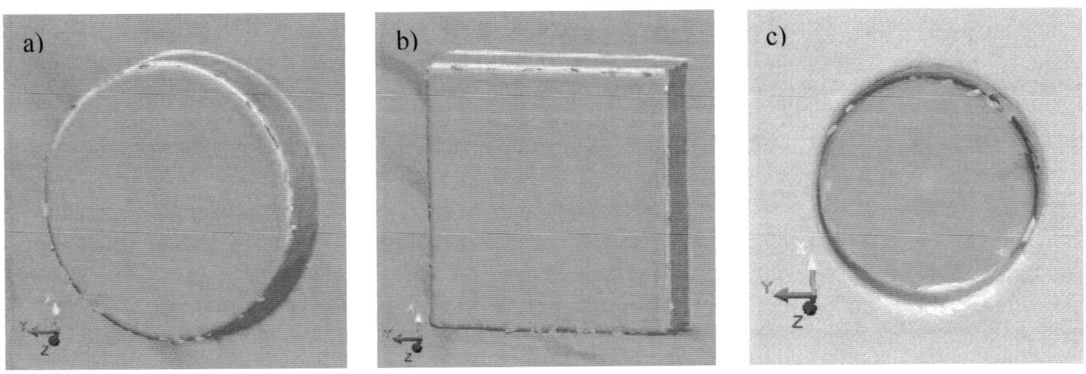

Figure 7. Thermally induced error during milling on selected geometries at time t = 0 min (grey) and t = 600 min (green); a) circle; b) square; c) hole

To assess the performance of the implemented heating system, the described test series is carried out again. At the time t = 0 min a component is manufactured by the robot system as reference. Then the heating systems are started, the robot system is heated for t = 11 min, the system is switched off and the stationary state is reached after t = 20 min. After reaching the stationary state another component is manufactured. The manufactured components are measured and the drift ΔAP_t is determined. A comparison of the drift ΔAP_t through the heating system with the error occurring during normal operation is shown in Figure 8.

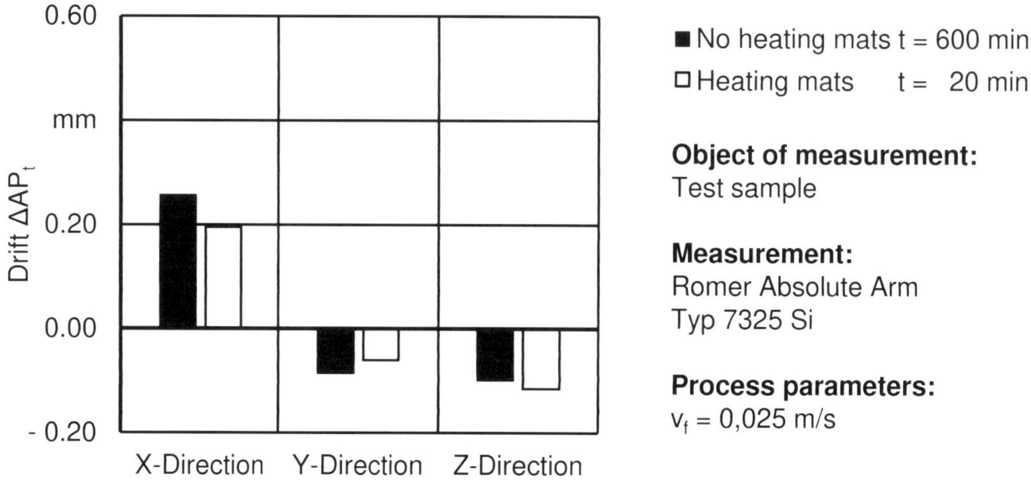

Figure 8. Comparison of drift ΔAP_t on sample component with and without heating system

It shows that the drift ΔAP_t of the heating system is comparable with the drift ΔAP_t of normal heating, see Figure 8. The component shows comparable drifts in all spatial directions. With the heating mats, a thermally induced error of $\Delta AP_t = 0.234$ mm can be generated within t = 20 min. For normal heating by the process, the system reaches this error value only after t = 600 min. For normal processing, a thermally stable robotic system can be generated after less than t = 30 min with the help of the solution. The achieved error deviation of the drift of the self-heating to the heated variants is with a value of $\Delta AP = 0.04$ mm only minimal.

5. Summary

Within the scope of the investigations, it was shown that a targeted heating of the robot structure enables faster and more economical precise manufacturing with a robot. The developed system can bring the robot within shortest time into a thermally stationary state and thus prevent thermal drift ΔAP_t. The drift ΔAP_t only needs to be corrected at the beginning of the process, no further corrections are necessary afterwards.

Especially for small and medium-sized companies, the developed system is a great opportunity to integrate robot systems into precise and flexible production. Whereas expensive systems such as laser trackers with a cost of over K = 100,000 € were previously necessary to compensate for thermal influences, the developed system with costs of up to K = 5,000 € is much cheaper and also easier to operate.

For normal machining, the solution can be used to create a thermally stable robot system after less than t = 30 min. With a voltage of U = 48 V and temperatures at $\vartheta_H = 90$ °C, the system poses only a low risk to users and robots. Furthermore, the achieved error deviation from the drift of the self-heating to the heated variants with a value of $\Delta AP = 0.04$ mm is only minimal.

It was shown that the system is slowed down by the low heating capacity P_H. Current systems do not allow for larger heating capacities P_H for this area of application. Flexible heating systems are reduced to small line cross-sections d_L and thus limited in heating power P_H. Development work would therefore be needed in the area of heating systems and the achievable heating power P_H at low line cross-sections d_L.

References

[1] Uhlmann, E.; Wacinski, M.; Dethlefs, A.: Robotergeführte Kantenbearbeitung von Turbinenteilen. Sonderdruck VDI-Z Integrierte Produktion 5 (2012).

[2] Uhlmann, E.; Heitmüller, F.; Manthei, M.; Reinkober, S.: Applicability of industrial robots for machining and repair processes. Procedia Cirp 11 (2013), S. 234 – 238.

[3] Abele, E.; Polley, W.; Ehm, A.; Troue, M.: Spanende Bearbeitung mit Industrierobotern - Thermische Einflüsse auf die Bearbeitungsgenauigkeit (English: Machining with industrial robots – thermal influences on the machining accuracy). wt Werkstattstechnik online 103 (2013) H.9, S. 706 – 711.

[4] Abele, E. et al.: Einsatz von Robotern in der spanenden Fertigung (English: Achieving higher accuracy of robots in milling applications). Wissenschaftsmagazin Nr. 1/2011, S. 44 – 49, Technische Universität Darmstadt, 2011.

[5] Rösch, O.; Zäh, M. F.: Steigerung der Arbeitsgenauigkeit von Fräsrobotern - Höhere Genauigkeit durch steuerungstechnische Kompensation der Roboter-Abdrängung (English: Increasing operational accuracy of milling robots). wt Werkstattstechnik online 103 (2013) H.9, S. 686 – 692.

[6] Ehm, A.: Mechanische Bearbeitung mit Industrierobotern in der automobilen Serienproduktion. Darmstadt: PT/TVM. 2014.

[7] Gräser, R.-G.: Ein Verfahren zur Kompensation temperaturinduzierter Verformungen an Industrierobotern. Forschungsberichte IWB. Hrsg.: Reinhart, G., München: Utz1999.

[8] Reinkober, S.: Fräsbearbeitung von Nickelbasislegierungen mit Industrierobotern. Berichte aus dem Produktionstechnischen Zentrum Berlin. Hrsg.: Uhlmann, E., Stuttgart: Fraunhofer Verlag, 2018.

[9] Richter, F.: Erfassung und Ausgleich thermisch bedingter Verformungen an Industrierobotern. Berichte aus dem Institut für Werkzeugmaschinen Konstruktion und Fertigung. Hrsg.: Heisel, U., Tübingen: MVK, 1998.

[10] Beyer, L.: Genauigkeitssteigerung von Industrierobotern. Forschungsberichte aus dem Laboratorium Fertigungstechnik. Hrsg.: Wulfsberg, J., Aachen: Shaker, 2005.

[11] Mohnke, C.; Reinkober, S.; Uhlmann, E.: Constructive methods to reduce thermal influences on the accuracy of industrial robots. . Procedia Manufacturing 33 (2019), S. 19 – 26.

[12] Uhlmann, E.; Mohnke, C.; Reinkober, S.: Konstruktive Maßnahmen zur Reduzierung des thermischen Einflusses auf die Genauigkeit von Industrierobotern. wt Werkstatt online 7/8 (2018), S. 486 – 492.

IOP Conf. Series: Materials Science and Engineering 1140 (2021) 012008 doi:10.1088/1757-899X/1140/1/012008

ROS middle-layer integration to Unity3D as an interface option for propulsion drive simulations of autonomous vehicles.

Vladimir Kuts[1*], Anton Rassõlkin[2], Andriy Partyshev[1], Sergei Jegorov[1], Viktor Rjabtšikov[2]

[1]Tallinn University of Technology, School of Engineering, Department of Mechanical and Industrial Engineering, Estonia, Harju, Tallinn, Ehitajate str. 5, 19086
[2]Tallinn University of Technology, School of Engineering, Department of Electrical Power Engineering and Mechatronics, Estonia, Harju, Tallinn, Ehitajate str. 5, 19086

[*]vladimir.kuts@taltech.ee

Abstract. As autonomous vehicle development continues at growing speeds, so does the need for optimization, diagnosis, and testing of various autonomous systems elements, under different conditions. However, since such processes should be carried out in parallel, it may result in bottlenecks in development and increased complexity. The trend for Digital Twins brings a promising option for the diagnosis and testing to be carried out separately from the physical devices, incl. Autonomous vehicles, in the virtual world. The idea of intercommunication between virtual and physical twins provides possibilities to estimate risks, drawbacks, physical damages to the vehicle's drive systems, and the physical one's critical conditions. Although the problem of providing communications between these systems arises, at the speed that will be adequate to represent the physical vehicle in the virtual world correctly, it is still a trending topic. The paper aims to demonstrate a way to solve this problem - by using ROS as a middleware interface between two twining systems on the autonomous vehicle propulsion drive example. Data gathered from the physical and virtual world can be exchanged in the middle to allow continuous training and optimization of the propulsion drive model, leading to more efficient path planning and energy-efficient drive of the autonomous vehicle itself.

1. Introduction

Simulation is an approximate or 1-to-1 imitation of a real process, often taking part in the virtual environment, troubleshooting, researching, testing, training, monitoring, controlling, or educating. In the past decade, simulations have been vital in production and development as they are capable of preventing many problems related to planning and reducing bottlenecks at early stages, also during the real-time maintenance of the process [1]–[5] and, especially with increasing technology complexity and rise in using fully autonomous systems, enforcing and changing work-safety features. One side of the simulation aspect - the concept of Digital Twin (DT) [6], [7] is being exploited in the related research to develop a precise dual-way synchronized simulation interface for the propulsion drives [8], [9] to be ready to be integrated into the electrical vehicles [10].
Physics simulations are very common and critical nowadays. They are used enormously in such applications as MATLAB Simulink, Simscape, CAD design, SolidWorks, etc., and gaming physical

processes simulations. They should be considered in the mechatronic systems' planning stage [11]. Of course, they all have approximation and simplifications, while not all possible physical laws can be yet simulated simultaneously; however, such simulations provide considerable benefit in research and testing.

In a previous study done by authors of the related paper, which was on the on electrical motors simulation under development of DT for propulsion drive of an autonomous electric vehicle [12], Unity3D was used for simulations of DT that was exchanging messages with Robot Operation System (ROS) node through a ROS bridge [13]. However, ROS is not being used only for robots but also for various drones, self-driving vehicles, and autonomous systems. ROS enables inter-process communication; it is believed to be a quality method of interconnecting a digital twin propulsion drive system with its real counterpart. ROS was used for performance calculation using an empirical performance model for induction motor (IM). As a visualization tool in the related research is being used Unity3D which is connected with ROS directly [14]. Even though Unity3D simulated most of the motor's physical behavior (torque and rotation), the response and received numerical values, unfortunately, do not suit the DT development in the long run. The reason for this is the complexity of the overall system of physics of IM. Moreover, to make the system transferable and usable with other models (not the ones present in Unity3D but also in Gazebo or elsewhere) the physics handling has to be close to standalone.

The research's main aim is to develop a framework and a toolkit, including a middle-layer ROS interface connected with the physical propulsion drive workbench and its DT, which can be visualized in various simulation engines. The related paper aims to develop a methodology to connect the interface with Unity3D for the visualization, considering data exchange and feedback.

2. Methodology

2.1. Working principle of a test bench on a digital twin

For the current case study, the DT operates on the simulated data generated based on real data measured and gathered from the 7,5 kW IM (ABB 3GAA132214-ADE). The data was gathered using the data acquisition system (DAS) Dewetron Dewe 2 and saved into files with a different extension (*.mat, *.xlsx, *.csv, *.txt). The measured data can be anything regarding the motor's operation, namely input currents and voltages, consumed and shaft powers, torque and angular velocity on data acquisition, and other side data calculated from them. According to DAS tuning (16Hz - 100kHz), the parameters can be measured with different frequencies, and received data is relative to time. This feature enables to recreate of the motor's behavior precisely as it happened in the real case scenario with the help of ROS Server. An example of such can be seen in Figure 1, where the input current from frequency converter to IM was recorded and now can be simulated in ROS (graph from ROS package *rqt plot* we were not included to the related paper because it could not handle plotting messages at such high frequency).

In the proposed DT system, ROS Server acts as a data server and physics simulator. The idea behind it is the following: the server is a standalone subsystem of a TB DT that is responsible for processing real, measured data of the motor, calculating other motor parameters based on the processed data, and streaming to the ROS topics available for models.

Figure 2 features the architecture of the DT setup for TB. The real data is fetched to appropriate ROS Nodes (components of ROS server that are performing calculations, real data processing, and streaming of data) present in the server, processed and translated into ROS messages, and finally, sent to the DT model over ROS Bridge. The real data can be based on the empirical model/map of the motor (or its part) or the actual raw data.

Upon receiving ROS messages, the model can perform the necessary actions to simulate the mechanical/electrical/thermal behavior. Models can be present in any simulation environment. They are subscribed to ROS Server's topics over API or ROS Bridge and configured to perform the necessary operations based on the subscribed ROS topic (for example, rotation based on received angular speed). Furthermore, the module can feature simulated 'measurement' devices/sensors that can send back the data over the ROS bridge. In this case, the ROS Nodes can process and calculate other required values, as it would happen in the real TB.

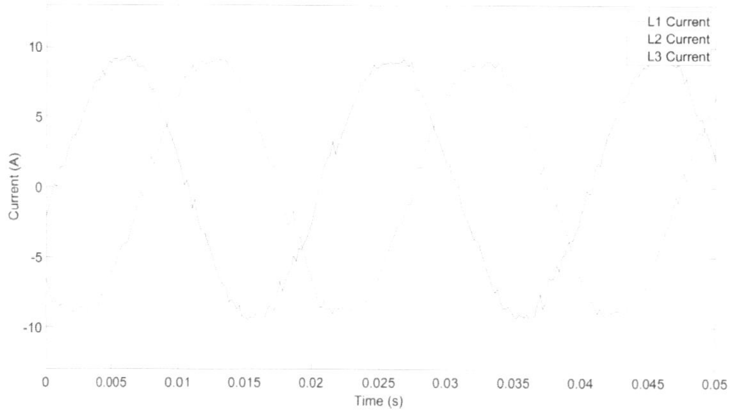

Figure 1. Input current measurements sampled at 5kHz frequency

The current DT of TB consists of the Unity 3D model and ROS Server. ROS Server streams simulated values regarding input power (3-phase current and voltages), efficiency calculated based on measured torque and angular velocity. The torque is calculated by the physics engine of the Unity3D, whereas other values are based on the real ones. This creates a problem of incorrect data calculation because Unity does not focus on calculating correct values on physics laws, as it is more for games, allowing developers to adjust the physics laws to the game setup. This is why the shift from the physics engine of the model environment to ROS was introduced. ROS server would serve physical parameters based on the real TB data and independent of the modeling environment.

Additionally, ROS can record *rosbags* – files with recorded values from topics/servers that can be played back to repeat the behavior. Such a feature would allow us additional analytical features from the DT side.

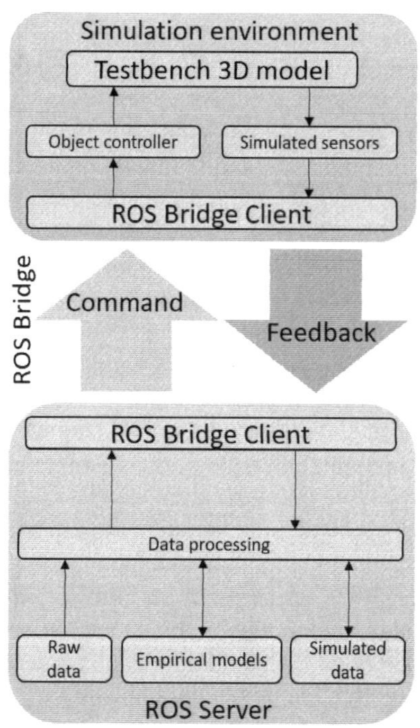

Figure 2. The generic architecture of TB DT

2.1.1. ROS Interfacing

To allow easy interfacing of ROS with other systems, a ROS Bridge node has to be used. It converts ROS communications into a JSON file format and sends them outside of the ROS ecosystem. JSON is used because of its universal format with existing libraries that support its serialization and deserialization in almost every modern programming language. Taking it one step further, ROS Bridge can be used to port specific ROS topics to and out of Message Queuing Telemetry Transport (MQTT) protocol to upscale the system and allow it to run on multiple machines around the world. This so-called MQTT Bridge sends data to the remote server by taking the serialized message on a specified ROS topic and publishes it into a specified MQTT topic. MQTT Bridge is also capable of the inverse - it receives a JSON-serialized message and attempts to deserialize it into a specified ROS topic in a specific message type. Together these systems make interfacing of ROS with any visualization solution much simpler to develop. To further simplify the deserialization process, classes that match ROS message types were created in C# for Unity3D implementation of the ROS interface. This approach can be considered the most efficient because, in this case, a ROS message delivered in the serialized form via MQTT can be directly deserialized into an object of a matching type. This approach can be implemented in similar ways on the majority of existing programming languages, making it the most straightforward and most versatile option.

Visualization is being done in Unity3D (See Fig. 3) engine connected to the physics simulator via ROS Interface, where it is a 1 to 1 scale propulsion drive model with the transmission, wheel parts, and non-visible gears. Model is being assembled as the physical one, and each part is being controlled by a related script, where data is being fed from the middle layer.

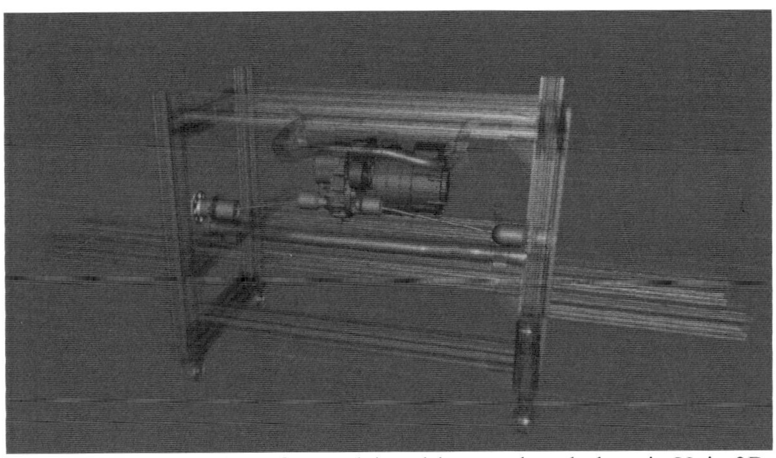

Figure 3. Visualization of propulsion drive test bench done in Unity3D

3. Discussion

The primary outcome of the related part of the more extensive research in developing the fully synchronized DT of the propulsion drive is that the ROS interface was developed. It is possible to feed it with gathered from the physical data and give to the visual simulated, which in related use-case is being Unity3D. The given data simulation runs and gives logged feedback about physical interactions back to the ROS middle layer, where the model is being improved and sent back to the visual side, improving it after each data movement loop. However, some limitations were met during the development of methodology, and more developments go to reach the final aim of the stated research aim (See Table 1).

Table 1. Limitations and further steps

Limitations	Future steps
The model was tested with only one type of visual simulation tool. Possible additional integrations should be done in the middle layer to be suitable for additional software tool packages.	To establish correct torque calculations based on the real values collected from the physical TB.
	To implement a two-way connection between physical TB and its DT.
If DT and TB work simultaneously over the internet, the frequency of data acquisition may be too high to send on time, the possibility of lags	The injection process flow of new components of TB into the DT.
	To create unpredicted behaviors in the system, trigger points, and try to make the system respond to the unpredicted change making it more adaptive to changes

4. Conclusions

The ROS interface connected with the Digital Twin of the propulsion drive workbench visualized in Unity3D was introduced during the related work. Raw and simulated data and empirical models can be post-processed and fed to the visual simulation, where additional data is being logged and given as feedback to the middleware to improve the model and physical simulation itself. The next crucial step is to feed physical simulation directly with data from the physical drive, enabling synchronization between the real and virtual worlds through the developed interface.

5. Acknowledgments

The research has been supported by the Estonian Research Council under grant PSG453 "Digital Twin for Propulsion Drive of Autonomous Electric Vehicle."

References

[1] R. AhmadiAhangar, A. Rosin, A. N. Niaki, I. Palu, and T. Korõtko, "A review on real-time simulation and analysis methods of microgrids," *International Transactions on Electrical Energy Systems*. 2019, doi: 10.1002/2050-7038.12106.

[2] S. Venkatesan, K. Manickavasagam, N. Tengenkai, and N. Vijayalakshmi, "Health monitoring and prognosis of electric vehicle motor using intelligent-digital twin," *IET Electr. Power Appl.*, 2019, doi: 10.1049/iet-epa.2018.5732.

[3] G. Turner, "Soaring through virtual aviation: The role of VR in aerospace manufacturing," *Manufacturing Global*, 2020. .

[4] L. Gevorkov, A. Rassolkin, A. Kallaste, and T. Vaimann, "Simulink based model of electric drive for throttle valve in pumping application," in *2018 19th International Scientific Conference on Electric Power Engineering, EPE 2018 - Proceedings*, Jun. 2018, pp. 1–4, doi: 10.1109/EPE.2018.8395996.

[5] I. Rasheed *et al.*, "Fast Numerical Techniques Based Analysis of Electromagnetic Problems Using MATLAB," in *Proceedings - 12th International Conference on Frontiers of Information Technology, FIT 2014*, Jun. 2015, pp. 115–120, doi: 10.1109/FIT.2014.30.

[6] N. Kousi, C. Gkournelos, S. Aivaliotis, C. Giannoulis, G. Michalos, and S. Makris, "Digital twin for adaptation of robots' behavior in flexible robotic assembly lines," in *Procedia Manufacturing*, 2019, doi: 10.1016/j.promfg.2018.12.020.

[7] V. Kuts, M. Sarkans, T. Otto, T. Tähemaa, and Y. Bondarenko, "Digital Twin: Concept of hybrid programming for industrial robots – Use case," in *ASME International Mechanical Engineering Congress and Exposition, Proceedings (IMECE)*, 2019, vol. 2B-2019, doi: 10.1115/IMECE2019-10583.

[8] V. Kuts *et al.*, "Synchronizing physical factory and its digital twin through an IIoT middleware: a case study," *Proc. Est. Acad. Sci.*, vol. 68, no. 4, pp. 364–370, 2019, doi: 10.3176/proc.2019.4.03.

[9] A. Rassõlkin, T. Vaimann, A. Kallaste, and V. Kuts, "Digital twin for propulsion drive of autonomous electric vehicle," in *60th International Scientific Conference on Power and Electrical Engineering of Riga Technical University (RTUCON)*, 2019.

[10] N. Khaled, B. Pattel, and A. Siddiqui, "Digital Twin development and cloud deployment for a Hybrid Electric Vehicle," in *Digital Twin Development and Deployment on the Cloud*, 2020.

[11] R. Sell, E. Coatanéa, and F. Christophe, "Important aspects of early design in mechatronic," in *Proceedings of the International Conference of DAAAM Baltic* , 2008.

[12] A. Rassõlkin *et al.*, "Digital Twin for Propulsion Drive of Autonomous Electric Vehicle," in *60th International Scientific Conference on Power and Electrical Engineering of Riga Technical University (RTUCON)*, Oct. 2019, pp. 1–4, doi: 10.1109/RTUCON48111.2019.8982326.

[13] A. Rassõlkin, V. Rjabtšikov, T. Vaimann, A. Kallaste, V. Kuts, and A. Partyshev, "Digital Twin of an Electrical Motor Based on Empirical Performance Model," in *Proceeding of 2020 XI International Conference on Electrical Power Drive Systems (ICEPDS)*, 2020.

[14] E. Sita, C. M. Horváth, T. Thomessen, P. Korondi, and A. G. Pipe, "ROS-Unity3D based system for monitoring of an industrial robotic process," in *SII 2017 - 2017 IEEE/SICE International Symposium on System Integration*, 2018, doi: 10.1109/SII.2017.8279361.

IOP Conf. Series: Materials Science and Engineering 1140 (2021) 012009 doi:10.1088/1757-899X/1140/1/012009

Green eco-friendly acoustic materials

J Lavrentjev

AcouTech Lab, Department of Mechanical and Industrial Engineering, Tallinn University of Technology, Ehitajate tee 5, 19086 Tallinn ESTONIA

juri.lavrentjev@taltech.ee

Abstract. Green soundproofing acoustic porous materials are a valid alternative to traditional synthetic materials. Green eco-friendly acoustic panels and other soundproofing solutions are made at least in part recycled or natural materials that are safer for the environment. They do not contribute to the environmental and toxic waste problem since at the end of their useful life, green materials can be disposed of without polluting the environment. They are increasingly being used in the automotive industries and also in construction improving the acoustic comfort inside buildings as well as mitigate reverberation effects and reduce the transmission of noise between rooms. In this paper, the acoustic properties of different but typical green acoustic materials are analysed and their acoustic characteristics experimentally determined. The problems of designing of green acoustic materials have been analyzed and the main solutions to the design and use in acoustics have been proposed. The effect of the structure and arrangement of materials on the acoustic properties of materials has been analyzed.

1. Introduction

The green eco-friendly acoustic materials are usually defined as acoustical sustainable materials, either natural or made from recycled materials. They are more and more a valid alternative to traditional synthetic materials. People recognise more that both construction of new homes and living in a home can create a big carbon footprint, and people are looking for green-built homes to reduce that footprint. The production and using of these materials generally has a much lower environmental impact, but a proper analysis of their sustainability, through Life Cycle Assessment has to be assessed. In fact many natural materials (bamboo, straws, wood industry waste of different fractions) show good sound absorbing performances [1]. Not all currently used recycled acoustic materials can be considered sustainable in terms of the energy consumption and impact to the human health. For instance, mineral wools are widely used for thermal and sound insulation, because of their good performance and low cost. However, mineral wool fibres, can cause skin irritation and dangerous when inhaled. [2, 3]. In the last years a great attention has been focused on green eco-friendly materials and many research centres have developed new sustainable materials, in many cases with interesting acoustical properties [4, 5]. However, many materials do not reach their potential acoustic properties, mainly due to the lack of general design principles. The aim of this work is to summarize the main failures in the design of these materials, based on the experience gained from the research of these materials in the AcouTech Lab of Tallinn University of Technology.

2. Experimental investigation

The most common parameter to characterize acoustic properties of materials is the absorption coefficient There are two techniques to experimentally determine the absorption coefficient: the reverberant room method usually following ISO 354:2003 [6] and the impedance tube transfer function method which follows ISO 10534-2:1998 [7]. In the first case (see Figure 1a) the absorption coefficient is determined in random incident sound field conditions inside a test chamber while in the impedance tube (see Figure 1b) only a direct incident field is considered. The results obtained from these two methods are identical only if the acoustic material is acoustically isotropic which condition is not satisfied in the case of porous acoustic materials.

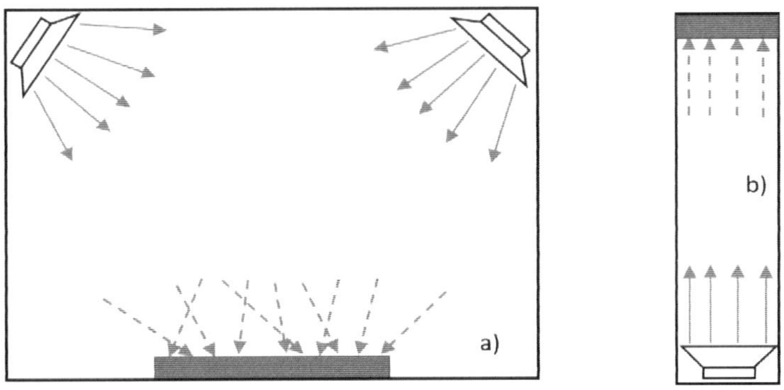

Figure 1. The techniques to experimentally determine the absorption coefficient, a) the reverberant room method, b) the impedance tube method.

The main advantage of the impedance duct method is the requirement of only a small test specimen size, equal to the tube cross-section area, usually few square centimeters. The measurable frequency range is determined by the inner diameter of the tube, setting the upper limit by the cut-on frequency of the higher-order modes. Typical measurement frequency range for impedance tube method is 100-8000 Hz. The standard reverberation room method demanding a minimum specimen size of $10m^2$. The method is suitable for a frequency range from 100 to 5000 Hz. In this paper the absorption coefficient of the acoustic materials is determined in reverberation room and in some cases the results obtained from impedance tube method for comparison are presented.

2.1. Reverberation room method

In reverberation room method the averaged reverberation time is measured inside the room with and without the test specimen fitted into: determining T_1 and T_2 accordingly. From the difference between the reverberation times an equivalent sound absorption area of the test specimen is calculated and the sound absorption coefficient is obtained by dividing the equivalent sound absorption area by the absorbing area S of the test specimen. The experimental investigation was carried out at Tallinn University of Technology, in AcouTech Lab [8, 9], see Figure 2. The reverberation room has volume of 66 m^3 which is less than specified in standard but it allows to use 2.5 m^2 specimens for absorption coefficient determination. The tests were performed by applying the interrupted noise technique, with pink noise emitted from an omnidirectional loudspeaker Brüel & Kjær 4292-L driven by an amplifier Brüel & Kjær 2734. The additional equipment used for this test includes the following: 2-channel noise level meter Brüel & Kjær 2270, condenser microphones Brüel & Kjær 4189, scanning microphone boom Brüel & Kjær 3923, a calibrator Brüel & Kjær 4231.

3. The measurement results

First typical acoustic material and the experimentally determined absorption coefficient is presented in Figure 3. It is a natural material made from leather waste. It is light and soft reticulated foam with open porosity about 90%. Due to technological restrictions it can be produced only in 10 mm thickness.

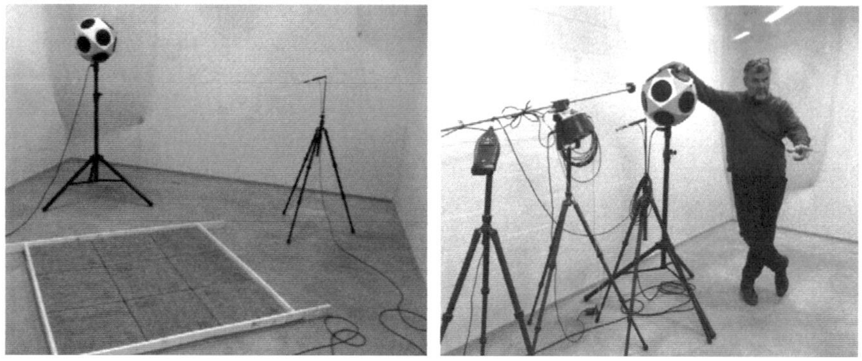

Figure 2. The reverberation room for absorption measurements in AcouTech Lab. a) measurement set-up, b) the equipment used in experiments.

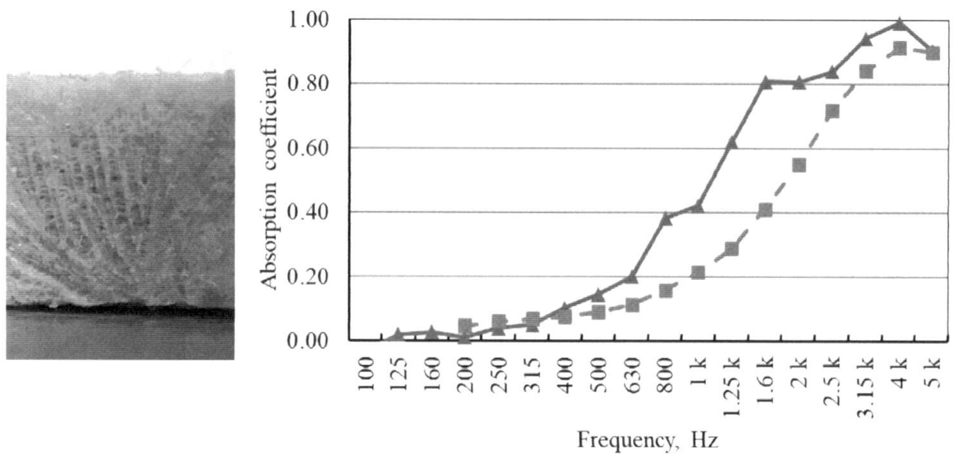

Figure 3. Reticulated leather waste foam, thickness 10 mm; blue line – measured in reverberation room, red line – measured in impedance tube.

First it is seen, that due to anisotropic material, the absorption coefficient is different weather measured in random incident sound field or direct incident sound field. Generally it is true that in random incident field the porous acoustic material exhibits larger sound absorption coefficient, since for the sound front coming oblique to the material surface. It is also clear that the small thickness of the material only has reasonable absorption coefficient only from 1000 Hz and above, which is not satisfactory for most of the applications. The maximum value reaches value 1, which shows that the porosity is good and has open pore system. Using the same material, experiments were performed with multiply layers of the material. In order to save material, the experiments were done in impedance tube. Then only small multiple specimens were needed, see Figure 4. It can be seen from the results, that with the adding layers, the impedance improves at the lower frequencies and retain the value at middle and higher frequencies. The four-layer case (40 mm thickness) is comparable with the industrial glass-fibre material covered with perforated textile (42 mm thickness). So the thickness matters and the technology should support production of thicker materials.

Figure 5 shows the measurement results for solid plate, having sub-millimeter size holes forming perforation. Typically these plates have thickness of 0.5 to 2 mm and becoming popular in different applications [10, 11]. In order to achieve reasonable absorption coefficient they are placed at the distance from solid surface. As it is seen from Figure 5, the increasing distance shifts the maximum values to the lower frequency region.

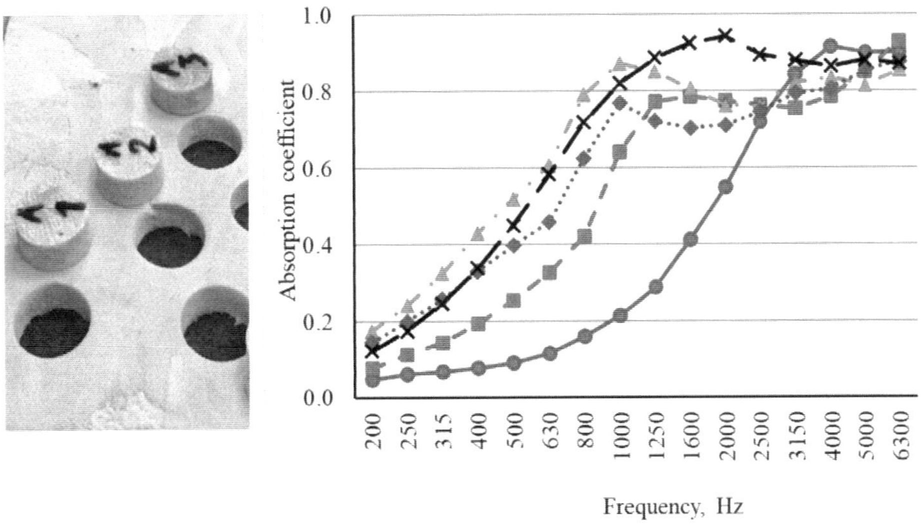

Figure 4. Reticulated leather waste foam, measured in impedance tube. Blue solid line - one layer, red dashed line – two layers, brown dotted line – three layers, green dashed-dotted line – four layers. Black long dashed line – reference material.

Figure 6 shows the result of thin carpet with the thickness 5 mm. Due to soft structure it cannot be placed at the distance from solid surface. Small thickness causes low absorption coefficient at low and middle frequency range. Due to big pores it has average absorption at higher frequencies. Figure 7 shows the results for panels made from wool residues. The panels have formed shape after applying pressure. The solid shape keeps the panel material at the certain distance from solid surface. As a result the absorption exhibits good values already at the low and middle frequency range. The maximum value however reaches only value 0.8. It is caused by the bad pore structure typical to fibre type porous materials. The measurement result for wheat straw pressurized panels (thickness 28 mm) are shown in Figure 8. Low values of the absorption coefficient are mostly caused by the big and partly closed pores of the straws.

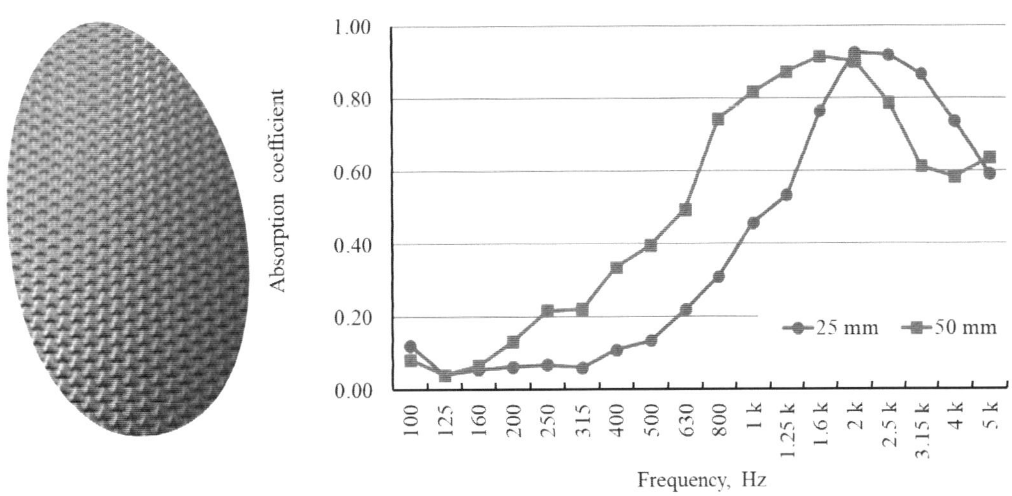

Figure 5. Microperforated plate, made from recycled metal. Placed at 25 mm and 50 mm from solid surface.

IOP Conf. Series: Materials Science and Engineering 1140 (2021) 012009 doi:10.1088/1757-899X/1140/1/012009

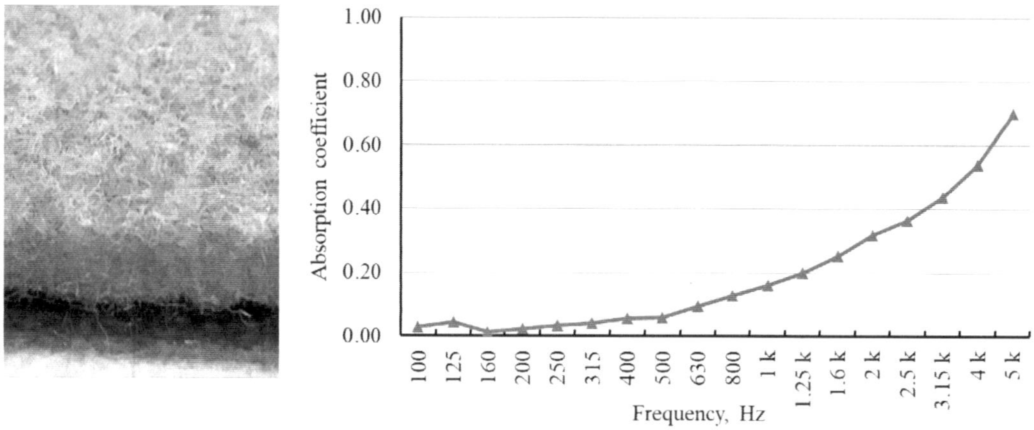

Figure 6. Thin (5mm) carpet made from recycled plastics

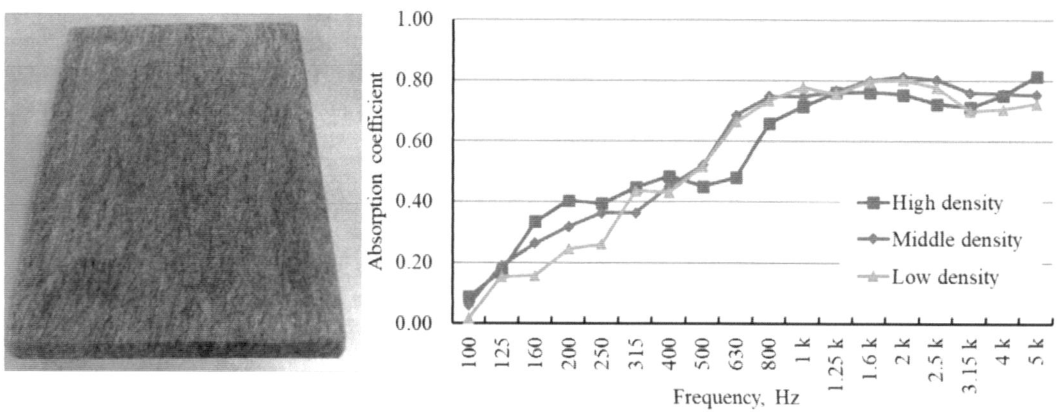

Figure 7. Panels made from wool industry residues. Thickness 5 mm, pressurized together and rigid.

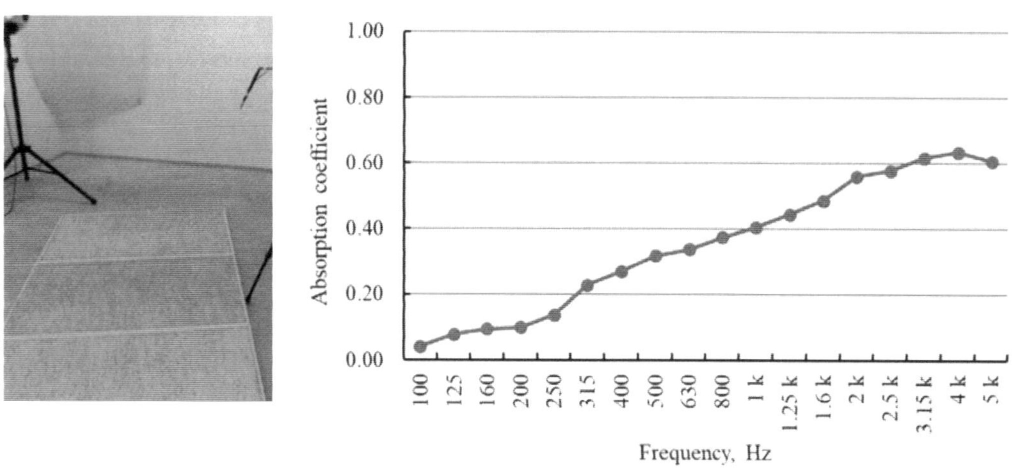

Figure 8. Wheat straw pressurized panels, thickness 28 mm.

Conclusions

In this paper the acoustic properties for green recycled and natural materials are investigated. The absorption coefficient is determined for different type of materials and their combinations. The absorption coefficient has been measured by implementing two standard techniques in comparison: the impedance tube method and the reverberation room method. Generally it has been demonstrated that an eco-friendly and recyclable novel porous material can offer an affordable replacement for synthetic foams and fibrous materials if the right design principles are applied.

The general design principles of green acoustic materials in order to achieve noise attenuation potential can be listed as follows:

- the best results exhibit porous materials with open pore structure, the pores should have small size
- the core material should have high elasticity
- the thickness of the material should be large enough to offer reasonable attenuation at lower frequencies
- in the case of thin material it should be placed at least one quarter of sound wave wavelength of the frequency range interested

References

[1] Asdrubali F, Schiavoni S and Horoshenkov K. V 2012 A Review of Sustainable Materials for Acoustic Applications *J of Building Acoustics* **19** 4, 283–312.

[2] WHO 2002 Man-made Vitreous Fibres *IARC Monographs on the Evaluation of Carcinogenic Risks to Humans* **81**.

[3] WHO 1988 Man-made Mineral Fibres and Radon *IARC Monographs on the Evaluation of Carcinogenic Risks to Humans* **43**.

[4] Novais R.M, Pullar R.C, Labrincha J.A, 2020 Geopolymer foams: An overview of recent advancements *Progress in Materials Science* **109**, article number 100621.

[5] Dissanayake D.G.K, Weerasinghe D.U, Thebuwanage L.M, Bandara U.A.A.N 2021 An environmentally friendly sound insulation material from post-industrial textile waste and natural rubber *Journal of Building Engineering* **33**, 2021, article number 101606.

[6] ISO 354:2003 Acoustics - Measurement of sound absorption in a reverberation room.

[7] ISO 10534-2:1998 Acoustics - Determination of sound absorption coefficient and impedance in impedance tubes - Part 2: Transfer-function method.

[8] Kabral R, Rämmal H, Lavrentjev J and Auriemma F 2011 Acoustic Studies on Small Engine Silencer Elements *SAE Technical Paper Series* 2011-32-0514.

[9] Lavrentjev J, Rämmal H, Tiikoja H, 2011 The passive acoustic effect of automotive catalytic converters *SAE Technical Papers* 2011-24-0219.

[10] Kabral R, Rämmal H, and Lavrentjev, J 2012 Acoustic Studies of Micro-Perforates for Small Engine Silencers *SAE Technical Paper Series* 2012-32-0107.

[11] Kabral R, Rämmal H, Auriemma F, Luppin J, Koiv R, Tiikoja H, Lavrentjev, J. 2012 A Novel Design for Cruiser Type Motorcycle Silencer Based on Micro-Perforated Elements *SAE Technical Paper Series*, 2012-32-0109.

Acknowledgments

This research was supported by Estonian Research Council Proof-of-concept grant EAG28

IOP Conf. Series: Materials Science and Engineering 1140 (2021) 012010 doi:10.1088/1757-899X/1140/1/012010

Adhesion of AlCrN coating deposited on TiB_2/Ti composites sintered by SPS dedicated for high temperature tribological applications

R Michalczewski[1], M Kalbarczyk[1,2], Z Słomka[1], I Hussainova[3], L Liu[3], M. Antonov[3]

[1] Łukasiewicz Research Network - Institute for Sustainable Technologies, ul. K. Pułaskiego 6/10, 26-600 Radom, Poland
[2] Faculty of Mechanical Engineering, Kazimierz Pułaski University of Technology and Humanities in Radom, Poland
[3] Department of Industrial and Mechanical Engineering, Tallinn University of Technology, Ehitajate 5, 19180 Tallinn, Estonia

Corresponding author: remigiusz.michalczewski@itee.lukasiewicz.gov.pl

Abstract. The aim of the work was to investigate the adhesion of AlCrN coating deposited by PVD method on TiB_2/Ti composites manufactured by Spark Plasma Sintering (SPS). The composites were produced from three kinds of mixtures of powders: (1) Ti and TiB_2, (2) Ti6Al4V and TiB_2 and (3) Ti and B. Each of the mixtures was prepared at 50/50 wt. % ratio of components. Before the coating deposition the specimens were polished to obtain the surface roughness typically required for machine parts. For each set of specimens an individual surface preparation procedure was developed. The coating adhesion was measured using a scratch tester. The process of scratching was carried out to observe the characteristic moments of adhesive and cohesive failures. The highest critical loads were obtained for the coating deposited on composites produced from the mixture of two powders containing pure Ti and TiB_2.

1. Introduction

A vast amount of manufacturing processes takes place at high temperatures and depend on the durability and reliability of components capable to cope with this harsh environment [1]. The overall objective of the work is to extend the lifetime of components operating under relative motion at high temperatures (HT), in the range from 400 to1000°C, by the development and application of novel lubrication concept based on the double-lubrication approach which relies on the design of innovative high temperature self-lubricating materials and lubricous multifunctional coatings [2]. To optimize the benefits achievable from coatings, it is vital to consider their implementation in the design stage [4, 5, 6]. The aim of this study was to select the input powder materials and methods of preparation for TiB_2/Ti ceramic-based bulk composites intended for PVD coatings deposition [7, 8]. One of the main factors enabling the successful application of PVD coating is the ability to remain attached to the substrate under required operating conditions. The adhesion between a coating and a substrate depends on many factors, among others on material structure and surface finishing, so in practice, it can be evaluated only in the experimental manner [9, 10]. So, the paper presents the results of research, aimed at the selection of the most suitable technology of TiB_2/Ti composites manufacturing and surface preparation for the deposition of the PVD coating.

2. Materials and Methods

2.1. TiB_2/Ti composites

Samples were designed to achieve target Ti-TiB composite with Ti/TiB=15/85 wt.% in fully reacted state. Ceramic titanium diboride plays the role of the hard phase interconnected by metallic titanium. Target 15/85 weight ratio of Ti/TiB corresponds to Ti:B=1.2167:1 molar ratio, hence, the raw powders resembles approximately 50/50 weight ratio for Ti/TiB_2 and Ti6Al4V/TiB_2 compositions (exact numbers are mentioned in Table 1). For powder preparation by self-propagating high-temperature synthesis method (SHS) the elemental titanium and boron powders were used, and the respective Ti/B ratio was calculated as 84.3/15.7 wt.%.

Table 1. Samples preparation procedure

Name	Powder development					SPS parameters		
	Raw powders				Preparation procedure	Pressure	Time	Temp.
	Ti	Ti6Al4V	TiB_2	B		P	t	T
	[wt.%]					[MPa]	[min]	[°C]
Group I	50.33		49.67		Conventional mixing	50	15	1250
								1450
Group II		52.3	47.7		Disintegrator mixing	50	15	1050
								1150
								1250
Group III	84.3			15.7	SHS synthesis, disintegrator mixing	50	15	1050
								1150
								1350

All samples are arrayed into three groups according to the different manufacturing strategies. Group I contains samples fabricated from the powder mixture prepared by the conventional rotative mixing of the pure titanium (99,98% purity) and titanium diboride (98,5% purity) raw powders. Powder development for samples listed in Group II was performed by the help of laboratory disintegrator milling system DSL-175, while titanium alloy (Ti6Al4V Grade 5) raw powder instead of pure titanium was used because of the spherical shape of particles.

Manufacturing routine for samples combined in Group III consists of two steps. Firstly titanium-boron mixture was used for combustion synthesis conducted in CPR3.5L constant pressure SHS reactor. Secondly, SHS product was broken into pieces by rolling mill and crushed to powder by disintegrator. The manufactured samples (discs) are presented in Figure 1. SEM images of developed powders are represented in Figure 2 (first row).

a) b)

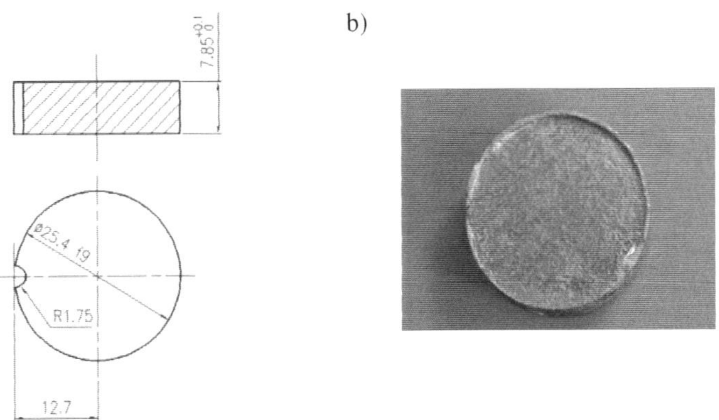

Figure 1. The sample (disc) made of TiB_2/Ti composite: a) a scheme, b) the state after sintering.

The developed powders were consolidated by spark plasma sintering (SPS) device (FCT Systeme GmbH, Germany) at 50 MPa pressure with the dwell time of 15 min, heating rate 100 Celsius per minute, applying different sintering temperatures, as mentioned in Table 1. Images of typical of SPS produced samples of each group are shown in Figure 2 (second row).

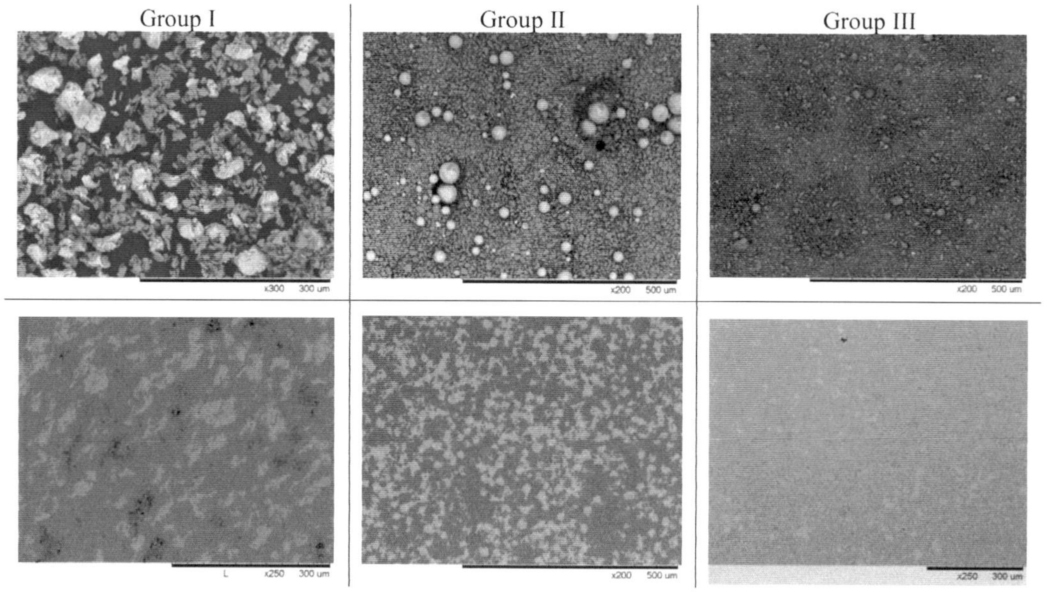

Figure 2. SEM images of the developed powders (first row), and the prepared samples (second row).

2.2. *Surface preparation*

The samples preparation was divided into the following steps. The first step was to remove the post-sintering scale from the surfaces of the specimens and the preliminary examination of samples surface. The second step was to put the number on the side of each sample using laser engraving.

The third step was to adjust the thickness of the samples to the test requirements using a grinder equipped with a grinding wheel (1A1 200x20x5x51 D151 K75 MBm). As the selected grinding wheel is intended for sinters and carbides, it contains diamond particles with a grain size of 126-150 μm, at K75 concentration and a soft resin binder. During the process a cooling-cutting fluid was used.

All of the grinding processes were done at the linear speed of 35 m/s. The grinding time (approx. 1h needed for thickness reduction of 0,5 mm per side), was adjusted according to the samples surface quality and hardness.

After obtaining the required thickness, the samples were polished by means of ATM Saphir 550 automatic polisher using the procedure presented in the Table 2.

Table 2. Samples polishing procedure.

Parameter	Stage 1	Stage 2	Stage 3	Stage 4
Polishing plate	Blue disc	Silver disc	White cloth	Red cloth
Lubricant	Water	Bio Diamant	Bio Diamant	Bio Diamant
Diamond particle size [μm]	-	9	3	1
Plate speed [rpm]	150**	150**	150**	150**
Head speed [rpm]	125*	125**	125**	125**
Load per sample [N]	30	25	20	15
Time [min]	60	30	60	60

* clockwise rotation direction, ** counterclockwise rotation direction

The given time is understood as a maximal required for the samples from the first group having the highest hardness, to reach the average roughness Ra < 0,1 μm.

2.3. *Coating deposition*

Nowadays, there is a wide range of either single layer coatings (e.g. TiN, CrN, ZrN, TiB_2), or multilayer coatings, with combined properties of several materials, pioneered by TiC/Al_2O_3/TiN, or multi-element coatings such as TiAlN, BCN, TiBCN, SiNC, AlCrN and TiAlYN-systems [11]. The modification of a coating composition leads to increase its area of application. The addition of Al to the TiN and CrN coatings results in increase in the wear resistance and thermal stability at temperatures up to 1000°C. For this study, from numerous coatings of interest, the widely available one was selected. The commercial coating AlCrN-based (BALINIT® ALCRONA PRO made by Oerlikon Balzers) was deposited on discs made of various types of TiB_2/Ti. The detailed deposition process parameters are not provided by the coating supplier. The selected coating exhibits excellent tribological characteristics at extremally high temperatures [12].The chemical composition of the coating, measured by X-ray microprobe EDS, is summarized in Table 3, while the other properties can be found in Table 4.

Table 3. The chemical composition of the coating (measured by EDS).

Coating name	Chemical composition [at. %]		
	Al	Cr	N
AlCrN	36.9±0.2	39.1±0.2	21.2±0.3

Table 4. The properties of the selected coating (manufacturer data).

Parameter	Value
Maximum work temp. [°C]	1100
Hardness Hv (0.05)	3200
Residual compressive stress [GPa]	-3
Thickness (calo-test) [μm]	1.2–1.3

The coating was deposited on a flat surface of the discs (samples) manufactured according to technology described in section 2.1 and finished following procedure presented in section 2.2.

2.4. *Adhesion measurement*

Adhesion of AlCrN coating was measured by means of the scratch-test method using REVETEST Scratch-Tester, manufactured by Anton Paar. A Rockwell C type intender with the angle of 120° and tip radius of 0.2 mm was used. The following test parameters were applied: maximum force of 100 N, progressive load from 0 to 100 N, incremental load rate of 10 N/mm, 10 mm/min scratch speed, and 10 mm scratch length. The values of the critical load Lc, causing characteristic damage of the coating, were determined through the analysis of the tangential force (Ft) and changes in the level of acoustic signal emission (AE), as well as optical observation of the scratch.

3. Results and discussion

3.1. *Surface characteristics*

The aim of the polishing process was to obtain a sample surface without extreme irregularities. For composite materials the surface preparation is much more complex than it is for steel alloys because of the presence of at least two phases of different mechanical properties and high porosity. There is no exact values of roughness parameters required by PVD technology, but it is obvious that the surface character has a strong influence on the performance of the coating-substrate systems. The finishing process was performed using the procedure described in section 2.2. The values of R_a parameter for each group of the samples are summarized in Figure 3.

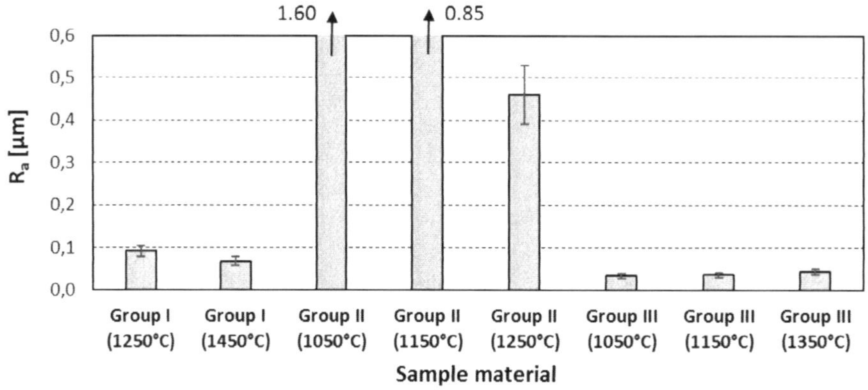

Figure 3. The average roughness of the polished samples (without coating).

The roughest surfaces, with the Ra values, even 10-fold higher than in other cases were obtained for the composite material of group II. The high roughness of samples from group II resulted from a numerous pores present in the surface, which are visible in 2D profiles measured using stylus profilometry – Figure 4. It was observed that sintering temperature affected the structure of the composite and in consequence the surface roughness. In most cases (group I and II) the increase in temperature caused the decrease in final surface roughness. The maximum depth of pores for the samples of group I is c.a. 2.5 µm, of group III is c.a. 1.5 µm, while for group II it reaches more than 30 µm.

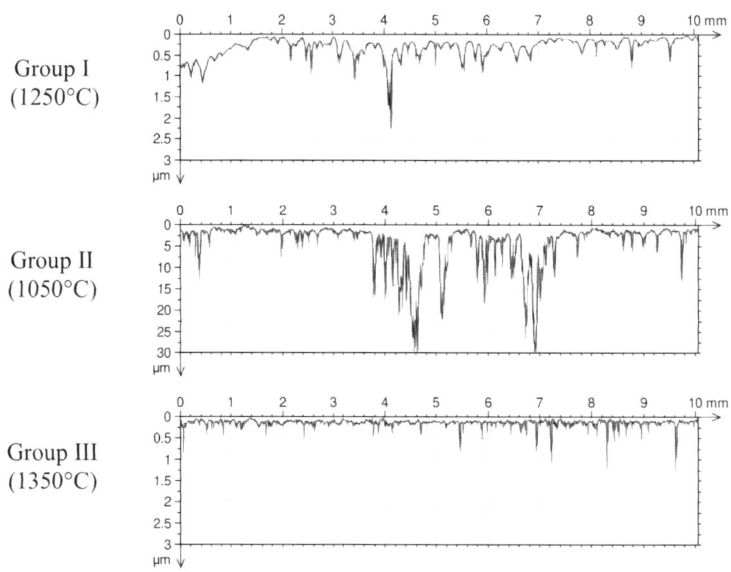

Figure 4. 2D profiles of the composite surfaces after polishing process.

The distribution of phases present in the surface is shown in Figure 5. The composite surface is not homogeneous, thus it can be expected that the adhesion between the coating and the substrate may vary depending on the particular phase present on the surface.

Figure 5. Porosities and phase differentiation of the composite samples revealed by polishing process.

Because of the porosity and the difference in particular phases mechanical properties it is very difficult to receive a surface which would be smoother than the obtained one. The continuation of the polishing process leads not only to flattening the existing asperities, but also opens the new holes, which is well visible in 3D images presented in Figure 6, especially for the composites from group I and II.

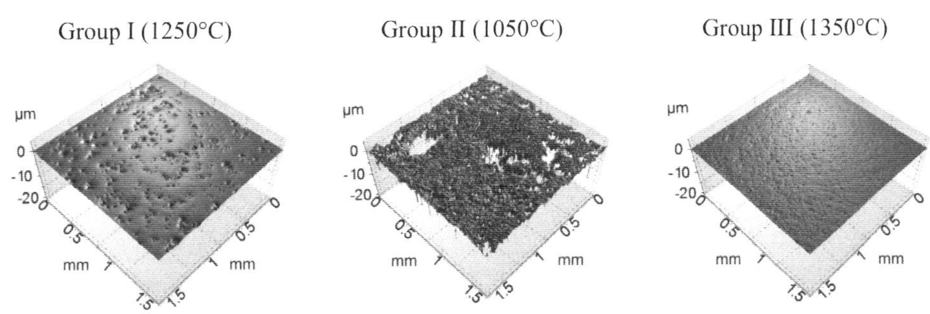

Figure 6. 3D images of the composite surfaces after polishing process

3.2. Coating adhesion

The scratch test was used to assess the thin coating adhesion to the substrates. Some of the types of failures depend on adhesion i.e. failure at the coating-substrate interface. Others depend on plastic deformation and fracture within the coating. There is no unique critical load defining the adhesion of a coating. So, for each coating three critical loads were determined: Lc1, Lc2 and Lc3. The typical failure for Lc1 was brittle tensile cracking. It is assumed that Lc1 critical load is considered to be a resistance to crack initiation. Lc2 represents the load at which first signs of chipping and delamination is visible on the scratch. When a gross spallation or a substrate exposure takes place the Lc3 critical load is reached. The critical loads measured for each coating-substate system are presented in Figure 7.

The highest values of critical loads were obtained for the samples with the coating deposited on materials of group I - samples fabricated from the powder mixture prepared by the conventional rotative mixing of the pure titanium and titanium diboride raw powders. The first brittle cracks were found at loads a bit above 30 N, the partial delamination in the range of 36-40 N, and exposure of the substrate (total removal of the coatings) at loads slightly above 75 N.

For the samples of group II it turned out to be impossible in most cases to measure the critical loads. Extremally rough surface caused that the first symptoms of coating failure were hidden in the background of numerous surface defects and pores of the composite. The total exposure of the substrate took place for loads of about 20 N. Because of the difficulty of precise determination of the critical loads and their low values, Lc1 and Lc2 for samples of group II were not presented in Figure 7.

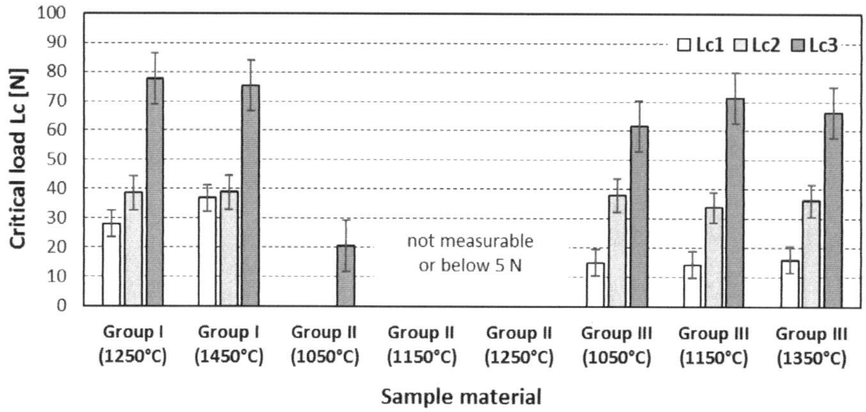

Figure 7. Scratch test results of AlCrN coating: Lc1 (brittle cracking), Lc2 (chipping, delamination), Lc3 (gross spallation, exposure of substrate).

For the coating deposited on substrate of group III, - samples produced from the powder mixture of pure titanium and boride powders the first brittle cracking was determined at the loads much lower than in the case of the group I samples. In case of the coated samples of group III, the first brittle cracking was observed at approx.15 N, partial delamination below 40 N, and exposure of the substrate (total removal of the coatings) at the loads between 60 and 75 N.

The micrographs of the scratches on the coated samples from group I, at the Lc1, Lc2 and Lc3, are presented in Figure 8.

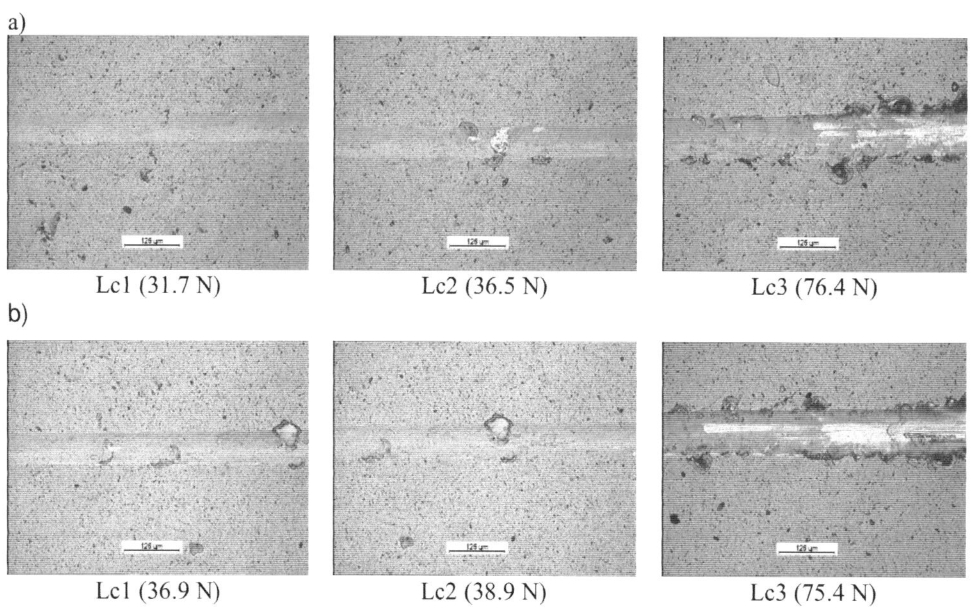

Figure 8. Evolution of the coating failure during the scratch test – Group I substrate, sintering temperature a) 1250°C, b) 1450°C.

The increase of sintering temperature from 1250 to 1450°C lead to increase of the adhesion of the AlCrN coating. Lc1 and Lc2 critical loads were higher for higher temperature, but the point the coating total removal was not changed. So, there can be clearly observed a dependence between sintering temperature and coating resistance to brittle cracking.

Evolution of the coating failure during the scratch test on the coating deposited on samples from group II is presented in Figure 9. Because of high roughness and porosity of the substrate it was not possible to perform precise estimation of the Lc1 and Lc2 critical loads. The group II composites requires further optimization., in the direction of obtaining a less porous bulk material, which would make it possible to investigate the aspect of coating-substrate adhesion.

5.4 N 15.6 N Lc3 (22.8 N)

Figure 9. Evolution of the coating failure during the scratch test – Group II substrate (sintering temperature 1050°C).

The critical loads obtained for the coating deposited on the surface of composites from group III depend on sintering temperature. Evolution of the coating failure during the scratch test is presented in Figure 10.

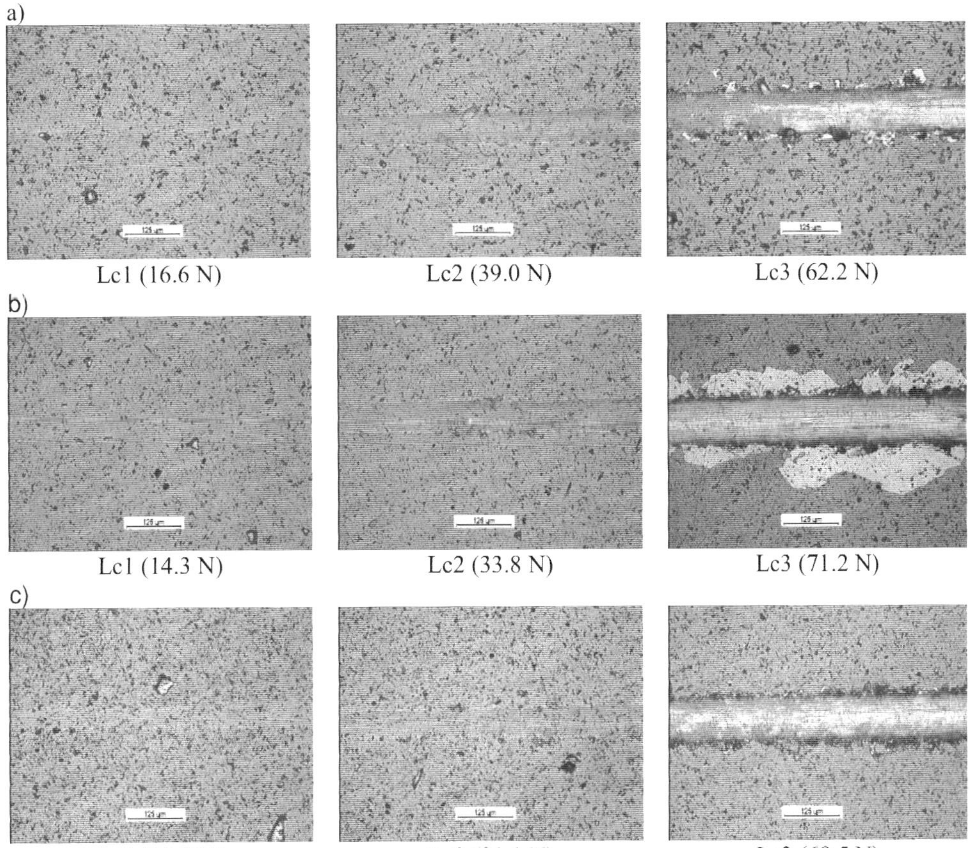

a)
Lc1 (16.6 N) Lc2 (39.0 N) Lc3 (62.2 N)
b)
Lc1 (14.3 N) Lc2 (33.8 N) Lc3 (71.2 N)
c)
Lc1 (15.5 N) Lc2 (31.1 N) Lc3 (69.5 N)

Figure 10. Evolution of the coating failure during the scratch test – Group III substrate, sintering temperature a) 1050°C, b) 1150°C, c) 1350°C.

The highest Lc3 (exposure of substrate) was observed for the substrate sintered at the temperature of 1150°C, however the first symptoms of coating failure was visible at relatively low Lc1 critical load (14.3 N).

4. Conclusions
On the basis of obtained results, the selection of input powder materials and methods of preparation for ceramic-based bulk TiB_2/Ti composites, intended for high temperature PVD coatings was done. The most suitable are the composites of group I, i.e. produced from a mixture of two powders: pure Ti and TiB_2, with 50/50 wt.% ratio. For this composite it is possible to obtain satisfactory quality of the surface and high adhesion of PVD coating. It could be assumed, that the selected substrate material and preparation method will be also suitable for other coatings. This is an important step forward to elaborate materials and technology towards a novel lubrication concept relying on the double-lubrication approach i.e. lubricious substrates and lubricious coatings.

5. Acknowledgments
Project HOTselflub is supported by the National Science Centre, Poland, under the M-ERA.NET 2, which has received funding from the European Union's Horizon 2020 research and innovation programme, under grant agreement no 685451.

References
[1] Varga M et al. Influence of velocity on high-temperature fundamental abrasive contact: A numerical and experimental approach 2019 *Wear* **426** 370-377
[2] Torres H et al. Tribological behaviour of self-lubricating materials at high temperatures 2018 *Int. Mat. Rev.* **63** 309-340.
[3] Torres H et al. Microstructural design of self-lubricating laser claddings for use in high temperature sliding applications 2018 *Surf. & Coat. Technol.* **337** 24-34.
[4] Michalczewski R et al. The rolling contact fatigue of WC/C-coated spur gears 2013 *Proceedings of the Institution of Mechanical Engineers, Part J: Journal of Engineering Tribology* **8** (227) 850-860.
[5] Michalczewski R et al. The Effect of a Gear Oil on Abrasion, Scuffing, and Pitting of the DLC-Coated 18CrNiMo7-6 Steel. 2019 *Coatings* **9** (1).
[6] Tuszyński W et al. The effect of WC/C coating on the wear of bevel gears used in coal mines. 2015 *Materials Science (Medžiagotyra)* **21** (3) 358-363.
[7] Basu B et al. Processing and properties of monolithic TiB_2 based materials 2006 *Int. Mat. Rev.* **51** 352-374.
[8] Liu L et al. SHS Produced TiB2-Si Powders for Selective Laser Melting of Ceramic-Based Composite 2020 Applied sciences-basel. 10 (9)
[9] Michalak M. et al. The method of tribotesting of PVD coated elements in oscillation motion at high temperature 2015 *Tribologia* **1** (259) 77-94.
[10] Antonov et al. Assessment of gradient and nanogradient PVD coatings behaviour under erosive, abrasive and impact wear conditions 209 Wear **267** (5-8) 898-906.
[11] Mayrhofer PH et al. Microstructural design of hard coatings 2006 *Materials Science* **51** 1032–1114.
[12] Michalak M et al. The Effect of Temperature on Wear Mechanism of the AlCrN Coated Components 2016 *Key Engineering Materials* **674** 233-238.

IOP Conf. Series: Materials Science and Engineering 1140 (2021) 012011 doi:10.1088/1757-899X/1140/1/012011

User-centered design for Human-Robot Collaboration systems

S L Pizzagalli, V Kuts and T Otto

Department of Mechanical Engineering, Tallinn University of Technology, Ehitajate tee 5, 19086, Tallinn, Estonia

simone.pizzagalli@taltech.ee

Abstract. Autonomous robotic systems are one of the pillars of Industry 4.0, together with Digital Twin (DT) simulations and advanced Human Machine Interfaces (HMI). Customization requirements in modern manufacturing demand a closer collaboration between operators and automated technologies, leading to a novel Human-Robot Collaboration (HRC) and interaction paradigm aimed at the augmentation of human capabilities in the workplace. This scenario calls for a new definition of HRC standards and design for safety, programming, and an overall assessment of modern cyber-physical systems. A comprehensive design process aiming to formulate a common framework of requirements based on human, organizational and production needs is missing. A user-centered design approach described may be a solution to address the open challenges of future HRC. DT and Augmented and Virtual reality technologies (AR/VR) are implemented in elaborated model as the necessary tools to assign the operator a central position in the design, control, and assessment of modern industrial collaborative scenarios.

1. Introduction

The advancements in industrial automation and the need for flexible, adaptable production systems aimed at mass customization lead to closer collaboration between humans and robots in several industrial fields. Big data and analytics, autonomous robots, simulation, system integration, industrial internet of things, cyber physical systems, the cloud, additive manufacturing, Augmented and Virtual reality (AR/VR) are among the leading technology drivers for the industry 4.0 (I4.0) revolution [1]. Human-Robot Collaboration (HRC), in particular, is a crucial aspect in the evolution of modern industry into a scenario that will see human workers being empowered by sensing and perceptive technologies supporting and perfecting human skills throughout the production process [2]. Natural, adaptive, and intuitive Human Machine Interfaces (HMI) are necessary to enhance safe and efficient teaming processes [3], favoring well-being, and technology acceptance in the upcoming industrial scenario. Recent studies show an increased interest in Digital twins (DT) AR/VR interfaces with examples of assessment and evaluation of safety [4] programming methods [5] and commissioning of HRC systems. A few studies attempt the combination of DT AR/VR technologies in a User Centered Design (UCD) approach focusing on specific use cases. Businesses adopting I4.0 driving technologies are aware of the risks and impacts these have on the operators [6]. No clear design and assessment methods are yet available or enough to provide a robust framework that would include human factors in design and decision-making strategies related to HRC. This work aims at addressing the relationships and synergies occurring between the HRC process and a UCD approach for collaborative robotic scenarios. The main research question is whether DT AR/VR interfaces are valuable and reliable tools for the development of modern user-centered robot collaboration systems improving the efficiency and safety of interaction between operators and machines in the manufacturing field.

2. Evolution of HRC in I4.0

Teaming skilled workers and robots in complex tasks' execution combines the advantages of automation with soft human skills and decision-making [7]. Safety, intuitive interfaces, including input modes, outputs, feedback, and programming approaches, together with innovative design and control methods, are the main ingredients and challenges for the upcoming development and implementation of collaborative production technologies [8]. The forthcoming massive exploitation of collaborative robots will lead to a series of organizational, human-centered issues, and production process adaptations [9]. Close collaboration between humans and robots should be addressed by evaluating the impact these technologies have on the industrial environment and the operators. Individual, organizational and robotic agent factors influence the design and organization of the future workspace with open challenges such as operator monitoring, stress level, workload assessment, robotic system acceptance, efficient reprogramming, data management, ethics, and privacy related to the collected data [10]. The change in production paradigms will transform the role and type of relationship and interaction that human workers have with automated technology from mere cooperation to an augmented and symbiotic paradigm. In their study, Romero et al. [11] describe how the coming generation of industrial Operators 4.0 (O4.0) will be enhanced in their physical, sensorial, and cognitive capabilities, being at the center of a symbiotic cyber-physical system. The agents involved in this new production paradigm will need to be intelligent; therefore, purposeful, perceptive, aware, autonomous, able to act, reflective, adaptable, learning, and conversational. While many of these qualities are part of human skills, industrial robots seem to be far from achieving these characteristics.

The popularity of DT solutions and research applications is constantly increasing [12]. The plasticity of this technology is demonstrated by examples spanning from manufacturing applications, electric engine optimization [13] to building construction [14], aviation, and healthcare [15]. The scalability of simulation driven solutions allow for applications supporting maintenance and production planning [16] to specific realtimed data driven model for product monitoring [17]. AR/VR HMI for HRC offer the advantages of including the user as an active agent in a robotic cell's control interface, which is itself a DT of the controlled system and environment. Many recent studies focus on experimental approaches to DT AR/VR interfaces with applications aimed at programming [18], control [19], design of collaborative industrial cells [20], and HRC safety assessment [21]. Other examples address the potential of these technologies in becoming UCD design and evaluation tools for advanced industrial systems and workstations [6] [22]. The rising HRC industrial paradigms need to be tested and validated both in terms of HRI, interface efficiency for robot control and system impact on the user. What is missing is a systematic approach that would allow the evaluation of all aspects involved in the HRC process and make DT AR/VR interfaces the main tool bridging human and robots in a UCD approach.

3. Collaboration levels, safety and programming in HRC

Human-robot levels of collaboration can be defined in different ways. Helms et al. [23] describe four types of human-robot operations: independent or parallel work, synchronized work simultaneous work and assisted or collaborative scenarios. The last scenario involves humans and robots having a common task, sharing workpieces, tools and workspace while performing the required task at the same time. De Luca et al. [24] further specify the relation between safety, coexistence and collaboration defining them as nested levels of human-robot interaction (HRI). Safety is defined as the core feature of robots involved in collaborative working scenarios. Collaborative industrial robots' requirements are described in [25] and include four main safety modes: Safety-rated Monitored Stop, Hand Guiding, Speed and Separation Monitoring, and Power and Force Limiting. Regardless of the characteristics of each specific method it is crucial to keep in mind the type of involvement and level of physical contact that the operator has in each mode and the implications for both collaborative task performance and robot programming operations. Different interaction modalities need a clear definition of robot context-awareness by sensors and artificial intelligence and have a different impact on the operator. Programming is another central aspect of modern collaborative robots as it involves a variety of dedicated interfaces or direct manipulation methods such as in walk-through or programming by demonstration. Virtual reality (VR) might offer a possible solution for some of the issues related to programming procedures by providing a repeatable controlled and safe environment for kinesthetic teaching procedures [26] also on the old generation of robotics systems.

4. UCD for HRC

The contemporary industrial production and O4.0 requirements in HRC need to be allocated in a design model able to organize collaboration, therefore task allocation and organization, safety and programming on the same design scope. A user centered design approach seems to be a valuable solution for this challenge. End-users within a UCD [27] must be active in an iterative design process involving planning, testing, commissioning, and assessment of the system. The design process should clearly allocate and define tasks between user and system and include multidisciplinary resources and skills both in defining the requirements and in the assessment phase. Examples of open and multidisciplinary approaches for the design of complex, intelligent systems, for example, can be found in autonomous driving vehicle design [28]. Moreover, requirements should be drawn by analyzing all involved stakeholders the context of use, including personal and organizational needs and objectives. This can be done by means of qualitative based descriptions and user models ethnographic research methods or quantitative data collection and analysis.

The new approach shown in Figure 1 aims at a comprehensive design solution for HRC by efficiently allocating safety, collaboration modes, interfaces, managerial and task-oriented issues and requirements in an iterative design and evaluation process revolving around the user and its context. The proposed method matches requirements and technologies typical of HRC systems with the UCD design cycle and specifications.

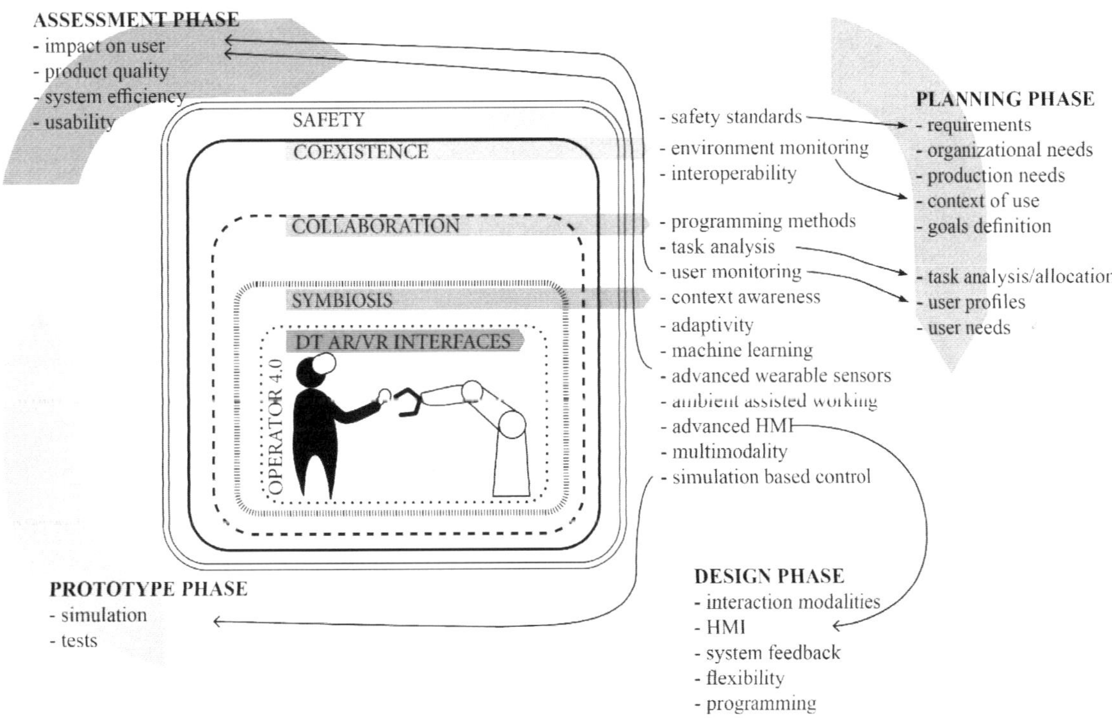

Figure 1. A conceptual view to synergies between modern HRC systems and User Centered Design

We propose that DT AR/VR interfaces will be the main technology driver for this new design method, the tool empowering the O4.0 paradigm and mediating human and robotic systems interaction. Real-time simulations and interaction technologies can assist in evaluating users, environmental, and organizational aspects by providing insights into the state of the system in all its aspects. DT AR/VR interfaces are grounding on IoT, advanced sensors, artificial intelligence, and data analysis technologies which are

IOP Conf. Series: Materials Science and Engineering 1140 (2021) 012011 doi:10.1088/1757-899X/1140/1/012011

promoting smarter safety systems and learning approaches in robotic collaboration and programming by grounding their advanced capabilities precisely on human and environment-based data and behavior analysis. Figure 2 shows the state of realization of the presented approach in the study case discussed in [29] demonstrating the centrality of DT AR/VR interfaces in assessing an efficient HRC system along a UCD workflow. In particular the prototype and assessment phases show the O4.0 paradigm of augmentation of control capabilities in the virtual simulation.

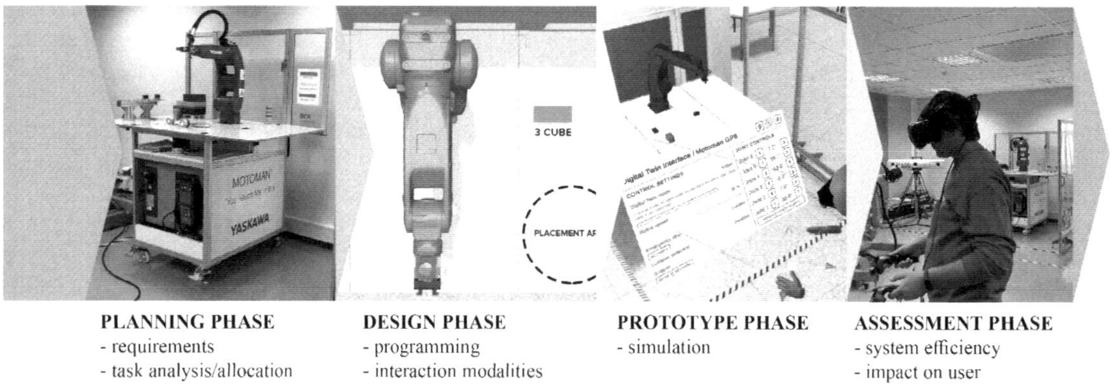

Figure 2. Workflow for human robot collaboration

5. User-centered DT AR/VR architecture for HRC

Considering the variability of collaborative design tasks, automated systems features, interfaces, safety, and programming methods, DT AR/VR interfaces allow for the efficient adaptation and scalability of these variables, including the human operator and the context of use as main players in the configuration of modem cyber-physical systems. VR supports the simulation of reproducible and controlled contexts of use and could potentially be adopted to integrate existing ethnographic research methods [30]. AR/VR technologies include motion and eye-tracking capabilities and can easily be integrated with environmental, physiological monitoring sensors, and motion tracking systems aimed at the description and evaluation of human factors. The use of DT based interaction methods offers a safe and effective approach for experiments on real machinery. The software integration model for a DT AR/VR based HRC systems supported by a user-centric design approach is presented in Figure 3.

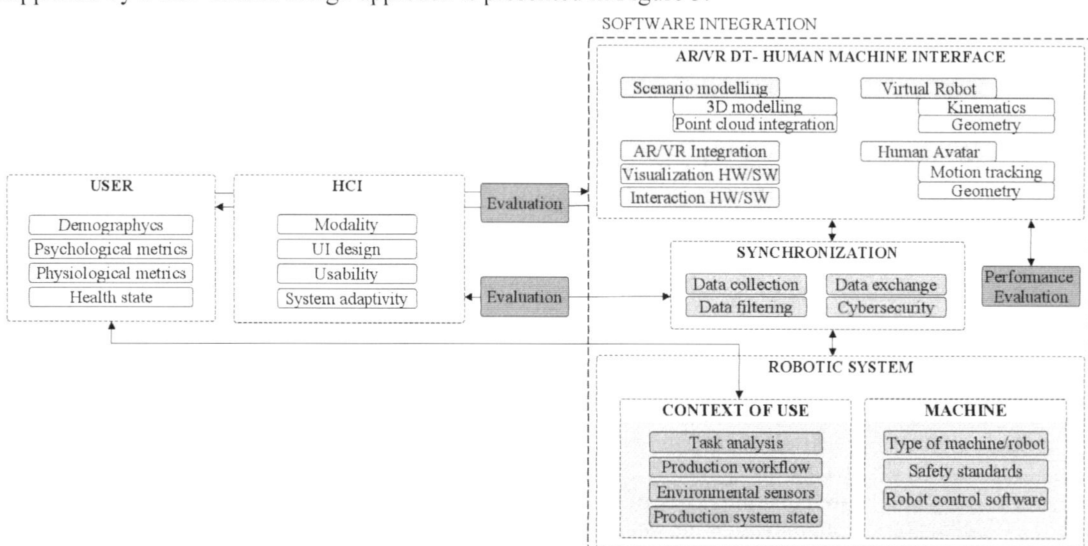

Figure 3. Software integration for HRC systems based on DT AR/VR interfaces

The model contributes to a truly immersive VR/AR training and control system, which provides an agent with higher-level involvement than customary training and thus being more efficient in the HRC process. Moreover, there is a need to develop metrics for assessment and validation of HRI standards in DT AR/VR interfaces of manufacturing systems. The central point of the research is that the test methods and metrics for human-robot teaming should be developed first in DT immersive interfaces to increase the safety level of physical industrial robotic and overall manufacturing systems. The proposed architecture allocates evaluation of DT system performance, HMI efficacy and usability, impact on the user and HRI assessment.

6. Discussion

The evolution of advanced industrial systems, where the operator gains a central position in collaborative production scenarios, encourages more comprehensive and holistic design solutions. UCD can be adopted in supporting the optimization and evaluation of HRC cells by organizing and arranging its components and characteristic along an iterative process that would include human stakeholders, context, and organizational issues in the same design scope. DT AR/VR interfaces offer a new paradigm of interaction with robotic systems and support O4.0 capabilities augmentation in the workplace. Moreover, by including the operator in the interface and simulation itself, AR/VR technologies support the planning and evaluation of advanced HMI in HRC within a user-centric design method. UCD based on DT AR/VR interfaces could lead to more efficient production processes, industrial cell deployment, evaluation of psychological, physiological well-being, and ergonomics factors in HRI.

Extensive research is necessary to clarify the state of the art in DT AR/VR interfaces by addressing use cases, type of interaction, and evaluation metrics. Based on the DT AR/VR for robotic arm control and programming presented in Figure 2, future works will understand how AR/VR is linked to the DT loop and HMI process. User based experiments will try to understand how humans are involved in each interaction, with the robot, with the AR/VR technologies, and with the DT simulated interfaces, and which could be shared metrics of evaluation between these processes. Future research will address cybersecurity, connectivity methods (between physical and virtual worlds) optimization, and development of virtual distributed infrastructure with open access for experimental design environment with various machinery for system integrators and end-user as well researchers.

7. Acknowledgments

This research was supported by project AR16077 Smart Industry Centre (SmartIC) No. 2014-2020.4.01.16-0183, supported by the EU Regional Development Fund.

References

[1] Vaidya S, Ambad P and Bhosle S 2018 Industry 4.0 – A Glimpse *Procedia Manuf.* **20** 233–8
[2] Romero D, Bernus P, Noran O, Stahre J and Fast-Berglund Å 2016 The Operator 4.0: Human Cyber-Physical Systems & Adaptive Automation Towards Human-Automation Symbiosis Work Systems *IFIP Advances in Information and Communication Technology* pp 677–86
[3] Berg J and Lu S 2020 Review of Interfaces for Industrial Human-Robot Interaction *Curr. Robot. Reports* **1** 27–34
[4] Oyekan J O, Hutabarat W, Tiwari A, Grech R, Aung M H, Mariani M P, López-Dávalos L, Ricaud T, Singh S and Dupuis C 2019 The effectiveness of virtual environments in developing collaborative strategies between industrial robots and humans *Robot. Comput. Integr. Manuf.* **55** 41–54
[5] Burghardt A, Szybicki D, Gierlak P, Kurc K, Pietruś P and Cygan R 2020 Programming of Industrial Robots Using Virtual Reality and Digital Twins *Appl. Sci.* **10** 486
[6] Longo F, Padovano A and Umbrello S 2020 Value-oriented and ethical technology engineering in industry 5.0: A human-centric perspective for the design of the factory of the future *Appl. Sci.* **10** 1–25
[7] Nahavandi S 2019 Industry 5.0—A Human-Centric Solution *Sustainability* **11** 4371
[8] Villani V, Pini F, Leali F and Secchi C 2018 Survey on human–robot collaboration in industrial settings: Safety, intuitive interfaces and applications *Mechatronics* **55** 248–66

[9] Demir K A, Döven G and Sezen B 2019 Industry 5.0 and Human-Robot Co-working *Procedia Comput. Sci.* **158** 688–95
[10] Bagdasarov Z, Martin A A and Buckley M R 2020 Working with robots *Organ. Dyn.* **49** 100679
[11] Romero D, Stahre J and Taisch M 2020 The Operator 4.0: Towards socially sustainable factories of the future *Comput. Ind. Eng.* **139** 106128
[12] Barricelli B R, Casiraghi E and Fogli D 2019 A Survey on Digital Twin: Definitions, Characteristics, Applications, and Design Implications *IEEE Access* **7** 167653–71
[13] Rassolkin A, Vaimann T, Kallaste A and Kuts V 2019 Digital twin for propulsion drive of autonomous electric vehicle *2019 IEEE 60th Annu. Int. Sci. Conf. Power Electr. Eng. Riga Tech. Univ. RTUCON 2019 - Proc.* 5–8
[14] Khajavi S H, Motlagh N H, Jaribion A, Werner L C and Holmstrom J 2019 Digital Twin: Vision, benefits, boundaries, and creation for buildings *IEEE Access* **7** 147406–19
[15] Bruynseels K, Santoni de Sio F and van den Hoven J 2018 Digital Twins in Health Care: Ethical Implications of an Emerging Engineering Paradigm *Front. Genet.* **9** 31
[16] Terkaj W, Tolio T and Urgo M 2015 A virtual factory approach for in situ simulation to support production and maintenance planning *CIRP Ann.* **64** 451–4
[17] Tšukrejev P, Kruuser K and Karjust K 2019 Production monitoring system development for manufacturing processes of photovoltaic modules *Proc. Est. Acad. Sci.* **68** 401–6
[18] Kuts V, Sarkans M, Otto T, Tähemaa T and Bondarenko Y 2019 Digital Twin: Concept of Hybrid Programming for Industrial Robots — Use Case *Volume 2B: Advanced Manufacturing* vol 2B-2019 (American Society of Mechanical Engineers)
[19] Horváth G and Erdős G 2017 Gesture Control of Cyber Physical Systems *Procedia CIRP* **63** 184–8
[20] Pérez L, Rodríguez-Jiménez S, Rodríguez N, Usamentiaga R and García D F 2020 Digital Twin and Virtual Reality Based Methodology for Multi-Robot Manufacturing Cell Commissioning *Appl. Sci.* **10** 3633
[21] Vosniakos G-C, Ouillon L and Matsas E 2019 Exploration of two safety strategies in human-robot collaborative manufacturing using Virtual Reality *Procedia Manuf.* **38** 524–31
[22] Peruzzini M, Grandi F and Pellicciari M 2020 Exploring the potential of Operator 4.0 interface and monitoring *Comput. Ind. Eng.* **139** 105600
[23] Helms E, Sehraft R D and Hägele M 2002 Rob@work: Robot assistant in industrial environments *Proceedings - IEEE International Workshop on Robot and Human Interactive Communication* (IEEE) pp 399–404
[24] De Luca A and Flacco F 2012 Integrated control for pHRI: Collision avoidance, detection, reaction and collaboration *2012 4th IEEE RAS & EMBS International Conference on Biomedical Robotics and Biomechatronics (BioRob)* (IEEE) pp 288–95
[25] 2016 ISO/TS 15066, Robots and robotic devices – Collaborative robots
[26] DelPreto J, Lipton J I, Sanneman L, Fay A J, Fourie C, Choi C and Rus D 2020 Helping Robots Learn: A Human-Robot Master-Apprentice Model Using Demonstrations via Virtual Reality Teleoperation *2020 IEEE International Conference on Robotics and Automation (ICRA)* (IEEE) pp 10226–33
[27] 2010 ISO 9241-210, Ergonomics of Human-system Interaction - Human-centered Design for Interactive Systems
[28] Sell R, Leier M, Rassolkin A and Ernits J P 2018 Self-driving car ISEAUTO for research and education *Proc. 2018 19th Int. Conf. Res. Educ. Mechatronics, REM 2018* 111–6
[29] Kuts V, Otto T, Tähemaa T and Bondarenko Y 2019 Digital twin based synchronised control and simulation of the industrial robotic cell using virtual reality *J. Mach. Eng.* **19** 128–44
[30] Rebelo F, Noriega P, Duarte E and Soares M 2012 Using virtual reality to assess user experience *Hum. Factors* **54** 964–82

Analysis of the tube piercing process types in terms of final product properties

Saeed Darki[1], Evgeniy Yurevich Raskatov[2]

[1&2] New Material and Technology Institute, Ural Federal University, Ekaterinburg, Russia

saeed.darki@yahoo.com

Abstract. The rotary tube piercing (RTP) is the first process of making seamless tube after producing the desired alloy ingot. There are several ways to make a seamless tube, one of the most common being RTP. This approach covers a wide range of processes that are categorized according to the number and shape of the rollers. On the other hand, each of these types has designed guides in the output and input of the piercing process. In this article, a new design of input and output guide for all types of rollers have been examined and simulated. Thus, three specimens including Diescher and Conical rollers were considered with 3 and 2 numbers, respectively. Results including torque, total force, temperature distribution and strain were extracted using FEM simulation. The results obtained through simulation are more in line with the experimental results obtained from previous research. While showing the successful performance of the output and input guides of the RTP process, the results indicate that the 3-roller Diescher type RTP has the most suitable arrangement for seamless tube production.
Keywords: Seamless tube, FEM, rotary tube piercing,

1. Introduction

The rotary tube piercing (RTP) process is one of the most common and efficient methods for producing seamless tube. On the other hand, this process is the first stage of production of the seamless tube, after the extraction of the desired material ingots. Therefore, examining the final product properties of the process can significantly shorten and facilitate other stages. Thus, the RTP types are selected according to the requirements for the final product and other operational parameters. There are two types of RTP: two rollers and three rollers. Each type has specially designed components, which include different shapes of the guides at the before or after or during the rolling zone. Also, mainly the shape of rollers includes: Conical and Diescher. Due to the wide range of types and components used in RTP, investigations have been done by other scientists and researchers.

Ding et al. (2018) implemented RTP as a new technology for manufacturing magnesium seamless tube. Also, they claimed the 400 ° C is the optimum temperature in terms of microstructure, uniformity of deformation, the penetration rate in the workpiece, and stress concentration. *Gamin et al. (2020)* studied the benefits and defects of the two rollers and three rollers for different application. The obtained results suggest that the process via two rolls can create a high accuracy in terms of the surface roughness of the tube. *Komori and Mizuno (2009)* investigated the angles within the piercing process and also the shear strain through the experimental and numerical model. Besides, they applied for the cone type roller within the process. Thus, they showed the shear strain ($\gamma_{\vartheta z}$) is independent on the feed angle at the cross angle 20 degrees. *Mao et al. (2020)* implemented two sets of the process tandemly: the piercing and the rolling. Based on the experimental outcomes, the maximum piercing force belongs to the 42CrMo between steel 1045 and AZ31. *Pater et al. (2019)* performed a comparison between three rollers piercing and two rollers. They claimed that the three-roller process has more favourable outcomes to produce the seamless tube. But still, the challenges of examining three rolls process have been left in the functional condition in terms of the longevity of the process components. *Romantsev et al. (2018)* implemented the new technology, based on the piercing process for the seamless tubes with high accuracy in terms of wall thickness. Consequently, they applied two types of the mandrel for piercing; high elongating with the conical shape and low elongating with the cylindrical shape. *Zhang et al. (2020)* developed the nickel-base thick wall tube piercing regards to optimization of parameters variations of process and solution investigation of layers separation phenomena in both surfaces of the tube. They expressed that high temperature and large strain are factors which can increase the external separation in range of r/R=0.8~0.95. *Murillo-Marrodán et al (2020)* carried out the investigation of the thick-wall seamless tubes. The extracted results from experimental and modelling displayed that the main reason for the eccentricity is the interaction of non-uniform temperature distribution in billet and the bending phenomenon of the plug and mandrel. In this case, it seems that the extreme eccentricity has been occurred by the geometry dimensions.

With taking into the previous investigation, in the current study has been examined the impacts of basic parameters for each type of rotary tube piercing. Thus, obtained results lead to select the most suitable approach to manufacture of the seamless tube. Based on operating conditions and facilities, the most suitable method is selected. The results have been obtained based on a considered mathematical model and taking into account the heat transfer conditions between the workpiece and other process components.

2. Materials and methods

In order to investigate RPT, two main types of this process have been modelled through three samples. Therefore, commercial software is used to simulate RTP. Also, the mathematical model has been implemented based on the finite element method. In terms of the mesh study of simulation, due to the increase in the precision of results and the decrease the running time, the most accurate type and element size have been implemented. Take into the nature of the process, the viscoplastic model of the material, heat transfer, and forces and moments have been considered as boundary conditions. Accordingly, the demanded results have been obtained via the three considered sample. On the other hand, in order to validate the mathematical model, a comparison between the experimental results and the simulation based on the total force has been performed.

Based on the number and shape of the rollers used, the process analysis is performed via FEM. The number and shape of the rollers of the simulation samples are as follows:

Table 1. The number and the shape of the simulated specimen

№ specimens	Number of rolls	Shape of rolls
1	2	Conical
2	3	Conical
3	3	Diescher

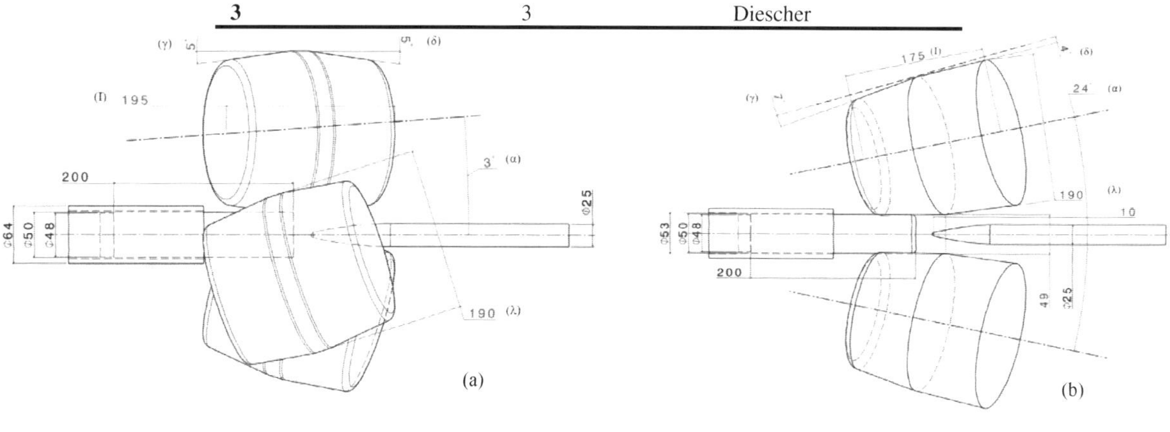

Figure 1. The schematic of piercing process. (a): Third specimen with three Diescher rollers. (b): Second Specimen with three Conical

IOP Conf. Series: Materials Science and Engineering 1140 (2021) 012012 doi:10.1088/1757-899X/1140/1/012012

While discussing the geometry of the produced tube and influence of the other process parameters; the feasibility of changing the shape of a conventional disk guide for a two-roller RTP process has been investigated. For this purpose, only one guide was placed at the input and at the output of the process, and a combination of plug and long mandrel was designed. Therefore, the possibility of simplifying the two-roller process was investigated and implemented by removing the guide discs. The schematics of the RTP for two types of process are shown in the figure 1.

In this investigation, the 3D-Deform software has been implemented as commercial software in order to the simulation of piercing rotary tube. Due to the facility of the software, it is possible to simulate this process only by using precise boundary conditions and executing practical parameters. Otherwise, the simulation of the process will not be complete or the manufactured tube will not have the desired geometry. It should be noted the size of ratio, the element type and the number of the elements are respectively: 3.5, Tetrahedral mesh, 25000.

Also, in Table 2, the value of each of the parameters have been stated for each type. In this process, the material used for the workpiece is DIN-C-15 and for other members of the AISI-H-13 process. Also, the temperature of the workpiece at the beginning of the process is 1200°C and other components are assumed to be 150°C. the material properties are as follows:

Table 2. The Geometry parameters of simulated specimen

Geometry parameters	Number specimen 1&2	Number specimen 3
Cross Angle, α°	12	3
Feed Angle, β°	20	20
Roll entrance angle,γ°	7	5
Roll output angle,δ°	4	5
Roll Diameter,λ[mm]	190	190
Roll length,l [mm]	175	195
Angular velocity ω [Rad/s]	8	8

Table 3. The properties of material in simulation

Material Property	Steel C15 (DIN)	AISI-H-13
Density	7.85 g/cm³	7.76 g/cm³
Poisson's Ratio	0.303	0.3
Young Modulus	210 GPa	210 GPa
Thermal Conductivity	64 W/m.K	28.6W/m.K
Thermal Expansion	12 E-6/C	10 E-6/C
Melting Temperature	1540 °C	-

Therefore, the behavior of viscoplastic material flow stress Equation (1) is described as follows [8]:

$$\sigma_f = 2707.11\, e^{-0.003\dot{2}5T}\ \varepsilon^{-0.13502}\ e^{0.05494/\varepsilon}\ \varepsilon^{0.1529} \quad (1)$$

Where $\sigma_f, \varepsilon, T\ and\ \varepsilon$ respectively, are: flow stress [MPa], effective strain - , temperature [C°] and strain rate $[S^{-1}]$

For friction, the shear stress model is used to describe the friction between the rollers, other components, and the workpiece. The choice of this model from the friction equation is based on large deformation of geometry for hot and dry processes. The Equation 2 has described the fraction in RTP [1].

$\tau = m.K$ (2) and $K = {}^{\bar{\sigma}}/_{\sqrt{3}}$ (3)

Where $\tau, m, k\ and\ \bar{\sigma}$ respectively, are : shear stress, Friction factor, material consistency and equivalent stress.

In the mathematical model section, the ALE approach is implemented in FEM simulation. Moreover, in the Mesh section, the tetrahedral mesh is used for meshing the geometry, and due to the large deformation of the geometry during the process, as well as the plug geometry, determining the optimal size ratio will be effective in increasing the simulation accuracy and reducing the simulation time.

In this simulation, the material in question is considered viscoplastic and for other components of the process, the rigid state of the material is assumed. In addition, from the viewpoint of the distribution temperature analysis of the workpiece, heat transfer has been considered between the components connected to the workpiece. Thus, the temperature distribution is obtained at the end of the process. The initial boundary conditions have been assumed constant for all three specimens due to the compare basic process parameters such as torque and force. The Alpha angle is one of the most important parameters in this process that determines the rotational and axial velocity of the workpiece. As well as, it has a significant effect on temperature distribution by reducing the processing time. Also, the temperature distribution is directly proportional to the torque and force of the rollers. In mathematical model, while examining the feasibility of the two-roller process without a disc guide and replacing it with an input guide, a long mandrel is used at the output to guide the pipe path, out of the rolling area. The new design, while simplifying the production process of seamless pipes, the parameters affecting the heterogeneous deformation of the workpiece, such as disc guides, have been removed. Therefore, the quality of the workpiece has been increased after rolling, and the process has become more controllable. It should also be noted that due to the complexity of the RTP process, the components of the process should be designed to fit within each other. The mismatch between the geometric and kinetic parameters of the process components can lead to the reduction in the speed, failure or cessation of the piercing process. Therefore, the design of the geometry parameters of the rollers and the mandrel and the inlet guide must all fit together.

3. Results and discussions

In this section, have been simulated and analyzed the most appropriate dimensions and adjustment of the RTP process while showing the feasibility of new design and simplification of the process at the RTP process. As mentioned in the previous section, to avoid trial and error in operating conditions and cost loss, as well as due to the advanced dynamics of the process, all dimensions, coordinates, and dynamic and kinetic parameters of the process components need to be simulated. Changes in process parameters such as roller rotation speed, angles and distance of the plug from the billet and the applied force of pusher can lead to cracks in the workpiece during the piercing process at the tip of the plug or non-uniformity of the wall thickness. It should also be noted that this process is only suitable for creating the initial shape of the seamless pipe from the produced ingot and only the uniformity of the pipe geometry is one of the most important design goals of this process. Therefore, to change the size and increase the accuracy of geometric dimensions or pipe tension and change the pipe diameters, more controllable processes such as radial forging are recommended.

In order to the validation of the mathematical model, the RTP process was simulated with three Diescher rollers at a 10° angle as the feed angle. Therefore, in the first 5 seconds of the process, a comparing was carried out between simulation and experimental results. The exerted forces in the simulation have an average of 53.199 KN and the experimental results have an average of 48 KN [6]. Thus, comparing the force values and the fluctuations during the process allow confirming the accuracy of the FEM simulation.

After mathematical modelling and carrying out simulation via FEM software, the results extracted from all three types of assumed processes are shown. Table 4 shows the comparison between the production pipe parameters of all three types of RTP processes. In addition, it expresses the amount of transfer torque to the rollers and the average force applied between the rollers and the workpiece.

Table 4. The results comparison of each simulated specimen

Results	Number specimen 1	Number specimen 2	Number specimen 3
Tube length [mm]	204.2	206.8	213.9
Outlet diameter [mm]	66.07	64.9	62.51
Inlet diameter [mm]	25.37	27.84	28.153
Wall thickness average [mm]	20.22	18.13	17.35
Final surface area [mm^2]	53028	57064	60915
Force average[KN]	75.80465658	39.4119268	42.67810227
Torque average[N/m]	1276.14919	917.9283054	1084.09141

IOP Conf. Series: Materials Science and Engineering 1140 (2021) 012012 doi:10.1088/1757-899X/1140/1/012012

Figure 2 shows the amplitude of the applied force for all 3 specimens. According to the results, the highest force has been exerted by first specimen rolls. Besides, the lowest average is related to the forces which exerted by the Diescher rollers. On the other hand, the amplitude variations of force at the third specimen is less than other samples; which can indicate low vibration the piercing machine and consequently the uniformity of the surface of the production pipe. The application of the less force by the rollers can indicate less frictional force between the part and the rollers. This phenomenon can help to increase the life of rollers and reduce the wear in them.

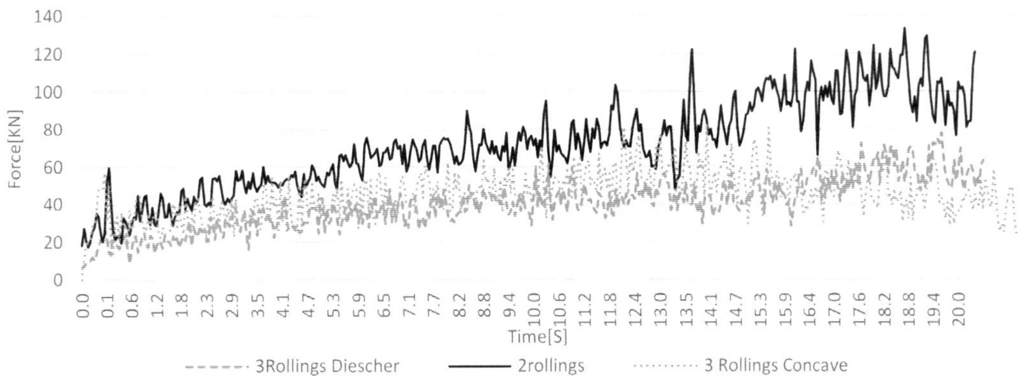

Figure 2. The variation applied force during RTP for each 3 types

Figure 3 shows the strain distribution during the piercing process at the final 15 mm. As the strain distribution in the longitudinal section of the workpiece shows, the process with two rollers has the highest strain, which is in accordance with the other results of previous experimental studies [7]. On the other hand, the strain distribution in sample three has a smaller value while it has a more uniform distribution.

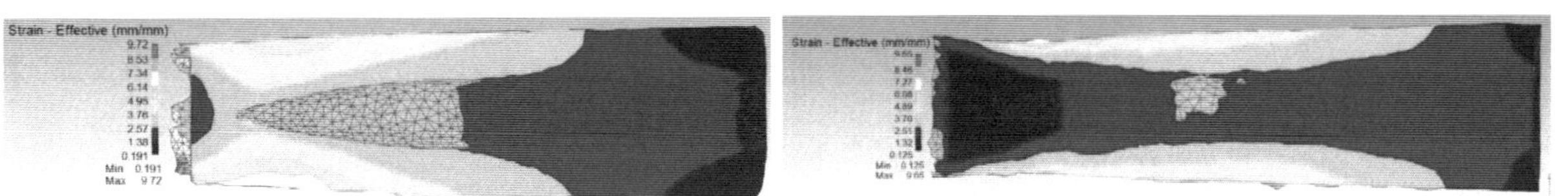

(a) (b)

Figure 3. The distribution of effective strain at the end 15 mm, on the longitudinal section of workpiece . (a) Three Conical, (b) three Diescher

Figure 4 shows the temperature distribution of the workpiece during the process at the final 15 mm. Based on the obtained results, the uniformity of its temperature distribution can be a factor to prevent the eccentricity of the workpiece and the production of seamless pipes with uniform thickness in the wall. Therefore, the homogeneous temperature distribution in sample No. 3 (Diescher) produces a tube with a better wall thickness uniformity.

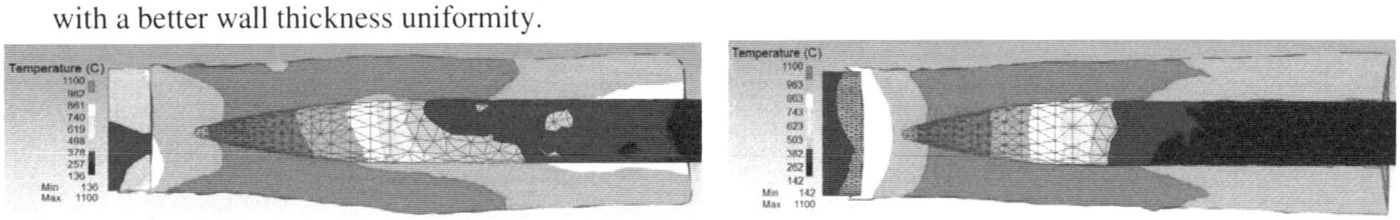

(a) (b)

Figure 4. The distribution of Temperature the end 15 mm, on the longitudinal section of workpiece . (a) Three Conical rollers, (b) three Diescher

Figure 5 exhibits the total velocity on the axial section of the workpiece at the final 15mm. Regarding the velocity distribution, the three rollers conical type include the less penetration rate in the material. Accordingly, it can be expressed the less contact surface between workpiece and rollers lead to the less total velocity in the workpiece and obviously the time of process increases.

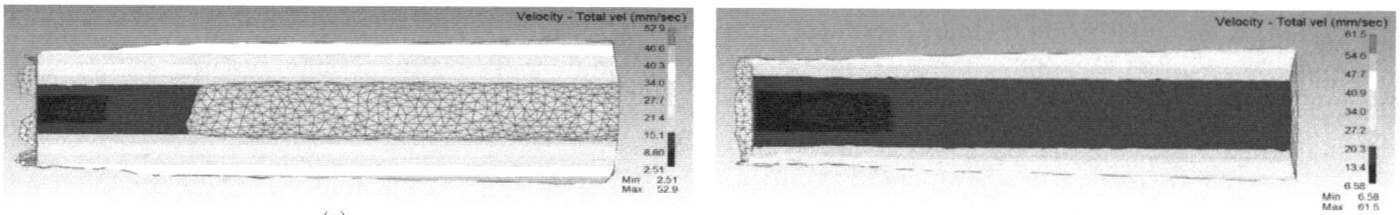

(a)

(b)

Figure 5. The distribution of Temperature the end 15 mm, on the longitudinal section of workpiece . (a) Three Conical, (b) Three Diescher

4. Conclusions

In this paper, while examining three models of the most common types of RTP processes, the optimization in the design of the input guide in order to simplify and feasibility the manufacturing process was investigated. On the other hand, a comparison was made between the properties of the pipe produced from the three samples, the results which can be used to select the most appropriate type of piercing process. Based on this, the extracted results can be expressed as follows:

1. While implementing the new design, through the designed guide, the o-wall deformation of the workpiece was eliminated and the seamless tube was obtained with greater uniformity in thickness and surface as a result of the simulation output.
2. According to the results, the highest average force has been applied by the conical type with two rollers and the lowest average force belonged to the Diescher type with three rollers with the value of 42.67 KN. In terms of average torque applied during the process, the conical type with three rollers has a lower value than the Diescher type due to the reduction of the contact surface in this type.
3. The uniform distribution of strain in the wall of the produced tube and on the other hand the uniform distribution of temperature along the tube, indicate the higher quality of the tube produced by specimen three. Besides, the higher penetration velocity in the third specimen leads to the manufacturing of the seamless tube at a shorter time. Thus, based on these parameters it can be claimed that the three-roller sample with Diescher type has more advantage for the production of seamless pipes in these dimensions.

5. References

[1] Design of Tools for Deformation Processes. 1986. doi:10.1007/978-94-009-4177-9.

[2] Barco J. Analysis of Wall Thickness Eccentricity in the Rotary Tube Piercing Process Using a Strain Correlated FE Model 2020. doi:10.3390/met10081045.

[3] Ding X feng, Shuang Y hua, Liu Q zu, Zhao C jiang. New rotary piercing process for an AZ31 magnesium alloy seamless tube. Mater Sci Technol (United Kingdom) 2018;34:408–18. doi:10.1080/02670836.2017.1393998.

[4] Gamin Y V., Romantsev BA, Pashkov AN, Patrin P V., Bystrov IA, Fomin A V., et al. Obtaining Hollow Semifinished Products Based on Copper Alloys for Electrical Purposes by Means of Screw Rolling. Russ J Non-Ferrous Met 2020;61:162–71. doi:10.3103/S1067821220020054.

[5] Komori K, Mizuno K. Study on plastic deformation in cone-type rotary piercing process using model piercing mill for modeling clay. J Mater Process Technol 2009;209:4994–5001. doi:10.1016/j.jmatprotec.2009.01.022.

[6] Mao F, Wang F, Shuang Y, Hu J, Chen J. Deformation behavior and experiments on a light alloy seamless tube via a tandem skew rolling process. Metals (Basel) 2020;10. doi:10.3390/met10010059.

[7] Murillo-Marrodán, A., García, E., Barco, J., Cortés, F., 2020. Analysis of wall thickness eccentricity in the rotary tube piercing process using a strain correlated FE model. Metals (Basel). 10, 1–18. https://doi.org/10.3390/met10081045.

[8] Pater Z, Wójcik Ł, Walczuk P. Comparative Analysis of Tube Piercing Processes in the Two-Roll and Three-Roll Mills. Adv Sci Technol Res J 2019;13:37–45. doi:10.12913/22998624/102766.

[9] Romantsev B, Goncharuk A, Aleshchenko A, Gamin Y, Mintakhanov M. Development of multipass skew rolling technology for stainless steel and alloy pipes' production. Int J Adv Manuf Technol 2018;97:3223–30. doi:10.1007/s00170-018-2134-3.

[10] Zhang Z, Liu D, Zhang R, Yang Y, Pang Y, Wang J, et al. Experimental and numerical analysis of rotary tube piercing process for producing thick-walled tubes of nickel-base superalloy. J Mater Process Technol 2020;279:116557. doi:10.1016/j.jmatprotec.2019.116557.

Modelling FGM materials. An accurate function approximation algorithms

J Majak[1], **M Mikola**[1], **M Pohlak**[1], **M Eerme**[1] **R Karunanidhi**[1,2]

[1]Department of Mechanical and Industrial Engineering, Tallinn University of Technology, Ehitajate tee 5, Tallinn, Estonia
[2]Department of Automotive and Aeronautical Engineering Hamburg University of Applied Sciences, Berliner Tor 9, 20099, Hamburg, Germany

juri.majak@taltech.ee

Abstract. The study is focused on development of an accurate and cost effective function approximation techniques for modelling functionally graded materials. Different grading functions (exponential, power law) are expanded into Haar wavelet series based on higher order Haar wavelet approach. The proposed techniques can be utilized also for modelling load cases, complex boundary conditions, grading functions etc.

1. Introduction

Recently, the Haar wavelet method (HWM) and higher order Haar wavelet method (HOHWM) have been utilized with success for solving wide class of engineering problems [1-7]. The differential governing equations of the structures, solved using HWM or HOHWM, involve often set of functions describing loading, boundary conditions, varying stiffness, geometry, density and material properties. In order to provide comprehensive approach, it is reasonable to treat these function in terms of Haar wavelets i.e. to expand these functions into Haar wavelet series [8-10]. Most commonly the functions accompanying the governing equations are expanded into Haar wavelets. However, here is reason to be careful, because it is shown in [11] that in the case where function is expanded into Haar wavelets the order of convergence with respect to mesh/resolution is equal to one. Since the order of convergence of the HWM is equal to two [12] and HOHWM higher, such an approach may lead to loss of accuracy of the solution, especially in the case of use of HOHWM. Thus, more accurate Haar wavelet based approach is needed to overcome possible loss of accuracy of the solution.
Current study is focused on approximation of grading functions of the functionally graded (FG) materials using higher order Haar wavelet method based function approximation. The HOHWM was introduced by authors for solving differential equations as improvement of the HWM [13]. Based on HOHWM, the higher order approach for functions approximation was introduced in [14]. Herein the latter higher order function approximation technique is extended for modelling grading functions of the FG materials. As result the generalized algorithm can be used for solving governing equations covering different grading functions (exponential, power law, four parameter functions, etc.). In the current study the second order derivatives of the grading functions are expanded into Haar wavelets providing fourth order convergence with respect to mesh.

2. Haar wavelets

The Haar wavelets are introduced as [1]

$$h_i(x) = \begin{cases} 1 & \text{for} \quad x \in [\xi_1(i), \xi_2(i)) \\ -1 & \text{for} \quad x \in [\xi_2(i), \xi_3(i)), \\ 0 & \text{elsewere} \end{cases} \tag{1}$$

where $i = m + k + 1$, $m = 2^j$ is a maximum number of square waves deployed in interval $[A, B]$ and the parameter k indicates the location of the particular square wave,

$$\xi_1(i) = A + 2k\mu\Delta x, \quad \xi_2(i) = A + (2k+1)\mu\Delta x, \quad \xi_3(i) = A + 2(k+1)\mu\Delta x,$$
$$\mu = M / m, \ \Delta x = (B - A)/(2M), \ M = 2^J . \tag{2}$$

The parameters j and J stand for the resolution and maximum resolution, respectively. Any square integrable and finite function in the interval $[A, B]$ can expanded into Haar wavelets.

3. Approximation of grading functions of the FG materials

Based on idea of higher order Haar wavelet method introduced by authors in [13] an accurate function approximation is derived for four simple and widely used grading functions described in the following subsections.

3.1. Exponential grading function

Due to its simplicity, the exponential grading function is widely used. In the case of axially graded materials the exponential functions are utilized commonly for describing the elasticity modulus $E(x)$ and the density $\rho(x)$ as

$$E(x) = E(0) * e^{2\beta x/L}, \ \rho(x) = \rho(0) * e^{2\beta x/L}. \tag{3}$$

In (3) L is a length of the beam and β is a grading parameter, $E(0)$ and $\rho(0)$ stand for the reference values of the elasticity modulus and density at $x = 0$. However, in order to provide unique approach for handing different grading functions, herein the volume fraction of the first constituent V_1 is expanded into Haar wavelet. The modulus of elasticity and the density can be expressed in terms of volume fraction V_1 as

$$E = (E_1 - E_2)V_1 + E_2 , \quad \rho = (\rho_1 - \rho_2)V_1 + \rho_2, \tag{4}$$

where the indexes 1 and 2 refers to the constituents (materials) 1 and 2, respectively. In the case of exponential grading function (3), the volume fractions of the constituents (materials) V_1 and V_2 can be derived as

$$V_1 = \frac{e^{2\beta} - e^{2\beta x/L}}{e^{2\beta} - 1}, V_2 = \frac{e^{2\beta x/L} - 1}{e^{2\beta} - 1}. \tag{5}$$

The higher order Haar wavelet expansion can be introduced as

$$f^{(n)}(x) = \sum_{i=1}^{2M} b_i h_i(x), \quad n = 0,1,2,..., \tag{6}$$

where b_i and $h_i(x)$ stand for the wavelet coefficients and Haar functions, respectively. In the case of $n = 0,1,2...$ the functions itself, its first derivative, second derivative, etc. is expanded into series of the Haar functions. In the following the value $n = 2$ is utilized in order to provide fourth order convergence

$$\frac{e^{2\beta} - e^{2\beta x/L}}{e^{2\beta} - 1} = \sum_{i=1}^{2M} b_i p_{2,i} + d_1 x + d_2. \tag{7}$$

In (7) $p_{2,i}$ stand for second order integrals of the Haar functions (1). Two integration constants d_1 and d_2 can be determined by satisfying equation (5) in boundary points $x = 0$ and $x = L$, respectively. The Haar wavelet expansion coefficients b_i can be determined by satisfying the equation (7) in grid points. In the case of uniform mesh used the grid points x_l are given as

$$x_l = \frac{2l-1}{4M}, \quad l = 1, \dots, 2M. \tag{8}$$

The volume fraction $V_1(x)$ be calculated for each coordinate x value by substituting the coefficients b_i, integration constants d_1 and d_2 in (7). The volume fraction $V_2(x)$ can be evaluated as $1 - V_1(x)$. The detailed expressions of the coefficients b_i, integration constants d_1 and d_2 are omitted for conciseness sake.

3.2. Power law grading function

The power law relation for describing FG materials is given as [15]

$$E = (E_1 - E_2)\left(1 - \frac{x}{L}\right)^k + E_2, \quad \rho = (\rho_1 - \rho_2)(1 - \frac{x}{L})^k + \rho_2. \tag{9}$$

In (9) k is a grading parameter. Despite to presence of one parameter in both, exponential and power law grading function, the power law function is more flexible/general. In exponential grading function (3) the parameter β value is determined by mechanical characteristic value at the point $x = L$. However, in the case of power law grading function (9) the values of the mechanical characteristics in boundary points are provided and the parameter k remains as design parameter. Varying the value of the parameter k allow to obtain different distributions of the mechanical characteristics. The higher order Haar wavelet expansion can be employed for the volume fraction of the constituent (material) 1 as

$$\left(1 - \frac{x}{L}\right)^k = \sum_{i=1}^{2M} b_i p_{2,i} + d_1 x + d_2. \tag{10}$$

Similarly, to above, the second order derivative of the volume fraction function is expanded into Haar wavelet and grading function is expressed in terms of second integrals of Haar functions. The elasticity modulus and density can be evaluated by substituting (10) in (9).

3.3. Four parameter power law grading function

In order to provide higher flexibility for describing the distribution of the FGM, the following four parameter model is considered for approximation of the volume fraction V_1 [15]

$$V_1 = C * \left[1 - \frac{x}{L} + \alpha\left(\frac{x}{L}\right)^\beta\right]^\gamma, \tag{11}$$

where parameters C, α , β and γ control the volume fraction V_1variation through the length of the axially graded structure. The higher order Haar wavelet expansion is given as

$$C * \left[1 - \frac{x}{L} + \alpha\left(\frac{x}{L}\right)^\beta\right]^\gamma = \sum_{i=1}^{2M} b_i p_{2,i} + d_1 x + d_2. \tag{12}$$

Note, that four parameter grading function can be utilized for design optimization of FGM structures.

IOP Conf. Series: Materials Science and Engineering 1140 (2021) 012013 doi:10.1088/1757-899X/1140/1/012013

3.4. Four parameter trigonometric grading function

Four parameter trigonometric grading function can be considered as an alternative to four parameter power law grading function. The corresponding volume fraction V_1 can be expressed as [15]

$$V_1 = C * \left[\frac{1}{2} - \frac{\alpha}{2} sin\left(\frac{n\pi x}{L} + \phi\right)\right]^{\gamma}. \tag{13}$$

Expanding the second order derivative of the volume fraction (13) into Haar wavelet series and integrating twice one obtains approximation of the function V_1 as

$$C * \left[\frac{1}{2} - \frac{\alpha}{2} sin\left(\frac{n\pi x}{L} + \phi\right)\right]^{\gamma} = \sum_{i=1}^{2M} b_i p_{2,i} + d_1 x + d_2. \tag{14}$$

The approximations used for above four grading functions are second order polynomials including global and local terms.

4. Numerical results

The values of the grading functions, its absolute errors and convergence rates corresponding to the exponential, power law and four parameter grading functions are given in tables 1 and 2, respectively. In Table 1 the approximation results are given for exponential and power law grading functions.

Table 1. Approximation for exponential and power law grading functions ($\beta = -0.549306, k = 1.5$).

	$V_1(x) = \frac{e^{2\beta} - e^{2\beta x/L}}{e^{2\beta} - 1}$			$V_1(x) = \left(1 - \frac{x}{L}\right)^k$		
2M	Function value at point x=L/2	Converg. rate	Absolute error	Function value at point x=L/2	Absolute error	Converg. rate
4	0.365989778		3.57E-05	0.353652500		9.91E-05
8	0.366023040	3.8946	2.40E-06	0.353549500	4.6608	3.92E-06
16	0.366025287	3.9954	1.50E-07	0.353553000	3.3869	3.74E-07
32	0.366025428	3.9996	9.40E-09	0.353553400	3.9854	2.36E-08
64	0.366025437	3.9999	5.87E-10	0.353553400	3.9969	1.48E-09
128	0.366025437	3.9994	3.67E-11	0.353553400	3.9990	9.26E-11

Table 2. Approximation for four parameter grading functions ($C = 1, \alpha = 1, \beta = \theta = 2, \phi = \frac{\pi}{2}, \gamma = 1, \eta = 1.2$).

	$V_1(x) = C * \left[1 - \frac{x}{L} + \alpha\left(\frac{x}{L}\right)^{\beta}\right]^{\theta}$			$V_1(x) = C * \left[\frac{1}{2} - \frac{\alpha}{2} sin\left(\frac{n\pi x}{L} + \phi\right)\right]^{\gamma}$		
2M	Function value at point x=L/2	Converg. rate	Absolute error	Function value at point x=L/2	Absolute error	Converg. rate
4	0.561810700		6.89E-04	0.653466700		1.04E-03
8	0.562454300	3.9150	4.57E-05	0.654447800	4.1021	6.07E-05
16	0.562497100	3.9975	2.86E-06	0.654504800	4.0206	3.74E-06
32	0.562499800	4.0000	1.79E-07	0.654508300	4.0050	2.33E-07
64	0.562500000	4.0000	1.12E-08	0.654508500	4.0013	1.45E-08
128	0.562500000	3.9999	6.99E-10	0.654508500	4.0003	9.08E-10

It can be observed from Table 1 that the order of convergence tends to four as expected and the absolute error reduces up to 10^{-11}. The results in Tables 2 are similar, high accuracy and fourth order rate of convergence is achieved. Thus, the obtained results can be used for HWM and also HOHWM (s=1) where convergence is not higher than four (convergence rate of the HOHWM is equal to 2+2s, where s is method parameter).
In future study the functions approximations developed, are planned to apply for structural analysis and design optimization of engineering structures [16-21] and production processes [22-25]. An another challenging research area is fractional calculus [26]. From one side, fractional calculus allows to describe a number of real world problems more naturally/objectively. For example, modelling the behavior of the viscoelastic material. From other side, the mainstream numerical methods cannot by applied directly for fractional calculus, but need adaption, refinement. The proposed function approximation technique can be applied for expansion of the fractional derivatives included in differential equations of integer order derivatives used for approximation of fractional derivatives. Solution of fractional differential and integro-differential equations is foreseen.

5. Conclusions

An accurate function approximation technique is proposed for modelling grading functions of the FG materials, based on idea of higher order Haar wavelet method. According to proposed approach the second order derivative of the grading function is expanded into Haar wavelet series. The two complementary integration constants are determined by using function values on boundary. As result fourth order convergence with respect to mesh was achieved. The obtained results are accurate enough utilizing with HWM and HOHWM (s=1). The proposed approach covers obviously further increase of accuracy by increasing the order of derivative expended into Haar wavelet series in formula (6).

6. References

[1] Lepik Ü, Hein H 2014 *Haar wavelets: with applications* (International Publishing Switzerland: Springer)
[2] Majak J, Pohlak M, Eerme M 2009 Application of the Haar Wavelet-based discretization technique to problems of orthotropic plates and shells. *Mechanics of Composite Materials* **45**(6) pp 631-642.
[3] Hein H, Feklistova L 2011 Computationally efficient delamination detection in composite beams using Haar wavelets. *Mechanical Systems and Signal Processing* **25**(6) pp 2257-2270
[4] Xie X, Jin G, Ye T, Liu Z 2014 Free vibration analysis of functionally graded conical shells and annular plates using the Haar wavelet method. *Applied Acoustics* **85** pp 130-142.
[5] Jin G, Xie X, Liu Z 2014 The Haar wavelet method for free vibration analysis of functionally graded cylindrical shells based on the shear deformation theory, *Composite Structures* **108** pp 435–448.
[6] Islam SU, Aziz I, Al-Fhaid AS 2014 An improved method based on Haar wavelets for numerical solution of nonlinear integral and integro-differential equations of first and higher orders. *Journal of Computational and Applied Mathematics* **260** pp 449–469.
[7] Aziz I, Islam SU, Khana F 2014 A new method based on Haar wavelet for the numerical solution of two-dimensional nonlinear integral equations. *Journal of Computational and Applied Mathematics* **272** pp 70–80
[8] Xie X, Jin G, Li W, Liu Z 2014 A numerical solution for vibration analysis of composite laminated conical, cylindrical shell and annular plate structures *Composite Structures* **111**, pp 20-30
[9] L. Wang, Y.Ma, Z. Meng 2014 Haar wavelet method for solving fractional partial differential equations numerically *Applied Mathematics and Computation* **227** pp 66–76
[10] Hsiao CH, 2015 A Haar wavelets method of solving differential equations characterizing the dynamics of a current collectionsystem for an electric locomotive *Applied Mathematics and Computation* **265** pp 928–935

[11] E. Babolian, A. Shahsavaran 2009 Numerical solution of nonlinear Fredholm integral equations of the second kind using Haar wavelets *Journal of Computational and Applied Mathematics* **225** pp 87–95
[12] Majak J, Shvartsman BS, Kirs M, Pohlak M, Herranen H 2015 Convergence theorem for the Haar wavelet based discretization method. *Composite Structures* **126** pp 227-232
[13] Majak J, Pohlak M, Karjust K, Eerme M, Kurnitski J, Shvartsman BS 2018 New higher order Haar wavelet method: Application to FGM structures. *Composite Structures*, **201**, pp 72-78.
[14] Majak, J, Eerme M, Haavajõe A, Karunanidhi R, Scholz D, Lepik A 2020 Function approximation using haar wavelets *AIP Conference Proceedings* **2293** Article number 230004 International Conference on Numerical Analysis and Applied Mathematics 2019 Greece; 23-28 September 2019 Code 165330 DOI 10.1063/5.0026543
[15] Alshabatat NT and Naghshineh K 2014 Optimization of Natural Frequencies and Sound Power of Beams Using Functionally Graded Material *Advances in Acoustics and Vibration*, Article ID 752361, https://doi.org/10.1155/2014/752361
[16] Lellep J, Majak J 1997 On optimal orientation of nonlinear elastic orthotropic materials *Structural Optimization* **14**(2-3) pp 116-120
[17] Zhang WH, Domaszewski M, Bassir H 1999, Developments of sizing sensitivity analysis with the ABAQUS code. *Structural Optimization* 17, 2 pp 219-225.
[18] Guessasma S, Bassir D, Hedjazi L 2015 Influence of interphase properties on the effective behaviour of a starch-hemp composite. *Materials and Design* 65, pp 1053-1063.
[19] Nahas M, Alzahrani M 2020 Optimal Stochastic Distribution of CNTS in a Cantilever Polymer Microbeam Using Artificial Neural Networks. *Mechanics of Composite Materials* **56**, pp 665–672.
[20] Cui D, Li DK 2019 Optimization of Extension-Shear Coupled Laminates Based on the Differential Evolution Algorithm. *Mechanics of Composite Materials* **54**, pp 799–814.
[21] Plotnikova SV, Kulikov GM, Shape Control of Composite Plates with Distributed Piezoelectric Actuators in a Three-Dimensional Formulation. *Mechanics of Composite Materials* **56** pp 557–572.
[22] Sell R, Otto T 2008. Remotely controlled multi robot environment. *Proc. of 19th EAEEIE Annual Conf.*: 19th EAEEIE Annual Conference, Tallinn, Estonia, June 29 - July 2, 2008. Tallinn, pp 20−25. DOI: 10.1109/EAEEIE.2008.4610152.
[23] Kuts V, Otto T, Tähemaa T, Bukhari K, Pataraia T 2018. Adaptive industrial robots using machine vision. ASME 2018 International Mechanical Engineering Congress and Exposition, 2: IMECE2018, November 9-15, Pittsburgh, PA, USA. ASME. DOI: 10.1115/IMECE2018-86720.
[24] Kaganski S, Majak J, Karjust K, Toompalu S 2017 Implementation of Key Performance Indicators Selection Model as Part of the Enterprise Analysis. *Procedia CIRP* **63**, pp 283-288, 50th CIRP Conference on Manufacturing Systems, CIRP CMS 2017 Taiwan; Code 137078
[25] Snatkin A, Eiskop T, Karjust K, Majak J 2015 Production monitoring system development and modification *Proceedings of the Estonian Academy of Sciences*, **64** 4 pp 567-580
[26] Majak J, Shvartsman B, Pohlak M, Karjust K, Eerme M, Tungel, E 2017 Solution of fractional order differential equation by the Haar Wavelet method. Numerical convergence analysis for most commonly used approach *AIP Conference Proceedings* **1738**, 8 June 2016, Article number 480110 International Conference of Numerical Analysis and Applied Mathematics 2015, ICNAAM 2015 Greece 23-29 September 2015, Code 122034.

Acknowledgments
The research was supported by Estonian Centre of Excellence in Zero Energy and Resource Efficient Smart Buildings and Districts, ZEBE, TK146 funded by the European Regional Development Fund (grant 2014-2020.4.01.15-0016).

Benchmark Concept for Industrial Pick&Place Applications

Axel Vick[1], Martin Rudorfer [2], Vojtech Vonasek[3]

[1] PhD Student, Process Automation and Robotics, Fraunhofer IPK Berlin, Germany
[2] Postdoctoral Research Fellow, University of Birmingham, United Kingdom
[3] Postdoctoral Research Fellow, Czech Technical University in Prague, Czech Republic

E-mail: axel.vick@ipk.fraunhofer.de

Abstract. Robotic grasping and manipulation is a highly active research field. Typical solutions are usually composed of several modules, e.g. object detection, grasp selection and motion planning. However, from an industrial point of view, it is not clear which solutions can be readily used and how individual components affect each other. Benchmarks used in research are often designed with simplified settings in a very specific scenario, disregarding the peculiarities of the industrial environment. Performance in real-world applications is therefore likely to differ from benchmark results. In this paper, we present a concept for the design of general Pick&Place benchmarks, which help practitioners to evaluate the system and its components for an industrial scenario. The user specifies the workspace (obstacles, movable objects), the robot (kinematics, etc.) and chooses from a set of methods to realize a desired task. Our proposed framework executes the workflow in a physics simulation to determine a range of system-level performance measures. Furthermore, it provides introspective insights for the performance of individual components.

1. Introduction

There is a high demand for robotic solutions particularly for industrial applications such as assembly, machine loading or unloading, and handling in general [1]. Yet, creating intelligent robots that can handle objects at a similar level of dexterity as humans is still an unsolved research problem [2]. The ongoing trend towards high-mix low-volume production makes this even more challenging, because potential solutions must generalize well or quickly adapt to novel manipulated objects. The Pick&Place task is one of the most studied problems especially as it is needed in many industry applications. Generally, the task is to move an object from one place to another. The main phases of Pick&Place are a) pre-grasping, where the end-effector moves to a position suitable for grasping, b) grasping, where the end-effector attempts to grasp the object and detach it from its initial position; c) transport, where the object is moved above the target place, and d) placement, where the object is intentionally released at the target zone [3]. There is a large variety of approaches to tackle this task, from engineered solutions which are specifically tailored towards certain objects, to learning-based solutions which might be more general but hard to replicate [4]. From the perspective of industrial engineers, it is hard to assess which methods or components will be suitable to implement a certain manipulation task.

Benchmarks and competitions are essential tools to make research progress measurable and different approaches comparable. They usually define a certain protocol, which comprises scenario and task descriptions, and a set of metrics to measure the performance of the method in question. Examples of such benchmarks for Pick&Place tasks are [3, 5]. By resorting to a specific protocol, the benchmark results become repeatable, i.e. they should yield the same results when the same method is tested again. However, the defined scenarios are often very specific, considering e.g. picking of certain foods [3] or simple blocks [5]. The measurements therefore have a poor validity for assessing the methods' suitability for real-world manipulation applications in the manufacturing environment, and the results of these benchmarks can be at most vaguely indicative of the expected performance. Moreover, the ranking of particular methods could even be reversed under different task and environment conditions.

We tackle this problem by proposing a benchmark that directly evaluates the applicability of given methods to specific industrial Pick&Place tasks. Crucially, we allow the user to define scenarios and therefore improve the validity of the measured performance indicators. The benchmark provides a set of given methods and evaluates them on the user-defined scenario in a simulation environment.

2. Components of a Pick&Place System

There are various approaches to tackle Pick&Place tasks, as indicated by Figure 1. All approaches start with a perception of a given scene or scenario including the dynamic object to be manipulated as well as other fixed objects considered as obstacles. One approach is to first estimate the object's pose with respect to the robot and then select feasible grasping points based on pre-determined (e.g. annotated) grasps followed by motion planning for picking and placing the object. The advantage is that the object's pose is known during grasping which facilitates precision placement, however, reliable object pose estimation is still difficult [6]. In contrast, other approaches circumvent the need for object pose estimation and try to directly identify suitable grasps based on the observed shape, e.g. by fitting shape primitives [7] or learning affordance models [8]. The grasp prediction is often split up in two parts: a generative model from which grasps are sampled and a discriminative model which ranks them w.r.t. success probability [9, 10]. After a grasp has been selected, these methods plan a trajectory which is then executed with open-loop control.

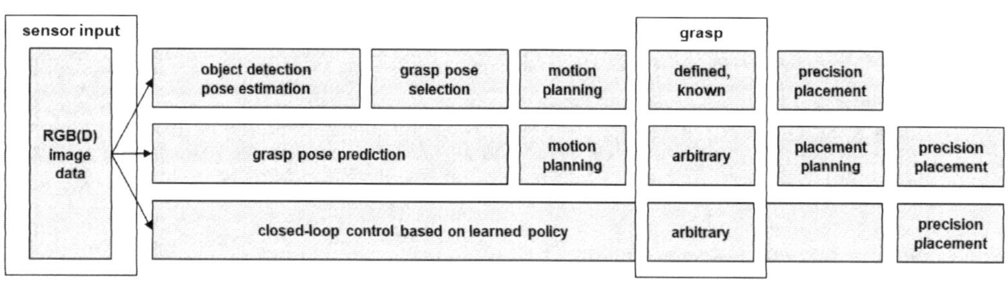

Figure 1. Different pipelines for Pick&Place tasks.

Closed-loop control can be achieved with methods such as vision-based reinforcement learning [11], but these approaches typically require extensive training data (580.000 real world grasp attempts in the case of [11]) and substantial computing resources. For more in-depth reviews of robotic grasping and manipulation we refer the reader to [2, 12, 13].

In order to create a long-lasting, meaningful benchmark we need to support evaluation of all the different approaches and their roughly four phases or components of object pose estimation, path planning, grasping and placing.

3. Existing Benchmarks and Metrics

The isolated performance of the individual components of Pick&Place tasks can be measured using dedicated component-specific benchmarks. Such benchmarks are available for object detection with pose estimation [6], and grasping [14, 15]. Evaluating motion planning is more challenging as it strongly depends on the task being solved. General discussion about benchmarking motion planners can be found in [16]. In [15], evaluation of motion planners for grasping is proposed.

On the other hand, system-level benchmarks evaluate the full pipeline from perception and planning to control during autonomous execution of a predefined task. Competitions which evaluate at system level are e.g. the Amazon Picking Challenge [17], the IROS grasping and manipulation challenge [18], and the assembly challenge at the World Robot Summit 2018 [19]. Yet, these competitions are rare events with a limited number of participants. It is not always possible to re-stage the competition in the lab due to e.g. accessibility of involved objects.

Benchmarks particularly for Pick&Place tasks have been recently proposed by [3] and [5]. In [3], a supermarket logistics scenario is considered. Five different types of foods in three difficulty levels have to be cleared with Pick&Place. Metrics are success rates on object and task level, mean picks per hour, and average duration of successful Pick&Place cycles with some introspection by phase.

In contrast, [5] consider a box-and-blocks test, in which simple wooden blocks have to be handled. In one sub-task they must be placed at pre-defined positions which also puts an emphasis on the placement accuracy. The benchmark evaluates grasping from very cluttered boxes with up to 100 blocks, but using only simple blocks limits the validity for applications in manufacturing environments. Besides the picking success rate as main metric, they also measure the planning and execution times and the distance traveled by the end effector, which makes the contribution of the motion planner more traceable.

4. Concept

A robotic scenario consists of four main parts: a) the environment including objects to be grasped, b) robot, c) the work flow, d) performance measures. The environment consists of static and dynamic objects that are described by shape, appearance, positions, including their geometry for collision detection. The static objects define obstacles which always have to be avoided by the robot. The dynamic objects are considered as movable and the robot can touch, grasp, and move them. The objects are described using the SDFormat specification [20]. The robot is described by its shape, kinematics and/or dynamics, position in the environment, etc. We utilize the URDF format to define the robot [21]. The task is described by the user by defining an initial state and a desired target state. The work flow defines what actions have to be performed in which order by the robot to fulfill the desired task.

The work flow is described by a directed graph, where nodes are the actions and edges connect preceding and following actions. The nodes can be considered as programs, e.g. standalone applications, services, ROS services, etc. The preconditions of a node define input data and state required for executing the given node. For example, the node 'Grasping' requires the positions of the object relatively to the robot end-effector (input data) and also requires the robot to be near the object (state). Each node also defines a group of methods that are suitable for realizing the node. For example, the node "Motion planning" in Fig. 2 accepts all programs from group "MP". The user can therefore prepare multiple motion planners (each will belong to the group "MP") and test all of them to identify which planner performs best.

Based on the scenario and the given work flow, performance measures can be computed in simulation. We use both system-level and introspective performance measures. Because some components are stochastic, the task must be executed repeatedly to obtain statistics. System-level metrics include: success rate of task (task completions over number of attempts), success

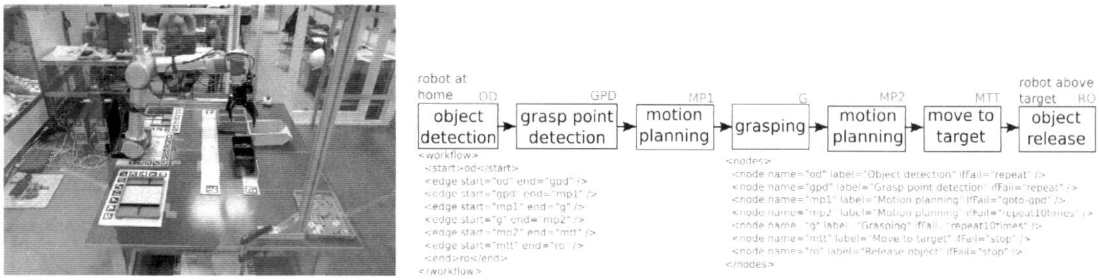

Figure 2. Example of a work flow graph. The preconditions are blue, the red labels denote group of methods.

rate of Pick&Place operation (successful operations over number of attempts), average duration of successful task completions, average duration of a successful Pick&Place cycle. We further evaluate the distance traveled by end-effector, times required for planning (robot stands still) and execution (robot moves) in successful Pick&Place operations.

While these metrics give a good overview of the general performance, they do not reveal much information on how to improve the control system. Therefore, we introduce further introspection metrics/protocols. In case of system failure, the failure is mapped to a node of the work flow graph, i.e. we can report the component-wise number of failures. Also, we can test multiple algorithms of a certain group (e.g. different motion planners). The task is then executed 10 times for each variant while all other components remain the same. This way, we can identify how one single component affects the overall scores and whether it works well in conjunction with other components. It is even possible to include a ground-truth solver in the comparison, e.g. for object pose estimation one "algorithm" could cheat and access the actual object poses. This gives insight into how much a better object pose estimation could improve the system performance, or if we should rather focus on improving other components instead.

The calculation of metrics is organized in two stages, the planning stage and the execution stage. In the planning stage, all parameters of object detection, grasping point selection and path planning are calculated w.r.t. the already known performance indicators. After successful planning (or using ground truth in case of failures) all the plans are executed in a physical simulation environment evaluating other performance indicators like grasping accuracy, path planning robustness in terms of collision-free manipulation of uncertain grasped objects.

The work flow graph can be organized in any complexity, mixing planning and execution stages, e.g. repeating object pose estimation while grasped, for improvements of the overall process.

5. Conclusions

We presented a concept for a scenario-driven benchmark system for industrial Pick&Place applications where users can define arbitrary tasks in arbitrary environments evaluating arbitrary planning methods, rather than e.g. evaluating only path planners in fixed settings. The methods are used in a service-based manner, allowing easy integration of new algorithms and exchange of individual components. Besides regular metrics for performance assessment, we specify introspection metrics that allow investigation of the interplay between different components (e.g. object detection and grasp synthesis) also while interacting with a physical simulation environment, representing the real-world execution stage of existing system-level benchmarks like picking challenges and competitions. This allows the practitioner not only to assess the suitability of available methods for their specific use case, but also to introduce changes to the scenario and observe the effect on the performance measurements.

Acknowledgments
This work was partially supported by the Engineering and Physical Sciences Research Council (EPSRC) under standard research grant No. EP/S032487/1 and by the Czech Science Foundation (GAČR) under research project No. 19-22555Y.

References

[1] R. Gruninger et al. "Market study on adaptive robots for flexible manufacturing systems". In: *2009 IEEE International Conference on Mechatronics*. IEEE, 2009.

[2] K. Kleeberger et al. "A Survey on Learning-Based Robotic Grasping". In: *Current Robotics Reports* (Sept. 2020).

[3] H. Mnyusiwalla et al. "A Bin-Picking Benchmark for Systematic Evaluation of Robotic Pick-and-Place Systems". In: *IEEE Robotics and Automation Letters* 5.2 (Apr. 2020), pp. 1389–1396.

[4] F. Bonsignorio and A. P. del Pobil. "Toward Replicable and Measurable Robotics Research". In: *IEEE Robotics & Automation Magazine* 22.3 (Sept. 2015), pp. 32–35.

[5] A. S. Morgan et al. "Benchmarking Cluttered Robot Pick-and-Place Manipulation With the Box and Blocks Test". In: *IEEE Robotics and Automation Letters* 5.2 (Apr. 2020), pp. 454–461.

[6] T. Hodaň et al. "BOP Challenge 2020 on 6D Object Localization". In: *European Conference on Computer Vision Workshops (ECCVW)* (2020).

[7] D. Pavlichenko et al. "KittingBot: A Mobile Manipulation Robot for Collaborative Kitting in Automotive Logistics". In: *Intelligent Autonomous Systems 15*. Springer International Publishing, Dec. 2018, pp. 849–864.

[8] L. Yen-Chen et al. "Learning to See before Learning to Act: Visual Pre-training for Manipulation". In: *2020 IEEE International Conference on Robotics and Automation (ICRA)*. IEEE, May 2020.

[9] Ü. R. Aktas et al. "Deep Dexterous Grasping of Novel Objects from a Single View". In: *CoRR* abs/1908.04293 (2019).

[10] J. Mahler et al. "Dex-Net 2.0: Deep Learning to Plan Robust Grasps with Synthetic Point Clouds and Analytic Grasp Metrics". In: *CoRR* abs/1703.09312 (2017).

[11] D. Kalashnikov et al. "QT-Opt: Scalable Deep Reinforcement Learning for Vision-Based Robotic Manipulation". In: *CoRR* abs/1806.10293 (2018).

[12] S. Caldera et al. "Review of Deep Learning Methods in Robotic Grasp Detection". In: *Multimodal Technologies and Interaction* 2.3 (Sept. 2018), p. 57.

[13] R. Li and H. Qiao. "A Survey of Methods and Strategies for High-Precision Robotic Grasping and Assembly Tasks—Some New Trends". In: *IEEE/ASME Transactions on Mechatronics* 24.6 (Dec. 2019), pp. 2718–2732.

[14] J. Falco et al. "Benchmarking Protocols for Evaluating Grasp Strength, Grasp Cycle Time, Finger Strength, and Finger Repeatability of Robot End-Effectors". In: *IEEE Robotics and Automation Letters* 5.2 (2020), pp. 644–651.

[15] Y. Bekiroglu et al. "Benchmarking Protocol for Grasp Planning Algorithms". In: *IEEE Robotics and Automation Letters* 5.2 (2019), pp. 315–322.

[16] M. Moll et al. "Benchmarking motion planning algorithms: An extensible infrastructure for analysis and visualization". In: *IEEE Robotics & Automation Magazine* 22.3 (2015), pp. 96–102.

[17] N. Correll et al. "Analysis and Observations From the First Amazon Picking Challenge". In: *IEEE Transactions on Automation Science and Engineering* 15.1 (Jan. 2018), pp. 172–188.

[18] Y. Sun et al. "Robotic Grasping and Manipulation Competition: Task Pool". In: *Communications in Computer and Information Science*. Springer International Publishing, 2018, pp. 1–18.

[19] F. von Drigalski et al. "Robots assembling machines: learning from the World Robot Summit 2018 Assembly Challenge". In: *Advanced Robotics* (Dec. 2019), pp. 1–14.

[20] O. S. R. Foundation. *SDFormat, a description language for Scientific Robotic Simulation*. http://sdformat.org/spec http://sdformat.org. 2016.

[21] ROS. *Accessed 26/11/2020*. http://wiki.ros.org/urdf.

Acoustic study of multi-layered microperforated elements for fibreless noise control applications

M. Villau[1], H. Rämmal[1] and J. Lavrentjev[1]

[1] Department of Mechanical and Industrial Engineering, TalTech, ESTONIA

Email: margus.villau@ttu.ee

Abstract. Recently fibreless solutions have become a trend in the design of sustainable and environmentally friendly duct silencers. Hereby microperforated panels have been proven to provide adequate performance for the substitution of unfavourable fibrous material layers commonly used in mass-produced silencers. This paper presents an acoustic study of microperforated elements aimed for an effective and eco-friendly heating, ventilation, and air conditioning (HVAC) duct silencer. Several microperforated sheet metal panels have been experimentally tested in a variety of layered configurations in order to maximize the sound absorption coefficient in operational frequency range. The optimal solution is implemented in the design concept of a novel fibreless HVAC silencer presented in this research. The results of the study demonstrate the appropriateness of the double-layered microperforates for a duct silencer as well as for a variety of Noise, vibration and harshness (NVH) implementations were potent noise absorption is aimed.

Keywords: microperforated element, absorption coefficient, HVAC silencer, fibreless.

1. Introduction

Noise control devices have been extensively used in various technical fields: e.g. automotive and aeronautical industries, construction engineering, HVAC systems, etc. In the recent decade, a requirement for eco-friendly noise cancellation materials in flow ducts (e.g. HVAC ducts, automotive exhaust silencers etc.) has been highlighted. The replacement of fibreless materials with non-fibreless materials has become a trend. One of the proposed materials with great potential is microperforated element typically provided in sheet metal panels. Variety of microperforated panels (MPP) have been widely applied for almost two decades being first introduced by Maa already in 1970s [1-3]. Maa proposed that the perforations are to be reduced to submillimetre level in order to provide sufficient acoustic resistance. He also developed analytical models for predicting acoustical properties (equivalent electrical circuit model) and pointed out that MPP's can be tuned for wide frequency band, thus suitable for variety of applications.

The properties and application of microperforations are studied extensively also in TalTech laboratory of technical acoustics [4-8]. A number of studies are mainly focused on applications for internal combustion engine exhaust silencers. Different microperforations have been studied with varying aperture geometry and pattern as well as porosity variations. Also, the authors have performed

investigations on the endurance and reliability for clogging of the microperforated tube sections in operating IC engine applications [6-7], successfully demonstrating the suitability of the solutions.
Multy-layer MPP set-up has been investigated also in past [9]. A few alternative configurations of aperture for MPP are investigated and presented in article [10].
The current study is focusing on the experimental determination of the properties of single and double layer MPP configurations using commercially available Acustimet® sheets produced by Swedish company Sontech International AB [11]. A recently published study [12] on the acoustical performance of Acustimet® aiming vehicle cabin noise suppression exhibits the potential. The main interest hereby is to investigate the opportunities to replace the commonly used fibrous material layers in mass-produced HVAC silencers with the MPP to achieve acoustically effective, fibreless and eco-friendly silencer concept.

2. Experimental studies

The experimental studies were carried out in TalTech laboratory of technical acoustics. The key parameters of the silencer elements were acoustically determined by using dedicated test facilities. The aim of this study is to determine sound absorption performance characteristics in variety of layouts for two different commercially available materials. A third MPP material sample was also included into the study for reference purposes. The materials tested, are shown in figure 1.

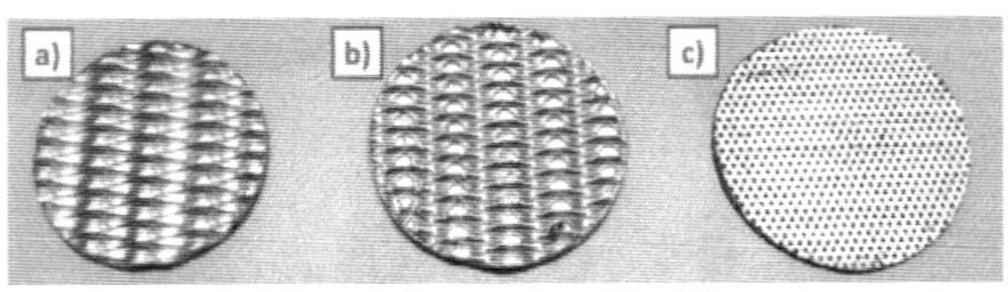

Figure 1. Material samples (D=28 mm) tested.
a) material A; b) material B; c) material C

Material A and B are commercially available under the brand name Acustimet® produced by Swedish company Sontech International AB, the third specimen is custom-made microperforated panel. The characteristic parameters for the different materials are shown in table 1.

Table 1. Properties of tested materials

	Material A	Material B	Material C
Aperture dimensions	1,7x0,1 mm	4x0,1 mm	d 0,5 mm
Aperture shape	Slit shape	Slit shape	Orifice
Material used	Aluminium alloy	Aluminium alloy	Steel alloy
Specific weight	2,6 kg/m^2	1,67 kg/m^2	3,54 kg/m^2
Porosity	~1%	~3%	~18%

Materials A and B were also studied by microscopy (see figure 3) to determine the aperture geometry and porosity. For Material C it was not necessary because of its simple orifice shaped apertures. As we can observe from figure 3, the materials A and B have a slit type aperture.

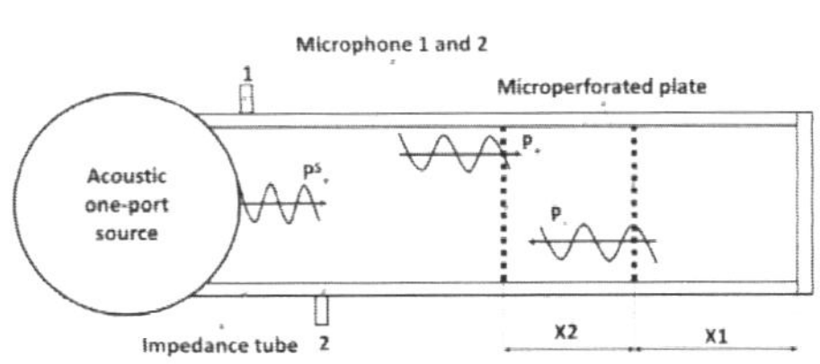

Figure 2. One-port measurement set-up to experimentally determine absorption coefficient of the materials [14].

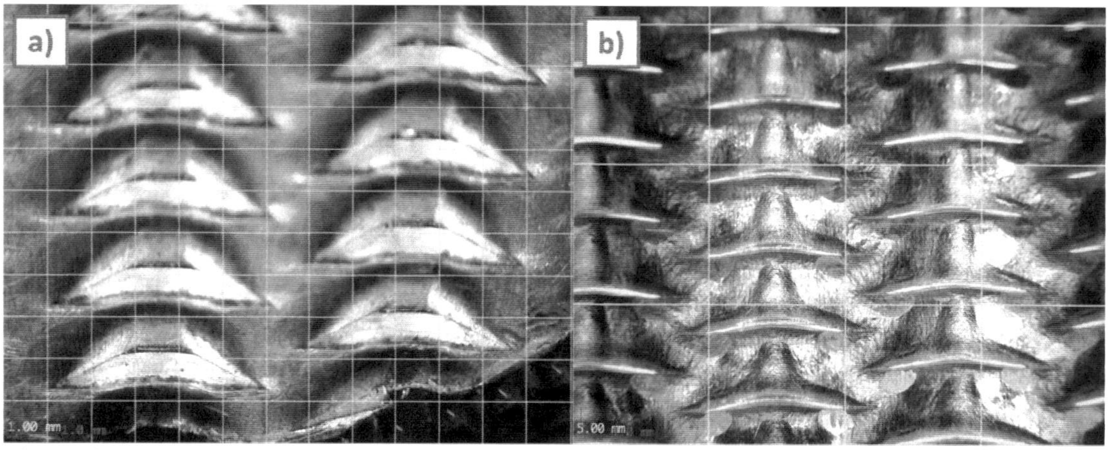

Figure 3. Microscopy images for the determination of aperture shape and size of materials A (left) and B (b, right)).

One of the most important parameters for characterization of noise attenuation performance is acoustic absorption coefficient which determination is discussed below.
The widely used method to determine the absorption coefficient of materials is the impedance tube method following ISO 10534-2:1998. The specimen size is only determined by the cross-sectional area of the test tube (see figure 2) [13].
The material specimen is fitted into the rigidly closed end of the tube. By applying excitation noise from the electro-dynamic driver from the other end of the tube and by measuring the acoustic signals from two microphones, it is possible to calculate the absorption coefficient of the material tested. The plane wave acoustic pressures at two microphones separated by distance s can be expressed as [13]:

$$p_1(f) = p_{1+}(f) + p_{1-}(f) \text{ and}$$

$$p_2(f) = p_{1+}(f)exp(-ik_+s) + p_{1-}(f)exp(ik_-s),$$

where p is the acoustic pressure, f is the frequency, $k = 2\pi f/c$ is the wave number, c is the speed of sound, - and + denote the pressure waves propagating in negative and positive direction along the tube axis. The reflection coefficient at microphone 1 is defined as [13]:

$$R(f) = p_{1-}(f)/p_{1+}(f).$$

It can be shown that:

$$R(f) = \left(H_{12}^e - exp(-ik_+s)\right)/(exp(ik_-s) - H_{12}^e),$$

where $H_{12} = p_2/p_1$ is the transfer function between induct mounted measurement microphones 1 to 2. Then the absorption coefficient of the material is calculated as [13]:

$$\propto_s (f) = |1 - R(f)|^2.$$

According to the impedance tube diameter of 24 mm, the cut-of frequency is 8376 Hz, and the observation range is chosen from 200 Hz to 8000 Hz which is determined by microphone separation (s=16mm). For the experiment, the investigated material is placed on to the tube with varying distance from back wall X1. The distance X1 is set from 0 mm to 50 mm with single layer set-up (see figure 3). With double layer set-up, there were placed two MPP in sequence with distances X1 and X2, where distance X1 was set 30 mm and 50 mm and distance X2 was varying from 5 mm 50 mm. Experimental equipment and set-up is shown in figure 4.

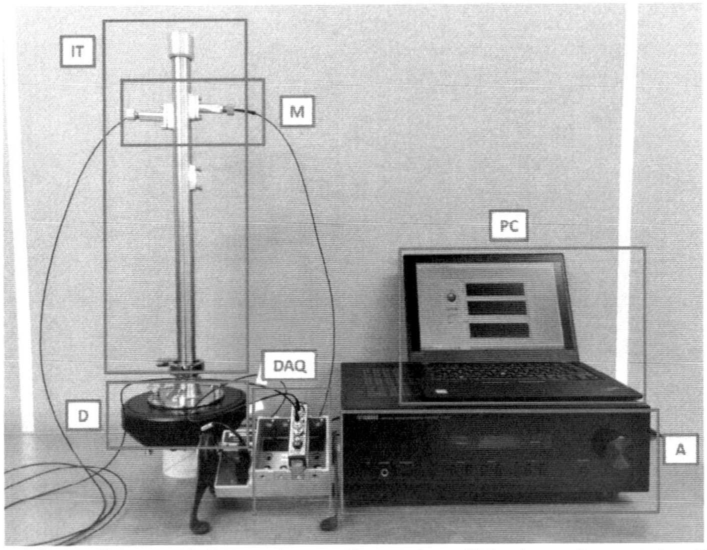

Figure 4. One-port (impedance tube) test rig configuration: IT – impedance tube (inner diameter 28 mm); M – two 1/4" microphones; D – electro-dynamic drive (sound source); DAQ – data acquisition system; PC – computer; A – excitation signal amplifier.

Experimental tests were carried out using the following equipment:

- two 1/4" prepolarized pressure microphones (G.R.A.S. 40BD),
- four channel DAQ analyser (NI 9234) and USB Carrier (NI USB-9162),
- electro-dynamic driver (BMS 4591),
- excitation signal amplifier (Yamaha AS201).

3. Results

Absorption coefficient tests with MPP were first conducted with varying distance X1 from back wall from 0 mm to 50 mm (see figure 2). The tests were repeated for all three different MPP materials, as shown in figure 1. Comparison results from the impedance tube test are presented in figure 5, where the distance X1 was set to 15 mm from back wall. It is clearly seen that material A and B exhibit higher absorption characteristic than material C. In figure 6 and figure 7 one can observe that by increasing the distance from the back wall, the absorption peak will be moved towards the lower frequencies. The initial back wall distance was set to 0 mm, as that perforated plate is at the far end of the test tube, and in this case the maximum absorption coefficient occurs at 3600 Hz. With distance X1 set to 15 mm, the maximum absorption shifts to around 2100 Hz.

Followingly, two extreme cases were tested, the distance X1 was first set to 30 mm then after to 50 mm and respective results are shown in figure 7, black dotted line and red solid line, respectively. We can observe that the maximum absorption shifts even more to the lower frequencies, 1200Hz and 900 Hz, respectively. Also, we can see that the second peak appears at higher frequencies with distance set to 30 mm. By setting the X1 to 50 mm a third peak appears in the frequency range studied.

For material C and for single layer set-up, the results are presented in figure 8. Distance X1 was varied between 0 to 15 mm. We can observe much lower absorption performance with higher porosity figures (18% respectively). For other set-ups, material C was excluded from further test with now matching performances with material A and B.

Secondly, the double layer set-up test was conducted with material A and B. The distance X1 was set to 30 mm and was held constant. Hereby the varying parameter was the distance X2 for which the

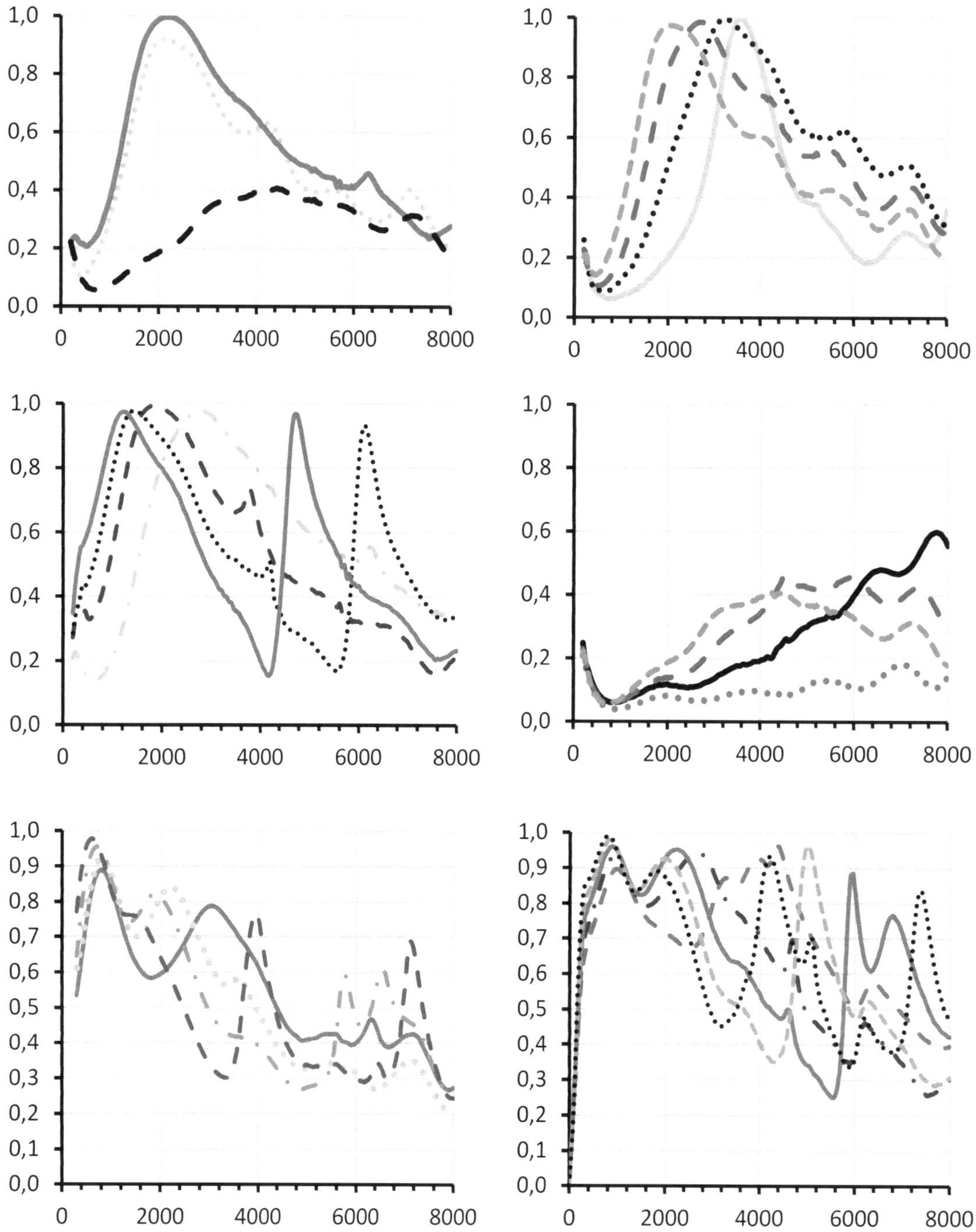

Figure 5-10. Comparison of test result in impedance tube (x-axis is Frequency [Hz]; y-axis is Absorption Coefficient [-]). In first graph (top left) are compared Materials A (yellow dot line), B (red solid line) and Material C (black dashed line). Absorption coefficient for one micro-perforated plate (material A). Distance X1 is set from 0 mm to 15 mm. Absorption coefficient for one micro-perforated plate (material B). Distance X1 is set from 0 mm to 15 mm. Fourth graph (middle right) is absorption coefficient for one micro-perforated plate (material C). For double layer set-up for material A (lower left) and material B (lower right). Distance X1 was set to 30 mm and distance X2 was varying from 5 to 50 mm.

absorption coefficient results are shown in figure 9. First absorption peak shifts even more to the lower frequencies, appearing at around 900 Hz. Considering in comparison, that the peak-frequency for single panel was at 1200 Hz with distance 30 mm and 900 Hz with 50 mm, we can conclude that with double layer set-up the overall distance (X1+X2) also appears to play key role in the silencing performance. Result for material B are presented in figure 10, and we can see even better sound apportioning performance within much wider frequency range.

4. Conclusions

Sound absorption properties of three fibreless microperforated sheet metal panels in a variety of single- and double-layer configurations are experimentally studied in this paper. The results exhibit the multi-layered material's suitability for sustainable noise control applications. Further investigation is planned for study different HVAC silencer layouts and configurations, e.g. for concentrically mounted microperforated tubes and for a varied number of expansion chambers.

Acknowledgements

This research was supported by Estonian Development Grant EAG28 (Smart micro-perforated silencer for heat, ventilation and air conditioning systems).

References

[1] Maa D Y 1975 Theory and design of micro perforated - Panel sound absorbing Construction, *Sci. Sin., XVIII, pp. 55-71*

[2] Maa D Y 1987 Micro perforated panel wide band absorbers, *Noise Control Engineering Journal, 29(3), pp. 77-84*

[3] Maa D Y 1998 Potential of microperforated panel absorber, *Journal of Acoustic Society of America, 104(5), pp. 2861-2866*

[4] Kabral R, Rämmal H, Lavrentjev J 2012 Acoustic Studies of Micro-Perforates for Small Engine Silencers, *SAE Technical Paper 2012-32-0107*

[5] Lavrentjev J, Rämmal H 2009 On experimental techniques to determine acoustic performance of small exhaust silencers, *SAE Technical Paper Series, 2009-32-0015*

[6] Rämmal H, Lavrentjev J 2019 Endurance of Micro-Perforated Elements in Unmanned Ground Vechicle's Small Diesel Engine Silencer Application, *SAE Technical Paper 2019-32-9533*

[7] Rämmal H, Lavrentjev J 2017 Reliability Study of Micro-Perforated Elements in Small Engine Silencer Application, *SAE Tech. paper 2017-32-0075*

[8] Kabral R, Rämmal H, Auriemma F, Luppin J 2012 A Novel Design for Cruiser Type Motorcycle Silencer Based on Micro-Perforated Elements, *SAE Technical Paper 2012-32-0109*

[9] Sakagami K, Yairi M, Morimoto M 2010 Multiple-leaf sound absorbers with microperforated panels: An Overview, Acoustics Australia, 76 – Vol. 38 August (2010) No. 2, pp 76-81

[10] Auriemma F, Rämmal H, Lavrentjev J 2013 Micro-Grooved Elements - A Novel Solution for Noise Control, *SAE Int. J. Mater. Manuf.* ***6****(3), 599-610*

[11] www.sontech.se

[12] Thaveedu J, Padavala P, Medisetti J 2020 Application of Micro-Perforated Panel in Automotive Vehicle Noise Refinement, *SAE Technical Paper 2020-01-5078*

[13] Rämmal H, Lavrentjev J 2020 Acoustic study of novel eco-friendly material for vehicle NVH applications, *Materials Today: Proceedings 10.1016/j.matpr.2020.04.632*

[14] Lavrentjev J, Rämmal H 2020 Experimental study of noise barrier boards with increased acoustic performance by utilizing Helmholtz resonator effects, *Materials Today: Proceedings 10.1016/j.matpr.2020.05.402*

The optimized electrochemical deposition of bismuth-bismuth telluride layered crystal structures

Aliaksei Bakavets[1], Yauhen Aniskevich[2], Genady Ragoisha[1], Natalia Tsyntsaru[3,4], Henrikas Cesiulis[3,5], Eugene Streltsov[2]

[1]Research Institute for Physical Chemical Problems, Belarusian State University, Minsk, Belarus
[2]Faculty of Chemistry, Belarusian State University, Minsk, Belarus
[3]Vilnius University, Faculty of Chemistry and Geosciences, Vilnius, Lithuania
[4]Institute of Applied Physics, Chisinau, Moldova
[5]JSC Elektronikos Perdirbimo Technologijos, Vilnius, Lithuania

E-mail: alexeibokovets@gmail.com

Abstract. Underpotential deposition, i.e. the cathodic deposition above reversible potential $E(Me^{n+}/Me)$, produces an atomic layer of a metal on a semiconductor electrode, such as e.g. bismuth telluride. This phenomenon allows electrodeposition of superlattices formed of building blocks of a layered semiconductor structure joined by biatomic metal interlayer. This work outlines the optimized pulse potential controlled electrodeposition of $(Bi_2)_m(Bi_2Te_3)_n$ films produced under mentioned above technique. The influence on the morphology of the electrodeposited films of key-parameters as applied pulse frequency, duty cycle, a routine of sodium dodecyl sulfate introduction in the electrolyte is discussed. The optimized procedure comprises a short (about 10 s) cathodic pre-treatment at high overpotential of the cathodic reaction, the subsequent periodic switching for 120 min between potentials of electrodeposition and refinement at 0.1 Hz and 5% duty cycle with addition of surfactant 60 min after the start of the electrodeposition.

1. Introduction

Nanostructures formed by an alteration of ultra-thin layers of a semiconductor and a metal are of the interest for various applications and are especially promising for the development of new thermoelectric materials [1-4]. The major characteristic of a thermoelectric material efficiency, its figure of merit, depends on ratio of electric and thermal conductivities. Variations of the two latter characteristics are coupled in usual materials and the main route to thermoelectric materials with low thermal conductivity uncoupled from electric conductivity implies application of nanostructures [5]. This stimulates much work on nanostructuring of bismuth telluride, one of the best thermoelectric materials at room temperature [6-8], including investigation of Bi_2Te_3 nanoparticles electrodeposition [9-20]. Superlattices of $(Bi_2)_m(Bi_2Te_3)_n$ series formed by alteration of Te-Bi-Te-Bi-Te quintuples and bismuth biatomic layers [1-3, 21-27] and similar $(Bi_2)_m(Bi_2Se_3)_n$ [28-32] and $(Sb_2)_m(Sb_2Te_3)_n$ [33] layered materials are especially attractive as their specific structure which determines low thermal conductivity [1, 2, 22, 24, 31] is characteristic of ordinary crystalline state, not only nanoparticles. We have presented recently [3] the electrochemical approach for such superlattice formation by in situ insertion of metal interlayers into semiconductor layered crystal structure during the semiconductor pulsed electrodeposition.

The prerequisites of the technique are the layered crystal structure of a semiconductor, such as of bismuth telluride crystal which is formed by stacks of Te-Bi-Te-Bi-Te quintuples joined by weak van der Waals bonds, and the capacity of given semiconductor to acquire metal adlayer on its surface above the reversible potential $E(Me^{n+}/Me_{bulk})$ of the corresponding bulk metal. The latter phenomenon, i.e. the underpotential deposition (upd) of metal adlayer on a semiconductor, has been reviewed lately in [34], electrochemical properties of metal monolayers (adlayers) on bismuth telluride were reported in [35, 36].

The goal of this work was the optimization of parameters of $(Bi_2)_m(Bi_2Te_3)_n$ pulse potential controlled electrodeposition for preparation of compact uniform films with monitoring of the electrodeposition by scanning electron microscopy (SEM). The scope of the experiment in this work was limited to the investigation of parameters of the pulsed electrodeposition effect on morphology in the electrodeposition of $(Bi_2)_m(Bi_2Te_3)_n$ films, as the proof of $(Bi_2)_m(Bi_2Te_3)_n$ formation at the condition of our experiments was obtained earlier [3] by numerous experimental and theoretical methods.

2. Experimental

Electrodeposition of $(Bi_2)_m(Bi_2Te_3)_n$ films was performed in a three-electrode cell with stainless steel foil as a working electrode, Ag|AgCl|KCl(sat) reference electrode, and Pt counter electrode. Pulse potential control was provided by Gamry Series 300 potentiostat. The morphology of the films in the electrodeposition was monitored using Hitachi TM3000 Tabletop microscope. Substrates were subsequently rinsed in the hot nitric acid, polished with 0.05 μm alumina suspension, and electrochemically treated by cyclic polarization in the potential range between –0.4 V and 0.8 V in 3M HNO_3 solution. The electroactive area was isolated from the rest of the electrode by acid-resistant varnish. Additional experimental details of $(Bi_2)_m(Bi_2Te_3)_n$ electrodeposition were presented in [3].

Bismuth adlayer (Bi_{ad}) is electrodeposited onto bismuth telluride in the electrode potentials region of bulk bismuth anodic oxidation above $E(Bi^{3+}/Bi)$ (figure 1). Thus, the electrodeposition of bismuth adlayer on Bi_2Te_3 is efficiently prevented from nucleation and growth of Bi metallic phase by the potential control. Bismuth telluride electrodeposition proceeds in a more negative electrode potential region concurrently with bulk bismuth electrodeposition. In order to enable alternating deposition of Te-Bi-Te-Bi-Te quintuples and Bi_{ad} with exclusion of metallic bismuth, the potential is alternatively switched between the both regions (figure 2). The two bismuth adlayers thus formed on adjacent Te-Bi-Te-Bi-Te quintuples combine into a bismuth bilayer, which results in the electrochemical formation of $(Bi_2)_m(Bi_2Te_3)_n$ superlattice structure [3], in which bismuth bilayers occupy places at van der Waals planes between quintuples of bismuth telluride structure. The $m : n$ ratio in the superlattice structure was controlled in [3] by the concentrations of bismuth and tellurium precursors in the electrolyte, $Bi(NO_3)_3$ and TeO_2 dissolved in an aqueous HNO_3 solution.

Parameters of the pulse electrodeposition were varied in each experiment. In the original procedure, each electrode was pretreated at –400 mV for 10 seconds in the electrolyte containing 8.75 mM $Bi(NO_3)_3$, 1.25 mM TeO_2, 1 M HNO_3, then the electrodeposition with rectangular wave pulse potential control at 1.0 Hz pulse frequency (f) and 10% pulse duty cycle (γ) was carried out for 30 minutes. The duty cycle is the fraction of each period attributed to the active electrodeposition stage. The cathodic pre-treatment stage was found to facilitate uniformity in the subsequent electrodeposition. The nucleation of bismuth telluride on the bare substrate proceeds at high overpotential during the cathodic pre-treatment stage, thus providing uniform coverage of the electrode surface by centers for the following active electrodeposition stage. The potential of active electrodeposition, E_{dep} in the region of Bi_2Te_3 electrodeposition, and also the refinement potential, E_{ref} in the region of bulk bismuth anodic oxidation and Bi_{ad} cathodic deposition (figure 2) were determined from cyclic voltammetry and were set to –100 mV and 60 mV respectively for 8.75 mM $Bi(NO_3)_3$,

1.25 mM TeO_2, 1 M HNO_3 electrolyte composition. Sodium dodecyl sulfate (SDS) was used as a surfactant.

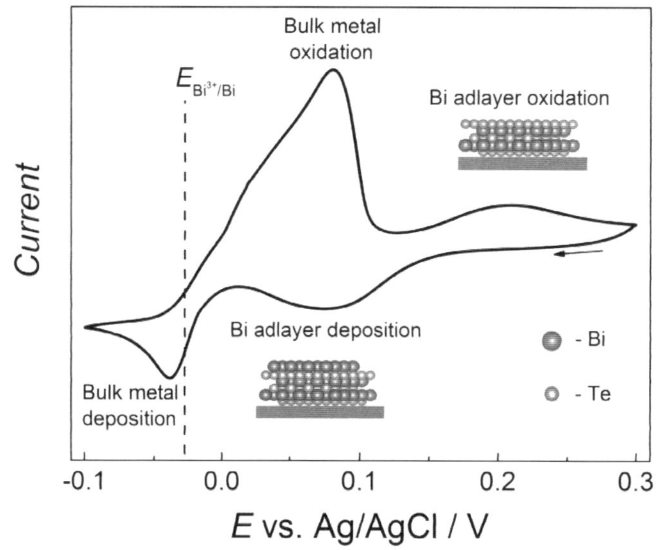

Figure 1. Cyclic voltammogram of Bi_2Te_3 electrode in 10 mM $Bi(NO_3)_3$, 1 M HNO_3 electrolyte solution.

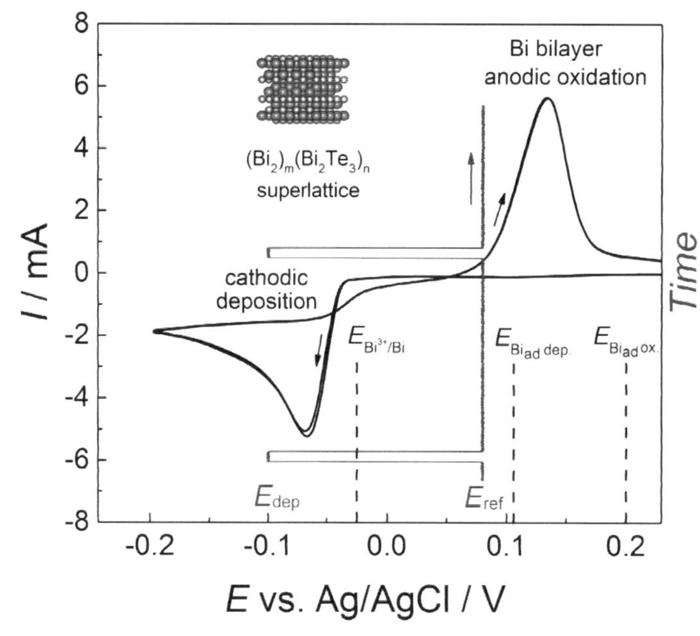

Figure 2. Pulse potential control profile of $(Bi_2)_m(Bi_2Te_3)_n$ electrodeposition (blue pulses) overlaid on a cyclic voltammogram (black) of $(Bi_2)_m(Bi_2Te_3)_n$ electrode on stainless steel substrate in 8.75 mM $Bi(NO_3)_3$, 1.25 mM TeO_2, 1 M HNO_3 electrolyte solution.

In the optimization procedure, γ (in %) was set to 5, 10, 20, 30, 40, 50, and 80, while other parameters were kept constant. Then, the pulse frequency, *f* (in Hz) was varied as 0.1, 0.5, 1.0, 5.0, and 10.0. The lowest frequency was restricted by 0.1 Hz, because of the required alteration of the electrodeposition and refinement stages. In the third optimization step, the time of the cathodic pre-treatment was varied as 10, 30, 60, and 120 seconds. The fourth variable parameter was time of the deposition: the films were deposited for 30, 60, 120, and 240 minutes. In addition, the effect of surfactant (sodium dodecyl sulfate) and the method of its introduction were studied.

3. Results and discussion

3.1 Pulse frequency and duty cycle

Pulse frequency and duty cycle control the time of the electrodeposition and refinement stages of the periodic process, and also replenishment of the electrolyte solution near the working electrode by electroactive species. Pulse frequency and duty cycle influence on the film morphology were studied independently. Figures 3 and 4 summarize the main effects of these parameters variation. A short time of the electrodeposition was used in those series since the formation of non-uniformities can be best observed at early stages.

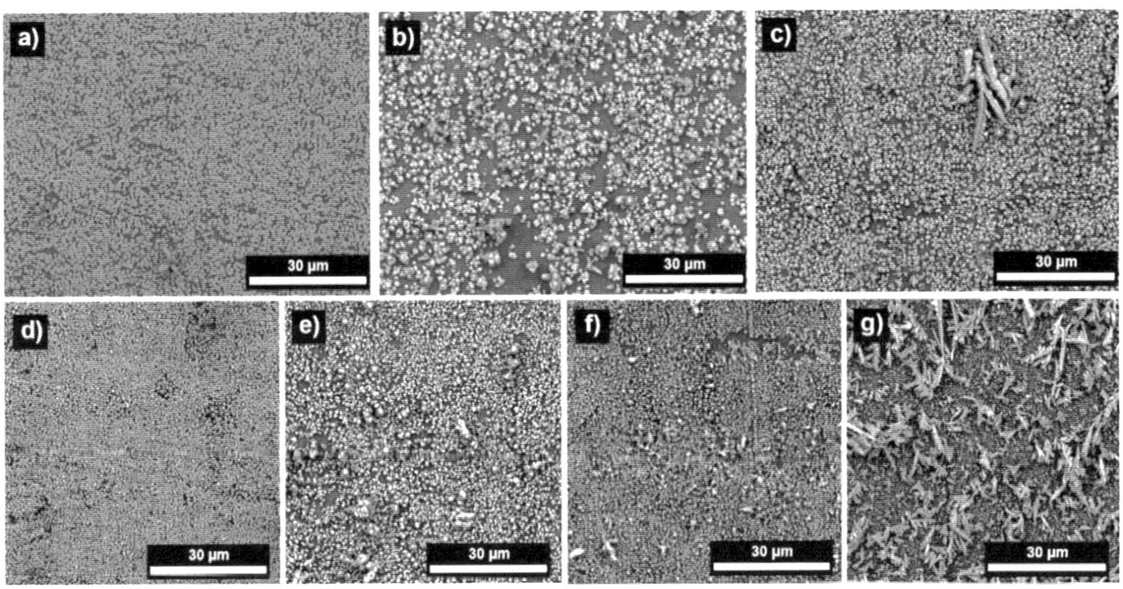

Figure 3. SEM images of $(Bi_2)_m(Bi_2Te_3)_n$ films electrodeposited at various γ (in %): a) 5; b) 10; c) 20; d) 30; e) 40; f) 50; g) 80. Other parameters are: $f = 1$ Hz, $E_{dep} = -100$ mV, $E_{ref} = 60$ mV.

At high duty cycle, the duration of the active electrodeposition stage increases, whereas the time available for diffusion and accumulation of electroactive species near the electrode surface and also the time of refinement stage decrease. As figure 3 shows, short pulses facilitate the formation of uniform films. However, when the pulses were short due to high frequency, the substrate was not uniformly covered by the $(Bi_2)_m(Bi_2Te_3)_n$ particles. Rare pulses favour the deposition of uniform-size crystallites (figure 4).

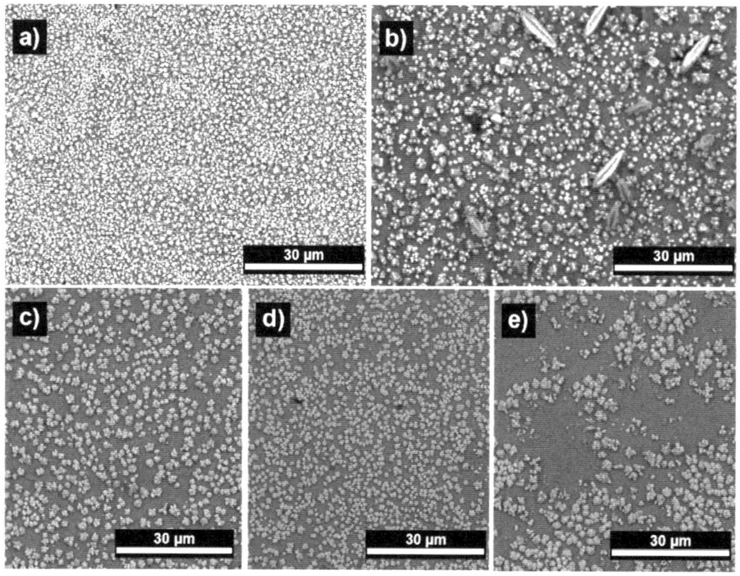

Figure 4. SEM images of $(Bi_2)_m(Bi_2Te_3)_n$ films electrodeposited at various pulse frequencies (in Hz): a) 0.1; b) 0.5; c) 1.0; d) 5.0; e) 10. Other deposition parameters are: γ = 10%, E_{dep} = −100 mV, E_{ref} = 60 mV.

3.2. Surface-active substance

Pulse deposition gives 3 µm thick film at 120 minutes, the thickness was typically sufficient for preparation of flexible thermoelectric devices with dissent thermoelectric parameters in the case of bismuth telluride [37]. However, the $(Bi_2)_m(Bi_2Te_3)_n$ films deposited without surfactant are rough, non-uniform, and porous, due to dendrite growth. In the case of bismuth telluride electrodeposition similar problem was mitigated with the use of sodium lignosulfonate surfactant [38, 39]. SDS has shown good results in our case. Its optimal concentration is approx. 55 mg L^{-1}; below and above this concentration the film roughness and formation of non-uniformly sized crystallites increases (figure 5).

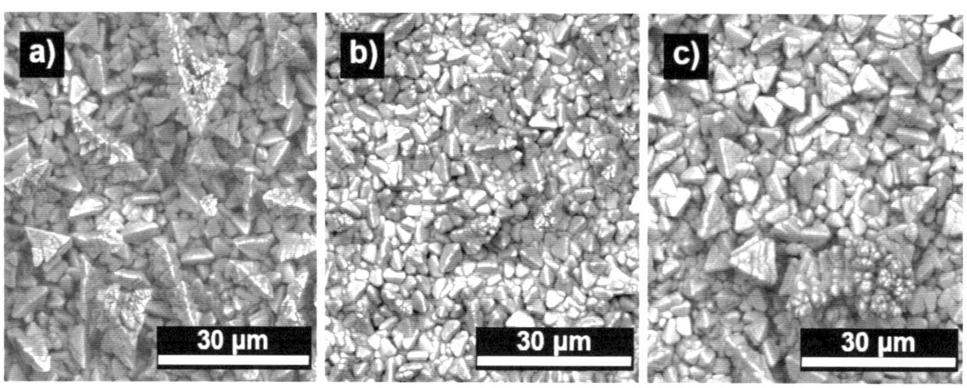

Figure 5. SEM images of $(Bi_2)_m(Bi_2Te_3)_n$ films electrodeposited at various SDS concentrations (in mg L^{-1}) in the electrolyte: a) 40; b) 55; c) 70. Other deposition parameters are: f = 0.1 Hz, γ = 5%, E_{dep} = −100 mV, E_{ref} = 60 mV.

The films electrodeposited from the acidic solution have originally poor adhesion to the stainless steel substrate. The addition of a surface-active agent at the beginning of the deposition process reduces the adhesion even greater and hinders the formation of the uniform film so that only rare agglomerates are formed (figures 6a and 6b). The solution of this problem consists in the addition of surfactant after the nucleation stage, when the film covers the electrode with the first layer of crystallites. The addition of the surfactant at this stage has provided the formation of a very compact uniform film with tapered crystallite shape morphology (figure 6c).

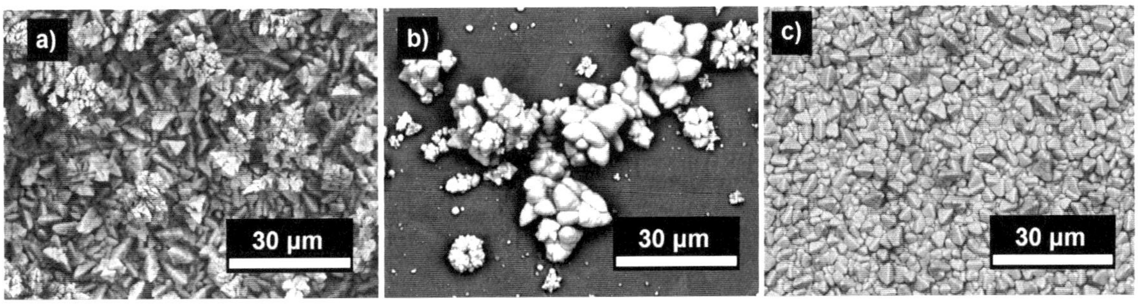

Figure 6. SEM images of $(Bi_2)_m(Bi_2Te_3)_n$ films electrodeposited: a) without SDS; b) with SDS added at the beginning of the deposition; c) with SDS added 60 min. after the start of the deposition. Other deposition parameters: $f = 0.1$ Hz, $\gamma = 5\%$, $E_{dep} = -100$ mV, $E_{ref} = 60$ mV.

Conclusion

The role of pulse frequency, duty cycle, cathodic pre-treatment and the way of SDS introduction in the electrolyte as factors affecting the morphology in the periodic pulse potential controlled electrodeposition of $(Bi_2)_m(Bi_2Te_3)_n$ films was investigated. The optimized procedure for the electrodeposition of compact films should include:

1) the short (about 10 s) cathodic pre-treatment at high overpotential of the cathodic reaction;
2) the subsequent periodic switching for 120 min of the potential between −100 mV and 60 mV at $f = 0.1$ Hz and $\gamma = 5\%$ in the active electrodeposition stage;
3) the surfactant addition after 60 min from the beginning of the electrodeposition.

Acknowledgments

Authors acknowledge funding from H2020 project MSCA-RISE-2017-778357-SMARTELECTRODES.

References

[1] Bos J W G, Zandbergen H W, Lee M-H, Ong N P and Cava R J 2007 *Phys. Rev.* B **75** 195203.
[2] Sharma P A, Lima Sharma A L, Medlin D L, Morales A M, Yang N, Barney M, He J, Drymiotis F, Turner J and Tritt T M 2011 *Phys. Rev.* B **83** 235209.
[3] Bakavets A, Aniskevich Y, Yakimenko O, Jo J H, Vernickaite E, Tsyntsaru N, Cesiulis H, Kuo L-Y, Kaghazchi P, Ragoisha G, Myung S-T and Streltsov E 2020 *J. Power Sources* **450** 227605.
[4] Termentzidis K, Pokropyvnyy O, Woda M, Xiong S, Chumakov Y, Cortona P and Volz S 2013 *J. Appl. Phys.* **113** 013506.
[5] Biswas K, He J, Blum I D, Wu C I, Hogan T P, Seidman D N, Dravid V P and Kanatzidis M G 2012 *Nature* **489** 414-8.
[6] Dresselhaus M S, Chen G, Tang M Y, Yang R, Lee H, Wang D, Ren Z, Fleurial J P and Gogna P 2007 *Adv. Mater.* **19** 1043–53.
[7] Heremans J P, Dresselhaus M S, Bell L E and Morelli D T 2013 *Nat. Nanotechnol.* **8** 471–3.
[8] Mamur H, Bhuiyan M R A, Korkmaz F and Nil M 2018 *Renew. Sustain. Energy Rev.* **82** 4159–69.
[9] Pinisetty D, Davis D, Podlaha-Murphy E J, Murphy M C, Karki A B, Young D P and Devireddy R V 2011 *Acta Materialia* **59** 2455–61.

[10] Prieto A L, Sander M S, Martin-Gonzalez M S, Gronsky R, Sands T and Stacy A M 2001 *J. Am. Chem. Soc.* **123** 7160–1.
[11] Sander M S, Gronsky R, Sands T and Stacy A M 2002 *Chem. Mater.* **15** 335–9.
[12] Menke E J, Li Q and Penner R M 2004 *Nano Lett.* **4** 2009–14.
[13] Yu H, Gibbons P C and Buhro W E 2004 *J. Mater. Chem.* **14** 595–602.
[14] Jin C, Xiang X, Jia C, Liu W, Cai W, Yao L and Li X 2004 *J. Phys. Chem.* B **108** 1844–7.
[15] Zhou J, Jin C, Seol J H, Li X and Shi L 2005 *Appl. Phys. Lett.* **87** 133109.
[16] Lee J, Farhangfar S, Lee J, Cagnon L, Scholz R, Gösele U and Nielsch K 2008 *Nanotechnology* **19** 365701.
[17] Mavrokefalos A, Moore A L, Pettes M T, Shi L, Wang W and Li X 2009 *J. Appl. Phys.* **105** 104318.
[18] Bejenari I, Kantser V and Balandin A A 2010 *Phys. Rev.* B **81** 075316.
[19] Frantz C, Stein N, Gravier L, Granville S and Boulanger C 2010 *J. Electron. Mater.* **39** 2043–8.
[20] Frantz C, Stein N, Zhang Y, Bouzy E, Picht O, Toimil-Molares M E and Boulanger C 2012 *Electrochim. Acta* **69** 30–7.
[21] Bos J W G, Faucheux F, Downie R A and Marcinkova A 2012 *J. Solid State Chem.* **193** 13–8.
[22] Thakur V, Upadhyay K, Kaur R, Goyal N and Gautam S 2020 *Mater. Today Adv.* **8** 100082.
[23] Concepción O, Galván-Arellano M, Torres-Costa V, Climent-Font A, Bahena D, Manso Silván M, Escobosa A and De Melo O 2018 *Inorg. Chem.* **57** 10090–9.
[24] Li Bassi A, Bailini A, Casari C S, Donati F, Mantegazza A, Passoni M, Russo V and Bottani C E 2009 *J. Appl. Phys.* **105** 124307.
[25] Eschbach M, Lanius M, Niu C, Młyńczak E, Gospodarič P, Kellner J, Schüffelgen P, Gehlmann M, Döring S, Neumann E, Luysberg M, Mussler G, Plucinski L, Morgenstern M, Grützmacher D, Bihlmayer G, Blügel S and Schneider C M 2017 *Nat. Commun.* **8** 14979.
[26] Chagas T, Ribeiro G A S, Gonçalves P H R, Calil L, Silva W S, Malachias Â, Mazzoni M S C and Magalhães-Paniago R 2020 *Electron. Struct.* **2** 015002.
[27] Stillwell R L, Jenei Z, Weir S T, Vohra Y K and Jeffries J R 2016 *Phys. Rev.* B **93** 094511
[28] Lind H and Lidin S 2003 *Solid State Sci.* **5** 47–57.
[29] Valla T, Ji H, Schoop L M, Weber A P, Pan Z H, Sadowski J T, Vescovo E, Fedorov A V., Caruso A N, Gibson Q D, Müchler L, Felser C and Cava R J 2012 *Phys. Rev.* B **86** 3–7.
[30] Gibson Q D, Schoop L M, Weber A P, Ji H, Nadj-Perge S, Drozdov I K, Beidenkopf H, Sadowski J T, Fedorov A, Yazdani A, Valla T and Cava R J 2013 *Phys. Rev.* B **88** 081108.
[31] Samanta M and Biswas K 2020 *Chem. Mater.* **32** 8819–26.
[32] Lind H, Lidin S and Häussermann U 2005 *Phys. Rev.* B **72** 184101.
[33] Johannsen J C, Autès G, Crepaldi A, Moser S, Casarin B, Cilento F, Zacchigna M, Berger H, Magrez A, Bugnon Ph, Avila J, Asensio M C, Parmigiani F, Yazyev O V and Grioni M 2015 *Phys. Rev.* B **91** 201101.
[34] Ragoisha G A, Aniskevich Y M, Bakavets A S and Streltsov E A 2020 *J. Solid State Electrochem.* **24** 2585–94.
[35] Bakavets A S, Aniskevich Y M, Ragoisha G A and Streltsov E A 2017 *J. Belarusian State Univ. Chem.* **1**(2) 3–13.
[36] Bakavets A S, Aniskevich Y M, Ragoisha G A and Streltsov E A 2018 *Electrochem. Commun.* **94** 23–6.
[37] Na J, Kim Y, Park T, Park C and Kim E 2016 *ACS Appl. Mater. Interfaces* **8** 32392-400.
[38] Caballero-Calero O, Díaz-Chao P, Abad B, Manzano C V, Ynsa M D, Romero J J, Muñoz Rojo M and Martín-González M S, 2014, *Electrochimica Acta* **123** 117-26.
[39] Naylor A J, Koukharenko E, Nandhakumar I S and White N M 2012, *Langmuir* **28** 8296-9.

Reduction of annotation efforts for multiclass object detection by using a domain awareness data combination strategy

C. Briese [1], B. Hummel[2], V. Hoang[1], M. Schlüter[1], J. Krüger[2]
[1]Fraunhofer IPK - Division of Automation Technology, Pascalstraße 8-9, 10587 Berlin, Germany, https://www.ipk.fraunhofer.de/logic
[2]Technische Universität Berlin - Division of Industrial Automation Technology, Straße des 17. Juni 135, 10623 Berlin, Germany, https://www.iat.tu-berlin.de/

E-mail: clemens.briese@ipk.faunhofer.de

Abstract. To train convolutional neural networks (CNN) it is common practise to collect a huge amount of data. This is cost intensive and often not applicable. Up to date several studies have investigated the concept of few shoot learning, e.g. 1-3 samples per class. Suboptimal is still the over fitting resulting from the gap between training data and representative test data in the application. Since this is still a field of intensive research, an alternative and common approach is transfer learning with data- and image augmented pictures. However, collecting and labelling data for fine-tuning can still take an enormous amount of time, when it comes to multiclass pictures in industrial applications like assembly kit verification. The kits often contain stock lists with a small interclass and a high intraclass-distance. A specific characteristic of stock lists is that parts are easily adaptable and exchangeable. To bring object detection closer to the industry, we successfully show a dataset driven approach that combines a single class collection of pictures, which we call single class (SC) dataset and adapt with a few samples the specific multiclass use case. In result, we use a model trained on a huge SC dataset that can easily and fast be adapted to specific industrial use cases.

1. INTRODUCTION

In last couple of years Deep Learning made a steady progress. Today's algorithms perform better than a human.[1][2] Despite the current high performance of neural networks (NN), the problem of the data basis still exists. Especially with regard to multiclass detection, the effort for labeling data is still immense. In order to achieve applicable detection accuracy, hundreds of images with the corresponding number of annotations are required. For the partly critical application in industry and industry-like contexts, the requirements on classification algorithms are much higher than in the consumer area. Accordingly, more data is needed to train algorithms. This problem is one of the key barriers for the use of machine learning, especially deep learning in many companies.

The completeness check of assembly kits is a promising application for the use of object detection algorithms. The goal is to detect and count the required components and sub-assembly parts in the run-up. A major challenge is the detection of very similar objects (e.g. different types of screws). If there is a high variance or change in the composition of these kits, image data must always be extracted and labeled again. This effort rarely justifies a possible conversion of the

technology.
This paper presents an extension of the transfer-learning approach by a step-by-step learning strategy. First the objects are trained and then the complete assembly kit is trained with a small amount of additional image data. This approach provides applicable results and can reduce effort significantly with a high variety of different or changing use cases. The approach is evaluated in an experimental study and analyzed with respect to potential reduction of annotation effort.

2. RELATED WORK

One goal of computer vision and deep learning is to deploy models in a production environment. Different papers show good results in industrial application for quality control task, e.g. surface damages [3].
In recent years, one of the main challenges has been the reducing the time needed to create a dataset and to train the model. Regarding the dataset, there are different approaches like using synthetic data or transfer learning with pretrained models. Pretraining is a widely used approach to learn a 'universal' feature representation using large-scale datasets, like ImageNet [4] COCO-Dataset [5] or Pascal VOC [6] .[7] This opinion was disproved, when Mahajan et. al. showed, that the improvements on object detection are relatively small. With a 3000 larger pretraining data set compared to ImageNet, the result was on the scale of +1.5 % average Precision on COCO [8].
Another area of research is the use of data augmentation for the artificial enlargement of data sets. [9]. According Zop et. al. an optimized data augmentation policy can improve the detection accuracy by more than +2.3 % mean average precision (mAP) on the COCO dataset [10].
The effect from huge datasets is limited and in spite of image and data augmentation, the acquisition of many images according to the use case remains necessary for pretraining. Different algorithms recommend typically hundreds or thousands of example representative images per class [11].
This is were a new approach in reducing dataset size comes in, called few shot learning (FSL), e.g. 1-3 samples per class. The idea is that a given set of classes with sufficient training data, "is used to improve the performance on another set of classes with very few labelled examples". [12] While some papers demonstrated promising preliminary results [13], Barz et.al.[12] noted, even though the field makes progress, this is still a field of research because they do not work well.
All these presented approaches are helpful and optimize the application of those models, but are rarely tested in the assembly industry. The industry has to deal often with slightly different, but fast changing, highly specialized use cases. Making detection algorithms applicable, the main effort is not computational side, but labelling those multiclass use cases.
Lim et. al. showed that borrowing and augmenting images with similar visual appearances to the target class can significantly improve accuracy especially, when only some examples for the class are available [14].
Additionally Chen et. al. noted that domain adaptation problem between different use cases can be reduced with specialized algorithms, whereby also more training data could possibly alleviate the impact of domain shift [15].
Learning that different classes can produce useful synergies and more data can alleviate the domain shift, collecting a dataset with all products of a stock list, as representative as possible, seems useful.

3. PROPOSED METHOD

An extended transfer-learning approach for object detection is developed. It consists of three training steps:

IOP Conf. Series: Materials Science and Engineering 1140 (2021) 012017 doi:10.1088/1757-899X/1140/1/012017

Diagram 1: Proposed process of training a multiclass detection algorithm

With the new intermediate step, the detection algorithm learns to extract and classify the features of possible objects of an assembly kit. For this, image data of the possible objects (single class images) must be generated and annotated once. Simultaneous labeling of several identical objects in an image is much faster than labeling different objects. The last step closes the resulting domain gap to multiclass detection using "small dataset learning". Due to the small amount of image data required for the respective application, the effort for annotation is reduced in the long-term. Since the annotation of single-class images is much simpler and faster compared to multiclass annotation, most of the effort is shifted to this area. It also can be easier supported by the use of data-driven algorithms.
As neural-network a YOLOv5-Framework is being used. It was firstly published in May 2020 by Jocher et al. [16] which is an improved PyTorch implementation of YOLOv3 by Redmon et al. [17] YOLOv5 offers four different models of which the largest one, YOLOv5x consisting of 9 million parameters and 284 layers, has been used.

4. EXPERIMENTS

As an experimental study, data of different complexities of assembly kits (MC data) and their individual parts (SC data) are recorded. With these data different strategies are applied to the training of a Yolov5x-NN. The images are acquired with the help of a static image setup with one consumer webcam and static lighting. The SC dataset consists out of 24 objects with 1000 labels each split with a training/validation from 2/1 ratio. The investigated complexities are:

(i) least complex: four different objects, two visually being similar
(ii) more complex: eight different objects, two times three visually being similar
(iii) most complex: twelve different objects, two times five being visually similar

Additionally, two training scenarios with different amounts of data are tested.
In the first training scenario each complexity-setup of the assembly training kit dataset consists of 50 images. The data is split up into 30 training-, 10 validation- and 10 test images. For the second training scenario only 20 images were used. Only the best approaches from the first scenario are used in the second.
Five strategies are compared, the new method presented in this paper (see chapter 3) and the possible variations (see tab. 1).
As evaluation metric two versions of mean average precision (mAP) is used. The mAP metric describes how many of the predictions correspond to reality, i.e. how accurate the predictions are. Mostly the value is considered at different IoU (intersection over unit), the value describes how well the object was localized (Eq: (1)). One Metric is the mAP at a IoU-value of 50 % (mAP@.5). and the second is the mAP at is at different points of the IoU from 50% to 95% in 5% steps (mAP@.5:.95). So the second metric is the more stringent one.

$$mAP = \sum_{classes} \frac{True\ positives}{True\ positives + False\ positives}, \quad IoU = \frac{area\ overlap}{area\ of\ union} \tag{1}$$

Google Colab was employed, with a Tesla T4 as GPU. For training the batch size was set to 16 and the epochs to 100. The training took about one hour.

Table 1: mAP@.5 and map@.5:.95 values for the different trials. SC-Data being the abbreviation for single class data

Inital	Pre-Trained						Scratch			
Trainsteps	SC-Data				None		SC-Data			
	Kit		No Kit		Kit		Kit		No Kit	
mAP@	.5	.5:.95	.5	.5:.95	.5	.5:.95	.5	.5:.95	.5	.5:.95
LeastComplex										
@ 50 samples	**0.983**	**0.812**	0.671	0.548	0.741	0.451	**0.983**	0.779	0.703	0.487
@ 20 samples	**0.938**	**0.742**	-	-	-	-	0.776	0.569	-	-
MoreComplex										
@ 50 samples	0.954	**0.682**	0.497	0.352	0.529	0.286	**0.969**	0.673	0.523	0.310
@ 20 samples	**0.875**	**0.613**	-	-	-	-	0.675	0.556	-	-
MostComplex										
@ 50 samples	**0.783**	**0.547**	0.310	0.189	0.45	0.269	0.734	0.464	0.446	, 0.252
@ 20 samples	**0.591**	**0.571**	-	-	-	-	**0.591**	0.348	-	-

5. EVALUATION

The test results of the Experiments are shown in Table 1. The first observation is that all experiments, with the new pretrained step on SC-Data, have reached the highest values for both mAP variants. In the highest complexity, the mAPs are around 80% for the simple metric variant and over 50% for the very stringent variant. The more complex the kit, the better the performance of the models trained on the SC-Data. The mAP values differ at highest complexity by almost the double. It is interesting that the experiments with a randomly initialized NN performed slightly better than the pretrained initial variant for the less complex kits and the simpler mAP-metric. This shows that the influence of a pretrained initial neural network is only really noticeable above a certain variance of objects.

The time needed to label an assembly kit image, averaged over all three complexities, is about

Table 2: Effort amortisation of the described approach compared to the standard procedure

number of kits	effort with SC [min]	effort without SC [min]
0 (inital)	800	0
1	831.67	158.33
...	...	...
7	**1021.67**	1108.3

95 s. Correspondingly with 20 images $\approx$ 32 min. Added to this is the one-time effort for the single-class data. It took about four seconds per SC-Label with about 1000 labels per object and 24 objects, which makes $\approx$ 800 min $\approx$ 13,33 h initial effort. Assuming that without the SC-training step about 100 labelled images of the kit are needed, this results in an effort of $\approx$ 158 min $\approx$ 2,6h per kit. Even with the conservative assumption of the four seconds labeling-time per object in the SC dataset, the associated initial effort is amortized from a kit number of 7 (see Tab. 2). This can be greatly reduced if the annotation of the single-class dataset is supported by intelligent algorithms like Background-subtraction or (semi-)automatic labeling.

6. CONCLUSION

This paper has shown that adding a training step with single-class data can improve the performance of detection algorithms for the visual check of assembly kits. We achieved mAP values twice as high compared to the training without the intermediate. With the approach presented here, the number of images required per assembly kit was reduced to about 20. This significantly reduces the effort for data collection. Despite a conservative estimate of the initial effort, the approach amortizes from about the sixth kit to be annotated.

REFERENCES

[1] John Markoff. A learning advance in artificial intelligence rivals human abilities. *The New York Times*, 2015.
[2] Licheng et. al. Jiao. A survey of deep learning-based object detection. *IEEE Access 7*, page 128837–128868, 2019.
[3] Qianwen et. al. Lv. Few-shot learning combine attention mechanism-based defect detection in bar surface. *ISIJ International*, 59:1089–1097, 06 2019.
[4] J. Deng and W. et. al. Imagenet: A large-scale hierarchical image database. In *2009 IEEE Conference on Computer Vision and Pattern Recognition*, pages 248–255, 2009.
[5] Tsung-Yi Lin et. al. Microsoft coco: Common objects in context. 2015.
[6] M. et. al. Everingham. The pascal visual object classes challenge: A retrospective. *International Journal of Computer Vision*, 111(1):98–136, jan 2015.
[7] Kaiming He et. al. Rethinking imagenet pre-training. *CoRR*, abs/1811.08883, 2018.
[8] Dhruv et. al. Mahajan. Exploring the limits of weakly supervised pretraining. In *Proceedings of the European Conference on Computer Vision (ECCV)*, September 2018.
[9] Luis Perez et. al. The effectiveness of data augmentation in image classification using deep learning. *CoRR*, abs/1712.04621, 2017.
[10] Barret Zoph et. al. Learning data augmentation strategies for object detection. *CoRR*, abs/1906.11172, 2019.
[11] C. H. Lampert et. al. Learning to detect unseen object classes by between-class attribute transfer. In *2009 IEEE Conference on Computer Vision and Pattern Recognition*, pages 951–958, 2009.
[12] Björn Barz et. al. Deep learning on small datasets without pre-training using cosine loss. *CoRR*, abs/1901.09054, 2019.
[13] Eli Schwartz et. al. Repmet: Representative-based metric learning for classification and one-shot object detection. *CoRR*, abs/1806.04728, 2018.
[14] Joseph Lim. Transfer learning by borrowing examples for multiclass object detection. 04 2013.
[15] Yuhua Chen et. al. Domain adaptive faster R-CNN for object detection in the wild. *CoRR*, abs/1803.03243, 2018.
[16] Glenn Jocher et. al. ultralytics/yolov5: v3.1 - Bug Fixes and Performance Improvements. October 2020.
[17] Joseph Redmon et. al. Yolov3: An incremental improvement. *CoRR*, abs/1804.02767, 2018.

IOP Conf. Series: Materials Science and Engineering 1140 (2021) 012018 doi:10.1088/1757-899X/1140/1/012018

Assistance Method for the Application-Driven Design of Machine Learning Algorithms

Adalbert Fono[1], **Gregor Thiele**[1], **Max Klein**[1] **and Jörg Krüger**[2]

[1]Fraunhofer Institute for Production Systems and Design Technology, 10587 Berlin, Germany

[2]Head of the Department for Industrial Automation at the Institute for Machine Tools and Factory Management, Technical University of Berlin, 10587 Berlin, Germany

E-mail: adalbert.fono@ipk.fraunhofer.de

Abstract. Machine learning (ML) offers a lot of potential for applications in Industry 4.0. By applying ML many processes can be improved. Possible benefits in production are a higher accuracy, an early detection of failures, a better resource efficiency or improvements in quantity control. The use of ML in industrial production systems is currently not widespread. There are several reasons for this, among others the different expertise of data scientists and automation engineers. There are no specific tools to apply ML to industrial facilities neither guidelines for setting up, tuning and validating ML implementations. In this paper we present a taxonomy structure and according method which assist the design of ML architectures and the tuning of involved parameters. As this is a very huge and complex field, we concentrate on a ML algorithm for time series forecast, as this can be used in many industrial applications. There are multiple possibilities to approach this problem ranging from basic feed-forward neural networks to recurrent networks and (temporal) convolutional networks. These different approaches will be discussed and basic guidelines regarding the model selection will be presented.The introduced assistance method will be validated on a industrial dataset.

1. Introduction

Driven by faster computer hardware and more available data, ML approaches - in particular Deep Learning methods - reached a superior state of capability. Nowadays, we can see ML applied in a broad spectrum of applications, e.g. in the medical sector, in consumer electronics, and in fundamental research in natural science, technology, engineering and mathematics [1, 2, 3, 4]. In context of Industry 4.0 many processes can be improved by incorporating ML. Possible benefits in the context of production are pointed out in [5]. However, domain experts are rather unversed in concepts and techniques applied in ML. Our goal is to provide a guideline for finding the ML method suitable for given tasks in the domain of time series forecasting on industry data.

2. State of the art

The most popular classes of ML methods for time-series processing are based on artificial neural networks. We present a short overview with advantages and disadvantages with regard to time series forecasting.

Feedforward Neural Networks (NNs) can be applied in many different data domains - such as tabular data, image data, etc. However, they do not offer any intrinsic capability to deal with sequential data like time series. A possible workaround is to apply a sliding window and consider only the corresponding interval. In case of data with long-ranging time dependencies this approach becomes unfeasible as it requires large window sizes which lead to an intractable NN model. Hence, it can be preferable to solve the time dependencies by a customized feature vector which contains useful information like gradient, mean, variance and other properties for every involved time series; in order to use this vector as additional input data. However, recurrent neural networks (see below) seem like a more natural and better choice. A comparison on medical data was conducted in [6].

Recurrent Neural Networks (RNN) are inherently designed to handle sequential data and, thus, seem like the obvious choice to deal with time series data [7]. In practice, two slightly adapted specifications are very popular - the Long Short Term Memory (LSTM) [8] and the Gated Recurrent Unit (GRU) [9] architecture. Due to their construction the learning procedure runs more smoothly and yields superior results in comparison with the classical RNNs. There are only subtle differences in their design and in most cases they reach similar performance [10]. Therefore, in the remainder we focus on LSTM networks but the conclusions also hold for GRU nets. LSTMs can be used in multiple network types such as bidirectional [11] (where the input sequence is processed forward and backwards), stacked [12] (comprised of multiple LSTM layers) and encoder-decoder networks [9] (where the input sequence is encoded into a fixed length vector representation by the encoder part, and the decoder part decodes the representation into another sequence). In recent years LSTMs are often combined with an attention mechanism [13] which helps the network - as the name suggests - to focus its attention on the 'relevant' parts of the input sequence. Moreover, there are networks which solely rely on the attention mechanism and dispense with recurrence entirely [14]. The main limitation of LSTM nets is that they can merely process regular time series, i.e. sequences of evenly spaced (uniform sampled) data points. There have been made efforts to adapt LSTMs to this irregular case [15], however, to the authors' knowledge no straightforward and efficient solution exists. Instead of adapting the LSTM unit directly, many approaches introduce (complex) architectures which try to handle the irregularity by variable mechanisms like signal splitting, incorporating the time differences and learning dense representations for each time step [16, 17].

Convolutional Neural Networks (CNNs) were originally designed to cope with image data [18] but its strengths also transfer to sequential data [19]. Moreover, CNNs can also be combined with LSTMs [20] or even adapted to handle time series data naturally with so-called Temporal Convolutional Networks (TCNs) [21, 22]. The advantages of those networks are their ability to deal with long-ranging time dependencies in the input sequence and thereby allowing the processing of long sequences. Similar to RNNs, these architectures are ill-suited for irregular time series and additional effort is required for adaptation [23, 24].

3. Concept
Applying an ML method to a given task requires corresponding knowledge. Nowadays many solutions and frameworks exist to support the implementation. Nevertheless, many decisions - regarding model selection and architecture, preprocessing, etc. - have to be made and various parameter need to be tuned. We aim to assist the process of finding a suitable structure and parameters accordingly. Here, we propose a guideline, in particular for model selection, which facilitates this process by emphasizing critical steps and their underlying motivation.
With a look on industrial demands, we can shrink the scope of scenarios. So, we focus on data acquired/recorded from sensors and control devices. Generally, the data form a multivariate

irregular time series. Our assessment is based upon expert interviews and profound knowledge acquired from multiple projects in this domain.

3.1. Criteria for model selection

In Sec. 2 we have described multiple different models with their strengths and weaknesses. In general, it can be differentiated between two classes. The first category covers basic models like NNs and plain LSTMs that are straightforward to implement as beginner-friendly frameworks exist, e.g. Keras for Tensorflow [25]. The drawback is that these models either have limited capacities in relation to sequential data (NNs) or require specific input types (regular time series for LSTMs). Hence, in many cases additional effort is required in the data preparation process. The second category includes more complex models like evolved LSTMs or TCNs for irregular time series which offer an end-to-end solutions that require minimal data processing and enable superior precision. Unfortunately, implementation and learning processes are more complex and require in-depth ML knowledge. Following, we present relevant criteria for the model selection.

The nature of available data forms the starting point for the model selection. Is it favorable to use the original data or a re-sampled version as model input? Using the original data is mainly motivated by a straightforward data preparation step without any obstacles. However, this simpler pipeline comes at the cost of a more evolved ML model. In particular, for irregular time series the applied model must be capable of handling this data. Unfortunately, there is no straightforward model architecture in this case and it is still an open research challenge to design a suitable model (see Sec. 2). Design, implementation and training of these models requires advanced ML knowledge. Due to the non-uniform sampling, non-specialized models cannot grasp the underlying system dynamics, i.e. the training does not yield a useful model.
In contrast, inserting a resampling step in the data pipeline yields a straightforward model selection, even for data of uneven structure. In this case, most methods are applicable, i.e. even less complex models are sufficient which eases the training step as well as the implementation. The drawback is that a thorough analysis of the data is required. The main task is to find a suitable sampling rate. Inapt sampling rates may introduce different characteristics or neglect important information of the time series. The problem of non-uniform sampling is discussed in the signal processing community for a long time [26]. Especially in online applications, resampling needs to tackle all relevant phenomena in the data stream despite of the limited accessible interval.
If both the generic approach and the resampling fail, a customized feature engineering can offer a transformation of the temporal information. The data needs to be analyzed by experts in order to set-up a feature vector which incorporates characteristics of the time series, e.g. statistical parameters, time difference between points, and causalities. Plenty of transformations are explained in [27] including useful code snippets in python. These customized features may help to predict the system behavior based on non-uniform signals even though not performing a complete resampling.

The prediction horizon is the second important parameter to be set. Does the task require multi-step predictions or is a simple one-step prediction sufficient? Models and data are not equally appropriate for each scenario. Some models are not suited for multi-step predictions, e.g. basic NNs are less applicable in this scenario compared to LSTMs.
The data itself can be used as guidance to a suitable model. Are there long-ranging time-dependencies or rather immediate causalities? In the first case, rather complex models are required which can cope with long sequences because past events may still have important impact far later. For phenomena with short-term effects, a concise horizon can already adequately describe the state of the underlying system and, thereby, less complex models suffice.

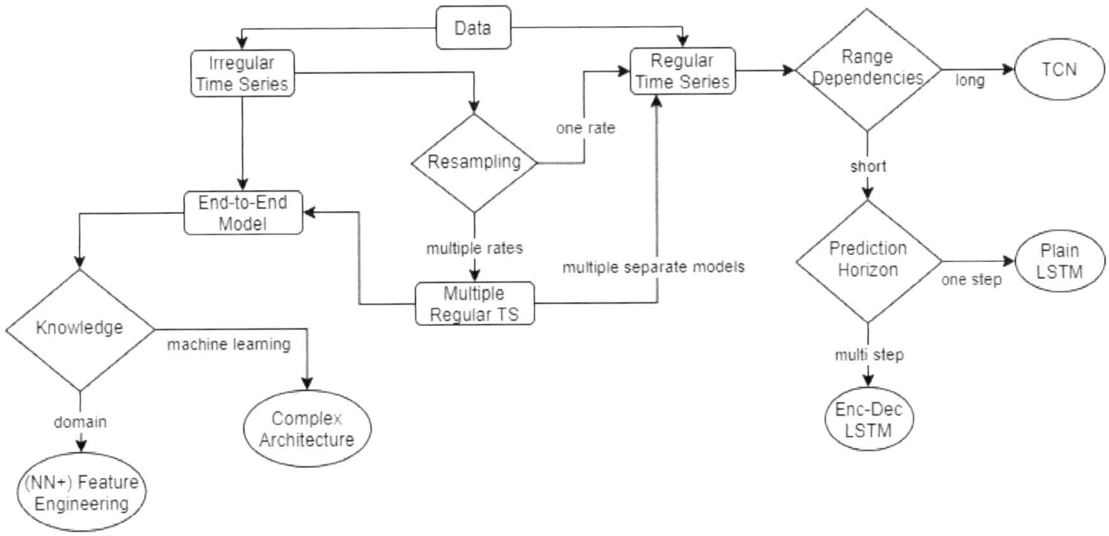

Figure 1. Taxonomy of the proposed concept

Model performance is a decision criterion that also needs to be considered. In an online environment with low latency requirements, computation time of the complete pipeline can be the bottleneck. In case of a large prediction horizon - compared with the latency - this problem diminishes. Similarly, if very high prediction accuracy is demanded, low complexity models may fail. In contrast, it is often easier to reach an acceptable floor using simpler models. Hence, it is worth to clearly weigh the options and demands beforehand.

3.2. Finding a suitable structure

How can the aimed ML pipeline be derived from the collected criteria? First, we can identify the following three main notions: 1) time series characteristics such as regular or irregular, long or short ranging dependencies, slow or fast dynamics; 2) addressed functionality of the model such as prediction horizon, one-step or multi-step prediction, required precision; 3) capabilities and effort of the developer such as domain knowledge, ML knowledge, development time. Then these notions and their evaluation for a specific task and data can serve as input arguments for a decision method which guides us to a proper structure of an ML model. The goal is to end up with the most reasonable starting point concerning performance as well as required effort. The taxonomy is presented in Fig. 1.

4. Practical application

We want to demonstrate the capabilities of our taxonomy on a practical example. Therefore, the process data from a combined heat and power plant was considered, in particular multiple heat sensors. Every sensor sends a base signal to the control unit in fixed time intervals, in our case 60 seconds. Additionally, the flexible send-on-delta sampling triggers the sensors in case of changes in the underlying signals exceeding a individual customized threshold. Hence, the data can be consider as a multivariate irregular time series. For checking the plausibility of process data, the 60 seconds base measurements are satisfying. Thus, focusing only on these measurements yields a regular time series with approximately the same information content which grasps the underlying process dynamics quite well. Therefore, an additional resampling step appears unnecessary. This leaves us with two options, either considering the irregular time

IOP Conf. Series: Materials Science and Engineering 1140 (2021) 012018 doi:10.1088/1757-899X/1140/1/012018

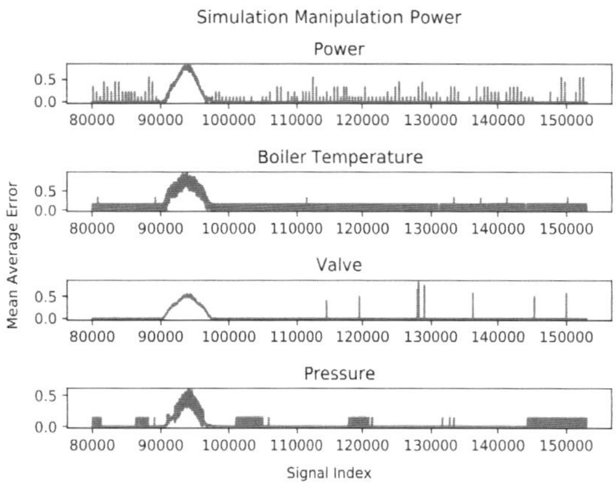

Figure 2. Simulation Manipulation Power, NN predictor

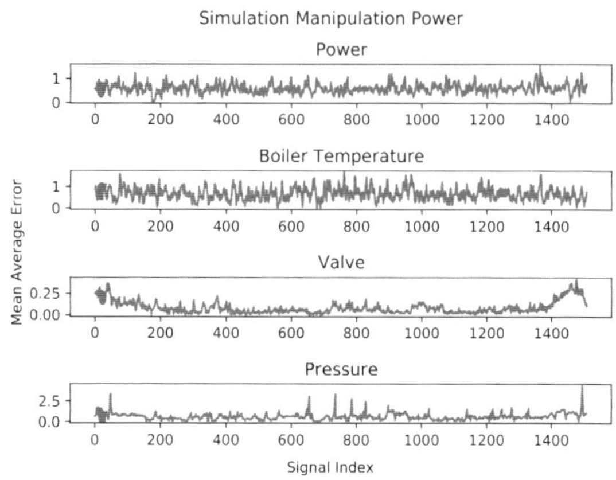

Figure 3. Simulation Manipulation Power, LSTM predictor

series with a complex model or involving only the regular time series using a simple model accordingly. Now, our decision method enables us to quickly determine a fitting model based on the desired use case.

Case-Study: Anomaly detection in power generation In [28], we compared two different ML algorithms developed for anomaly detection in distributed power generation. Both aim to handle time signals; the first one combines regular NN with customized features, the second one utilizes a plain LSTM. From a real power plant, we acquired data from multiple sensors. As explained above, they were sampled with both a uniform 60 seconds clock and triggered by relative changes which results which results in a non-uniform distribution.
We examined two artificial anomalies: scaling of a rotation speed measurement and overriding a power sensor in a calm operation point with a constant, i.e. signals are only generated uniformly every 60 seconds.. The first one leads to a contradiction in the underlying physics, e.g. every single instant is invalid. The second one results in unreasonable sampling which is only visible in the time information. While the LSTM approach performs superior in the first case, it fails to detect the odd sampling in the second case. In contrast, the combination of NN and a customized feature engineering detects the first manipulation less robust but achieves respectable results in the second case. Results of the second manipulation are shown in Fig. 2 and Fig. 3. This little experiments illustrates how design decision in machine learning can affect the performance.

5. Discussion and outlook

Confronted with plenty of possible ML configuration, we suggest a taxonomy to derive a suitable algorithm based on a methodical analysis of the use-case. The advantage of our decision method is that it enables the (inexperienced) user to choose a fitting model based on a quick analysis without in-depth knowledge of the whole field.
This approach poses the question if and how the covered design decisions can have decisive impact on the performance respectively. An input-output-schema could show how each architecture decision and parameter effects on the desired functionality, especially under consideration of the industrial context.

Acknowledgment

This research was proportionate supported in the project EnerSec (project number 16KIS0847) by the German Federal Ministry of Education and Research and in the project *Reinforcement Learning for complex automation technology* by Investitionsbank Berlin (IBB), co-financed by the European Regional Development Fund.

References

[1] Krizhevsky A, Sutskever I and Hinton G E 2012 Imagenet classification with deep convolutional neural networks *Proc. of the 25th Int. Conf. on Neural Information Processing Systems - Volume 1* NIPS'12 (Red Hook, NY, USA: Curran Associates Inc.) p 1097–1105
[2] Hinton G, Deng L, Yu D, Dahl G E, Mohamed A, Jaitly N, Senior A, Vanhoucke V, Nguyen P, Sainath T N and Kingsbury B 2012 *IEEE Signal Processing Magazine* **29** 82–97
[3] AlQuraishi M 2019 *Bioinformatics* **35** 4862–4865
[4] Silver D *et al.* 2016 *Nature* **529** 484–503
[5] Fahle S, Prinz C and Kuhlenkötter B 2020 *Procedia CIRP* **93** 413 – 418 ISSN 2212-8271 53rd CIRP Conference on Manufacturing Systems 2020
[6] Lipton Z C, Kale D C, Elkan C and Wetzel R 2015 *arXiv e-prints* arXiv:1511.03677 (*Preprint* 1511.03677)
[7] Rumelhart D E, Hinton G E and Williams R J 1986 Learning internal representations by error propagation *Parallel Distributed Processing – Explorations in the Microstructure of Cognition* (MIT Press) chap 8, pp 318–362
[8] Hochreiter S and Schmidhuber J 1997 *Neural computation* **9** 1735–80
[9] Cho K, van Merriënboer B, Gulcehre C, Bahdanau D, Bougares F, Schwenk H and Bengio Y 2014 Learning phrase representations using RNN encoder–decoder for statistical machine translation *Proc. of the 2014 Conf. on Empirical Methods in Natural Language Processing (EMNLP)* (Doha, Qatar: Association for Computational Linguistics) pp 1724–1734
[10] Chung J, Gulcehre C, Cho K and Bengio Y 2014 Empirical evaluation of gated recurrent neural networks on sequence modeling (*Preprint* 1412.3555)
[11] Schuster M and Paliwal K 1997 *Signal Processing, IEEE Transactions on* **45** 2673 – 2681
[12] Malhotra P, Vig L, Shroff G and Agarwal P 2015 Long short term memory networks for anomaly detection in time series *Proc. of 23rd European Symp. on Artificial Neural Networks, Computational Intelligence and Machine Learning, ESANN 2015* pp 89–94
[13] Graves A, Wayne G and Danihelka I 2014 Neural turing machines (*Preprint* 1410.5401)
[14] Vaswani A, Shazeer N, Parmar N, Uszkoreit J, Jones L, Gomez A N, Kaiser L and Polosukhin I 2017 Attention is all you need *Proc. of the 31st Int. Conf. on Neural Information Processing Systems* NIPS'17 (Red Hook, NY, USA: Curran Associates Inc.) p 6000–6010
[15] Neil D, Pfeiffer M and Liu S C 2016 Phased lstm: Accelerating recurrent network training for long or event-based sequences *Proc. of the 30th Int. Conf. on Neural Information Processing Systems* vol 29 pp 3882–3890
[16] Srivastava S, Sen P and Reinwald B 2020 *arXiv e-prints* arXiv:2004.03398 (*Preprint* 2004.03398)
[17] Singh B P, Deznabi I, Narasimhan B, Kucharski B, Uppaal R, Josyula A and Fiterau M 2019 *arXiv e-prints* (*Preprint* 1905.00125)
[18] LeCun Y and Bengio Y 1998 *Convolutional Networks for Images, Speech, and Time Series* (Cambridge, MA, USA: MIT Press) p 255–258
[19] Zhao B, Lu H, Chen S, Liu J and Wu D 2017 *J. of Systems Engineering and Electronics* **28** 162–169
[20] Livieris I, Pintelas E and Pintelas P 2020 *Neural Computing and Applications* **32** 1–10
[21] Lea C, Flynn M, Vidal R, Reiter A and Hager G 2017 Temporal convolutional networks for action segmentation and detection *2017 IEEE Conf. on Computer Vision and Pattern Recognition (CVPR)* pp 1003–1012
[22] Bai S, Zico Kolter J and Koltun V 2018 *arXiv e-prints* (*Preprint* 1803.01271)
[23] He Y and Zhao J 2019 *J. of Physics: Conf. Series* **1213** 042050
[24] Wan R, Mei S, Wang J, Liu M and Yang F 2019 *Electronics* **8** 876
[25] Chollet F *et al.* 2015 Keras https://keras.io
[26] Marvasti F A 2001 *Nonuniform sampling : theory and practice* (New York (N.Y.) : Kluwer academic/Plenum)
[27] Subasi A 2020 Chapter 2 - data preprocessing *Practical Machine Learning for Data Analysis Using Python* ed Subasi A (Academic Press) pp 27 – 89 ISBN 978-0-12-821379-7
[28] Klein M, Thiele G, Fono A, Khorsandi N, Schade D and Krüger J 2020 Process data based anomaly detection in distributed energy generation using neural networks *2020 International Conference on Control, Automation and Diagnosis (ICCAD)* pp 1–5

Industry 4.0 readiness of manufacturing sector in the Baltic Region

SM Bazaz[1], S Penttilä[2], M Ollikainen[3], J Ratava[4], J Varis[5]

[1] Research Assistant, LUT University, Lappeenranta, Finland
[2] Junior Researcher, LUT University, Lappeenranta, Finland
[3] Project Researcher, LUT University, Lappeenranta, Finland
[4] Associate Professor, LUT University, Lappeenranta, Finland
[5] Professor, LUT University, Lappeenranta, Finland

Email: Sara.Moghadaszadeh.Bazaz@lut.fi

Abstract. The aim of the research is to provide a sample data and analysis to present the state of digitalization in the Baltic sea region. A matrix contains seven operation fields and ten services and technologies mapped available digital services in six countries like Denmark, Estonia, Finland, Latvia, Lithuania, and Poland. A comparison analysis describes the level of digitalization strength and weakness areas by country in Baltic sea region. Support structure analysis is a reference to select the suitable partner to enhance the level of digitalization in countries. The results show that there are differences in the strength and weakness areas in Baltic sea region. Further the strength areas of each country are compared to weakness areas of digitalization to develop a roadmap to improve the readiness to implement and use Industry 4.0 functionality and tools.

1. Introduction

Manufacturing industry is rapidly digitalizing. Integrating manufacturing with modern communication technologies, automation, and robotized boost availability, flexibility, reliability and maintainability in manufacturing industry. [1,2] In addition to various production technologies utilized as a part of a production chain, the complete computer-aided process of delivering affordable high-quality products requires several digitalized systems. [3,4] Digitalization assists companies in measuring performance, efficiency to arrange a plan to reach goals of the company. [5,6] Key benefits of digital manufacturing are improved employee collaboration, connectivity, productivity, and potentially autonomous process analysis and improvement capabilities, known as Industry 4.0. [7,8] These together allow enhancing the supply chain to the point of enabling customized lots of a single product. However, onboarding new technologies in the dominantly SME manufacturing industry can be challenging. [9,10]

In this study, a dataset collected by questionnaire is combined with manufacturing companies' capabilities collected by interviews or listed on public sources such as web pages.[1] From this data, it is possible to see the existing level of digitalization in each of the studied countries, and in the case of potential improvement, see if the relevant technologies are available in the country. For specialized maintenance, it is also important that the services are provided nearby.

[1] This work has been done as a part of the Interreg Baltic Sea Region project Innovation Framework for Challenge Oriented Intelligent Manufacturing (INforM), grant number #R078.

2. Methods

The digitalization structure and readiness for the Industry 4.0 in metalwork manufacturing is evaluated with six countries in Baltic region: Denmark, Finland, Estonia, Latvia, Lithuania and Poland. First, method to gather the state of digitalization in companies -questionnaire and the support structure matrix survey in each country is introduced. Further the method to analyse the differences in the digital readiness and need for support structures in different categories between countries is explained.

2.1. State of digitalization in companies in each country

The state of digitalization in each country in the survey by web questionnaire and company interviews. The level of digitalization in companies' questionnaire was based on the ANSI/ISA-95 classification of systems. Thus, ten main topics was derived in the questionnaire including five questions each topic. The questionnaire consisted following topics:

1. Digitalization in Customer Relationship Management (CRM);
2. Digitalization in technology planning, Product Lifecycle Management (PLM);
3. Vertical and horizontal value chain integration through Computer networks;
4. Digitalization in a Computer-Aided Quality Control (CAQC);
5. Digitalization in production monitoring, using Manufacturing Execution System (MES);
6. Digitalization in production planning, Enterprise Resource Planning (ERP);
7. Using hardware for digitizing the manufacturing (towards Cyber Physical Systems – CPS);
8. Digitizing the inbound logistics and Warehouse Management Systems (WMS);
9. Digitizing equipment maintenance process via Computerized Maintenance Management System (CMMS);
10. Digitizing quality assurance and Laboratory Information Management System (LIMS).

The questionnaire questions consisted of three answer options: "Not existing", "Partly used", "Totally used" giving percentage value of 0%, 50% and 100%. The total cumulative results of each topic were calculated by using average % of the total answers separately for each country. While this scale is coarse, with a sufficient amount of answers, a quantitative analysis can be performed.

2.2. State of digitalization support in each country

The level of the support structure of the companies in the study was gathered in the form of survey. The survey consisted information of service providers and operational digital technologies in manufacturing industries gathered with publicly available sources. The operators in the survey were divided in seven groups as follows:

1. Engineering includes manufacturers or vendors, and manufacturing or mechatronic systems designers;
2. IT comprises entities providing supporting software or hardware, including software as a service (SaaS) and cloud models;
3. Competence centers provide expertise for project or program support and typically are skill-based or network-based centers;
4. Educational institutions include universities, universities of applied science, and vocational schools;
5. Digital innovation hubs refer to social communities or research centers that provide expertise on technology trends, knowledge and strategic innovation management, and industry-specific insights;
6. Industrial associations support employees and industry business.
7. Public organizations include research organizations and various development organizations.

The support structure matrix indicates the distribution of the digitalization in each country, presenting the number of digital services available in each category. Each operator in the survey can fit in multiple groups depending on the operator's area of expertise.

2.3. Comparison of level of digitalization of companies in the Baltic sea region

The level of digitalization in each country in the study were compared between the categories provided in 2.1. The data of each country was plotted in the bar chart divided in the groups presented in 2.1. The data set consists of 64 entities in Denmark, 27 in Estonia, 85 in Finland, 62 in Latvia, 60 in Lithuania, and 71 in Poland. The collected data may be used means to explore and find the suitable solution for needed services. Some overlap and related specialization are apparent. For example, an IT company provides ERP, MES, and PLM services suggests a developer or vendor of a management software suite. An engineering and IT company providing PLM, CPS, LIMS, and maintenance suggests a manufacturer of CPS with services to support its main product such as PLM for product development, LIMS to test its product and provide maintenance for its customers. And there are educational institutes and research hubs which are active in several fields and their role in supporting the industry. The differences in the readiness for Industry 4.0 was presented and the lack of required digitalization level in each category between the companies was evaluated.

2.4. Support functionality analysis of the Baltic region

Support functionality analysis of each country in the study was evaluated by plotting the data of each group presented in 2.1 in the form of bar chart. The differences in the support structures between the countries were evaluated. The support structure matrix and level of digitalization in companies in each country were compared to find the improvement areas of digitalization. Further the improvement areas were compared to countries support structure matrix to find the support to category from other countries in the study. Finally, the roadmap to increase the level of digitalization and readiness for Industry 4.0 implementation in Baltic region is presented.

3. Results and Analysis

The collected data set, while not exhaustive, may be considered representative of the level of activities in each country, see Figure 1. It is readily apparent that, relatively speaking, Denmark, Estonia, and Poland have an engineering-heavy sample. Estonia and Denmark have relatively fewer IT companies, whereas the data set includes many Polish companies having activities on that field with the other countries having a good number of companies in the sample. Estonian part of the data set has a high representation of educational institutes, competence centres, and digital innovation hubs; many of these are public organizations suggesting a high level of communal support. For the supporting entities, Poland and Denmark had relatively few educational institutes listed in the sample; Finland, Poland and Denmark had few competence centres. No digital innovation hubs were listed in the study for Poland and few for Denmark. Estonia had an especially high share of public organizations, with Finland and Denmark having few. Latvia and Estonia listed a higher number of industrial associations, whereas Denmark had none in the sample.

The structure of the entities studied varies between countries. There is relatively little overlap in all countries. In Finland and Denmark, education institutes are listed to provide engineering and IT services In Estonia, Latvia, Lithuania, and Poland education institutes are listed to provide engineering, IT services, competence centers, and digital innovation hubs. The individual services or technologies supported are shown in Figure 2 which illustrates the distribution of services and technologies among in the six countries studied. Note that this illustrates the share of entities studied that offer services or products of this type relevant to their operating fields. We may roughly divide the services to more commonly used (PLM, VH, CAQ, MES, CPS) and less used (CRM, ERP, LIMS, maintenance).

Finland, Estonia, and Poland have a relatively high degree of services offered in Product Lifecycle Management. In vertical and horizontal value chain integration, Estonia has the highest share over 50%. with Poland and Denmark having the lowest level of adoption. In Computer-Aided Quality Control, Poland seems to be the only notable outlier, having relatively low level of availability. For manufacturing executions systems, Denmark has a high level of services offered, with other countries having a lower level and Finland coming in last. In Cyber-Physical Systems, Denmark appears to take the lead (study was conducted in the "Danish Robot Valley" in Fyn) with other countries enjoying a good level of adoption.

In Customer Relationship Management, Estonia and Finland have the highest level of support, with relatively little interest in Denmark. One might assume that such functions may be included in ERP systems, which appear to enjoy good support in Estonia and Finland, some support in Lithuania but surprisingly low support otherwise. For Warehouse Management Systems, all partners have a similar rate of adoption; some functionality may be included in ERPs. Denmark seems to have a lower rate of WMS services or products offered. In Laboratory Information Management, Finland and Estonia seem to have a higher level of digitalization than the other partners. Finally, maintenance systems appear to enjoy better level of support in Finland than elsewhere.

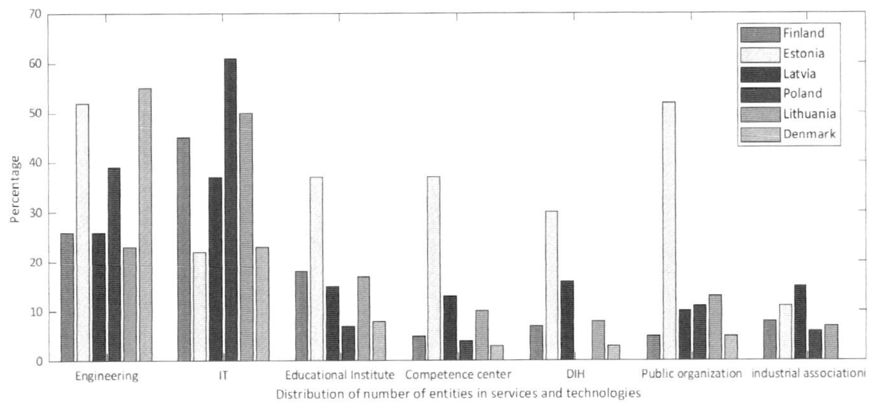

Fig 1. Distribution of manufacturing-relevant entities in each country. One entity having divisions in different fields means totals may add up over 100%.

Overall, Estonia and Finland seem to have a high level of readiness for Industry 4.0, with a comparative focus on customer and customer satisfaction-oriented systems, services and technologies such as ERP, CRM and LIMS. The Danish part of the study suggests significant competence in manufacturing automation. Lithuania appears to be having a slightly higher level at many of the less-adopted systems compared to its neighbors.

As a rough estimate, under 15% share of companies providing a service in a country might warrant a concern for availability, and under 30% indicate some room for improvement. These are apparent in the collected data. Overall, we may conclude that the availability of maintenance systems important for keeping high-tech equipment running smoothly is low. Outside of Estonia and Finland, ERP, CRM, PLM, and LIMS seem to need improvement. Denmark has a low share in most technologies outside of CPS, MES and CAQ, though it might be that the number of robotics companies is relatively high in the sample. Other notable opportunities for improvement include VH and CAQ systems for Poland and CAQ systems for Latvia.

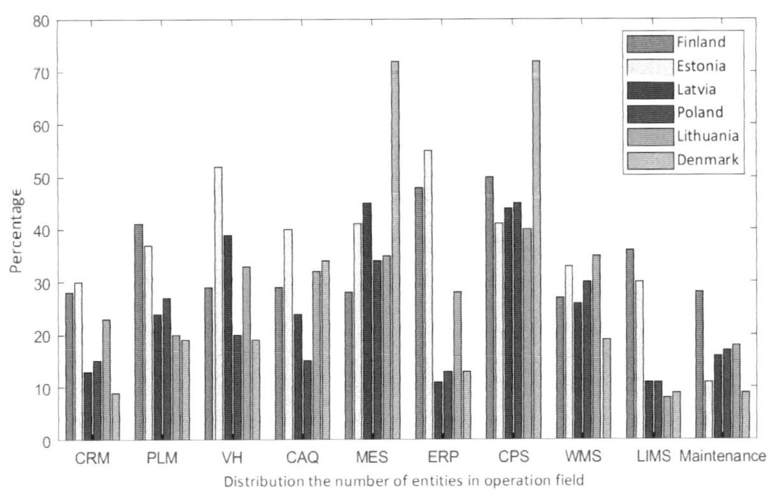

Fig 2. Country comparison - services and technologies supported (percent of entities listed)

4. Discussion

The reliability of the study is dependent on the data gathered. The data does not represent the whole view of the companies nor service providers and therefore the results of the study does not give absolute result of the digitalization level of the companies in each country. However, the experts in manufacturing have been doing the surveys and questionnaires in each country and therefore the study presents a brief but easily generalized overview of the level of digitalization and support structure in Baltic region.

The results of the study can be applied in the practice by acknowledging and sharing the information between the countries via e.g. digital platform offering easy access to companies for the information. Companies can find the suppliers, vendors and knowledge about the Industry 4.0 benefits as well as practical steps how to apply the theory in practice. Thus, the level of digitalization can be increased in the companies. By increasing the level of digitalization in companies the Industry 4.0 can be implemented further in practice. This can not only enhance the viability in the company level but also in national as well as EU level. However, the language barrier between the countries might affect the applicability as especially SME's in Latvia, Lithuania and Poland are not using English as a commonly used language in everyday work life. Hence requiring the translation process in between the service providers to increase the effectiveness. For further study, a wider database of levels of digitalization in whole EU could be applied to increase the collaboration and potentially competitiveness of EU.

5. Conclusions

The survey about the digitalization level of six countries in Baltic region has been done based on the surveys and questionnaires. A resulting support structure matrix consists of seven operation fields and ten services and technologies. Information has been collected of the popular and available entities in its industrial region. The full results may help the countries to develop policy and find out their weakness and strengths in ten important and useful digital technology area. The support functionality analysis assists countries and entities in the countries to collaborate with the other partners to enhance their digital level in industry. Overall, Finland and Estonia have good availability in most services. Danish robotics cluster is notable. Denmark, Poland, and Latvia need to improve some services and Lithuania has the average share among the countries. Along with maintenance automation, CRM, ERP, and LIMS have the most potential to be improved to reach higher level of digitalization.

References

[1] Mahmood, K.; Karaulova, T.; Otto, T.; Shevtshenko, E. (2019). Development of cyber-physical production systems based on modelling technologies. Proceedings of the Estonian Academy of Sciences, 68 (4), 348−355.

[2] Kangru, T.; Riives, J.; Mahmood, K.; Otto, T. (2019). Suitability analysis of using industrial robots in manufacturing. Proceedings of the Estonian Academy of Sciences, 68 (4), 383−388.

[3] Qi Q, Tao F 2018 Digital Twin and Big Data Towards Smart Manufacturing and Industry 4.0: 360 Degree Comparison *IEEE Access. 6 (2018) 3585-3593.*

[4] Lõun K, Riives J and Otto T 2011 Evaluation of the operation expedience of technological resources in a manufacturing network. *Est J Eng* 2011 **17** 1, 51-65.

[5] Riives J, Karjust K, Küttner R, Lemmik R, Koov K, Lavin J. 2012 Software development platform for integrated manufacturing engineering system. *Proceedings of the 8th International Conference of DAAAM Baltic Industrial Engineering, Tallinn, Estonia, 19-21 april 2012. Ed. Otto, T. Tallinn: Tallinn University of Technology, 555−560*

[6] Kaganski S, Majak J, Karjust K, Toompalu S. 2017 Implementation of key performance indicators selection model as part of the Enterprise Analysis Model. *Procedia CIRP, 63: The 50th CIRP Conference on Manufacturing Systems (CMS). 03.05-05.05 Taichung City, Taiwan. Elsevier, 283−288.10.1016/j.procir.2017.03.143.*

[7] Kolberg D, Zühlke D 2015 Lean Automation enabled by Industry 4.0 Technologies *IFAC-PapersOnLine, 2015 1870-1875.*

[8] De Carolis A, Macchi M, Negri E, Terzi S 2017 Guiding manufacturing companies towards digitalization a methodology for supporting manufacturing companies in defining their digitalization roadmap, in: 2017 *Int. Conf. Eng. Technol. Innov.* 487-495.

[9] Bazaz SM et al. MEs' Support Functionality Analysis Based on Statistical Analysis 2020 *Proc Manuf* 51 2020 960-966

[10] Roblek V, Meško M, Krapež A 2016 A Complex View of Industry 4.0 *SAGE Open. (2016).*

IOP Conf. Series: Materials Science and Engineering 1140 (2021) 012020 doi:10.1088/1757-899X/1140/1/012020

Inertial Measurement Unit based Human Action Recognition for Soft-Robotic Exoskeleton

Jan Kuschan, Moritz Burgdorff, Hristo Filaretov, Jörg Krüger

Fraunhofer Institute for Production Systems and Design Technology IPK, Germany

E-mail: jan.kuschan@ipk.fraunhofer.de

Abstract. Absence from work caused by overloading the musculoskeletal system lowers the life quality of the worker and gains unnecessary costs for both the employer and the health system. Exoskeletons can present a solution. Typically, such systems struggle with stiffness and discomfort and primarily a lack of battery lifetime. Soft-robotic exoskeletons offer a possibility to overcome these problems by increasing the system flexibility, not limiting the supported DoF and being actuator and joint together. Since soft-robotic exoskeletons can be designed only using power when supporting the wearer, it is possible to increase the battery lifetime by only acting on those actions for which the wearer needs support. Dealing with controls for soft-robotic exoskeleton one major difficulty is to find a compromise between saving energy and supporting the wearer. Having an action-depending control can reduce the supported actions to cover only relevant ones and increase the lifetime of the battery. The system conditions are to detect the user actions in real-time and distinguish between actions which require support and those which do not. We contribute an analysis and modification of human action recognition (HAR) benchmark algorithms from activities of the daily living, transferred them onto industrial use cases containing short and mid-term action and reduce the models to be compatible using embedded computers for real-time recognition on soft exoskeletons. We identified the most common challenges for inertial measurement units based HAR and compare the best-performing algorithms using a newly recorded data set overhead car assembly for industrial relevance. As a benchmark data set we focused on the "Opportunity" data set. By introducing orientation estimation, we were able to increase the F1 scores by up to 0.04. With an overall F1 score without a Null-class of up to 0.883, we were able to lay the foundation to use HAR for action dependent force support.

1. Introduction

Physically demanding tasks in manufacturing, logistics, handcraft and service are vital contributors to early damage of the musculo-skeletal system and especially the spine [11]. This stress leads to a reduced quality of life and also to a decreased capacity for work [10, 16]. Exoskeletons can present a solution. Typically, such systems struggle with stiffness and discomfort and primarily a lack of battery lifetime [9, 7]. Soft-robotic exoskeletons offer a possibility to overcome these problems by increasing the system flexibility, not limiting the supported Degree of Freedom (DoF) and being actuator and joint together [8, 20]. Since soft-robotic exoskeletons can be designed only using power when supporting the wearer, it is possible to increase the battery lifetime by only acting on those actions for which the wearer needs support. Human Activity Recognition (HAR) can be used to predict movements and support every individually classified action, but can also be constrained to only support task-related actions. The combination of Inertial Measurement Unit (IMU) based HAR and soft exoskeletons are therefore predestined to create an action-based prediction of the current or future activities as early and accurate as possible.

The idea behind HAR is that characteristic sensor signals directly correspond to specific body movements [18] which can, therefore, be detected and classified from a time series of sensor data. Traditional approaches to HAR rely on hand-crafted or heuristic information, where expert knowledge is used to identify relevant features. This method is highly restrictive and leads to difficulties with recognizing high-level behaviours, as engineered features are only "convenient mathematical operations" and "do not relate to units of behaviour" [18]. An additional problem is the general transfer of this explicit knowledge to different application domains. More recent methods however make use of machine learning with earlier works using techniques such as Deep Belief Networks (DBNs) ([19]) or Hidden Markov Models (HMMs) above Restricted Boltzmann Machines (RBMs) layers ([1]). While combining hand-crafted feature approaches with machine learning methods can lead to systems that perform well in certain scenarios ([6]), machine learning and more specifically deep learning methods still offer major advantages, such as more robustness to a large variety of actions and different people.

2. Related Work

In [14] an active textile upper arm, elbow and hand exoskeleton for workers support was presented. Kuschan et al. described how an angular control could be used to achieve a gravity compensation of the arm. Since this supports mainly static arm poses but due to the stiffness of the system also interferes with dynamic movements of the user, it is necessary to detect current and future movements. The system consists of five IMUs — one on the chest, two on the upper arm, one on the lower limb and one on the hand.

2.1. Datasets

While working with wearable systems like soft-robotic exoskeletons, it makes sense to use the existing sensors like IMUs to avoid the typical problems of working with vision-based HAR. Even if sensor-based HAR is not as common as video or image-based HAR, a large variety of datasets are available. These datasets are often specialized in one specific topic like recognizing sports activities [23], elderly care [2] or daily activities [15]. Chavarriaga et al. [5] created a challenge for HAR based on the publicly available Opportunity dataset. While Activity of Daily Living (ADL) often consists of long- (e.g. relaxing, running), mid- (e.g. cooking, washing dishes) and short-term (e.g. open doors) actions, in industrial use-cases we are often confronted with repetitive processes of mid- (e.g. mounting something) and short-term (e.g. picking up something) actions.

Opportunity Dataset Due to its vast documentation, the Opportunity dataset is a very common benchmark for HAR. It consist of $n_c = 18$ numbers of sporadic gesture classes, with the *Null* class occupying 72%. The activities from the Opportunity dataset were recorded in a home environment and comprised gestures performed during everyday activities. The recordings include four subjects each of which performed five ADL sessions, during which they only followed a high-level description of the task and therefore had free interpretation of how to achieve the goal rather than step-by-step instructions. Every subject also performed a drill session, during which they executed 20 repetitions of a sequence of nine activities. The dataset was recorded at a sample rate of 30 Hz and used a large number of sensors of different modalities that were shared between the environment, the objects and the subjects and is around 6 h long. The subjects' body-worn sensors included 7 IMUs and 12 3D acceleration sensors with a total number of relevant on-body sensor channels of $D_t = 133$. The IMUs provide the 3D acceleration, 3D angular velocity, 3D magnetic field and the sensor orientation in quaternions.

2.2. Classifier

In the domain of HAR, many different machine learning techniques have already been successfully employed. Choices range from shallow models, such as k-Nearest Neighbors (KNN) ([13]), Decision Trees ([3]) or Joint boosting ([4]) paired with automatic or hand-crafted feature extraction, to deep learning models, such as Deep Neural Networks (DNNs), Convolutional Neural Networks (CNNs), autoencoders or RBM. For this work, the models were selected based on their reported classification performance on the aforementioned "Opportunity" dataset and the perceived complexity of their structure. Three models were selected and will hereinafter be referred to by the names based on their respective publications: Deep Convolutional Long Short-Term Memory (DC-LSTM) [18], Cascaded Bidirectional and Unidirectional LSTM-based Deep Recurrent Neural Network (C-DRNN) [21] and CNN-2 [17].

DC-LSTM, as well as CNN-2, make use of multiple convolutional layers for feature extraction, but neither model uses padding, and as such, they produce increasingly smaller feature maps with every consecutive layer. Therefore, both models can only be applied if the length of the sliding window is above a certain threshold, which is dependent on the amount of convolutional and pooling layers.

C-DRNN The Cascaded Bidirectional and Unidirectional Long Short-Term Memory (LSTM)-based Deep Recurrent Neural Network (DRNN) from [17] is a cascaded structure, in which the first layer is a bidirectional LSTM layer, the output of which gets concatenated with a simple summation and is then followed by unidirectional LSTM layers (where bidirectional means that the layer consists of two LSTMs, one of which processes the data backwards). Even though the reported F_1 score indicates state-of-the-art performance, several observations about the methodology have to be considered for its interpretation. First, all datasets were divided with a simple 80/20 split between training and test data with no validation set. This proceeding did not adhere to the Opportunity challenge guidelines and as such did their test data optimizations as opposed to on a validation set. Next, there is no description of how multiple activity classes during one window were handled. And lastly, they present no detailed description of their final model structure, and only a few hyperparameters are mentioned.

#	Type	Layer Size	Parameters: Opportunity $D = 113$, $n_c = 18$	Parameters: Overhead Car Assembly (OCA) $D = 72$, $n_c = 7$
1	*Bidirectional LSTM*	64 units	91 648	70 656
2–3	*LSTM*	64 units	33 280	33 280
4	Fully Connected (Softmax)	n_c units	1 170	455
Total			159 378	137 671
Total (Relevant Layers)			**158 208**	**137 216**

Table 1: Layer configuration and number of parameters per layer in C-DRNN. Number of parameters can vary with the framework used for the implementation. D is the number of features, n_c is the number of classes. Relevant layers are italicised.

DC-LSTM DC-LSTM [18] is an architecture that combines multiple convolutional and recurrent layers. It has proven state-of-the-art performance in the related domain of speech recognition.

The model was tested on the Opportunity and Skoda datasets [22] and delivers state-of-the-art performance scores on both of them. Regarding the Opportunity dataset, they used all on-body sensors and data splits as outlined in the Opportunity challenge guidelines.

The model was imported from their publicly available repository without any major changes. It consists of four stacked convolutional layers with 64 5×1 filters, each of which operates along the time axis. Subsequently, the output is fed into two stacked LSTM layers with 128 units each and one softmax layer for classification, shown in table 2. For training, the publication proposes a learning rate of $10e^{-3}$, which, however, in their implementation is changed to $10e^{-4}$. The training was therefore performed with the RMSprop optimiser with learning rate of $10e^{-4}$, decay factor of $\rho = 0.9$ and a dropout probability of $p = 0.5$. Another inconsistency between their publication and implementation is the difference in the window length. The paper proposes a window length of 500 ms, for which they claim to have obtained the best test results, whereas the implementation uses a window length of 800 ms, which is used by the model that produces the reported F_1 score.

CNN-2 The CNN-2 from [21] can be seen as two consecutive blocks of two convolutional layers followed by one maxpooling layer, which results in two fully connected and one softmax layer for classification.

A detailed listing of the model architecture can be found in table 3. It was first proposed in [18] as the baseline model and then refined with the use of pooling layers by [21], although it still only served as a baseline model, it achieved state-of-the-art performance on the Opportunity dataset with $F_1 = 0.9152$. The tests adhere to the Opportunity guidelines. It was trained with the RMSprop optimiser with a fixed learning rate of $10e^{-3}$, a decay factor $\rho = 0.95$ and a dropout probability of $p = 0.5$.

3. Experiments

Overhead work is a typical problem in ergonomic analysis. For this dataset we present in the Experiments section different modifications of the benchmark algorithms we worked out from ADL datasets. We also introduce orientation estimation for IMU based HAR and other varieties of the input data.

3.1. OCA

To have a dataset that meets our requirements, we have recorded the OCA dataset in laboratory environments. OCA covers the mounting and dismounting of a heat capsule under a car, a cyclic process which needs approximately 75 s. We separated the cycle in six mid level classes: "Mount

#	Type	Filters @ Patch Size	Parameters	
			Opportunity $D = 113$, $n_c = 18$	OCA $D = 72$, $n_c = 7$
1	*Convolution*	64 @ 5×1	384	384
2–4	*Convolution*	64 @ 5×1	20 544	20 544
5	*Recurrent*	128 units	3 769 344	2 425 344
6	*Recurrent*	128 units	132 096	131 584
7	Fully Connected (Softmax)	n_c units	2 322	903
Total			3 965 778	2 619 847
Total (Relevant Layers)			**3 963 456**	**2 618 944**

Table 2: Layer configuration and number of parameters per layer in DC-LSTM. The number of parameters in the original publication is incorrect, due to an error in their calculation of the number of parameters of the recurrent layers. The number of parameters can vary with the framework used for the implementation. Relevant layers are italicised.

IOP Conf. Series: Materials Science and Engineering 1140 (2021) 012020 doi:10.1088/1757-899X/1140/1/012020

#	Type	Filters @ Patch Size	Parameters	
			Opportunity $D = 113$, $n_c = 18$	OCA $D = 72$, $n_c = 7$
1	*Convolution*	64 @ 5 × 1	384	384
2	*Convolution*	64 @ 5 × 1	20 544	20 544
3	MaxPooling	2 × 1		
4–5	*Convolution*	64 @ 5 × 1	20 544	20 544
6	MaxPooling	2 × 1		
7	Fully Connected	512 units	22 217 216	33 030 656
8	Fully Connected	256 units	131 328	131 328
9	Fully Connected (Softmax)	n_c units	4 626	1 799
Total			22 415 186	33 225 799
Total (Relevant Layers)			**62 016**	**62 016**

Table 3: Layer configuration and number of parameters per layer in CNN-2. Relevant layers are italicised.

Mount Cover Panel	Screw in Cover Panel	Take Screw-driver	Take Cover Panel Down	Unscrew Cover Panel	Place Screw-driver Down
Worker takes the cover panel and loosely mounts it.	Worker screws in all screws for the cover panel.	Worker picks the screw-driver up from the table.	Worker takes down the cover panel and puts it down.	Worker un-screws all screws that hold the cover panel in place.	Worker places his screw-driver on a table.

Table 4: High level description of the custom data set.

Cover Panel", "Take Panel Down", "Take Screwdriver", "Place Screwdriver", "Screw in Panel" and "Unscrew Panel", shown in table 5. Without line production in the lab it was necessary to dismount the heat panel after mounting it and since this is not part of the original process we decided to handle it as independent classes.

The choice of the actions is related to the wearable robotics use-case. OCA contains very dynamic short-term overhead movements, like mounting cover panel and take cover panel down, without the need of support, but also very static overhead actions like screwing in and unscrewing the panel, which requires substantial support. The better the classification rate of the static actions, the better the wearer is supported and the better the classification rate of the dynamic short-term actions the less energy is wasted.

For the recording process subjects were equipped with a sensor vest devised like in [24]. This vest consists of 12 IMUs, covering all major body segments (figure 1).

All IMUs provide 3D acceleration, 3D angular velocity and the sensor's orientation in

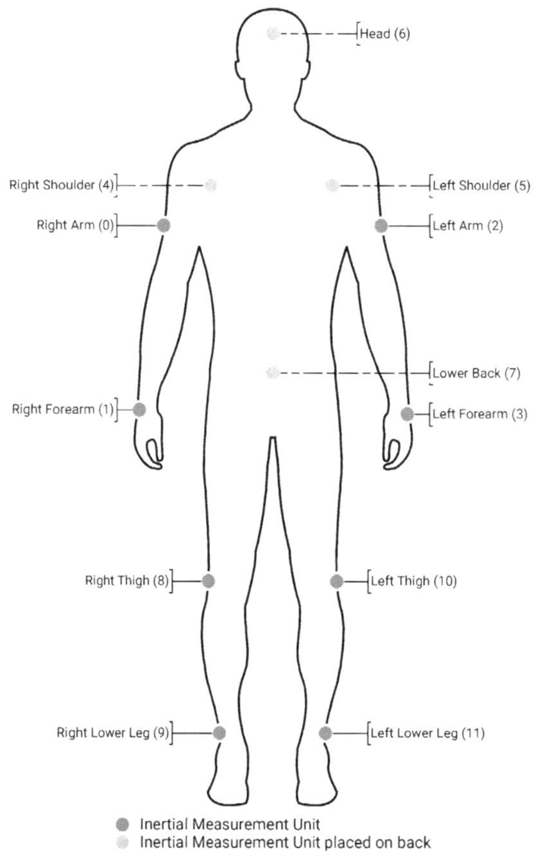

Figure 1: Front view; Sensor placement on the vest for the custom dataset. The numbers indicate the index of the sensor in the dataset. [24]

quaternions for a total number of $D_t = 120$ channels. The dataset was recorded at a constant sample rate of 40 Hz, is around 90 min long and covers 70 cycles. All sessions were video-recorded and then retroactively labelled with a synchronization tool. Finally, we had five test subjects where person 1, 2 and 3 each recorded two independent datasets, each approximately 10 minutes long, and person 4 and 5 each recorded one dataset, about 15 minutes each.

The short duration actions like "Take Panel Down", "Take Screwdriver" and "Place Screwdriver" comprise a tiny part of the total dataset, while the *Null* class is the most dominant with 42%. Still, this is much less than the 71.9% *Null* class presented in the Opportunity dataset. Since "Screw in Panel" and "Unscrew Panel" are very similar actions, we discuss their influence on the model if they are kept distinct and if they are combined into one class.

3.2. Training

The three models [18, 21, 17] were chosen because they have the best benchmark performance on the Opportunity dataset and also provide available code. Only the architecture of the models was used, all weights are initialized prior to training and affected only by our own training procedure.

For use in this work a few variations on the original C-DRNN have been made. The authors of [17] calculate a prediction for every sample $\boldsymbol{y}_c^t$, where $t = 1, \ldots, T$, and uses a merging step to obtain a class distribution for the entire input window. The prediction of the last sample

$\boldsymbol{y}_c^T$ was replaced similar to the final step in DC-LSTM to predict the most recent movement. Additionally, the handling of multiple classes in one window was changed, and the data split from the Opportunity challenge was used.

The final model is presented in table 1 and consists of three layers, where the first layer is the bidirectional LSTM layer, which is followed by two unidirectional LSTM layers and one softmax layer for classification. All LSTM layers have 64 units and use the tanh activation function. It was trained with the Adam optimiser [12], with a learning rate of $10e^{-3}$ and a dropout probability of $p = 0.2$.

For training and inference, all models receive the same input structure, a series of $T \times D$ matrices. Likewise, they all produce the same output, which is a probability distribution $\boldsymbol{y}_c = (y_1, \ldots, y_{n_c})$ over all n_c classes, with $\sum_i y_i = 1$ and $y_i \in [0, 1]$, for every one of those matrices.

In table 6, we present the variations of the input data, which we also use for the Result and Discussion section. The first models were trained only using the acceleration and angular velocity raw data captured by the IMUs. For the subsequent variations, we added an orientation estimation calculated according to

$$\begin{aligned} \boldsymbol{q}_t &= \boldsymbol{q}_i \otimes \boldsymbol{q}_{[x|y|z]} \otimes \boldsymbol{q}_i^* \\ \alpha_{i,[x|y|z]} &= \operatorname{acos}(q_{t,z}) \end{aligned} \tag{1}$$

where $\boldsymbol{q}_{[x|y|z]} = [0,1,0,0]^T_{[x]}, [0,0,1,0]^T_{[y]}, [0,0,0,1]^T_{[z]}$ and $\boldsymbol{q}_i$ is the orientation from the acceleration and angular velocity of IMU i.

For the third variation of the data, we removed the acceleration, because the orientation and angular velocity are enough to determine the configuration of a kinematic chain (in this case, the human body). Finally, we also removed the angular velocity because it is merely the first-order derivative of the orientation.

For the training, we split the dataset into training, test and validation parts without the leave one out method, following the rules of the Opportunity dataset [5]. But in the later results, we also discuss the influence of isolating one test subject as designated test data.

Name	**Repetitions**	**Instances**	**(%)**
Null	791	5 749	42.0
Mount Panel	68	1 188	8.7
Take Panel Down	67	535	3.9
Take Screwdriver	130	483	3.5
Place Screwdriver	127	497	3.6
Screw in Panel	199	2 593	18.9
Unscrew Panel	200	2 640	19.3
Total		**13685**	

Table 5: Class labels and their distributions for the OCA dataset. Repetitions and instances show the distribution of the individual classes (across all subjects and sessions) after application of the sliding window.

IOP Conf. Series: Materials Science and Engineering 1140 (2021) 012020 doi:10.1088/1757-899X/1140/1/012020

3.3. Performance

In the domain of HAR there are multiple performance measures with individual strengths, such as accuracy, precision, recall, F scores or Receiver Operating Characteristics (ROC) curves. The F_1 score belongs to the most commonly used metrics, as it provides a simple yet expressive value that can be used for easy comparison between multiple approaches. More in-depth analysis of a given model can be achieved through the use of confusion matrices, as they offer the ability to evaluate the classification outcome on a per-class level. Furthermore, it can also be employed as a similarity measure between different activities in a dataset, where more misclassifications between two classes can be interpreted as low interclass variability. However, the F_1 score suffers in expressiveness when datasets are heavily unbalanced because trivial classifiers that only predict the most prevalent class can achieve high scores. While confusion matrices can be normalized to handle this effect, the F_1 score has to be extended to the weighted F_1 score, which takes the prevalence of each class into account.

$$wF_1 = \sum_i 2w_i \frac{p_i \cdot r_i}{p_i + r_i} \tag{2}$$

where $w_i = n_i/N$ is the proportion of number of instances of class i (n_i) to the total number of instances (N), and p_i and r_i are the recall and precision per class respectively, with $p_i = \frac{TP_i}{TP_i+FP_i}$ and $r_i = \frac{TP_i}{TP_i+FN_i}$. For the interpretation of the results, these two metrics will be used.

3.4. Tests

The tests for this work were conducted in two parts. First, the models were trained and tested on the input variations 1–4 (table 6) on both the Opportunity and OCA datasets with the default architectures and hyperparameter settings. Those results serve as a baseline for both classification accuracy and inference time of the original models and explore whether the introduction of model-based features to the input of the unchanged model increases model classification performance. Secondly, all permutations of layer settings for the models were trained on the Opportunity dataset with the default input. Next, the best model was chosen according to the following formula:

$$\underset{m \in \mathcal{M}}{\arg\max} \Big(|f_t(m; \mathcal{M}) - f_{val\_acc}(m)| \cdot r(m) \Big) \tag{3}$$

where $\mathcal{M}$ is the set of all tested architectures, f_t is a logarithmic trend line for the given set of validation accuracy over the present reduction in model complexity, f_{val_acc} is the validation

Variant	Features added to the data	Features removed from the data
1	–	–
2	orientation estimation	–
3	orientation estimation	acceleration channels
4	orientation estimation	acceleration and angular velocity channels

Table 6: Variants of the input data. Orientation features were calculated as described in equation 1.

accuracy of the given model architecture and r is the average complexity reduction in % of the model m compared to the original model. The logarithmic trend line was chosen to have the form $f_t = -a \cdot e^{b \cdot x} + c$, because we expect the models to quickly rise in performance with higher model complexity but level out. Afterwards, the selected model was trained and tested on input variations 1–4 on both the Opportunity and OCA datasets.

4. Results

Table 7 shows the weighted F_1 scores (equation 2) for the three used networks of both the Opportunity dataset and the OCA dataset. Further, the different varieties of input data are presented, and the influence of reducing the model complexity is shown. Especially for the Opportunity dataset, the weighted F_1 score is significantly higher than the one without *Null* class. Since a high amount of samples has a direct influence on the weighted F_1 score, the difference to the OCA should relate to the class distribution with 79.2% for Opportunity and 42% for OCA. The F_1 scores without *Null* class performs better for both the default and reduced models of the C-DRNN and reduced models of DC-LSTM and CNN-2 better, than for the Opportunity data set. Interestingly, no distinct pattern for the different input data varieties is observable. The best classifiers seem to be either of variant two (acceleration, angular velocity and orientation estimation) or variant one (acceleration and angular velocity). The Opportunity dataset has a better classification of variant two for the default models and variant one for the reduced models. We assume that this is due to the ratio between reduced layers and an increased amount of input data. As a result, the networks are not able to find a suitable solution for the higher amount of data. Contrary to the expectation that the orientation estimation combined with angular velocity contains all information and the acceleration is not needed, there is a drop in the performance for most of the datasets and models using variant three. Still, there is a lower decrease in performance for the reduced models. Using only the orientation estimation as input data for the models results in the worst classification results.

The confusion matrices for the models trained on the OCA dataset were chosen based on the best F_1 presented in table 7. Having a look at the confusion matrices for the different models of the OCA dataset in figures 4, 2 and 3 one can see that all models suffer in distinguishing between the "screw in" and "unscrew cover panel" classes. At the current state, it appears more like a random decision between those two actions. The recognition error between the remaining classes is very low and mostly only the *Null* class has a bigger influence on the false detection rate.

The confusion matrix of the C-DRNN on the OCA dataset, shown in figure 2, presents the most robust distinguishing between "screw" and "unscrew cover panel" of the three trained models. One the other hand, only the C-DRNN seems to have problems to classify "take screwdriver" and "place screwdriver down" correctly.

The results of DC-LSTM shown in figure 3 and CNN-2 shown in figure 4 are very similar, but the DC-LSTM model differs by displaying the highest false positive rate for *Null* class.

Although the CNN-2 does not consider history, which seems to be an advantage for repetitive actions, it appears as a robust classifier. One can only guess that it is maybe due to the number of repetitions.

Having the results from the confusion matrices of the three trained models, we decided to train the models again, but this time with a merged class of "screw" and "unscrew cover panel". The results are showcased in figure 5 and table 8.

Table 9 shows the classification results for the original OCA dataset, but not using the defaults from the Opportunity dataset. In this test, we changed the training, validation and test data using leave one out. Therefore, the test subjects one, two and three are used for training, while test subject four is used for testing and five for validation. The weighted F_1 score shows poorer results for all models and nearly all variants. The best results comparing the

Model	Variant	C-DRNN		DC-LSTM		CNN-2	
		wF_1	wF_1^N	wF_1	wF_1^N	wF_1	wF_1^N
Default	1	0.874	0.549	0.886	0.596	**0.898**	**0.636**
	2	**0.880**	**0.573**	**0.895**	**0.623**	0.895	0.612
	3	0.864	0.517	0.879	0.560	0.893	0.597
	4	0.844	0.453	0.879	0.552	0.890	0.585
Reduced	1	**0.881**	**0.578**	**0.887**	0.575	**0.899**	**0.636**
	2	0.880	0.554	0.886	**0.580**	0.890	0.607
	3	0.875	0.545	0.883	0.562	0.889	0.589
	4	0.841	0.458	0.877	0.540	0.882	0.562

(a) Opportunity data set

Model	Variant	C-DRNN		DC-LSTM		CNN-2	
		wF_1	wF_1^N	wF_1	wF_1^N	wF_1	wF_1^N
Default	1	0.741	0.617	**0.695**	**0.569**	0.751	0.634
	2	**0.766**	**0.664**	0.673	0.517	**0.754**	**0.638**
	3	0.600	0.475	0.670	0.568	0.709	0.605
	4	0.425	0.256	0.439	0.247	0.478	0.291
Reduced	1	0.735	0.607	0.694	0.581	**0.781**	**0.684**
	2	**0.768**	**0.664**	0.708	0.579	0.727	0.596
	3	0.590	0.456	**0.723**	**0.624**	0.703	0.604
	4	0.432	0.264	0.477	0.299	0.494	0.322

(b) OCA data set

Table 7: Weighted F_1 score of default and reduced models with all input variants on the respective data set. Weighted F_1 without the *Null* class (wF_1^N).

Model	Variant	C-DRNN		DC-LSTM		CNN-2	
		wF_1	wF_1^N	wF_1	wF_1^N	wF_1	wF_1^N
Default	1	0.819	**0.771**	0.824	0.803	**0.883**	**0.873**
	2	**0.848**	0.824	**0.861**	**0.848**	0.873	0.854
	3	0.740	0.702	0.761	0.724	0.784	0.752
	4	0.530	0.436	0.661	0.610	0.659	0.592

Table 8: Weighted F_1 score of default model with all input variants of the OCA data set with combined screw and unscrew class. Weighted F_1 without the *Null* class (wF_1^N).

original configuration and leave one out validation dropping between 5 and 12 %. The confusion matrix in figure 6 presents an expected behaviour. Since the test subject from the validation set performs the actions a little bit different, it is not predicted as the correct action class, but instead as the Null class. The rest, for example, the classification problems between screwing actions, is comparable to the results from the original dataset.

5. Discussion

We analyzed and modified algorithms from different neural networks that performed best at HAR in ADL and figure out if they can be used to support wearable soft robotics control. In order to test the algorithms in a realistic industrial scenario, we recorded a dataset representing a

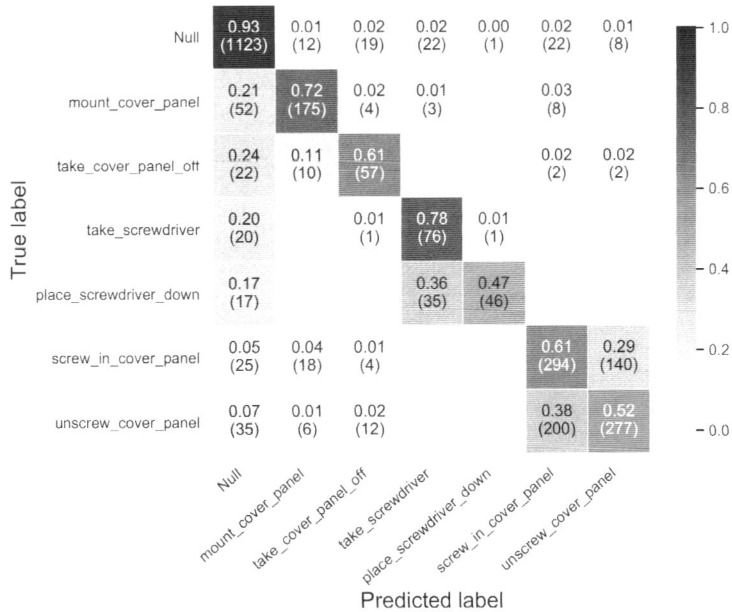

Figure 2: Confusion matrix of C-DRNN on the OCA dataset.

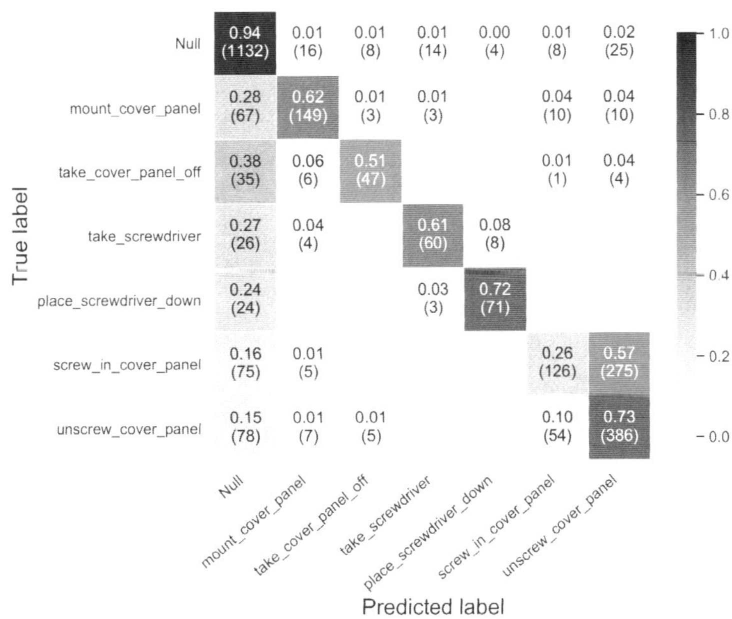

Figure 3: Confusion matrix of DC-LSTM on the OCA dataset.

common industrial assembly task. The actions in the dataset have significant kinetic similarities, which makes them difficult to classify. Nevertheless, trivial adaptation of HAR classifiers that exhibit state of the art performance on the Opportunity ADL datasets delivers promising results. Using our OCA dataset, which only consists of 90 minutes of recorded data, we were able to

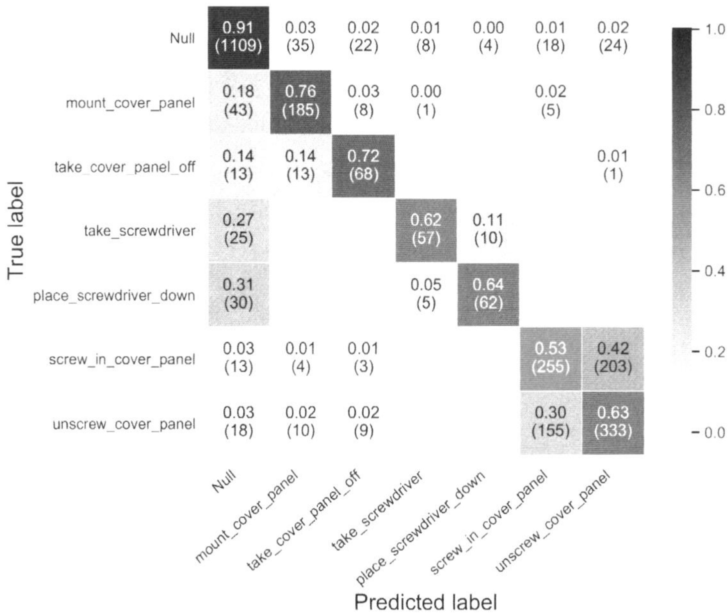

Figure 4: Confusion matrix of CNN-2 on the OCA dataset.

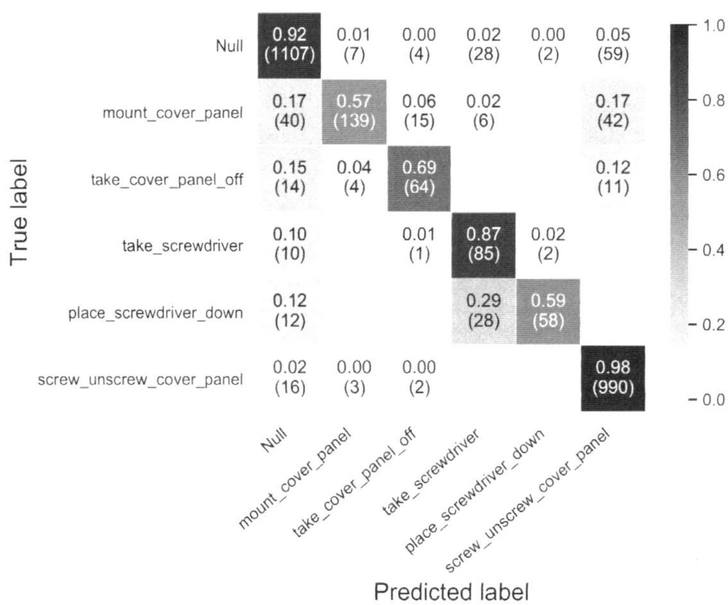

Figure 5: Confusion matrix of C-DRNN on the OCA dataset with combined screw and unscrew cover panel.

achieve a comparable F_1 score (for the variant with a combined "Screw" and "Unscrew" class). It must be noted, however, that our set of target classes is smaller than Opportunity's—we have 6 target classes, while Opportunity has 17 (both excluding the Null class). Furthermore, the confusion matrices make it clear that some of the remaining errors occur between kinetically

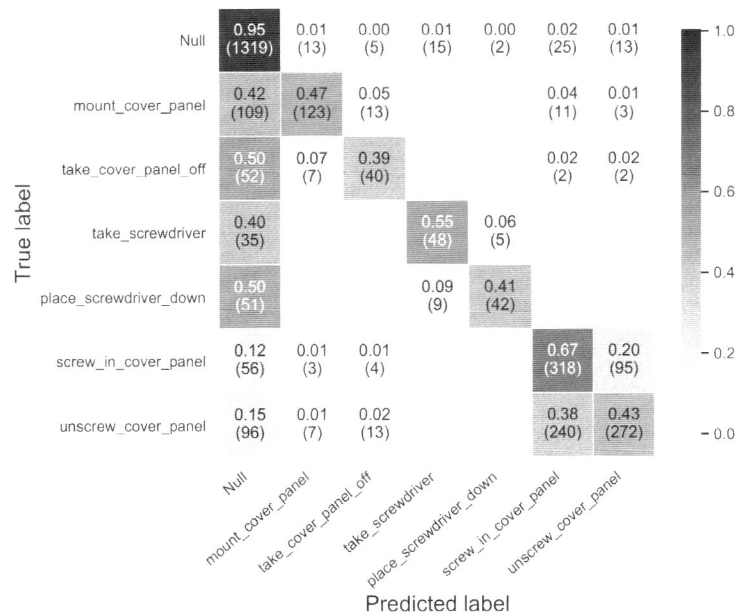

Figure 6: Confusion matrix of CNN-2 on the OCA dataset, using leave one out validation.

Model	Variant	C-DRNN		DC-LSTM		CNN-2	
		wF_1	wF_1^N	wF_1	wF_1^N	wF_1	wF_1^N
Default	1	0.626	0.485	**0.616**	0.431	**0.693**	**0.562**
	2	**0.655**	**0.512**	0.543	0.377	0.655	0.487
	3	0.586	0.445	0.607	**0.457**	0.286	0.000
	4	0.433	0.224	0.512	0.333	0.497	0.310

Table 9: Weighted F_1 score of default model with all input variants of the OCA data set, using leave one out validation. Weighted F_1 without the *Null* class (wF_1^N).

similar classes—taking/placing down the screwdriver, and mounting/unmounting the cover panel. The majority of the misclassifications are Null class false positives. The difference in performance between the three examined neural network architectures is far smaller than the difference caused by merging the very similar "Screw" and "Unscrew" classes. We therefore conclude that for similar classes of use-cases—HAR problems with target classes composed of a small set of full-body actions, the most important characteristic is a favorable data setup. That is, the largest improvements in performance can be achieved by first ensuring that the target classes contain minimal kinetic similarity, assuming that is not already irrevocably defined by the problem at hand.

In our use-case, the HAR classification scheme is meant to be used as part of a soft robotics control scheme and must therefore ensure that the system only supports the necessary actions and no Null class actions. Having only a very low false prediction rate at the Null class, this can be assured. But still having a true positive rate of 98% only for the screwing action when it is combined, shows that improvement on algorithms or amount of data is necessary. This result leads to efficient energy management for soft-robotic exoskeletons, because already with those algorithms, it is possible to cover nearly all relevant actions to support and ignore approximately

66% for external force support inappropriate actions. That more data is needed is also shown by the leave one out validation, where the F_1 score drops rapidly. Using the same setup for different people without specializing it for every wearer is an important requirement for industrial soft robotic support systems. The amount of data could also be a reason why the CNN-2 performs similarly to the other algorithms, which should be in advantage due to the repetitive task. Having a greater error for the short time actions like taking the screwdriver can be explained by having an overlapping class from the former window class. This affects short time actions more than long-term actions. Having the processing unit on the system and not on an edge cloud or the like, smaller models make sense to reduce calculations. Therefore, it is beneficial that the F_1 score of the reduced models shows nearly the same results as the default models. Nevertheless, this will change with an increase of data, which must be taken into account. The different input variants influence the results. But, using only acceleration and angular velocity and adding orientation estimation show comparable results, only CNN-2 seems to prefer variant 1. Variant 3 and 4 on the other can not match up to the first and second.

Having over 98% of mid-term actions seems suitable to set up a HAR based control for soft robotic wearables. For the more dynamic short-term actions, the results are okay, but not sufficient to set a robust control. But it is necessary to record more data from different persons to find a more robust model for people not included in the training set. Considering that the dataset contains only 90 minutes working material, it seems to be a manageable overhead for introducing such systems in industrial environments.

6. Future Work

HAR based control for soft-robotic wearables is due to different challenges not implemented yet. First, it is crucial to figure out at which sequential area of the action the classifiers perform worst and what is the reason for it. If it is at the beginning or end of the movements, perhaps changing the window sizes might alleviate the problem. Having the main errors in the middle of the action sequence, one can think about facing the issue through a more robust control. We now established a base to program a control for soft robotic wearables and will port the classifiers onto an embedded computer to use it as the main processing unit. To generate as much data as possible and not miss a vital body segment, we surely overbuilt the sensor system. This will be reduced in the future, in accordance with its impact on the classification results. In the end, exoskeletons only have a small number of sensors at the supportive areas. Last but not least, we were able to show the feasibility even with a small dataset, and surely we want to increase the amount of data, by different persons, more iterations and also more different actions.

References

[1] Mohammad Abu Alsheikh et al. “Deep Activity Recognition Models with Triaxial Accelerometers”. In: *CoRR* abs/1511.04664 (2015). arXiv: `1511.04664`. URL: `http://arxiv.org/abs/1511.04664`.

[2] Oresti Banos et al. “Design, implementation and validation of a novel open framework for agile development of mobile health applications”. In: *BioMedical Engineering OnLine* 14.Suppl 2 (2015), S6. DOI: `10.1186/1475-925X-14-S2-S6`.

[3] Ling Bao and Stephen S. Intille. “Activity Recognition from User-Annotated Acceleration Data”. In: *Pervasive Computing*. Ed. by Alois Ferscha and Friedemann Mattern. Berlin, Heidelberg: Springer Berlin Heidelberg, 2004, pp. 1–17. ISBN: 978-3-540-24646-6.

[4] Ulf Blanke and Bernt Schiele. “Daily Routine Recognition through Activity Spotting”. In: *Location and Context Awareness*. Ed. by Tanzeem Choudhury et al. Berlin, Heidelberg: Springer Berlin Heidelberg, 2009, pp. 192–206. ISBN: 978-3-642-01721-6.

[5] Ricardo Chavarriaga et al. "The Opportunity Challenge: A Benchmark Database for On-body Sensor-based Activity Recognition". In: *Pattern Recogn. Lett.* 34.15 (Nov. 2013), pp. 2033–2042. ISSN: 0167-8655. DOI: 10.1016/j.patrec.2012.12.014. URL: http://dx.doi.org/10.1016/j.patrec.2012.12.014.

[6] Z. Chen et al. "Distilling the Knowledge From Handcrafted Features for Human Activity Recognition". In: *IEEE Transactions on Industrial Informatics* 14.10 (Oct. 2018), pp. 4334–4342. ISSN: 1551-3203. DOI: 10.1109/TII.2018.2789925.

[7] Michiel P De Looze et al. "Exoskeletons for industrial application and their potential effects on physical work load". In: *Ergonomics* 59.5 (2016), pp. 671–681.

[8] Benjamin Gorissen et al. "Elastic Inflatable Actuators for Soft Robotic Applications". In: *Advanced Materials* 29.43 (2017), p. 1604977. DOI: 10.1002/adma.201604977. eprint: https://onlinelibrary.wiley.com/doi/pdf/10.1002/adma.201604977. URL: https://onlinelibrary.wiley.com/doi/abs/10.1002/adma.201604977.

[9] Karen N Gregorczyk et al. "Effects of a lower-body exoskeleton device on metabolic cost and gait biomechanics during load carriage". In: *Ergonomics* 53.10 (2010), pp. 1263–1275.

[10] Kåre B Hagen and Ola Thune. "Work incapacity from low back pain in the general population". In: *Spine* 23.19 (1998), pp. 2091–2095.

[11] Damian Hoy et al. "The epidemiology of low back pain". In: *Best practice & research Clinical rheumatology* 24.6 (2010), pp. 769–781.

[12] Diederik P. Kingma and Jimmy Ba. *Adam: A Method for Stochastic Optimization.* 2014. arXiv: 1412.6980 [cs.LG].

[13] K. Kunze et al. "Towards Recognizing Tai Chi - An Initial Experiment Using Wearable Sensors". In: *3rd International Forum on Applied Wearable Computing 2006.* Mar. 2006, pp. 1–6.

[14] J. Kuschan et al. "PowerGrasp: Concept for a novel Soft-Robotic Arm Support System". In: *ISR 2018; 50th International Symposium on Robotics.* June 2018, pp. 1–6.

[15] Heike Leutheuser, Dominik Schuldhaus, and Bjoern M. Eskofier. "Hierarchical, multi-sensor based classification of daily life activities: comparison with state-of-the-art algorithms using a benchmark dataset". In: *PloS one* 8.10 (2013), e75196. DOI: 10.1371/journal.pone.0075196.

[16] Nikolaos Maniadakis and Alastair Gray. "The economic burden of back pain in the UK". In: *Pain* 84.1 (2000), pp. 95–103.

[17] Abdulmajid Murad and Jae-Young Pyun. "Deep Recurrent Neural Networks for Human Activity Recognition". In: *Sensors* 17.11 (2017). ISSN: 1424-8220. DOI: 10.3390/s17112556. URL: https://www.mdpi.com/1424-8220/17/11/2556.

[18] Francisco Javier Ordóñez and Daniel Roggen. "Deep Convolutional and LSTM Recurrent Neural Networks for Multimodal Wearable Activity Recognition". In: *Sensors* 16.1 (2016). ISSN: 1424-8220. DOI: 10.3390/s16010115. URL: https://www.mdpi.com/1424-8220/16/1/115.

[19] Thomas Plötz, Nils Y. Hammerla, and Patrick Olivier. "Feature Learning for Activity Recognition in Ubiquitous Computing". In: *Proceedings of the Twenty-Second International Joint Conference on Artificial Intelligence - Volume Volume Two.* IJCAI'11. Barcelona, Catalonia, Spain: AAAI Press, 2011, pp. 1729–1734. ISBN: 978-1-57735-514-4. DOI: 10.5591/978-1-57735-516-8/IJCAI11-290. URL: http://dx.doi.org/10.5591/978-1-57735-516-8/IJCAI11-290.

[20] Panagiotis Polygerinos et al. "Soft Robotics: Review of Fluid-Driven Intrinsically Soft Devices; Manufacturing, Sensing, Control, and Applications in Human-Robot Interaction ". In: *Advanced Engineering Materials* 19.12 (2017), p. 1700016. DOI: 10.1002/adem.201700016. eprint: https://onlinelibrary.wiley.com/doi/pdf/10.1002/adem.201700016. URL: https://onlinelibrary.wiley.com/doi/abs/10.1002/adem.201700016.

[21] Fernando Moya Rueda et al. "Convolutional Neural Networks for Human Activity Recognition Using Body-Worn Sensors". In: *Informatics* 5 (2018), p. 26.

[22] Skoda Data Set. http://www.har-dataset.org/doku.php?id=wiki:dataset. Accessed: 2019-07-26. 2008.

[23] Muhammad Shoaib et al. "Fusion of Smartphone Motion Sensors for Physical Activity Recognition". In: *Sensors (Basel, Switzerland)* 14.6 (2014), pp. 10146–10176. ISSN: 1424-8220. DOI: 10.3390/s140610146.

[24] Mikołaj Walukiewicz. "Aufbau und Inbetriebnahme eines Sensoranzugs für die messtechnische Erfassung von Posen und Bewegungen am Menschen". Bachelor Thesis. Germany: Technical University of Berlin, 2019.

IOP Conf. Series: Materials Science and Engineering 1140 (2021) 012021 doi:10.1088/1757-899X/1140/1/012021

Laboratory stand for multi-axis control of stepper drives via EtherCAT fieldbus

Marcin Paprocki, Krystian Erwinski

Departament of Automatics and Measurement Systems, Institute of Engineering and Technology, Nicolaus Copernicus University in Torun, Grudziadzka 5/7, 87-100 Torun, Poland

E-mail: marcin.paprocki@umk.pl

Abstract.

In this paper a laboratory stand is presented that was developed for a multi-axis machine stepper drive control system. Each stepper drive utilizes an EtherCAT communication module based on a development board with software developed by the authors. The modules are used to exchange process data with a higher level controller. The EtherCAT communication standard is described including it's utilization for stepper drive control. A functional description of the developed communication module is also presented. The aim of the Laboratory stand is to conduct research of cyclic communication (1 milisecond or less) and synchronization between multiple drive slave devices and a supervisory controller. The main focus of the presented results was to determine EtherCAT communication cycle jitter on the developed modules. The main aim of the research was to verify whether the data exchange via EtherCAT bus has an impact on the synchronous operation of the multi-axis system.

1. Introduction

In many areas of industrial automation, communication buses are commonly used. In recent years, networks based on the Ethernet bus have been increasingly used. A specific version of standard Ethernet is real-time Ethernet, which ensures very low delays in the network data exchange and thus obtaining transmission time determinism. There are many varieties of this Ethernet, the most important are: EtherNet/IP, PROFINET IRT, EtherCAT, POWERLINK and SERCOS III [1, 2]. One of popular standards is EtherCAT. This standard ensures high transmission speed (100 Mbps), short minimum communication cycle time (even 12 µs) and short service time for a single network node (less than 1 µs). Moreover, it implements a distributed clocks mechanism that enables synchronization of network nodes with an accuracy of 1 µs. These features are particularly desirable in systems where synchronized control of drive systems is necessary, e.g. in CNC machines [3–6] and robots [7–10]. For this reason, EtherCAT bus is one of the most widely used in distributed motion control systems.

The article presents the EtherCAT slave device communication module designed for use in stepper drives. The EtherCAT communication standard was discussed along with a description of the slave device operation using the CANopen profile. The structure and operation of the communication module are discussed. A description of the laboratory test stand for multi-axis control of stepper drives was presented along with the research. The presented EtherCAT communication module is an extension of the authors' works in the field of drive control systems using real-time Ethernet [11–13]. The main aim of the research was to verify whether the data exchange via EtherCAT bus has an impact on the synchronous operation of the multi-axis system.

2. EtherCAT bus

There is one master and many slaves on the EtherCAT bus. The master device and slave devices are in the same network segment. A single network with one master device can support up to 65,535 slaves. EtherCAT uses Fast Ethernet standard (100 Mbps, UDP/IP) in full-duplex mode - one pair of wires (TX) is used to transmit data from the master to the slaves and the other (RX) to transmit in the opposite direction. A line, tree, star or ring topology can be used. The data section in an EtherCAT frame consists of datagrams, each of which contains data for individual slaves. The data is transferred between master and slaves in the form of process data (PDO - Process Data Object). Each PDO contains information that allows to verify that the data has been correctly read and processed. The standard Fast Ethernet network controller (network card) can be used in master devices (e.g. a PC). Unlike master devices, slave devices require to use a specialized controller in their construction.

The master device is responsible for initiating the data transfer to the slave nodes. First, the master device sends one Ethernet frame that reaches all slaves. The first slave then extracts the input data from the frame and inserts the output data into the frame. The modified frame is sent to the next slave again. The end slave device sends the modified frame to the master device. From the master's point of view, only one data frame is sent and received. The EtherCAT controller of the slave device at the time of receiving the frame only reads data from the frame and writes data to the frame. This is done "on the fly" - without the frame being analyzed and processed by the controller. The actual data processing by the protocol stack and the preparation of the feedback data takes place later. As a result, the delays introduced by the network node are minimal (several nanoseconds).

The CSMA/CD mechanism is not used in EtherCAT. The data transfer time is completely controlled by the master which should be capable of real-time deterministic operation to ensure cyclical data transfers with low latency and low jitter. In the EtherCAT bus, synchronization of network nodes is possible. For this purpose, the distributed clock mechanism (DC - Distributed Clock) is used. This mechanism allows all network nodes to be synchronized with an accuracy of less than 1 microsecond. In this mechanism, the clock of one of the slave devices in the network is a reference clock to which the clocks of the other devices are periodically synchronized. The process of clock synchronization measures network propagation delays during network initialization. Thanks to this mechanism, all devices on the EtherCAT network can be triggered simultaneously. This is important in multi-axis control, where the individual drives must move synchronously to each other.

3. EtherCAT slave communication module based on LAN9252

The EtherCAT protocol defines the physical, data link, and application layers of the OSI model. These layers are implemented differently for master and slave devices. The slave devices are equipped with a dedicated ESC (EtherCAT Slave Controller) to ensure very fast "on-the-fly" processing of data from the Ethernet frame.

In the physical layer, slave devices use the standard RJ45 (8P8C) connector and standard Ethernet PHY (physical layer) systems. For the data link layer, ESC controllers are used. Authors use the LAN9252 [14] chip from Microchip Technology, which has built-in Ethernet PHY circuits. The application layer is most often implemented by software in microprocessor systems. The authors implemented the EtherCAT application layer as dedicated software on a 32-bit microcontroller - STM32F411CE [15] from STMicroelectronics. Figure 1 shows a diagram of an EtherCAT slave communication module based on LAN9252.

The data exchange with the application layer on the microcontroller takes place using the local ESC interface, which in the EtherCAT standard is called PDI (Process Data Interface). From the point of view of the STM32F411CE microcontroller, the ESC chip is treated as an external memory that has the ability to generate interrupts. Through PDI, it is possible to read and write appropriate memory regions in ESC - data transmission from and to the microcontroller. In this work, the PDI in the SPI mode was used in LAN9252 (ESC) (Figure 1). The information exchange between devices through the SPI begins when the microcontroller sets the state "0" on the SCS_N line. Communication via PDI consists of the address sending phase and the data transfer phase to a given address in ESC. In the address phase, the master transmits the address to be accessed and the command. In the data phase, the output data sent by the slave device is presented and the input data is sent by the master device.

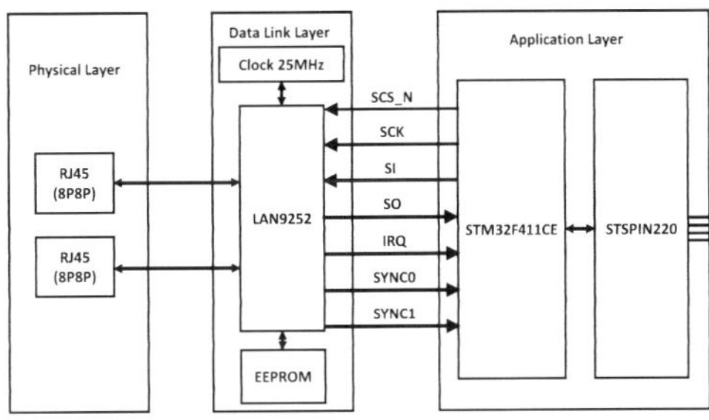

Figure 1. Block diagram of ESC LAN9252 connection with STM32F411CE microcontroller via SPI (PDI)

The PDI is responsible for the synchronization of work between the ESC and the microcontroller. There are three timing modes in the EtherCAT standard itself. They are FreeRun, SM-Synchron and DC-Synchron respectively. In FreeRun mode the slave device does not generate the IRQ interrupt signal via ESC. Data exchange via PDI is controlled and triggered directly by the microcontroller. In the SM-Synchronization mode, the synchronization takes place as a result of receiving an Ethernet frame by the slave device. The ESC will generate the IRQ interrupt signal to trigger the microcontroller to handle the process data. In the DC-Synchronization mode, synchronization occurs as a result of a state change on the SYNC0 line, controlled by ESC. The SYNC0 signal is generated simultaneously on all devices of the EtherCAT network. The synchronization of this signal between all slave devices was due to the operation of the distributed clocks mechanism.

There are several variants of EtherCAT application layer protocols available. The most commonly used protocol is CANOpen over EtherCAT (CoE) [16]. The class of the slave device (e.g. I/O module, electric drive, position transmitter) is defined by the so-called device profile. This profile defines an OD (Object Dictionary) structure that maps the process and configuration variables of a device. This enables the creation of a standard interface between the motion controller application and the communication layer (application layer), independent of the type and manufacturer of the master controller. In the presented communication module, the authors implemented a dedicated CoE profile designed to support a single stepper drive by each module.

4. Laboratory stand for multi-axis control of stepper drives via EtherCAT fieldbus

In order to verify the correct operation of the communication module, a laboratory stand was built with four STM32F411CE microcontroller systems, ESC LAN9252 modules and STSPIN220 stepper motor driver [17] with stepper motor (diagram of a one module in Figure 1). The EtherCAT stack application layer implemented on the STM32F411CE microcontroller uses the

Beckhoff EtherCAT Slave Stack Code Tool v.5.11 library and the authors' own software, needed for the operation of the PDI (SPI). A PC with Windows 10 with TwinCAT v.3.1 software was used as the master controller (master device). The TwinCAT application uses a real-time micro kernel that ensures that the software runs with time determinism. In the TwinCAT application, the control software has been implemented into a laboratory stand (multi-axis control). The diagram of the laboratory stand with the four-channel oscilloscope (RIGOL MSO5074) is shown in Figure 2.

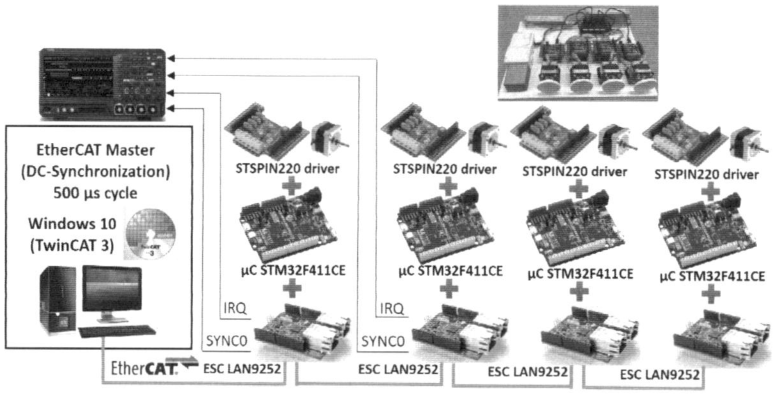

Figure 2. Diagram of a laboratory stand with research equipment

The correct operation of EtherCAT communication modules on a laboratory stand was tested during its operation (setting positions for stepper motors). All four modules exchanged information via the EtherCAT bus in the DC-Synchronization mode. The cycle time was 500 µs. The cyclicity of SYNC0 and IRQ signal generation was measured for two communication modules. The IRQ signal informs about the availability of data on the SPI bus (data from PC), and the SYNC0 signal signals the start of operation of the software controlling the STSPIN220 controller. The test results are shown in the Figure 3.

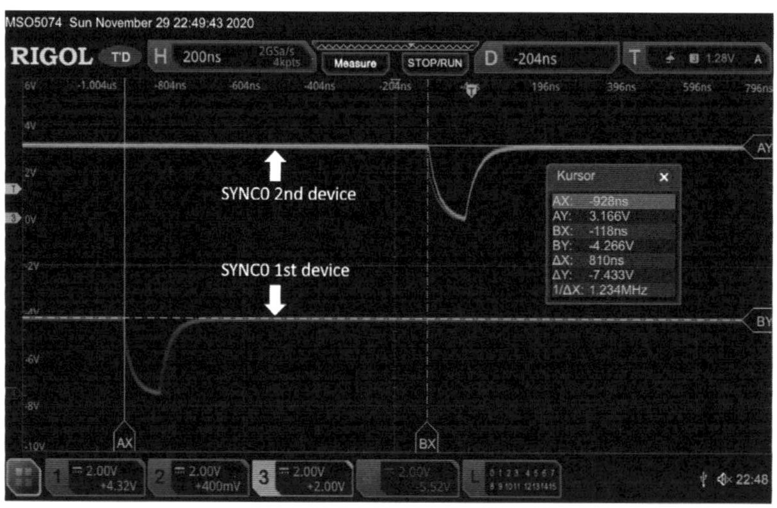

Figure 3. Time dependence measurements for SYNC0 signals

As expected, the SYNC0 signal change for both modules takes place every 500 µs, which confirms the correct operation of the communication system. The shift of the SYNC0 signal between the first module and the second module is 810 ns. Similar results are obtained for the IRQ signal for both modules. The time difference is 850 ns. The research confirms that the synchronization of the stepper drives operation at the ESC level (physical layer and data link layer) is very good (¡1 µs). Therefore, in some experimental solutions, the challenge may be to implement the software application layer of drive module. Taking into account the fact that this layer is most often implemented in a microcontroller, on which the drive control algorithm also works, it is important to maintain the timing of both tasks (communication stack and control algorithm).

At the laboratory stand, the data exchange cycle between communication modules was 500 µs. The EtherCAT standard provides for even shorter communication cycles (100 µs and less), where synchronization is likely to be kept (further research required). For such short communication cycles, communication modules can be used in servodrive solutions. This opens up new possibilities in which in the control processors of servodrives can only implement the basic functionality of such a drive (e.g. torque controller or algorithms for direct control of the power modules). The control systems themselves can then be placed in a supervisory system (e.g. at EtherCAT master system). The authors consider this an interesting direction for further research that will be continued in the future. In particular, when the communication module can be implemented in experimental drive systems for testing advanced control methods (including also for multi-axis servo systems). They can be, for example, adaptive, predictive, cross coupled, repeatable and other regulators [18, 19].

5. Conclusion

The article presents the EtherCAT bus communication module dedicated to electric drives. The module was implemented using the Microchip LAN9252 integrated circuit and the STM STM32F411CE microcontroller. The EtherCAT bus stack with the CANOpen application layer protocol has been implemented in the microcontroller. The module is designed for electric drives under development. This applies to both commercial devices as well as drives developed as part of research.

References

[1] Zurawski R 2014 *Industrial communication technology handbook* (CRC Press)

[2] Wilamowski B M and Irwin J D 2016 *Industrial communication systems* (CRC Press)

[3] Martinov G, Kozak N and Nezhmetdinov R 2017 *2017 International Conference on Industrial Engineering, Applications and Manufacturing (ICIEAM)* (IEEE) pp 1–4

[4] Jia H, Yao P, Li B and Tian X 2017 *2017 Chinese Automation Congress (CAC)* (IEEE) pp 2842–2846

[5] Grigoriev S N and Martinov G M 2016 *Procedia CIRP* **41** 858–863

[6] Li B, Lin H, Zheng L, Sun S and Yin Z 2016 *2016 IEEE Advanced Information Management, Communicates, Electronic and Automation Control Conference (IMCEC)* (IEEE) pp 795–801

[7] Liu Z, Liu N, Zhang T, Cui L and Li H 2015 *2015 IEEE International Conference on Information and Automation* (IEEE) pp 1708–1713

[8] Zhang G, Ni F, Li Z and Liu H 2018 *2018 IEEE International Conference on Mechatronics and Automation (ICMA)* (IEEE) pp 2169–2174

[9] Delgado R, Kim S Y, You B J and Choi B W 2016 *2016 13th International Conference on Ubiquitous Robots and Ambient Intelligence (URAI)* (IEEE) pp 710–715

[10] Yoon D K, Kim S Y, Cho J, Lee K K and You B J 2014 *2014 11th International Conference on Ubiquitous Robots and Ambient Intelligence (URAI)* (IEEE) pp 253–256

[11] Paprocki M, Wawrzak A, Erwiński K, Karwowski K and Kłosowiak M 2017 *Measurement Automation Monitoring* **63**

[12] Erwinski K, Paprocki M, Grzesiak L M, Karwowski K and Wawrzak A 2012 *IEEE Transactions on Industrial Electronics* **60** 628–636

[13] Paprocki M, Wawrzak A, Erwiński K and Kłosowiak M 2018 *Mechanik* **91** 299–303

[14] Microchip Tech. 2015 *2/3-Port EtherCAT Slave Controller with Integrated Ethernet PHYs*

[15] STMicroelectronics 2017 *Data Sheet - STM32F411CE* rev. 7

[16] Chen Y, Chen H, Zhang M and Li Y 2010 *2010 IEEE International Conference on Intelligent Computing and Intelligent Systems* vol 2 (IEEE) pp 67–70

[17] STMicroelectronics 2019 *Data Sheet - STSPIN220* rev. 4

[18] Szczepanski R, Tarczewski T and Grzesiak L M 2019 *Applied Soft Computing* **83** 105644

[19] Szczepanski R, Tarczewski T and Grzesiak L 2020 *Bulletin of the Polish Academy of Sciences. Technical Sciences* **68**

IOP Conf. Series: Materials Science and Engineering 1140 (2021) 012022 doi:10.1088/1757-899X/1140/1/012022

Automation of Life Data Analysis Processes

Jonathan Krämer[1], Gregor Thiele[2], Theresa Johanni[2] and Jörg Krüger[1]

[1] Department for Industrial automation at the Institute for Machine Tools and Factory Management, Technical University of Berlin

[2] Department for Process automation and Robotics, Fraunhofer Institute for Production Systems and Design Technology

E-mail: jonathan_kraemer@gmx.de

Abstract. Life data analysis is an effective statistical process to gain information on the reliability of technical facilities. The Weibull analysis is a widely-used method for modeling the probabilities of different functional failures. For its application in modern manufacturing and maintenance processes, an automation of this analysis is expedient. An exemplary process of data collection and fitting of a parametric model with continuous time scale is analyzed in this article. Its automation with different statistic software tools is described and comparatively evaluated. With the ambition of minimizing manual effort in the process, a highly automated solution was developed in this research. We describe this self-implemented solution in "R" and justify that this procedure automatizes the process more than the examined commercial implementations. The here presented work proves to be a practical real-time assistance, right now applied in aircraft maintenance.

1. Introduction

1.1. Life data analysis and Weibull distribution

A technical device fulfills its demanded functionality only for a certain time period, from its launching date to its end-of-life (EOL). For mounted parts the EOL is often caused by a malfunction and comes with the substitution of the component. Therefore, failure data is collected and analyzed to investigate the reasons and estimate the time of failure for identical or similar constructions. Life data analysis is the statistical field which provides the methods to model the failure data and gain insight on the behavior of technical facilities.

To get a more concrete picture of the kind of data this work is tackling, one can imagine this everyday world example: Lets say an overhaul worker works in aviation industry. To be able to better predict the EOL of all the technical devices he is responsible for, it is natural to collect some data and try to gain insights on this basis. So, if for example an engine fails and the worker identifies destroying corrosion on the engine blades, he makes a note to his failure data repository with the engine's specification and its individual life period (measurable e.g. in hours or hot-cold-cycles) and the corrosion as specific failure cause. All this collected data has to be analyzed by means of statistical methods. If done manually, this analysis is a time-consuming task. Especially if the workperson is not that well trained in these methods, it would be advantageous to have this procedure as automatized as possible. As a result, the workperson is able to reliably predict the EOL of engines due to corrosion and replace them in advanced at a suitable moment without endangering the function of the product during employment, meaning e.g. an aircraft during its flight.

The above-mentioned statistical methods build the scientific field of life data analysis. Here, the Weibull distribution is often used to model failure data. In 1951, Wallodi Weibull promised that this distribution serves for a wide variety of data [13]. In fact, it proves to be state-of-the-art for life data analysis on all kind of data from different industrial sectors. Especially, it is capable to approximate all typical failure frequencies of a technical facility during its whole utilization time, representing different modes of failure of a technical facility. Moreover, it serves to mimic other commonly used distributions like the exponential distribution. For more information on the Weibull distribution see the references given in Sec. 1.3.

1.2. Outline of this article

After the above given motivation and the short introduction to the theoretical background of this work, Sec. 1.3 summarizes the state-of-the-art of science and technology this research is based on. Sec. 2 describes the examined life data analysis process, its requirements and outputs. The considered software tools for the automation are introduced. Sec. 3 covers the realization of the life data analysis using different software solutions. Sec. 4 gives the conclusion and an outlook.

1.3. Related Work

Our understanding of life data analysis is based on [3], [8] and [10]. For more engineering orientated work, see [1]. The usage of software for life data analysis is described in [12], [4], [11] and [2]. One specific solution to automate a life data analysis, stepping in the same direction as this work, is mentioned in [9]. This solution motivated the alternative of self-implementing the process in the programming language "R" using the package [5] as discussed in Sec. 3. More information on this self-written implementation and its evaluation can be found in [6].

2. Concept

This section presents the industrial process that was automated in the course of this survey. We present features of the data that was provided for analysis, expound the requirements of the specific process and formulate the steps of its automation and its expected outputs. For more detailed information on the calculations that are part of these steps, see the listed literature for the process of life data analysis mentioned in Sec. 1.3.

2.1. Structure of the given process

In this work, we discuss an overhaul process of the aviation industry. Especially in the maintaining of aircraft, the reliability of all system components is of paramount importance. Plenty of technical entities are measured frequently, resulting in a comparatively rich data pool. Still, the hereon based analysis process does not constitute a fully variable but custom-designed compilation of the methods that are summarized under the name of life data analysis. Also, it is expected to create a specific output: a parametric model with continuous time scale that fits the available set of data points.

The life data analysis starts with sorting the data and calculating the median ranks. On this basis, the probability of failure for each data point is calculated. Subsequently, for each cumulative distribution function (CDF) we have to configure a probability paper (coordinate system) in which the respective distribution forms a straight line. Even this simple assistance by predefined probability papers helps automating tasks in overhaul processes.

After defining a suitable (still empty) plot, the data points used for the regression are drawn in the graph of the distribution followed by the regression line respectively. Fitting can be performed using two different methods which should be available in the software tool. The median rank regression has a graphical reference as it tries to minimize the residuals x on y or y on x. Alternatively, one can estimate the maximum likelihood function to find the most likely parameters for the distribution. After this calculation of the distribution parameters, the distribution can be plotted into the probability plot. For a deeper understanding of the fitted distribution, we add confidence intervals for point estimations as well as confidence bounds across the whole distribution. The calculation method of those intervals depends on the used method for the model estimation. Preferred methods for interval estimation are maximum likelihood bounds, fisher matrix bounds and beta-binomial bounds. In this specific application, extrapolation of the resulting model serves the prediction of future effects. On the basis of this model, different output values can be derived, as discussed in the following Sec. 2.2.

2.2. Structure of the result
The procedure of fitting the model to the data is described above. Here, we list the results and outputs that are needed for the user interface. The data and distribution should be visualized in a probability plot in an interactive way. This allows the user to retrieve the exact data and distribution parameters. Moreover, the region of confidence needs to be visible in the plot. For more detailed information about the model and the failure behavior, special values like the distribution parameters, the goodness-of-fit indicators and the important B-lifes should be accessible as well. For a user with a special interest in the value of a certain failure time or percentage of failure, a corresponding calculator is advantageous. To get a perspective of the model fit a comparison of all implemented distributions is useful.

2.3. Process requirements
Oriented on the considered use-case of the aviation industry with the described process structure we deduce a list of requirements that the automated solution must fulfill. They are divided into the categories technical/statistical and automatizing. Referring to these requirements, we assess the software tools according to their adequacy for implementation in the aimed life data analysis. Concluding from experience, the default distribution to fit the data should be the Weibull distribution. Still, different distributions that are additionally relevant in the practice of life time analysis should be available for modeling as well.

3. Practical realization
3.1. Software
Available software solutions for automating life data analysis are listed in [7]. This extensive table particularly provides the evaluation of the requirements given in Tab. 1 for each candidate. Like that, it helps with the decision which program is most suited to automate a given process. The results of the automation of the process with the most promising solutions for our case are laid out in the following.

3.2. Evaluation and Interpretation
Here, the automation of the given industrial process described in Sec. 2 with the different suitable software solutions of the previous Sec. 3.1 is presented. Four different solutions are promising to fit the requirements of our process. The commercially available tools "Minitab", "Weibull++" and "Relyence Weibull" and a self-written implementation in "R". For a detailed description of the implementations of the process with these solutions, see [6] or the related work in Sec. 1.3. The applications and results of the automation of the process are expounded and compared as displayed in Tab. 2. For all four solutions advantages and disadvantages are gathered for these aspects: data input, possibilities for individual configurations and availability of the process results. In the given table each row represents a software solution and the columns are the sections of interest.

Table 1. Requirements

Category	Criterion
Automating	graphical user interface
	structured results that are easy to understand
	user-friendly but extensive configurability
	maximized automation
Technical	database communication
	probability plot with confidence region
	processing of censored data: left, right and interval censored
	modelling the data with different distributions: exponential, normal, log-normal, logistic, log-logistic, smallest extreme value and Weibull distribution
	methods for estimating distribution parameters: median rank regression and maximum likelihood estimation
	methods for estimating confidence intervals: beta-binomial bounds, likelihood-ratio bounds, fisher-matrix bounds
	estimation quality of the different distributions based on goodness-of-fit parameters

Table 2. Interpretation

	data input	manipulation	process output
Minitab	+ Database connectivity − manual entry	+ many options manipulatable + three methods for analysis − no further option for automation	+ detailed information + used methods and options + easy export of results − no calculator available
Weibull++	+ request for censoring of data − no import option	+ options changeable when looking at results − not well described	+ used methods and options + easy export of results + calculator − plot and values representable only separately − default plot w/o confidence region
Relyence Weibull	− import only from software from Relyence − manual entry	+ options changeable when looking at results	+ interactable plot + easy export + plot is main output
R-solution	+ import from .csv-file + database connection programmable − no direct manual entry	+ options changeable at all time + default options given	+ interactable plot + overview over all available distributions + information over used methods and options + plot is main output − no export option

4. Conclusion and Outlook

This last section presents a conclusion of the results of this work. It also lists interesting research questions that came up during this work but were not pursued.

4.1. Contribution

From all discussed tools, the self-implemented solution in "R" reaches the highest degree of automation. This was achieved by a suitable preparation of the results which makes them understandable even for inexperienced users and by the generic design appropriate for many tasks in overhaul work.

A manual data import or an option to export results is not given, but just because this was not part of our specific requirements. Such interfaces can be added easily. This solution allows to perform the entire procedure of life data analysis by only one click. The automation is especially striking with regards to time-consuming tasks as the analysis of the data differentiated according to specific failure causes as described in our everyday example in Sec. 1. The software is designed user-friendly and easy to operate also for users with little statistical background, i.e. shop floor staff. Because of the focus on overhaul tasks, our solution is not designed for statisticians who may require more flexibility in the analysis. Despite of the extensive functionality, the results still need to be checked and interpreted manually.

4.2. Further Potentials

To further validate our results, it would be interesting to implement the different solutions in real overhaul plants and evaluate the impact of this innovation as well as the feedback on the user experience with the different software tools. For every use-case, the interested engineers need to evaluate whether it is more efficient to utilize commercial toolboxes or to implement own approaches as demonstrates in this article.

References

[1] Dr. R. B. Abernethy. *The New Weibull Handbook*. Dr. R. B. Abernethy, 1993. ISBN: 0-9653062-3-2.

[2] R. S. Amberti D. Kenett and S. Zacks. *Modern Industrial Statistics*. Wiley Academic, 2014. ISBN: 978-1-118-45606-4.

[3] L. A. Escobar and W. Q. Meeker. *Statistical Methods for Reliability Data*. Wiley-Interscience, 1998. ISBN: 978-0-471-14328-4.

[4] S. M. Gupta and K. K. Pochampally. *Reliability Analysis with Minitab*. Taylor & Francis Ltd., 2016. ISBN: 978-1-4987-3767-8.

[5] T.-G. Hensel. *cran/weibulltools*. 2019. URL: https://github.com/cran/weibulltools (visited on 05/24/2020).

[6] J. Krämer. "Automatisierte Lebensdatenanalyse mittels der Weibull-Verteilung zur Instandhaltung von Triebwerken". Technische Universität Berlin, 2020. DOI: https://dx.doi.org/10.14279/depositonce-10996.

[7] J. Krämer. *Software solutions for life data analysis*. DOI: https://dx.doi.org/10.14279/depositonce-11002.

[8] J. F. Lawless. *Statistical Models and Methods for Lifetime Data*. John Wiley & Sons, 2002. ISBN: 978-0471372158.

[9] K. McLain and R. Warren. "Automated Reliability Life Data Analysis of Missles in Storage and Flight". In: *Annual Proceedings on Reliability and Maintainability Symposium*. IEEE, 1990. DOI: 10.1109/ARMS.1990.68006.

[10] W. B. Nelson. *Applied Life Data Analysis*. John Wiley & Sons, 1982. ISBN: 9780471094586.

[11] K. Stephens. *Reliability data analysis with Excel and Minitab*. ASQ Quality Press, 2012. ISBN: 978-0-87389-831-7.

[12] A. Tobias and D. C. Trindade. *Applied Reliability*. Taylor & Francis Inc, 2009. ISBN: 978-1584884668.

[13] W. Weibull. "A Statistical Distribution Function of Wide Applicability". In: *ASME Journal of Applied Mechanics. Transactions of the American Society of Mechanical Engineers* 18 (1951), pp. 293–297.

IOP Conf. Series: Materials Science and Engineering 1140 (2021) 012023 doi:10.1088/1757-899X/1140/1/012023

Universal identification and control of industrial manufacturing equipment as a service

V. Tessaro[1], **A. Vick**[2] **and J. Krüger**[3]
[1] Technical University Berlin, Berlin, Germany
[2] Process Automation and Robotics, Fraunhofer IPK, Berlin, Germany
[3] Industrial Automation Technology, Technical University Berlin, Berlin, Germany

E-mail: axel.vick@ipk.fraunhofer.de

Abstract. Rapid prototyping as well as retrofitting and digitization of legacy manufacturing equipment often needs design and application of closed loop controllers. The analysis and modeling for such systems like energy-conversion or material transport devices is labor-intensive and needs process understanding. This paper presents a universal approach of identification and closed loop control of arbitrary linear systems delivered through web services using OPC UA as a standardized industrial communication interface. The identification service was used to model the dynamics of a 6-DOF industrial robot and a laboratory-scale water plant containing two separately controllable pumps. The control service successfully controlled the robot's linearizable axes and the water plant's pumps by using their respective identified state-space models. To evaluate the performance of the controllers in terms of stability, accuracy, and time response, target trajectories and disturbances like signal noise and latency in communication were introduced. Simulation and laboratory experiments show promising results for control of diverse systems with varying time-constants and imply broad applicability. So, this paper brings a guideline to efficiently implement model predictive control in manufacturing.

1. Introduction

With the transformation to industry 4.0 the automation and digitization of factories and production systems is increasing constantly. The development turning away from monolithic control structures and towards high connectivity in distributed systems opens up new opportunities of using cloud services. Aside from computing and storage capacities, cloud services can also offer complex functionalities like the control of industrial robots and other equipment [1].

Cloud service based control structures in conjunction with services for system identification have the potential to make the integration of conventional industrial equipment – independent of type or brand – and new prototypes into a factory cloud possible or easier. Due to the cloud services' universal applicability labor-intensive and knowledge requiring model synthesis and optimization of control algorithms can be omitted. Thus, research potential arises concerning the feasibility of the universal approach as well as the applicability to unknown dynamic systems. In this paper we focus on linear or linearizable systems for a start.

2. Related works

The outsourcing of computationally demanding functionalities to a cloud can be attractive and is researched intensively. Such a task with a high requirement for computing capacity is image processing. Rudorfer et al. [2] designed a cloud-service based system architecture within which image processing is offered as a web service. From their experiments they conclude the possibility of fast development of new image processing applications. Due to the usage of the communication protocol OPC UA different modular services from diverse sources can be employed. Though, the authors found the services to lack real-time capability.

Wassermann et al. [3] also designed a cloud-service based system architecture with the objective to offer path planning, robot control, virtualization and hardware services as modular, independent, relocatable web services. They demonstrated their approach with a pick-&-place task in an environment containing obstacles. The object recognition and the path planning were executed correctly and the authors found the processing time to be acceptable. For providing universal applicability they proposed the usage of OPC UA as a communication interface.

In a network environment with varying latency and band width Kilinc et al. [4] compared the performance of two cloud-service based controllers, a model predictive and a P-PI-controller, when controlling a 6-DOF industrial robot. Therefor, the model predictive control (MPC) was modified so that a once computed input sequence could be further used in the case of connection failures. The MPC was based on a state-space model that was derivated analytically and the controllers' parameters were tuned manually. In experiments the MPC turned out to be more stable and faster for increasing latencies, tested up to 160 ms. Additionally, it showed less overshoot and a smaller steady-state error than the P-PI-controller.

Instead of the derivation of a robot's state-space model Briese et al. [5] investigated the possibility to use a non-model based approach, the modified Switching Active Disturbance Rejection Control, for a cloud-service based, distributed robot control in a network with jitter. In experiments they achieved the stabilization of all six robot axes for cycle times up to 100 ms but physical disturbances could not be compensated completely and the stabilization at high cycle times could only be achieved with loss of performance. Though, one of the advantages of the authors' approach is the small number of parameters to tune and therefore an easier model synthesis than when using first-principle methods.

3. Methods for universal identification and control

3.1. OPC UA

For the communication between servers and clients we chose the OPC UA standard. OPC UA is a TCP/IP-based machine-to-machine protocol for industrial server-client applications and it offers access options for data and remote procedure calls on servers. We used the binary protocol version because of its smaller overhead and therefore higher speed and lower resource consumption.

Based on the principle presented by Vick at al. [1], we introduced an additional layer between logical functionalities and communication. We built all services as automation services that inherit from an OPC UA adapter class and contain an automation class object. The adapter class contains methods for communication whereas the automation class provides the more complex functionalities. Thus, we achieved a separation of these two functionalities which allows easier reuse of both. We built these classes using the python-opcua [6] module for our servers implemented in Python and using the open62541 library [7] for our servers in C++.

3.2. Subspace Identification

A state-space model is a transfer system's representation containing the relations of input, state, and output variables as first order differential equations. Its system specific matrices are obtained by performing a subspace identification on an input-output data set. With this data subspace identification methods try to find a subspace spanned by the columns of an extended observability matrix. One well known method is Multivariable Output Error State Space (MOESP) which is used in this paper.

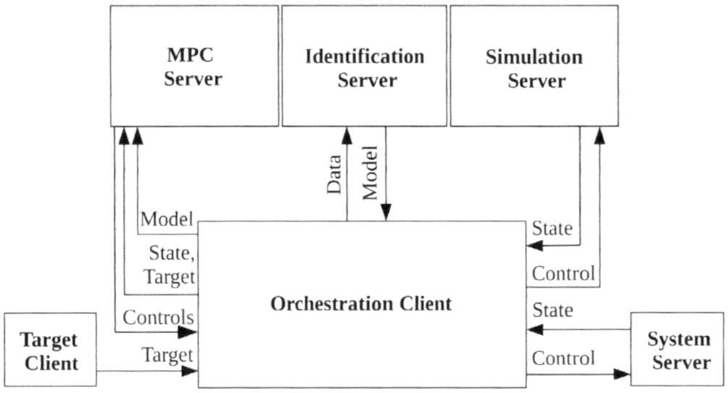

Figure 1. System architecture with communication between servers and clients.

The Python module SIPPY [8] provides MOESP and other methods for identifying dynamic SISO and MIMO systems based on their input-output data. If the system to be identified is a robot-like system, e.i. if the respective numbers of first and second order states and the number of inputs are equal, then we assume one independent system per axis, identify these systems individually and eventually combine them to the overall state-space matrices. Otherwise, if we cannot make an assumption about the system then the system order and the model structure are chosen automatically by the means of different models' Bayes Information Criterion. Subsequently, we transform the matrices so that the output matrix is an identity matrix and we discretize the state-space model.

3.3. Model Predictive Control

A discrete state-space model can be used in model predictive control (MPC) of processes for estimating future outputs. These future outputs depend on a sequence of future and past inputs. The deviation of the estimated outputs from a target trajectory and the control effort up to a control horizon are weighted in an objective function. This function is then minimized by iteratively modifying the input sequence with respect to the input limitations. Only the first signal of the input sequence is used to control the system and all other inputs are discarded because finally the control horizon is shifted one step to the future and the control cycle restarts.

Vick et al. [9] extended the MPC principle considering distributed systems with varying latencies in communication. Instead of discarding all input signals except the first, all inputs are buffered. This allows to use a shifted input sequence to compensate for the transfer time from the controller to the system as well as to reuse a once computed input sequence in case of disturbances of the cycle time or communication failures.

4. System Architecture

We designed a cloud-based system architecture where every service's task is processed independently of other services. The orchestration client [2] cycles through the process logic and forwards service's outputs to other services and thereby triggers server-side actions. The services and the orchestration client as well as their exchanged data are depicted in figure 1.

The *system service* is closest to the controlled system and provides access to its hardware e.g. actuators controlled by inputs and sensors outputting the current state. Additionally, there is a *simulation service* that can simulate such a system and its reactions to inputs. The *identification service* identifies a state-space model based on input and output data. The *simulation service* and the *MPC service* have to be initialized with this model. When provided the current state of the system, the MPC server can then compute an input sequence to reach a target state. Target states, trajectories and input sequences can be generated by the *target client*.

Table 1. Average steady-state errors, overshoots and rise times of the robot axes 1, 4, 5, and 6 and the water plant (W).

Achse	1	4	5	6	W
Steady-state error	-0.4%	-3.9%	-0.9%	-2.1%	0.2%
Overshoot	5.4%	1.9%	14.5%	9.4%	31%
Rise time	1.0 s	2.4 s	0.5 s	0.4 s	0.6%

5. Experimental setup and plan

We tested our approach using a 6-DOF robot and its simulation of which the axis positions and velocities are controlled through the axis motors' currents and a laboratory-scale water plant with two pumps that can be used to control the pressure difference from the system normal. Both systems provide access to their actuators and state variable measurements with the means of OPC UA servers. Firstly, random input sequences are used to record input-output data sets of the systems. Then, based on this data we identify one state-space model per data set. Finally, we use each of these models to control the systems on a target trajectory.

In the simulator experiments, we use two random trajectories, one for identification and one for validation. The identification trajectory is re-recorded with an increasing standard deviation of Gaussian noise ranging from 0 degrees to 3 degrees and with increasing latencies of 7 ms, 14 ms, 35 ms, 70 ms, 105 ms, and 140 ms. We check for model accuracy by simulating the validation trajectory with the identified models and calculating the normalized root mean square error (NRMSE) as a means of comparison of modeled and true state variables.

For the robot and water plant, we record six identification trajectories and one validation trajectory, respectively. For each model parameter, we calculate the parameter mean and certainty. Then we check again for model accuracy by comparing the validation trajectory simulated with the identified models to the real validation trajectory.

In order to test for controller performance, we control the all systems on a target trajectory and calculate the mean steady-state errors, overshoots and rise times.

6. Results and Discussion

In the simulations, the model accuracy for the axis velocities was higher than for their positions and decreased with increasing latency from a mean NRMSE of 0.73 to 0.025 and remains constant for increasing noise. As does the steady-state error which stays in the range from -1.5% to 1% for all tested noise levels and latencies. The overshoot increases from 0% to an average of 2% for growing latencies and remains constant for increasing noise. Rise times and overshoots depend significantly more on the MPC weights than on noise or latencies.

In the robot experiments the axes 1, 4, 6, and 5 turned out sufficiently linear so that a linear state-space model could be identified. Therefore, only those axes were considered in the experiments. At a mean latency of 70 ms, the NRMSE of the velocities was again better than the NRMSE of the positions with mean values of 0.83 and 0.23, respectively. Small-scale bootstrapping showed certainty of the model parameters with all certainties at least one magnitude smaller than their respective parameter's mean except for the parameters that were identified as zero. The steady-state error for axes 1 and 5 was lesser than 1% and for axes 4 and 6 lesser than 4%. The tuning of the MPC weights was more difficult for the real robot's axes leading to higher average values of overshoots as shown in table 1.

The water plant showed to be sufficiently linear in a range from 60% to 120% of pump power. At a mean cycle time of 300 ms the NRMSE of the modeled system pressure was 0.78 with a high certainty of the models' parameters. The steady-state error (see table 1) is lower than the actual value because the it is negative if the pressure differences are lesser than 0.08 Pa and positive if they are greater than 0.08 Pa (see figure 2). Thus, the error values cancel each other

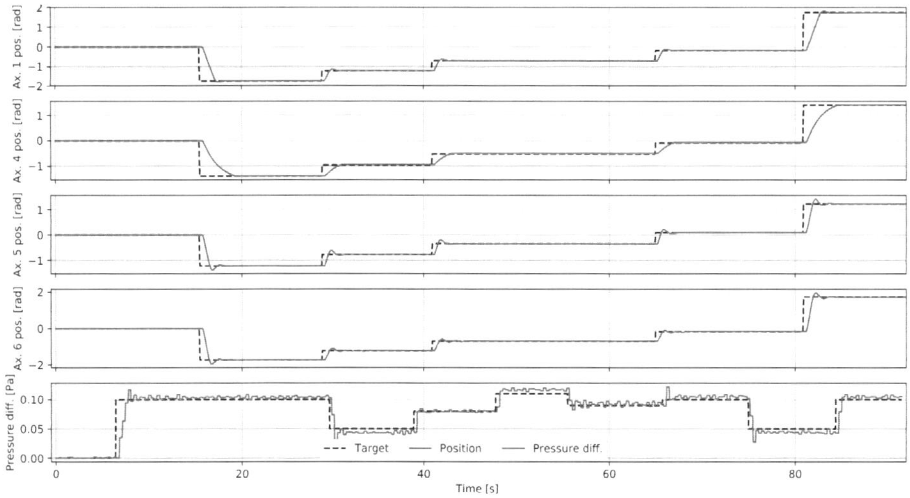

Figure 2. MPC performance experiments for the robot and the water plant.

out on average. After removing outliers considered to be singular non-linearities the average overshoot is still high with 31%. This is because no negative inputs are possible for the water plant and therefore the MPC weights should have been tuned to punish the control effort more.

7. Conclusion

For the MPC applies: the greater the discrepancy between a system and its model the less performant the MPC will be that is based on such model. Yet, the adequate tuning of the MPC weights has a significant impact on the control performance too which we found to exceed the influence of fluctuations in model quality caused by disturbances like noise or latency of the tested magnitudes. We identified the models of structurally and physically diverse systems and successfully controlled them by using these models in an MPC, thereby implying the feasibility of our approach. However, further research is needed to identify universal and automatable methods for tuning MPC weights so that a consistently high control quality for a multitude of systems can be achieved. Likewise, the extension of our linear modeling approach by a non-linear portion would allow for a greater variety of controllable systems.

References

[1] Vick A and Krüger J 2018 Using opc ua for distributed industrial robot control *ISR 2018; 50th Int. Symp. on Robotics* (VDE) pp 1–6
[2] Rudorfer M and Krüger J 2018 *Procedia CIRP* **76** 144–8
[3] Wassermann J, Vonásek V and Vick A 2018 Distributed industrial robot control using environment perception and parallel path planning cloud services *2018 IEEE 23rd Int. Conf. on Emerging Technologies and Factory Automation (ETFA)* vol 1 (IEEE) pp 1263–6
[4] Vick A, Kilinc A and Guhl J 2019 Variable-latency networked p-pi and mpc controller performance for industrial robots *2019 24th IEEE Int. Conf. on Emerging Technologies and Factory Automation (ETFA)* (IEEE) pp 1501–4
[5] Briese C, Vick A and Krüger J 2018 Cloud-based active disturbance rejection control for industrial robots *2018 IEEE 23rd Int. Conf. on Emerging Technologies and Factory Automation (ETFA)* vol 1 (IEEE) pp 559–65
[6] Free OPC-UA Library python-opcua `https://github.com/FreeOpcUa/python-opcua` [Accessed: 23 Jan 2021]
[7] Pfrommer J, Profanter S, Iatrou C P, Palm F and the open62541 community open62541 `https://open62541.org/` [Accessed: 23 Jan 2021]
[8] Armenise G, Vaccari M, Di Capaci R B and Pannocchia G 2018 An open-source system identification package for multivariable processes *2018 UKACC 12th International Conference on Control (CONTROL)* (IEEE) pp 152–157
[9] Vick A, Guhl J and Krüger J 2016 Model predictive control as a service – concept and architecture for use in cloud-based robot control *2016 21st Int. Conf. on Methods and Models in Automation and Robotics (MMAR)* (IEEE) pp 607–12

IOP Conf. Series: Materials Science and Engineering 1140 (2021) 012024 doi:10.1088/1757-899X/1140/1/012024

The method of tribotesting of PVD coated elements in oscillation motion at high temperature

R Michalczewski[1], M Kalbarczyk[1,2], D Maldonado-Cortés[3]

[1] Łukasiewicz Research Network – Institute for Sustainable Technologies, Radom, Poland

[2] Faculty of Mechanical Engineering, Kazimierz Pulaski University of Technology and Humanities in Radom, Poland

[3] Departamento de Ingeniería, Universidad de Monterrey, San Pedro Garza García, Mexico

Corresponding author: remigiusz.michalczewski@itee.lukasiewicz.gov.pl

Abstract. The development of new types of coatings is realised by many scientific and technological centres in numerous countries. To evaluate new material solutions, researchers use various methods with wide range of tribotesters. Consequently, the obtained results are incomparable. The barrier of new technologies development, especially surface modifications dedicated to work at high temperatures, is a lack of worldwide recognised tribological test methods. This paper presents the method for tribological evaluation of tribosystems with PVD/CVD coated elements, in oscillatory motion and at high temperature (up to 900 °C). The method is realised by using SRV Optimol Instruments Prüftechnik GmbH (Germany) device. The realised verification tests proved the correctness of the elaborated methodology.

1. Introduction

A vast amount of manufacturing processes operate at extreme temperatures and rely on the safety and reliability of components to be capable to cope with harsh environment. Tribological systems working under high operating temperatures (up to 1000 °C), have recently received a great deal of attention, because of the processes (such as oxidation, softening and creep), which although, at room temperature, can be treated as negligible, at elevated temperatures can play a dominant role, especially as their relation with wear is highly complex [1, 2].

Reliable performance of tools and devices working at elevated temperatures requires the use of proper materials with enhanced thermal and wear resistance, as well as the possibility to ensure low energy dissipation by low friction behavior in a wide range of temperature [3, 4].

Currently, one of the leading technologies for depositing the multifunctional coatings is physical vapour deposition (PVD) [5, 6]. Thin hard coatings are used in numerous applications to extend the life of the machine parts [7, 8]. The concept of using PVD coatings is based on previous works, in which it was proved that the presence of Al in the classical TiN and CrN coatings results in the increase in wear resistance and thermal stability at the temperatures up to 1000 °C [9, 10]

The important mechanism worth to be noted here is the formation of highly protective double oxide layer, including a very dense $(Cr,Al)_2O_3$ on top, providing high wear resistance and friction stability under severe stresses occurring at such temperatures [11].

Nowadays, there is a wide range of either single layered (e.g. TiN, CrN), or multilayered coatings with combined properties of several materials, beginning with $TiC/Al_2O_3/TiN$ in 1982, and also multi-element coatings such as TiAlN, BCN, AlCrN and TiAlYN-systems [12]. The modification of coating compositions leads to the extension of their area of application.

Despite the extraordinary wear resistance of the $Ti_{1-x}Al_xN$ coatings their frictional performance at high temperature is not sufficient, so one of possible ways to provide low friction behavior is the introduction of friction decreasing components to receive structures like: TiAlCrN/TiAlYN or TiAlN+WC/C (a-C:H:W). However, such modifications cause a drop of the temperature resistance by a few hundred degrees Celsius.

The promising results were obtained for TiAlN-based, AlCrTiN-based, TiBN, TiSiN, CrVN and nanocomposite CrAlSiN. In particular, the coatings containing vanadium have been identified as providing low friction at around 600 °C via the in situ formation of V_2O_5.

To evaluate the new material solutions, researchers use various methods and wide range of tribotesters [13 14]. Consequently, the obtained results are difficult to compare [15, 16]. One of the barriers to develop new technologies, especially surface modifications dedicated to work at high temperatures, is a lack of worldwide recognized and standardised tribological test methods [17]. In this paper the authors presents two methods for tribological evaluation of tribosystems with PVD/CVD coated elements in conditions of oscillatory motion and at high temperatures. The first concerns the wear assessment, and the second is focused on friction coefficient measurement at wide range of temperature from room up to 900 °C.

2. Test rig, materials and methods

2.1. Test rig

The wear tests were performed be means of ball-on-disc SRV friction and wear tester (Schwingungs Reibung und Verschleiss [SRV], Optimol Instruments Prüftechnik, Germany) using reciprocating motion of the ball sliding against coated disc (Figure 1). During the run the temperature is measured by the thermocouple situated in the base plate of the test disk.

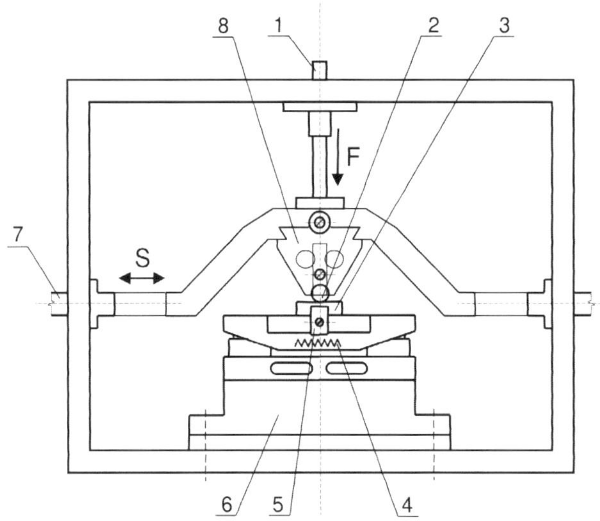

Figure 1. The scheme of the SRV machine test chamber: 1 – load rod (F – direction of normal force), 2 – test ball, 3 – test disc, 4 – electrical resistance heater, 5 – test disc holder, 6 – support, 7 – piezoelectric measuring device (S – direction of oscillating motion), 8 – test ball holder.

IOP Conf. Series: Materials Science and Engineering 1140 (2021) 012024 doi:10.1088/1757-899X/1140/1/012024

2.2. Specimens

The specimens (discs) were made of K10F sintered carbide (WC/Co) supplied by Sinter Sud. The drawing and photograph are presented in Figure 2. The properties of sintered carbide are presented in Table 1.

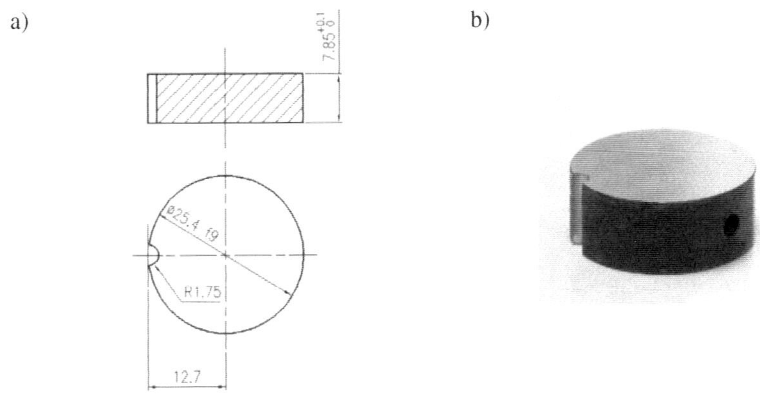

Figure 2. The specimen (disc) made of sintered carbide: a) the drawing, b) the photograph.

Table 1. The properties of K10F sintered carbide (according to the supplier specifications).

Parameter	Value
Type	K10F
Composition, wt%	91.5 WC, 8.5 Co
Density, g/cm^3	14.7
Hardness, HRA	92.0
Vickers hardness	1800 HV30
Grain size, μm	0.8

2.3. Substrate surface preparation

The samples preparation was divided into the following steps. The first step was to cut the samples from the rod, by means of electrical discharge machine using Cu wire (WEDM), with distilled water as a dielectric.

The next step was to obtain the required thickness of test samples using a grinder equipped with a grinding wheel (1A1 200x20x5x51 D151 K75 MBm) intended for sinters and carbides. The grinding wheel contains diamond particles with a grain size between 126 and 150 μm, at K75 concentration in a binder from a soft resin. During the process a cooling-cutting fluid was used. The grinding was done at the linear speed of 35 m/s.

After obtaining the required thickness and initial roughness, the samples were polished by means of ATM Saphir 550 automatic polisher using the procedure presented in the Table 2.

Table 2. The parameters used for specimens polishing.

Parameter	Stage 1	Stage 2
Polishing plate	Cameo Gold	Cameo Gold
Lubricant	BioDiamant 6 Mpe (6 μm diamond grain size)	BioDiamant 3 Mpe (3 μm diamond grain size)
Plate speed [rpm]	200	200
Load per sample [N]	15	15
Rotation direction	clockwise	clockwise
Time [s]	240	240

The results of surface roughness (R_a and R_z parameters), obtained after each of performed preparation steps, are presented in the Table 3.

Table 3. The roughness parameters of discs after surface preparation.

Surface	Roughness parameters	
	R_a [μm]	R_z [μm]
After grinding	0.0318	0.2233
After polishing (Stage 1)	0.0173	0.1412
After second polishing (Stage 2)	0.0028	0.0220

2.4. *Coatings*

The two types of commercially available coatings deposited by PVD (Physical Vapour Deposition) process were tested. The operational temperature upper limit of the AlTiN is 1000 °C and TiAlN coating is 900 °C. The detailed deposition process parameters are not provided by the coating supplier. These coatings maintain hardness even at high temperatures and are resistant to wear and thermal shock. The chemical composition and mechanical properties of the coatings are summarised in Tables 4 and 5.

Table 4. The chemical composition of the coatings (EDS, acc. voltage: 15 kV, take off angle: 30.2°).

	Concentration [wt%]			Concentration [at.%]		
Coating type	Al	Ti	N	Al	Ti	N
AlTiN (A)	37.1±0.3	40.1±0.2	22.3±0.4	36.1±0.4	22.3±0.1	41.7±0.6
TiAlN (B)	24.1±0.1	55.3±0.1	20.4±0.4	25.4±0.1	32.8±0.1	41.7±0.1

Table 5. The properties of investigated coatings.

Coating type	Hardness HV0.05	Thickness [μm]	Friction coefficient against dry steel	Surface roughness after deposition Ra [μm]
AlTiN (A)	3000	1.9 – 2.0	0.35	0.0643
TiAlN (B)	3300	1.4 – 1.5	0.30 – 0.35	0.0228

These coatings are used in hot forging and extrusion industries to protect dies and moulds operating at high temperatures. They are also applied to internal combustion engine components (e.g. nozzle needles and control valves of fuel injection systems). The coatings were deposited on the samples made of K10F sintered carbide.

2.5. *Wear test procedure*

The test parameters for wear assessment are summarized in Table 6. The parameters were selected in such a way, as to obtain measurable wear during the test but also to prevent the coating from being rubbed through (based on the experience gathered during previously conducted research).

On a single disc, 5 test runs are carried out in the places indicated in Figure 3. The test runs are performed in the order of increasing temperature. The volumetric wear was determined by means of Talysurf CCI – Lite Non-contact 3D Profiler.

Table 6. Test parameters for wear measurements.

Parameter	Value
Temperature, °C	25 (RT), 400, 600, 750, 900
Temperature gradient, °C/s	1
Time of temperature stabilisation, s	1200
Load, N	5
Stroke, μm	1000
Frequency, Hz	10
Run time, s	300
Ball material	Si_3N_4 (grade G20)
Ball diameter, mm	10
No. of repetitions	min. 3

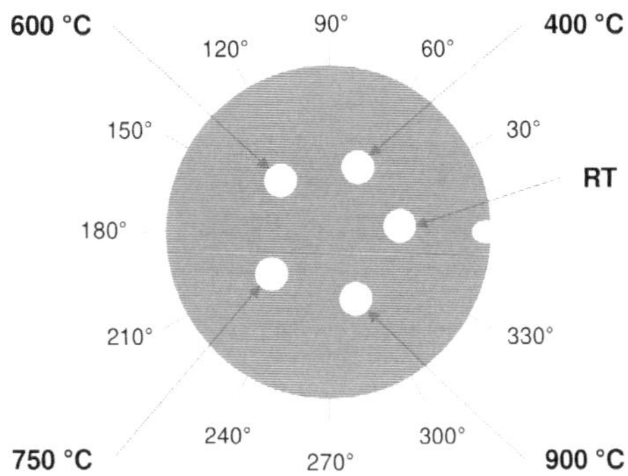

Figure 3. The location of wear scars on the disc.

In high-temperature tests, the method of heating the samples is important. Its full cycle consists of: heating from room temperature to the set point (e.g. 600 °C) at a constant rate of 1 °C/s, stabilization of the set temperature, and the 300 seconds test run. An example of the wear test step is shown in Figure 4.

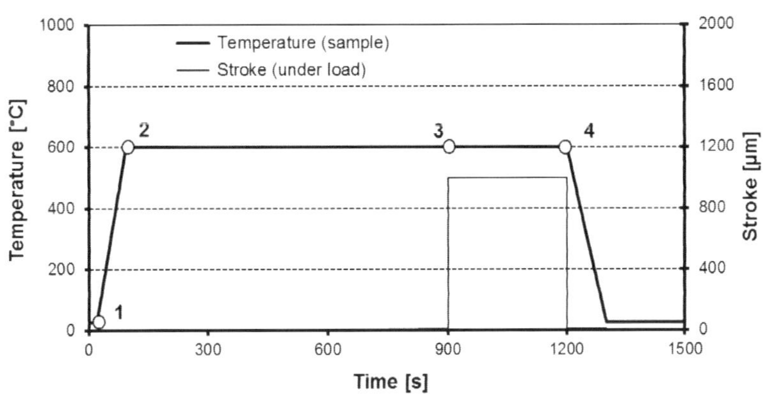

Figure 4. The example of the test procedure for wear measurement (at 600 °C) 1 - the beginning of the test, 1-2 – heating from RT up to 600 °C with 1 °C/sec. increase rate, 2-4 – stabilization at 600 °C, 3-4 – 300 seconds run, 4 – the end of the test.

After the test run, the wear scar is measured using a 3D profilograph. The run is treated as valid if the coating is not rubbed through.

2.6. The friction coefficient measurement procedure

The test parameters for the friction tests are summarized in Table 7.

Table 7. The parameters for friction measurements.

Parameter	Value
Temperature, °C	25 (RT), 100, 200, 300, 400, 500, 600, 700, 800, 900
Temperature gradient, °C/s	1
Time of temperature stabilisation, s	300
Load, N	20
Stroke, μm	1000
Frequency, Hz	10
Run time, s	10
Ball material	Si_3N_3
Ball diameter, mm	10
No. of repetitions	min. 3

The procedure for measuring the friction coefficient is shown in Figure 5a. The detailed course of the test for the 700 °C cycle is shown in Figure 5b. The friction coefficient is calculated as the arithmetic mean of the maximum quantities in the cycle.

a)

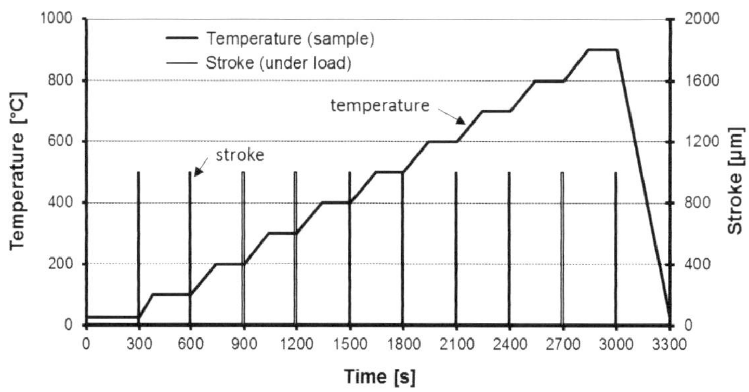

b)

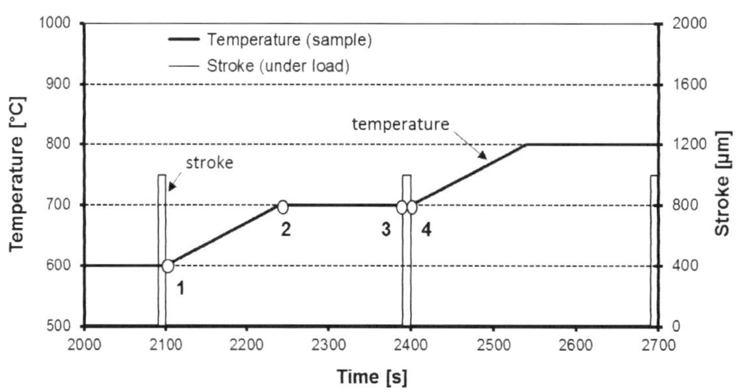

Figure 5. The friction coefficient measurement procedure: a) temperature curve with vertical lines representing each of measurement cycles, b) the detailed section of test between 600 and 800 °C: 1-2 – heating from 600 °C up to 700 °C with 1 °C/s rate; 2-4 – stabilization at 700°C; 3-4 – 10 s measurement cycle.

Aforementioned approach makes it possible to conduct one test run on a single disc, covering the whole selected temperature range: from RT up to 900 °C. Due to the fact that the friction process always takes place in the same area of the specimen, the test run time was reduced to 10 seconds. Between the test runs, heating and temperature stabilization of the sample is applied. After reaching the set temperature, another test run is carried out (in the same place, but at a higher temperature).

Results and discussion

In order to verify the developed research methodology, tests of wear resistance and friction coefficient were carried out for two selected coatings dedicated to high-temperature applications.

2.7. Wear test results

The experiments on coatings have shown a considerable dependency between tribological characteristics and temperature. In most cases the increase in temperature increases the wear. However this rule is not valid for coating intended for high temperature applications. For coating denoted as "B" the minimum wear was observed at 600 °C, while the wear of the coating denoted as "A" remained high.

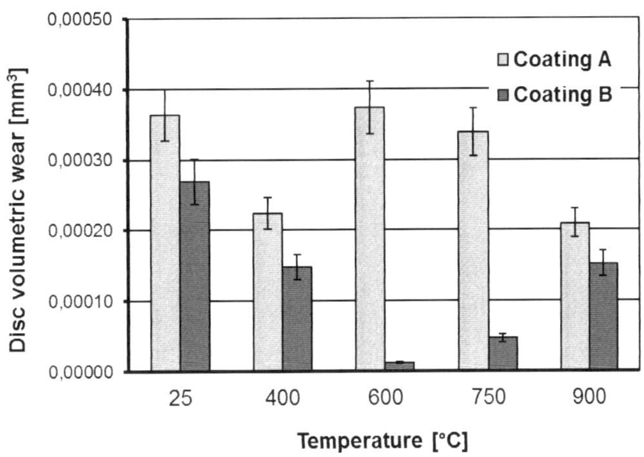

Figure 6. The results of discs volumetric wear at selected temperatures.

The developed method made it possible to precisely determine the differences between the wear characteristics of the A and B coatings. Despite the fact that both coatings are dedicated to high-temperature applications, their wear characteristics differ depending on the temperature. For coating A the wear at 600 and 750 °C was similar to the wear at room temperature (25 °C). The minimum wear was recorded for 400 and 900 °C. However, the result obtained for the temperature of 900 °C should be rejected, because the coating was completely removed during the test – the maximum depth of wear track, measured with the profilograph, exceeded 3 μm, which means that the ball rubbed through the coating and went into the substrate material.

At room temperature, the wear of the coating B was only slightly lower than that of the coating A. Research using the developed method allowed to find differences between these coatings at 600 °C and 750 °C. Contrary to the combination with coating A, for the combination with coating B, at 600 °C and 750 °C, the volumetric wear was several times lower than the one measured at room temperature.

The appearance of wear scars on the coated discs is shown in Figure 7. The maximum depth of wear scars is summarized in Table 8.

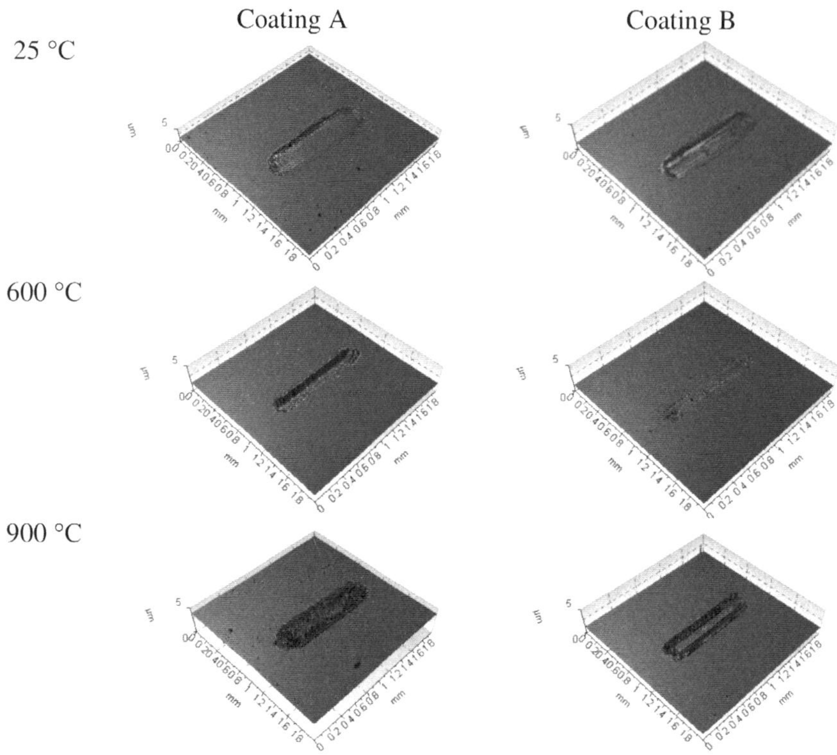

Figure 7. 3D images of wear scars on the coated discs.

Table 8. The maximum depth of wear scars measured at various temperatures.

Temperature [°C]	Coating A [μm]	Coating B [μm]
RT	1.6	1.1
400	2.2	1.6
600	2.2	1.5
750	2.2	1.1
900	3.1	1.5

The test conditions are set between two boarders: obtaining the wear high enough to be clearly measurable (while maintaining a satisfactory repeatability), and avoiding the total removal of the investigated coating. For the adopted parameters, the maximum depth of the wear scar for coating A slightly exceeds the thickness, but microscopic observations showed that the abrasion occurs only on a small part of the friction mark, which is considered as acceptable. However, for the B coating, no total removal of the coating was found in the entire test temperature range. Increasing the severity of the test conditions (time, frequency, load) would lead to the coating removal in each test run, which would be disqualifying for the test method. On the other hand, reducing the value of the test parameters would result in obtaining unmeasurable wear.

Summarizing, it can be stated that the results of the verification tests proved the correctness of the adopted test method for the assessment of the wear resistance of elements covered with coatings intended for operation at high temperatures.

2.8. Friction coefficient test results
The results of friction coefficient at selected temperatures are presented in Figure 8.

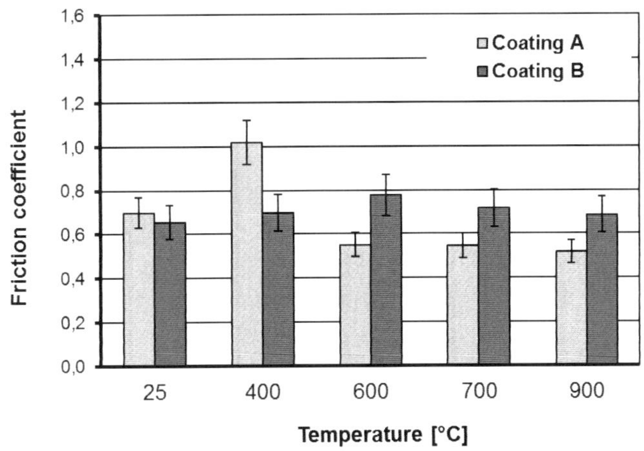

Figure 8. The results of friction coefficient at different temperatures.

At room temperature, the friction coefficients of the coatings A and B do not differ significantly. Only for the test temperature of 400 °C, the friction coefficient of the coating A is higher than for the coating B. The obtained results showed that using the developed methodology, it is possible to precisely measure the coefficient of friction in the entire temperature range from RT to 900 °C. The method has several significant positive attributes. The method is well repeatable as the obtained results are characterized by a scatter lower than 10%. It is also relatively low time and cost consuming. 3 sets of samples are sufficient for the entire measuring cycle, which can be conducted with a time schedule of approximately three hours.

3. Conclusion
The results of the verification tests proved the correctness of the adopted test method to assess the wear resistance and friction coefficient of the elements covered with coatings intended for operation at high temperatures.

4. Acknowledgments
The authors wish to express their thanks to Carlos Sánchez Villarreal from the University of Monterrey UDEM (Mexico) for his support with tribological experiments and helpful discussions.

Project HOTselflub is supported by the National Science Centre, Poland, under the M-ERA.NET 2, which has received funding from the European Union's Horizon 2020 research and innovation programme, under grant agreement no 685451.

References
[1] Yung D et al. Tribological performances of ZrC-Ni and TiC-Ni cermet reinforced PTA hardfacings at elevated temperatures 2017 *Surface and Coating Technology* **309** 497-505.
[2] Kolnes M et al. Wear of potential tool materials for aluminium alloys friction stir welding at weld temperatures 2019 *Proceedings of the Estonian Academy of Sciences* 68 (2) 198-206.
[3] Antonov M et al. Experimental setup for testing and mapping of high temperature abrasion and oxidation synergy 2009 Wear **267** (11) 1798-1803.
[4] Ngaile G et al. A tribo-testing method for high performance cold forging lubricants 2007 *Wear* **262** (5-6) 684-692.
[5] Michalczewski R et al. The Effect of a Gear Oil on Abrasion, Scuffing, and Pitting of the DLC-

Coated 18CrNiMo7-6 Steel. 2019 *Coatings* **9** (1).
[6] He Q et al An integrative approach to coating/carbide substrate design of CVD and PVD coated cutting tools during the machining of austenitic stainless steel 2020 Ceramics International **46** (4) 5149-5158.
[7] Tuszyński W et al. The effect of WC/C coating on the wear of bevel gears used in coal mines. 2015 *Materials Science (Medžiagotyra)* **21** (3) 358-363.
[8] Michalczewski R et al. The problems of resistance to scuffing of heavily loaded lubricated friction joints with WC/C coated parts. 2014 *Industrial Lubrication and Tribology* **66** (3) 434-442.
[9] Aouadi S et al. Lubricious oxide coatings for extreme temperature applications: A review 2014 *Surface and Coating Technology* **257** 266-277.
[10] Geng Z et al. Tribological behavior of patterned TiAlN coatings at elevated temperatures 2019 *Surface and Coating Technology* **364** 99-114.
[11] Michalak M et al. The Effect of Temperature on Wear Mechanism of the AlCrN Coated Components 2016 *Key Engineering Materials* **674** 233-238.
[12] Mayrhofer PH et al. Microstructural design of hard coatings 2006 *Materials Science* **51** 1032–1114.
[13] Minewitsch A A Some Developments in Triboanalysis of Coated Machine Components 2011 Tribology in Industry **33** (4) 153-158.
[14] Michalczewski R et al. A method for tribological testing of thin hard coatings 2002 *Tribotest Journal* **9** 117-130.
[15] Tuszynski W et al A new scuffing shock test method for the determination of the resistance to scuffing of coated gears 2012 *Archives of Civil and Mechanical Engineering* **12** (4) 436-445.
[16] Michalczewski R. et al. A method for the assessment of the rolling contact fatigue of modern engineering materials in lubricated contact 2012 *Transactions of FAMENA* **36** (4) 39-48.
[17] Michalak M et al. The method of tribotesting of PVD coated elements in oscillation motion at high temperature. 2015 *Tribologia* **1 (259)** 77-94.

IOP Conf. Series: Materials Science and Engineering 1140 (2021) 012025 doi:10.1088/1757-899X/1140/1/012025

Generic Architecture for the Automatic Parametrization of Production Machine Assembly Programs

Philipp Stephan[1]*, Jessica Fisch[1], Alperen Can[1], Oliver Heimann[2], Gregor Thiele[2], Jörg Krüger[3]

[1]Mercedenz-Benz AG, Berlin, Germany, [2]Fraunhofer IPK, Berlin, Germany, [3]Technische Universität Berlin, Berlin, Germany

* Corresponding author. E-mail address: philipp.stephan@daimler.com

Abstract. In a high-mix low volume production environment, time to market is a key factor. However, one bottleneck lies in the often times manual parametrization of machines for new or modified designs. A truly flexible manufacturing environment therefore requires a continuous data flow from the design stages to the shop floor. This paper presents a concept for the automated parameterization of machines at a large automotive plant. Therefore, this paper initially discusses the results of a stakeholder analysis. The stakeholders comprise of different departments related to the product design and manufacturing processes. The requirements resulting from the interviews conducted with the stakeholders are grouped and ordered by priority. Secondly a general architecture for the control interface is presented. It includes multiple submodules, which model the continuous data flow between the departments and the production systems. The first main step of the data flow is the transformation of the information which is presented in various styles depending on the source departments to structured and standardized data. Thereafter machine parameters are generated automatically by a submodule using the structured input data and inference rules. Finally, the architecture supports the automatic transfer of the machine readable output data to the assembly line. To test the architecture a prototype comprising of more than 100 robots in a live production environment is implemented. It allows for a continuous data flow from the design and productions planning department to the robots. This enables a flexible process control, which up to today has supported the fast roll out of more than 100 new product variants. In contrast to the conventional manual setup of the machines for operation, the prototype was able to show that the monetary and time expenditure could be reduced by 95 percent.

1. Introduction

The trend towards highly customized products leads to novel challenges for manufacturing. An ever growing number of variants needs to be manufactured on the same line. This requires a high flexibility of the involved machines with regards to their physical capabilities as well as their software parametrization [1]. Furthermore, this desired flexibility necessitates the integration of upstream design processes to provide the correct parameters. However, this vertical integration of processes and machines is aggravated by the current landscape of proprietary systems and controls [2], [3]. As a result, a lot of time is required to manually set up the production machines. This manual influence not only results in reduced transparency with regards to the current machine setup but also introduces additional delay and error sources between the design and manufacturing processes.

Therefore, to enable a continuous data flow between all involved parties a common, machine readable information model is needed. However, there currently is no such standard.

2. Contribution

This paper presents a conceptual architecture for a continuous data flow between all stakeholders involved in the design and production of mechanical components. The model is based on a stakeholder analysis conducted at an automotive plant.
Based on the stakeholder analysis an information model is derived that maps the information requirements of the manufacturing equipment onto the information sources in the upstream design processes. It considers three different priorities for the requirements which deal with 1) safety of the operator and equipment, 2) data processing and security and 3) ease of use. The model is then used for the design of a continuous data flow architecture that connects the involved departments. Based on the initial results with a prototype implementation, this architecture allows to reduce the cost and time needed for the addition of a new product variant by 95%.

3. State of the Art

Looking at the state of the art, there are different developments that deal with the networking of different subsystems across the different company levels. The piCASSO project [4], [5] deals with a cloud-based, modularized control architecture for dynamic robot programming at the control level. It provides an architecture that considers all machines as cyber physical devices which are addressed through standardized interfaces. This allows to interchange hardware with equal capabilities while providing a common interface for purely virtual control services which provide for the actual process control. This service based architecture leads to an increase in flexibility of the production system. While the piCASSO project focused on the benefits of a distributed control architecture it relied on standardized machine to machine communication frameworks. The most common framework is OPC-Unified Architecture which provides software interfaces for communication as well as a means for data modelling on a process and enterprise level [6]. The BaSys framework [7] builds upon OPC UA and provides a reference architecture for production systems, which allows a service-based networking of different technologies in the production environment. AutomationML is a highly standardized file format based on XML, which is intended to enable the standardized exchange of machine, production and product data in order to ensure the consistency of the information available [8]. Based on the ideas of these technologies the concept of the data-driven parameterization of production machines has been developed.

4. Concept

This section first presents the requirements engineering process. Thereupon the information model is derived and used to design a system architecture for the continuous data flow.

4.1. *Requirements*

To identify the requirements of the information model a stakeholder analysis with all parties involved in the design and production of the parts has been conducted. In particular the analysis includes internal production-related departments (PRD) and production employees (PD), as well as departments for standardization (DS), process optimization (PO), improvement management (IM) and information technology (IT). Furthermore, external service providers (ESP) are included in the basis for decision-making.
In an initial step, the involvement of the different parties and their influence on the process has been rated as shown in Figure 1. It shows that the design related departments and the production workers are the most important stakeholders, as they directly interact with the system. The production departments also have detailed information about their equipment and its data requirements. In addition, the production operators have an in-depth knowledge of the machines and their programming. DS engineers are trained to support projects of this kind and to provide standardized solutions or specifications for potential problems. Therefore, their influence on the project is very high and the interest is high. The department of process optimization has a high influence and interest, because important information

from already implemented improvements could be contributed and the implementation represents a considerable optimization in the production process. The IT department has a high influence because it ultimately provides the resources for implementation. The low influence of the IM results from the fact that this department only deals with already existing processes. On the other hand, the high level of interest is due to the fact that after the first prototype has been started, improvement measures can be implemented in other departments. External service providers would have little influence and little interest in the project at this point, since basic requirements must first be established in order to formulate concrete questions or demands on service providers. In the following, the requirements are analyzed and specified based on workshops and interviews conducted with the stakeholders, available documentation as well as information from existing and related systems.

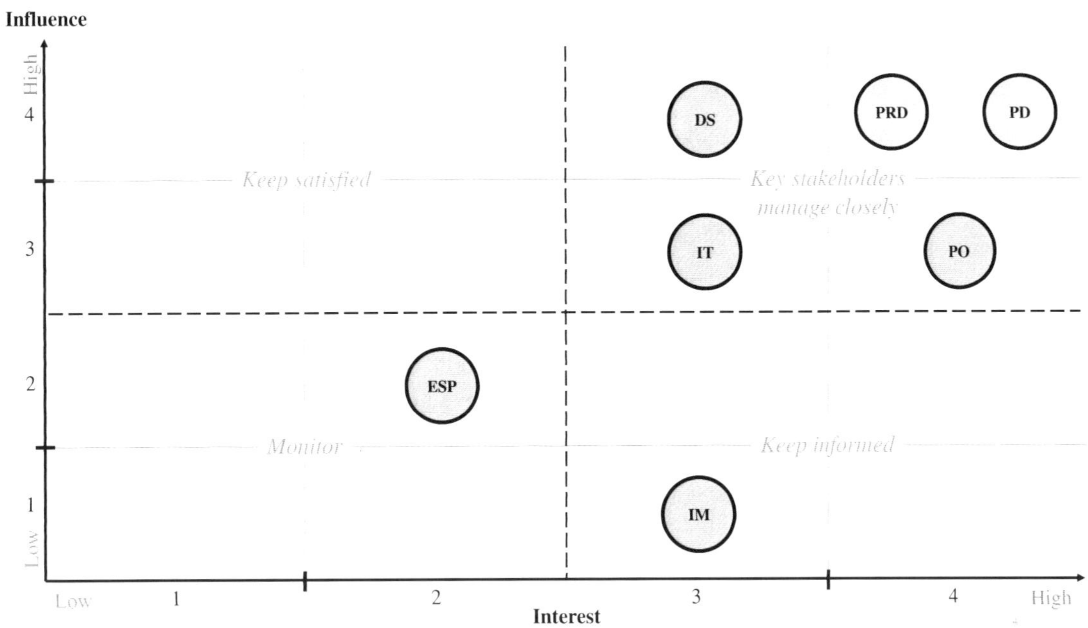

Figure 1: Interest and Influence of the possible stakeholders

During the analysis, the unstructured information obtained from the expert interviews were classified, assigned to one of the subsystems (shown in Figure 2) and prioritized. There are requirements for the following subsystems:

- the department(s) (Input Requirements - INR),
- the interface (Interface Requirement - IR) and
- the production machine(s) (Output Requirement - OUTR).

Aside this allocation, the requirements are further divided into the following three priority classes:

- Priority 1: Requirements for operational and functional safety as well as core functionality,
- Priority 2: Requirements which are decisive for the functionality of the automated parametrization and as well as the verifiability of production quality and finally
- Priority 3: Other requirements that are not necessary for the core functionality but contribute to the ease of use.

Table 1 presents the resulting catalog of priority 1 requirements. An example for a priority 2 requirement is the provision of metadata and process results, which provides necessary information for the verifiability of production quality, but does not contribute to safety or core functionality. An example for priority 3 is the overall usability, which neither contributes to safety or the core functionality nor provides functionality for the automated parametrization or data flow and yet is of importance for the acceptance and deployment of the system.

Table 1: Priority 1 requirements as found in the stakeholder analysis

ID: Keyword	Explanation
INR01: Input data format	Minimum semistructured, database convertible, human readable and modifiable
INR02: Standardization	Individual attributes defined by a uniform nomenclature
INR03: Data provision	Provision of input data in a defined structure
IR01: Interoperability	Bidirectional communication via standardized communication protocol
IR04: Security	(traceable) data access management
IR05: Safety	Safety checks prior to execution of a process with new parameters
IR20: Standardization	Uniform data types and structures, uniform naming schemas
IR21: Process data format	Minimum semistructured, machine-readable and human-readable
IR22: Data provision	(Automatic) provision of process data in a defined structure
OUTR01: Integration	Data processing and update of machine parameters between production cycles
OUTR02: Data Consumption	No computation within the machine program, Templated programs
OUTR03: Safety	Update notification, enforce testing of new motion sequences
OUTR04: Communication	Common protocol that is widely available on machines

4.2. Architecture and data flow

Finally, based on the requirements gathered through the stakeholder analysis an architecture for the data flow is designed as presented in Figure 2. The individual sub-systems, departments (grey), interface (green) and production (orange) are shown. The input data is generated and stored within a system of the corresponding department. Examples for input data are CAD data (i.e. geometry or material information), machine data, tool data, and quality data or logistics information. Afterwards the input

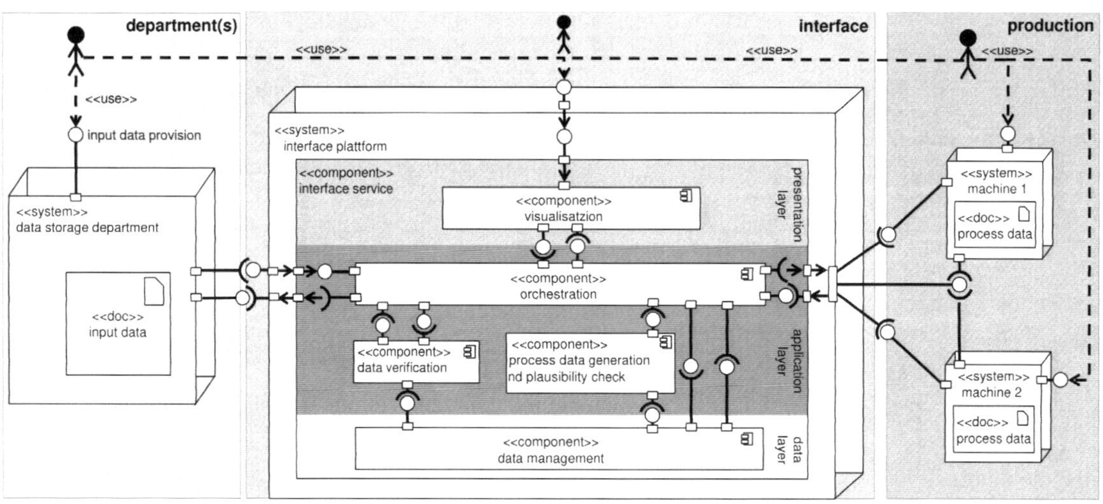

Figure 2: Architecture overall system

data is automatically transferred to the interface, where different services realize the data processing to generate machine-readable process data. The process data is then distributed to the respective production machines in a way that ensures data synchronization over the entire production line. The interface is divided into three sublayers: presentation, application and data layer. The presentation layer (blue)

provides the contents (such as current input and process data, archived data, change and error logs) of the service for different user groups.
The application layer (red) is the layer in which different algorithms for data processing and orchestration are implemented. In the data layer (yellow), the input and process data are stored. Depending on the data structure, persistent data storage in an SQL or NoSQL database is conceivable. Furthermore, a file-based storage for the process data is necessary. These must be machine-readable, at least semi-structured and stored in a defined file format (e.g. json, csv or xml) at a previously defined storage location. The data transfer to the production machines (orange) is carried out via a standardized communication protocol. The process data is then processed and persistently stored locally on the machine so that the machine program can access it. This means that in the event of a network failure, synchronized production data is stored on the machine. If the process data has an impact on the machine behaviour, a manual function check by a machine operator is necessary prior to clearance of the automatic program execution. If this check passes, series production is continued with updated process parameters otherwise the old parameters are kept in use until a new update is provided.

5. Conclusion and Outlook

The concept outlined in the paper provides standardized specifications to establish a continuous data flow between departments and shop floor via a central interface, thus enabling flexible production.
To test the architecture and the general data flow, a prototype has been implemented for an assembly line consisting of more than 100 6-axis articulated robots. The input data for more than 100 variants is automatically read from the construction and production planning departments and processed via a central interface. Robot program parameters are then automatically generated by inference rules and forwarded into program templates for the different machining processes. Initial results of the prototype implementation are very promising. As no external robot programmers were required to manually program all robots for all product variants, the costs and time taken for robot programming has been reduced by 95%. With the proposed automatic parameterization, the robots are ready for production in just 15 minutes. This includes the required manual safety check of the newly parametrized robot program. The previous process requires about eight hours of machine downtime. The more transparent change management further allows to identify error causes faster and has helped to reduce the number of rejected or faulty products.
In future work we like to expand the current prototype to a wider spectrum of machines (e.g. machine tools, measuring machines) and address limitations in the current implementation.

6. References

[1] Peschke and Eckardt. Flexible Produktion durch Digitalisierung. Carl Hanser Verlag; 2019.
[2] T. Strasser, A. Zoitl and G. Ebenhofer. 4diac – "Ein Open Source Framework für verteilte industrielle Automatisierungs- und Steuerungssysteme" in INFORMATIK 2010. Service Science – Neue Perspektiven für die Informatik. Band 1, K.-P. Fähnrich and B. Franczyk. Eds. Bonn: Gesellschaft für Informatik e.V.; 2010, pp. 435–440.
[3] V. M. Banholzer. Gestaltungsdiskurs Industrie 4.0: Akzeptanzaspekte, Frames. Institutionalisierung. Wiesbaden: Springer; 2018. p. 221–239.
[4] A. Vick, V. Vonásek, R. Pěnička and J. Krüger. Robot control as a service — Towards cloud-based motion planning and control for industrial robots. 2015 10th International Workshop on Robot Motion and Control (RoMoCo). Poznan; 2015, pp. 33-39, doi: 10.1109/RoMoCo.2015.7219710.
[5] A. Vick, C. Horn, M. Rudorfer and J. Krüger. Control of robots and machine tools with an extended factory cloud. 2015 IEEE World Conference on Factory Communication Systems (WFCS). Palma de Mallorca; 2015, pp. 1-4, doi: 10.1109/WFCS.2015.7160575.
[6] M., Schleipen, S.-S., Gilani, T., Bischoff, J. Pfrommer. OPC UA & Industrie 4.0 - Enabling Technology with High Diversity and Variability. Procedia CIRP, 2016;57:315–320.
[7] https://www.basys40.de/ (18.11.2020)
[8] AutomationML e.V. AutomationML and eCl@ss integration: Whitepaper. AutomationML e.V, 2015.

Simulation based Performance Analysis of Production Intralogistics

Jelena Golova[1], Kashif Mahmood[1*], Tõnis Raamets[1]

[1]Department of Mechanical and Industrial Engineering, Tallinn University of Technology, Tallinn, Estonia.

*E-mail: kashif.mahmood@taltech.ee

Abstract. Production planning and scheduling rely heavily on the efficient operations of production logistics and material handling equipment. Industry 4.0 technologies such as Internet of Things (IoT), Digital Twins, and Artificial Intelligence (AI) can be applied to production logistics in terms of autonomous mobile robots that facilitate to increase the flexibility and productivity of the whole production site. However, before the implementation of an automated production logistics systems, its feasibility must be analysed. This paper describes a simulation-based approach, including the definition of and comparative analysis of Key Performance Indicators (KPIs), to analyse the performance of production intralogistics applied to a selected use case. The presented approach offers a proof of concept on the basis of which decision-makers can implement mobile robots for intralogistics in their own production environments.

1. Introduction

In the scope of production management, the performance of activities such as obtaining raw materials to delivering finished goods to customers, need to be jointly studied and analysed. These activities are highly interconnected, and the analysis of the performance of those activities can help optimize manufacturing and logistics operations. The improvement of production intralogistics – the internal transportation of goods within a given manufacturing facility – has a major impact on the production efficiency of the whole site. As such, the requirement to optimize internal logistics systems in terms of operational performance, throughput and sustainability arises [1]. Although automation contributes a lot to business value creation and has already been to some extent introduced into the intralogistics of manufacturing facilities (e.g. conveyors, fork-lifters and pallet trucks), the aforementioned equipment allows only for a low degree of flexibility, whilst other tasks, such as loading & unloading, and the authorization of goods, are still mainly performed manually [2]. High level automation, such as the introduction of Autonomous Mobile Robots (AMR) into the intralogistics of the facility, offers a more flexible solution that can lead to a more efficient process of transportation.

Whilst intralogistics automation promises many benefits, any change within the production site introduces new challenges. For example, to ensure a smooth transition into the new workflow, a thorough change management course for line operators is recommended to be planned and carried out. Moreover, internal logistics systems are highly complex, with the deployment of AMRs requiring a thorough preliminary study and analysis. Therefore, the method of simulation and 3D visualization can be used to analyse and verify the change. Simulation modelling, paired with the Digital Twin concept and the setup of KPI (Key Performance Indicator) targets, has become a staple framework in operations management today, for the insights gained facilitate better decision-making in terms of financial, time oriented, material and energy savings, as well as the ability to streamline the process activities [3].

2. Literature Review

For the literature review, state-of-the-art articles relevant to the field of this study were analysed. Topics include the automation of production intralogistics through AMRs, the significance of simulation modelling and 3D visualization as decision-making tools, and a brief explanation of relevant Key Performance Indicators. Moreover, similar studies and related approaches are referred to in this section.

2.1. AMR in Intralogistics

Automation and the application of Internet of Things (IoT) have become widely associated with areas such as production, logistics and transportation. From the other side IoT applications in production and logistics should be seamlessly integrated into companies Manufacturing Engineering Systems (MES) [4]. AMRs for factory floor logistics persuade the smart factory concept with disruptive technologies such as Artificial Intelligence (AI), simulation and Digital Twin [5]. The use of AMRs as a type of material handling equipment creates a more ergonomic workspace for the production floor employees. Furthermore, proper deployment of AMRs may lead to an increase of production capacity and flexibility, while reducing transportation defects. Other automated solutions for factory logistics, such as conveyors, forklifts, pallet trucks, and automated guided vehicles, do not offer the same level of flexibility in terms of routing. In contrast, AMRs can be reprogrammed to be used in different applications, reacting to different data inputs. They are smaller in size and more agile than traditional automated guided vehicles; as such, they can access the production area more efficiently [6]. However, the implementation of AMRs for production intralogistics needs to be justified and verified before the physical set-up on the production floor.

2.2. Simulation Modelling and 3D Visualization

Simulation modelling is the creation of a digital model of a real-world system. Various what-if scenarios can be tested on a valid digital representation of a system to analyse, optimize, and predict the performance of processes based on set parameters [7]. After thorough experimentation in this risk-free environment, an optimal system configuration can be found and carried over into the real world. Potential problems and bottlenecks are discovered and reacted upon early in the process, thus leading to the improvement of set KPIs [8]. 3D visualization is an essential tool used to validate the simulation models´ feasibility by taking the geometry of the facility, line or process into account. 2D simulation, in comparison, offers only a low level of visual commissioning. The usage of Industrial Virtual Reality (IVR)-based 3D visualizations, which can be adapted to simulated 3D assembly layouts, product models, and process flows, may prove to be beneficial in such cases [9]. In this study, a production floor was simulated in a 3D simulation software; the assembly steps, including the mechanisms and tools on the factory floor were modelled, and the changes of the location of AMRs were analysed.

2.3. KPIs for Production Intralogistics

Performance indicators or KPIs aim to deliver information needed for the performance analysis of manufacturing operations, Intralogistics come under the discipline of operations management, and as a result, KPIs related to manufacturing operations are appropriate to production intralogistics as well, as defined in the ISO 22400 family of standards. The standard classifies KPIs based on their purpose of use, such as performance that can be measured in terms of cost, time, quality, flexibility and sustainability. Likewise, they are applicable to different types of operations, such as production, material handling, quality assurance, maintenance, and so on. [10]. Performance indicators not only showcase what has happened; they also indicate what will happen, as reactive steps will be taken by decision-makers to combat any weaknesses represented in the KPIs [11].

In this study, KPIs like *utilization*, *throughput*, and *cost* of AMRs, as well as transportation *defects* were chosen. The paper proposes a 3D simulation-based approach to analyse the performance of the production intralogistics process, though the suggested approach can be implemented to other processes as well. Several other research papers addressed the topics of intralogistics automation and the deployment of mobile robots for transportation on the factory floor [5], [12], [13]. The value of this study lies in the simplicity of the synchronized analysis approach, compared to the more difficult to construct and adopt procedures described in the aforementioned papers.

3. Simulation-based Approach for Analysis

The continuous 4-step approach (see figure 1) was adopted through the comprehensive literature review. The case study technique was used as a research method, and a use-case is introduced to validate the relevancy of the proposed approach.

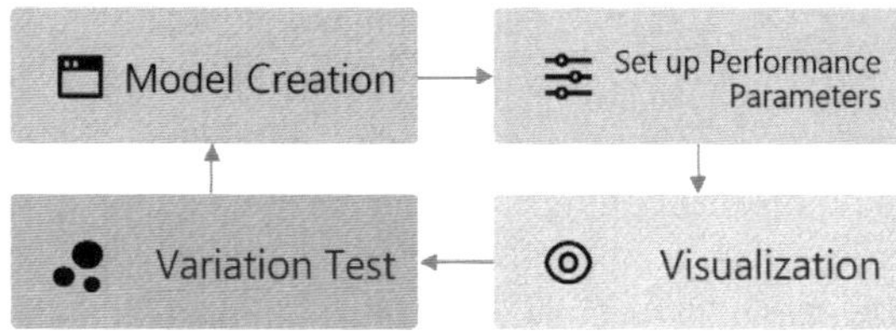

Figure 1. Simulation-based approach to analyse the performance of production intralogistics.

3.1. Model Creation

Creation of the model, a fundamental step in simulation-based analysis, facilitates to capture and describe the problem properly. A model is created by mapping the real environment to the virtual one through a specified computer-based application. A model should grasp and reveal the dynamics of a process such as the occurring of events, changes in activity timing, and resources´ state. Model parametrization also includes the selection of components (entities, source and sink, resources, etc.) relevant to the specific problem statement and system.

3.2. Performance Parameters

The acquisition of data related to the target process or system is crucial for the setup of the desired KPIs in this second step. Process parameter data, such as system specifications, input variables, and process performance metrics, are needed for the analysis. The aforementioned indicators numerically describe the behaviour of the resources, as well as activities' performance. Despite a wide selection of performance metrics being presented in various literature, KPIs must be selected depending on the underlying strategies of the company, for only then can the simulation model be built in response to the specific problem statement of the organization. For the intralogistics performance analysis and the objective of the test case, we incorporated *throughput*, *utilization*, *cost* and *defects* of the transportation activity as KPIs for the simulation analysis. Further details can be found in table 1.

Table 1. KPIs selected for this study.

KPI	Formula	Description
Throughput	$Throughput\ (R) = \frac{Average\ Inventory\ (I)}{Average\ Flow\ Time\ (T)}$	Shows the number of products transferred by an AMR from one station to another per unit time
Utilization	$Utilization = \frac{Task\ Performing\ Time}{Total\ Working\ Time} \times 100$	The percentage time that an AMR performs tasks out of the total working time or a shift duration
Cost	$Cost = Initial\ Investment + Operating\ Cost$	Shows the monetary impact of AMR in monetary terms
Defects	$Defects = \frac{Number\ of\ irregulaities}{Number\ of\ transportations} \times 100$	Expressed by irregularities in the transportation process (wrong number of goods, wrong type of goods, wrong destination)

3.3. Visualization

The exact-scale digital model of the production floor in 3D verifies the work of the real system, ensuring that the created model behaves as intended. 3D simulation assists users to visualize staff, equipment, building facility, and other items and processes in the virtual environment. The verification can be performed by providing real input data to the model and comparing the results with historical data.

Visualization also represents processing data in the form of a dashboard which helps to determine between strategic alternatives.

3.4. Variation Test
In this step, the simulation model can be allowed to test several tactical variations and scenarios that capture uncertainty. Sensitivity Analysis and Parameter Variation experiments are commonly used to reveal the effect of randomness and parameter change to the simulation model's behaviour.

4. Case Study
The proposed simulation-based approach was applied to the intralogistics process of a chemical manufacturer which produces detergents and hand sanitizer. The manually operated transportation of goods is a key operation in the production facility. Due to the high demand of products and, thus, the subsequent increase of production capacity and flexibility, the company decided to analyse and improve the intralogistics process with the intention to automate the production floor logistics by implementing AMRs. This solution is expected to reduce the transportation time and ultimately increase the process productivity, as well as cut down on workers' fatigue. The studied production facility consists of four production lines that fill empty bottles (in containers) of different sizes with liquid, label and cap them. The intralogistics related activities, planned to be executed on four different stations with the help of an AMR, are as follows:
1. Loading of products (empty bottles) in ***warehouse*** and transportation to production line
2. Unloading of empty bottles at the start of ***production line***
3. Loading of filled bottles at the end of ***production line*** and transportation to finished goods area
4. Unloading of filled bottles in ***Finished Goods*** (FG) area and moving back to the Warehouse (WH)

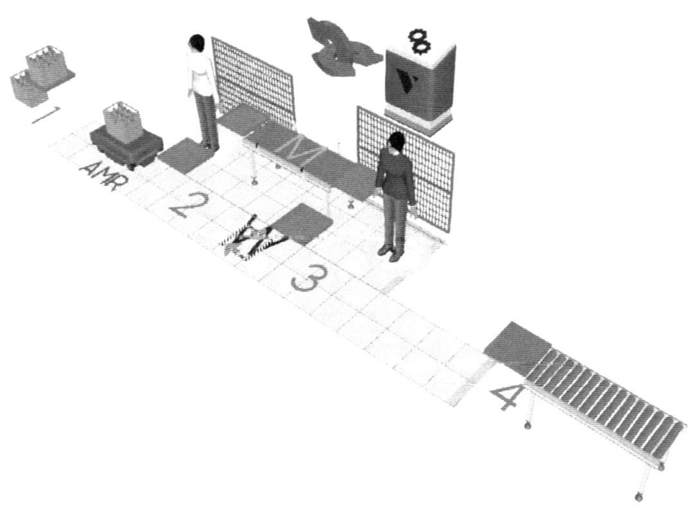

Figure 2. Simulation model of a single production line with AMR.

The 3D simulation models of the use case were created and analysed in Visual Components 4.2, a 3D manufacturing simulation software. The physical setup of the production lines and routes mapping of the AMR were constructed on the basis of full-scale production layout. Figure 2 gives a concise view and a single production line simulation model, where the intralogistics activities were marked and executed as defined above with the corresponding numbers. Figure 3 is a holistic view of the production facility and illustrates the transportation of goods using the AMR following the route WH → Production → FG → WH. The movement of the AMR was mapped and analysed during the simulation, with the green-coloured marking showing the movement of the AMR in the production area, the red-coloured one - to and from the FG area, and the yellow-coloured route – to WH and from WH.

IOP Conf. Series: Materials Science and Engineering 1140 (2021) 012026 doi:10.1088/1757-899X/1140/1/012026

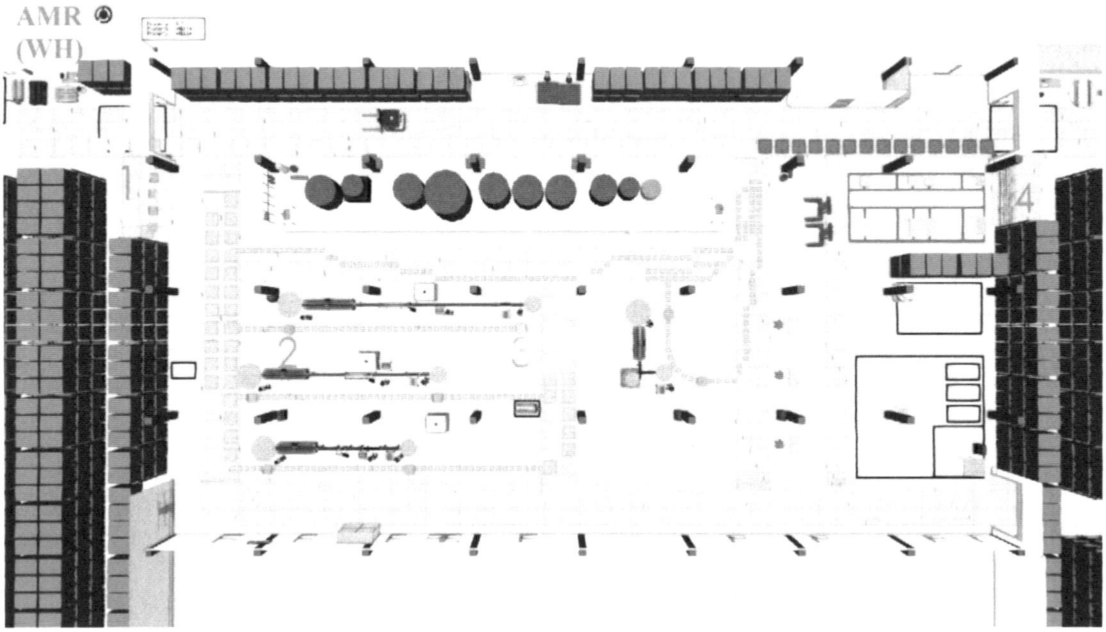

Figure 3. Detailed simulation model of the production facility (top-view).

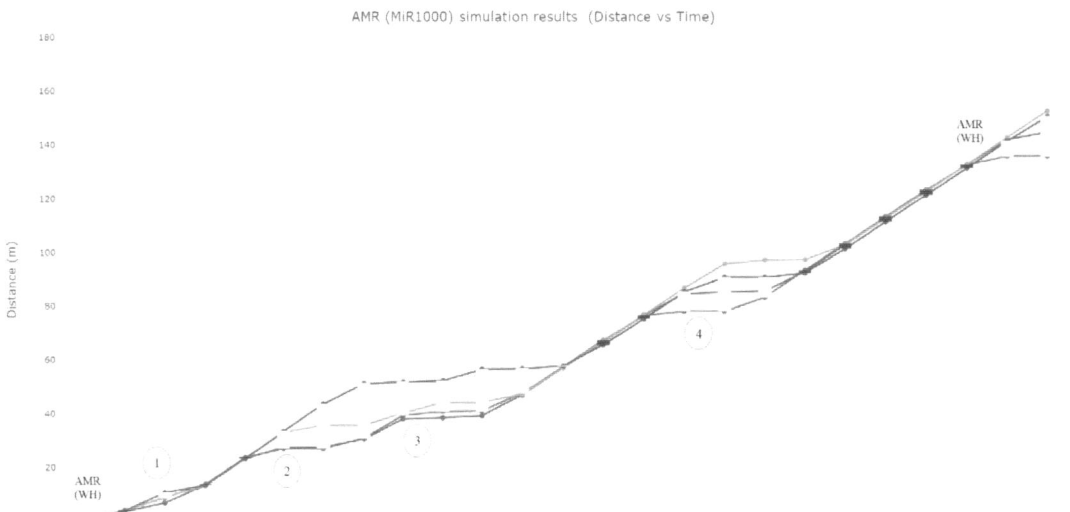

	0	10	20	30	40	50	60	70	80	90	100	110	120	130	140	150	160	170	180	190	200	210	220	230	240
Line 1	0	4,1	8,6	13,7	23,6	27,0	27,0	30,8	39,3	40,6	40,8	47,5	57,4	67,2	76,3	78,0	78,0	83,3	93,2	103,2	113,2	123,2	132,6	135,7	135,7
Line 2	0	3,6	6,7	13,2	23,3	27,4	27,4	30,5	37,9	38,4	39,0	46,9	57,0	66,7	76,6	84,6	85,2	85,5	92,8	102,8	112,8	122,8	132,8	142,0	144,7
Line 3	0	3,9	8,7	13,2	23,3	32,9	35,6	35,6	40,2	44,0	44,0	47,2	56,9	66,8	76,8	86,6	95,6	97,1	97,1	102,7	112,7	122,7	132,7	142,7	152,4
Line 4	0	3,9	10,6	13,5	23,4	33,4	43,4	51,1	51,7	52,1	56,3	56,6	57,5	65,5	75,3	85,3	90,7	90,7	92,3	101,3	111,3	121,3	131,3	141,3	150,4

Time (sec)

Figure 4: Time versus distance graph of an AMR (simulation outcome).

The results of the simulation analysis can be observed in figure 4. The graph, showing the time spent and distance covered by the AMR, helps to perceive the idleness and busyness of the robot. One AMR was used to feed and serve four production lines. For the 8-hour simulation run, performance metrics such as throughput and utilization were determined. By introducing variations in the simulation model (like the number of AMRs needed for the current production capacity), the effect of an AMR implementation to the transportation cost and defects was observed. The impact of the change, i.e. the automation of the production intralogistics operation, was monitored through previously defined KPIs; the results are shown in table 2. The deployment of an AMR shows a positive impact on every KPI.

Table 2. KPI observations of current real-world and simulated scenarios.

KPIs	Current Scenario (Manual Labor)	Automated Scenario (AMR - Simulated)	Remarks
Throughput	11 pallets per shift (8 hours)	11 pallets per shift (8 hours)	Same throughput, but AMR more flexible than manual process **(+ve)**
Utilization	Fully loaded	Half-loaded	Use of AMR → extra capacity to feed/serve more than four lines **(+ve)**
Cost	Manual transportation costs	Enables effective (human & robot) resource allocation	Use of AMR → less logistics employees → reduced transportation costs **(+ve)**
Defects	Irregularities due to disorganized corridors	Irregularities did not exist thanks to designated routes for AMR	Use of AMR → neat and clean routes → less irregularities **(+ve)**

5. Conclusion

The proposed simulation-based approach is intended to help analyse the feasibility of automation of intralogistics processes and the implementation of AMRs in production logistics. Due to the possibility of achieving a high level of accuracy in the representation of a real production facility in 3D modelling and simulation software, the authors of this study recommend using the aforementioned Industry 4.0 tools as part of the decision-making workflow when automating intralogistics processes. The case study ensured the effective use of 3D simulation and visualization which helped to reduce the installation time of AMRs and analyse the production capacity to figure out the number of AMRs needed to fulfil the current capacity requirement. Moreover, with the defined KPI analysis, it is technically feasible to use AMRs for intralogistics, and it may enhance the proactive decision making as well. Mobile robots are flexible tools which can be applied in different use cases as needed and can be introduced to a production facility stage-wise, first testing a solution with just one AMR, and then gradually increasing their number per required capacity. The simulation-based approach can be replicated in other companies in the future, especially those that are dealing with similar business processes and production environments.

6. References

[1] Mörth O, Emmanouilidis C, Hafner N and Schadler M, 2020, Cyber-physical systems for performance monitoring in production intralogistics, *J. of Computers & Industrial Engg.*, vol. 142, pp. 1-10.

[2] Granlund A, and Wiktorsson M, 2014, Automation in internal logistics: Strategies and operational challenges. *International Journal of Logistics Systems and Management*, 18(4), 538–558.

[3] Mahmood K, Karaulova T, Otto T and Shevtshenko E, 2017, Performance Analysis of a Flexible Manufacturing System, *Procedia CIRP*, vol. 63, 424 – 429.

[4] Riives, J.; Karjust, K.; Küttner, R.; Lemmik, R.; Koov, K.; Lavin, J. (2012). Software development platform for integrated manufacturing engineering system. *Proceedings of the 8th International Conference of DAAAM Baltic Industrial Engineering, Tallinn, Estonia*, 19-21 april 2012. Ed. Otto, T., pp. 555–560.

[5] Klumpp M, *et al*, 2019, Production logistics and human-computer interaction state of the art, challenges and requirements for the future, *The International Journal of Advanced Manufacturing Technology*, vol.105, 3691–3709

[6] Fragapane G, *et al*, 2020, Increasing flexibility and productivity in Industry 4.0 production networks with autonomous mobile robots and smart intralogistics, *Annals of Operations Research*, Springer.

[7] Laguna M and Marklund J, 2019, *Business Process Modeling, Simulation and Design*, 3rd Ed, CRC Press,

[8] Straka M, *et al*, 2017, Application of extendsim for improvement of production logistics' efficiency, *International Journal of Simulation Modelling*, vol.16 (3), 422-434.

[9] Masteamssik S, Schulze T, Raab M and Lemessi M, 2016, Comprehensive 3D Visualization of Simulated Processes in Virtual Factories, *International Conf. Modeling, Sim. and Vis. Methods*, pp. 50-56.

[10] Hwang G, Lee J, Park J and Chang T, 2017, Developing performance measurement system for Internet of Things and smart factory environment, *International Journal of Production Research*, vol. 55, 2590–2602.

[11] Mahmood K, Lanz M, Toivonen V and Otto T, 2018, A Performance Evaluation Concept for Production Systems in an SME Network, *Procedia CIRP*, vol. 72, 603 – 608.

[12] Michalos G, Kousi N, Makris S and Chryssolouris G, 2016, Performance assessment of production systems with mobile robots, *Procedia CIRP*, vol. 41, 195 – 200.

[13] Fischer M, *et al*, 2010, Automated 3D-Motion Planning for Ramps and Stairs in Intra-Logistics Material Flow Simulations, *Proceedings of the 2010 Winter Simulation Conference*, pp. 1648-1660.

IOP Conf. Series: Materials Science and Engineering 1140 (2021) 012027 doi:10.1088/1757-899X/1140/1/012027

High-temperature abrasion resistance and wear mechanisms of chilled high-chromium cast irons

Lukas Widder[a*], Harald Rojacz[a], Markus Buscher[b] and Markus Varga[a]

[a] AC2T research GmbH, Viktor Kaplan-Straße 2/C, 2700 Wiener Neustadt, Austria
[b] Kalenborn Kalprotect GmbH, Asbacher Str. 50, 53560 Vettelschoß, Germany

* lukas.widder@ac2t.at

Abstract. Hard and highly wear resistant chilled cast irons are of high interest in heavy industries like mining and steel production to achieve durable and sustainable wear protection. In this work a comprehensive view on abrasion mechanisms and the change in microstructure of these materials at elevated temperatures is targeted. Therefore, two different types of commercially available chilled high-chromium white cast irons were chosen. Detailed material characterization and hot hardness tests up to 1000°C were performed to get a first indication of temperature stability. For analysis of abrasive behaviour dry sand/steel wheel testing up to 700°C was conducted. A major part of the work was a thorough examination of the wear tracks and the influence on the microstructure. In the investigated materials chromium carbides $Cr(Fe)_7C_3$ and Mo-rich precipitations are present; the size and stoichiometry depend on the chemical composition. According to the differences in carbide size and type and the respective matrix composition, the hardness and abrasive behaviour changes, especially at higher temperatures. Incorporation of SiC abrasive in the wear zone leads to an in-situ wear protection, which was especially effective at 300°C increasing material's wear resistance compared to room temperature (RT). At higher temperatures up to 700°C it was possible to compensate the hardness loss of the cast iron materials.

1. Introduction

Many fields of heavy industry such as industrial mining and milling or the production of steel rely on applications with exceptional abrasion and wear resistance under elevated temperatures. High-chromium white cast irons are used in harsh abrasive environments as appear for mechanical components which somehow manipulate or process large quantities of raw and bulk materials [1,2]. For these highly abrasive applications mainly hypoeutectic alloys (*<4.3 % C*) with Cr-contents of up to *30 wt. %* are commonly applied [3]. Their high volume fraction of hard eutectic chromium-rich carbides in the microstructure makes them extremely abrasion resistant and leads to beneficial mechanical properties such as its high hardness [4-6]. Additionally, the toughness of the matrix exhibits substantial influence on the wear behaviour [7]. The applied chilled casting method improves wear resistance through defined directions of solidifications and finer and more homogeneously distributed carbide grain structure [8,9].

The main objective of the presented work is the comprehensive determination of wear behaviour [10] of two high-chromium white cast irons with distinct differences in chromium content. Abrasion resistance of said materials was tested at elevated temperatures up to 700°C. Furthermore, the influence of applied harsh abrasive environment on the subjacent microstructure was thoroughly examined.

2. Experimental section

2.1. *Materials investigated*

In the present study, two hypoeutectic cast irons commonly applied in high temperature applications were compared. The chemical composition of the samples can be seen in Table 1. Both materials are of relatively common composition but significantly differ in the amount of added Cr, which is of fundamental importance for hardness and wear resistance through carbide precipitation. KALCAST 153 contains *16 wt. % Cr* and is in the medium Cr-range for high-Cr white cast irons. In KALCAST 157, the amount of Cr is increased to *26 wt. %*. All specimens were cut directly from as-produced chill casted plates and surface-ground for further processing in tribological tests. According to their difference in Cr-content the materials reached hardness values of ~760 HV10 (KALCAST 153) and ~840 HV10 (KALCAST 157).

Materials were analysed by scanning electron microscopy (SEM – JEOL JIB-4700 F) equipped with an energy-dispersive X-ray spectroscope (EDS - Bruker X-flash 6/30) and an electron backscatter diffraction detector (EBSD – Bruker e-Flash HR). Phase analysis was performed pre- and post-testing, aiming on microstructural changes, carbide breaking and particle embedment.

Table 1: Chemical composition of applied cast iron samples.

wt. %	*C*	*Cr*	*Mo*	*Fe*
KALCAST 153	3	16	1	78
KALCAST 157	3	26	1.5	65

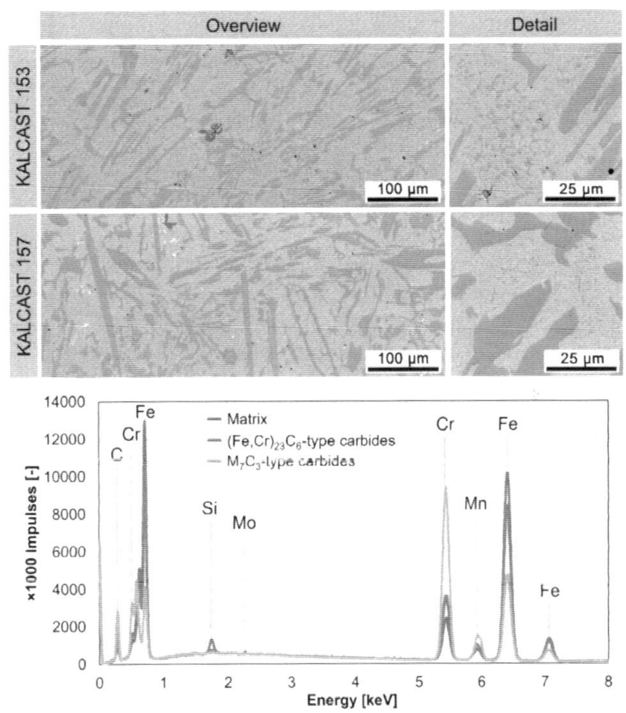

Figure 1: Microstructure and EDS data of the cast irons.

2.2. *Materials investigated*

Figure 1 depicts the images of examined cast irons microstructures. Both cast irons show fine martensitic matrix structure with precipitated carbides (dark areas). KAL CAST 153 tends to form fewer primary carbides than KALCAST 157. Combined EDS and EBSD spot analyses point out that those carbides mainly are of $(Cr, Fe)_xC_y$-type, whereas the EBSD measurements indicate a mixture of both, τ-carbides $(Cr_x\text{-}Fe_y)_{23}C_6$ and M_7C_3-type carbides, depending on local segregations within the carbides.
Secondary carbides (small dark gray phases within the martensitic matrix) are of Cr_7C_3-type and can be detected in KALCAST 153, where their appearance is polygonal, whereas in KALCAST 157 both carbide types can be found in the matrix as spikes. Mo-rich precipitates can be easily visualized as bright grey areas – the amount of those precipitations correlate with the amount of Mo present; in KALCAST 153 fewer precipitates can be seen than in KALCAST 157.

2.3. *Wear testing methods*

To evaluate the materials' abrasive wear behaviour three approaches were undertaken from RT to high temperatures: i) Low-stress abrasion testing according ASTM G65 procedure B (130 N normal load, 1436 m sliding distance, RT, according to the standard [11]), which gives a good indication of low-stress wear behaviour, e.g. loose material sliding over a chute.

ii) Usually, with increased temperatures the hardness of materials decreases, which gives a first indication on the abrasive wear resistance. Therefore, the hot hardness of the materials was measured up to 1000°C with a test device developed at AC²T [12]. Indents with a Vickers-diamond were made (HV10, 3 indents per temperature) from RT to 1000°C in steps of 100°C. To avoid surface deterioration by oxidation the test procedure was carried out in vacuum atmosphere. Measurement of the indents was done by means of optical microscopy after cooling of the samples. iii) Abrasive wear resistance at process temperature is of high importance in many industrial applications. Therefore, we made high-stress abrasive wear tests with a block-on-ring configuration, where abrasive is introduced into the wear gap [11]. For equal temperature conditions, both, the abrasive and the sample, are heated. Tests were done at RT, 300, 500, and 700°C with 10 N normal load. A martensitic steel wheel counter-body with 500 HV and broken SiC (size F36, 425-600 µm) at a flow rate of 150 g/min as abrasive were used, this entails much higher stress in the tribocontact, leading to high-stress conditions. The sliding speed was adjusted to 1 m/s and testing time was 10 min, resulting in 600 m sliding distance. The wear volume was calculated via weight loss and the indicated material density, normalized with the sliding distance resulting in the wear rate (mm³/m). Oxidation of the samples at elevated temperatures generally leads to an increase in sample weight, whereas herein this is negligible compared to the weight loss caused by the abrasion experiments [13]. Hence, the influence of oxidation on the sample weight was neglected for calculations of the wear rates.

3. Results and discussion

3.1. Hot hardness

Results of the hot hardness tests are shown in Figure 2. In the range from RT to ~500°C a significant difference in hardness can be observed: the cast iron containing more Cr-carbides (KALCAST 157) shows much higher hardness for this temperature-range. At 600°C the hardness of the high-Cr sample slightly changes to a similar level as for the lower-Cr cast iron. There, the influence of the softened matrix becomes predominant and the high amount of carbides cannot compensate the loss of mechanical resistance against the indentation. At high temperatures of >800°C, a higher hardness for the cast iron containing the higher Cr-amount can be measured, which can be attributed to the different appearance of the present carbides, since spike-like secondary carbides are present in the matrix at a higher amount.

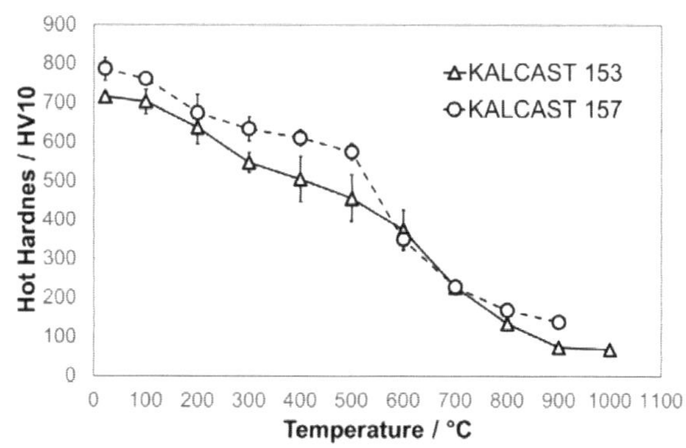

Figure 2: Hot hardness of used cast iron types.

3.2. Abrasion experiments

Results of the standard G65 experiments (see Figure 3) revealed substantial differences in wear resistance at low-stress conditions, which could mainly be caused by the difference in hardness and carbide content of the two materials. Containing low amounts of Cr, KALCAST 153 exhibited much higher wear rates as could be observed for KALCAST 157. The increased amount of hard Cr-carbides could resist the applied abrasive stress much more effectively. An increase of Cr content of *10 wt.%* could reduce the wear losses by over 30 % for the herein

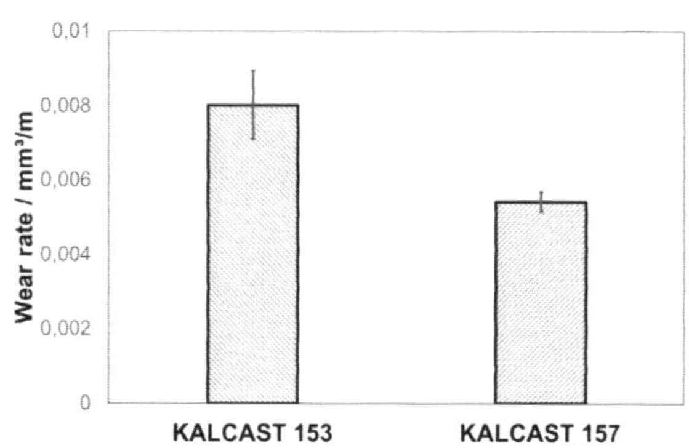

Figure 3: Standard ASTM G65 abrasion results.

tested low-stress environment. This can be attributed to the increased amount of coarse primary carbides of KALCAST 157 compared to KALCAST 153. Furthermore, spike-like sub-micron sized secondary τ-carbides of $(Cr_x\text{-}Fe_y)_{23}C_6$ and M_7C_3-type carbides in KALCAST 157 tend to increase the low-stress abrasion resistance of the matrix at a higher level, compared to the polygonal M_7C_3 secondary carbides within KALCAST 153.

For high-stress abrasion experiments (see Figure 4) at room-temperature a similar wear rate was observed for both cast irons – slightly higher for low-Cr cast iron, KALCAST 153. Interestingly, increasing the testing temperature to 300°C induced an extensive drop of wear losses for both materials, especially for KALCAST 153, resulting in a substantially higher wear resistance of this material compared to KALCAST 157. A further increase of temperature to 500°C lead to a raise in wear rates to the room temperature-level for KALCAST 157. Even for 700°C tests wear losses for this material remained at the same level and did not increase any further. From very low wear rates at 300°C, KALCAST 153 showed a linear increase when temperatures were raised to 500°C and 700°C. At 700°C both iron cast materials again showed similar wear rates, albeit results for the low-Cr KALCAST 153 were still slightly lower. Generally, wear rates of KALCAST 157 vary much less (~20%), compared to KALCAST 153 whereas in the medium range of 300-500°C KALCAST 153 exhibited substantially lower material losses (45%) compared to room temperature.

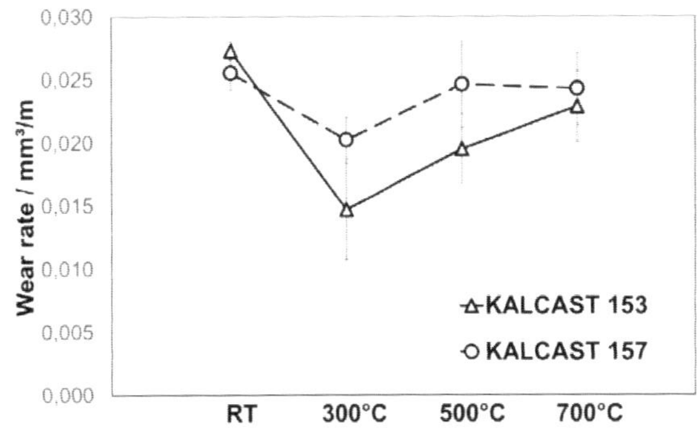

Figure 4: High-stress abrasion results.

Cross-sectional SEM analyses of the high-stress abrasion wear tracks provide a deeper understanding of the occurring wear mechanisms [14,15]. Figure 5 shows SEM images of metallographic cross-sections of wear tracks at RT, 300°C and 700°C to depict the correlation of the microstructure to wear mechanisms at different temperatures. As seen in these images, wear mechanisms change with increasing temperature. At RT only few SiC-abrasive particles are incorporated since the initial matrix hardness is at a high level. Increasing the samples' temperature, which decreases the matrix hardness, tends to have a positive effect on the in-situ formation of a tribolayer, more precisely a mechanically mixed layer (MML) like observed in previous studies [16,17].

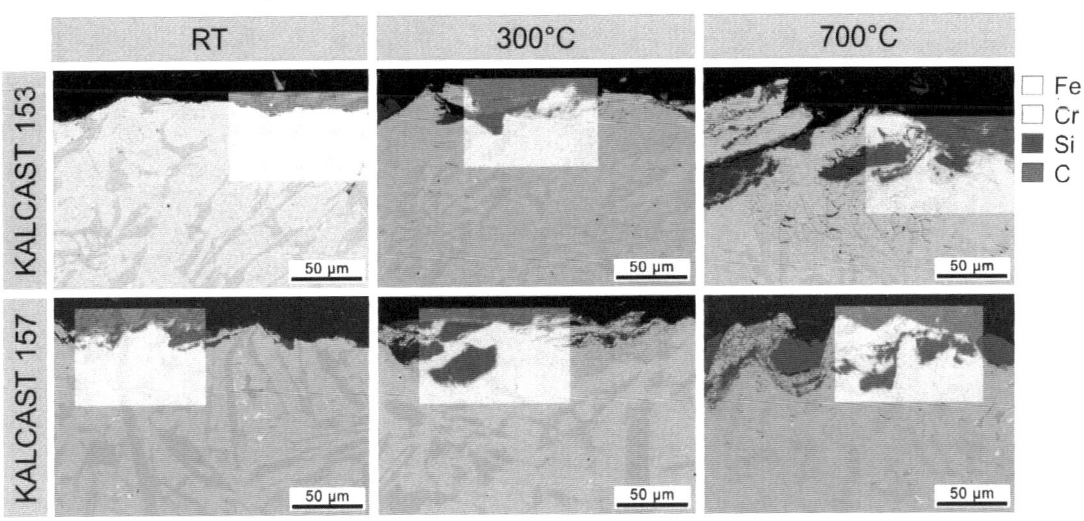

Figure 5: Cross-sectional SEM analysis of the wear scars including inserts of overlayed EDS-mappings.

At 300°C wear rates are the lowest, which may be attributed to a still sufficient hardness of the matrix against removal by abrasive material. But low enough to incorporate very hard SiC abrasives to form a MML, hindering further wear loss. At 500°C and 700°C KALCAST 153 exhibits increasing wear rates, which can be assigned to a lower matrix hardness. Leading to breaking of the coarse secondary carbides within the matrix, this impedes their effectiveness as wear protection. KALCAST 157 features finer secondary carbides in the matrix, reducing hardness loss and impeding primary carbide breakage, but also allowing for MML formation. In summary, both materials are capable of improving, their RT-wear resistance by in-situ MML-formation, which compensates for the hardness loss due to applied elevated temperatures.

4. Conclusions

High -temperature abrasion is a serious issue in many industrial applications. Within this work two white cast irons with high Cr-content (*16* vs. *26 wt. %*) were evaluated regarding their abrasive wear resistance and wear mechanisms up to 700°C. It was found, that the higher Cr-content led to the precipitation of larger primary carbides, but also to spike-like $(Cr_x\text{-}Fe_y)_{23}C_6$ and M_7C_3-type secondary precipitations, which were especially beneficial in low-stress abrasion at room temperature. The lower Cr alloy showed polyhedral M_7C_3 secondary precipitations, which allowed more incorporation of abrasive in high temperature high-stress testing, improving its wear resistance up to 45% compared to room-temperature.

Acknowledgments

This present study was funded by the accompanying partner Kalenborn Kalprotect GmbH and the Austrian COMET Programme (K2-Project InTribology, no. 872176) and carried out at the "Excellence Centre of Tribology", Wiener Neustadt, Austria (AC2T research GmbH).

References

[1] Ngqase M and Pan X 2020 *J. Phys. Conf. Ser.* **1495** (1), p. 012023). IOP Publishing
[2] Kvon S S, Kulikov V Y, Filippova T S and Omarova A E 2016 *Metalurgija* **55**(2) 206-08
[3] Laird G, Gundlach R and Röhrig K 2000 *Abrasion-resistant cast iron handbook* (Illinois: American Foundry Society) p. 72
[4] Kopyciński D, Kawalec M, Szczęsny A, Gilewski R and Piasny S 2013 *Arch. Metall. Mater.* **58**
[5] Pero-Sanz J A, Plaza D, Verdeja J I and Asensio J 1999 *Mater. Charact.* **43**(1), 33-39
[6] Lu B, Luo J and Chiovelli S 2006 *Metall. Mater. Trans.* A *37*(10), 3029-38
[7] Hussain N, Kumar A and Vijayanand P 2014 *Int. J. Eng. Res. Technol.* **3**(7), 2278-81
[8] Bedolla-Jacuinde A, Arias L and Hernández B 2003 *J. Mater. Eng. Perform.* **12**(4), 371-82
[9] Smith M 1994 *The influence of the degree of transformation of M7C3 carbides to M23C6 carbides on the properties of high chromium white cast irons* (Doctoral dissertation, University of the Witwatersrand, ZA)
[10] Widder L, Grafl A, Lebel A, Tomastik C and Brenner J 2011 *Tribol. und Schmierungstechnik*, **58**(2), 11-15
[11] ASTM International 2010 ASTM G65-04-Standard Test Method for Measuring Abrasion Using the Dry Sand/Rubber Wheel Apparatus
[12] Varga M, Flasch M and Badisch E 2017 *Proc. Inst. Mech. Eng.* J **231**(4), 469-78
[13] Liu H N, Nomura M, Ogi K and Sakamoto M 2001 *Wear* **250**(1-12), 71-75
[14] Rojacz H, Mozdzen G, Weigel F and Varga M 2016 *Mater Charact.* **118** 370-81
[15] Rojacz H, Premauer M and Varga M 2018 *Wear* **410** 173-80
[16] Varga M and Badisch E 2017 *Wear* **384-385** 114-23
[17] Rojacz H, Pahr H, Baumgartner S and Varga M 2017 *Trib. Int.* **113** 487-99

Thermal properties of calcium-aluminate based materials

P Kulu[1], D Goljandin[1*], R Traksmaa[1], T Kaljuvee[2] and A Gregor[2]

[1] Tallinn University of Technology, Department of Mechanical and Industrial Engineering, Ehitajate tee 5, 19086, Tallinn, Estonia.
[2] Tallinn University of Technology, Department of Material and Environmental Technology, Ehitajate tee 5, 19086, Tallinn, Estonia.

dmitri.goljandin@taltech.ee

Abstract. Good cementing properties, fast setting and strong thermal performance make calcium-aluminate a valuable raw material for use in the production of different types of new refractory materials, e.g., heat conductive/storage materials. The main aim of the study was to determine thermal properties of novel Nb-slag based materials with different fillers, and to clarify optimal composition and technology. Process of preparation of studied materials was following: mixing of components, casting to moulds and hardening of materials. To estimate potential application areas, we studied the following thermal properties of CA-based materials: thermal behaviour, coefficient of thermal expansion (CTE) and conductivity. For this small cylindrical specimens were cut out from produced materials, and plates sized 300 x 300 mm were used for conductivity studies. Different compositions of CA-based materials, the hardening process, and the influence of mechanical activation on the strength were analyzed. Best thermal properties similar to the analogous reference material were obtained by quartz sand and granite sand as filler materials. The thermal conductivity of the novel CA-based material is 1.5 times higher and the bending strength is about 3 times higher compared to commercial thermoplates.

1. Introduction

Slags from metallurgy composed primarily of silica with concentrations of constituents like alumina, magnesia and other components have the potential value for application, and technologies have emerged to recycle and reuse reprocessed slag in different building materials [1]. Niobium slag resulting from the aluminothermic process at Sillamäe metallurgy plant, Estonia, contains mainly calcium-consisting aluminate (CA) and due to non-full calcium-aluminothermic reduction, some amount of initial materials – mainly niobium oxide Nb_2O_5 (3-7%) and pure Nb up to 1% (depends on the batches) [2].

The technology of separation of pure Nb based on the disintegrator treatment of CA consists of the following steps: precrushing of big pieces, disintegrator milling and separation of milled product to fine (< 0.3 mm) and coarse fractions, followed by separative milling of coarse fraction to separate pure Nb [2, 3]. Preliminary experiments were conducted to clarify the potential application area of ground CA as the main component of heat accumulation and heat-transfer materials. Similar alumina-silica based ceramic materials are used for accumulation (slow heat release) of heat in fireplace stoves and fireplaces [4]. Due to analogous compositions of niobium slag, it may be a potential material for refractory products.

It was demonstrated by our previous experiments [2, 3] that CA-based materials have high potential of application as the main component of refractory materials, especially in the form of heat accumulation and heat transfer elements of stoves, fireplaces etc. Our preliminary experiments showed that CA-based materials have high mechanical properties and similar heat conductive properties (about 0.34 $W \cdot m^{-1}K^{-1}$ against 0.42 $W \cdot m^{-1}K^{-1}$ of commercial) [2].

2. Experimental

2.1. Initial materials

Niobium slag was subjected to impact milling and a coarse ground product was subjected to separative milling in disintegrator DSL-115. Used ground slag contained mainly calcium aluminates (CA); fineness was <125 μm. Old CA (aged more than some 10 days) was used in our experiments. To have fresh mechanically activated CA, it was milled immediately before using (during the next 5 h).

The following CA based mixtures were studied (Table 1):

- Pure CA (fresh and old);
- CA with additions of granite sand (GS), CEN standard quartz sand (QS) and polarstone sand (PS). The fillers were obtained by disintegrator milling of GS gravel, standard QS and PS wastes. It was shown the same chemical compositions after milling;
- CA with addition of fly ash cenospheres of alumina-silica (AC) (lightweight, inert, hollow sphere filled with inert air or gas – a typical byproduct of coal combustion at thermal power plants (Biotecha Latvia Ltd, Riga, Latvia), mainly consisting of SiO_2 and Al_2O_3). The chemical composition of AC was studied previously in [6]. Figure 1 shows the SEM images of the used filler materials.

As a reference material, analogous Cebud thermoplate CPA-K [5] were studied and concrete (cement + standard sand) of hardened materials was prepared and tested for comparison.

Figure 1. Filler materials: a – granite sand (GS), b – quartz sand (QS), c – polarstone sand (PS) and d – fly ash cenospheres (AC).

For the study of mechanical properties, bending strength specimens sized 40 x 40 x 160 mm were prepared. For the study of thermal properties, cylindrical specimens with ⌀7 mm from mechanical test pieces were cut out. For the study of thermal conductivity, the plates sized 300 x 300 x 25 mm were produced. Prepared mixtures of CA with fillers and addition of water (about 33%) were casted to molds and densified by vibration. The procedure was as follows: preparation of components → mixing of mixtures → hardening at air 24 h → drying at 100 °C for 24 h. Density of CA-based materials with filler GS and QS was about 2.0 g/cm^3, with PS was about 1.5 g/cm^3 and with AC was about 1.3 g/cm^3.

2.2. Characterization of mechanical properties

Prismatic specimens were tested on a universal testing machine according to standard EN 196-1 [7]. Bend strength of the specimens at three-point bending was determined. Universal testing machine P-5 (50 kN) was used and distance between the points of support was 100 mm.

2.3. Characterization of thermal properties

The thermomechanical characteristics of the samples from heat-conductive materials were determined on Setaram Setsys 1750 CS Evolution dilatometer. Cylindrical specimens with a diameter of 7 mm and a height of 15-16 mm in corundum sample holder were heated up to 1000 °C at a heating rate of 5 °C per minute and cooled down at the same rate in an Ar atmosphere. The diagrams DTDA – T and Δl% – T were plotted.

Thermal conductivity was measured according to the standard EVS-EN 12667 [8] using LaserComp FOX-304 heat flowmeter. The average heat flux was used to calculate the thermal conductivity (λ) and thermal resistance (R) according to Fourier's Law.

For thermal plates the optimal composition of CA : Filler (50:50) was selected keeping in the mind as technological as well strength properties of studied materials.

Table 1. Composition of studied materials.

Composition [wt%; conditions; fraction size]		Post-treatment
CA[a]	Filler	
[100; fresh and old]	–	Hardening at 20 °C / up to 54 days
[66, 50 and 34; fresh]	**granite sand** (GS) [34, 50 and 66; milled; < 125 μm]	
[66, 50 and 34; fresh]	**quartz sand** (QS) [33, 50 and 66; milled; < 125 μm]	
[66, 50 and 34; fresh]	**polarstone** (PS) [33, 50 and 66; milled; < 125 μm]	
[75 (vol% 50); fresh]	**ash cenospheres** (AC) [25 (vol% 50); < 300 μm;]	
–	portland **cement** [25] + standard **sand** [75; 2÷0 mm]	Hardening at 20 °C / 7 days

[a] – fraction size < 125 μm

3. Results and discussion

3.1. Thermal Analysis

As it follows from the DTDA-diagram belong to old CA (Figure 2), changes up to 100 °C are connected with the emission of physically bound water. Changes in the range from 200 up to 400 °C are probably connected with thermooxidation of the matter contained in Nb-slag as remains of the reduction process and change at 970 °C caused probably with formation of liquid phase.

The contraction-expansion behavior of samples of the studied materials by heating up to 1000 °C at the heating rate of 5 °C/min is presented in Figure 3.

As the graphs show, up to 180 – 200 °C, low shrinkage (up to –0.1%) can follow. Then, around 260 °C, deep shrinkage of bodies at 340 – 380 °C follows, which is most significant for pure CA (fresh -1.03%, old –0.85%). Starting from 400 °C, stabilization takes place.

Behavior of composites (except CA+QS) differs from the behavior of pure CA; shrinkage is less and more dependent on the temperature. Shrinkage increases up to 360 °C (up to –0.5%) and then stabilizes.

Composition containing quartz sand (CA+QS) behaves in a slightly different way: shrinkage is low (–0.2%) up to temperature 480 °C and then expansion up to 580 °C (up to +0.5%) follows and it is stabilized. The thermal behavior of CA+AC body is characterized in an analogous way: starting from 550 °C notable elongation (+0.6%) and stabilize from 600 °C up to 1000 °C.

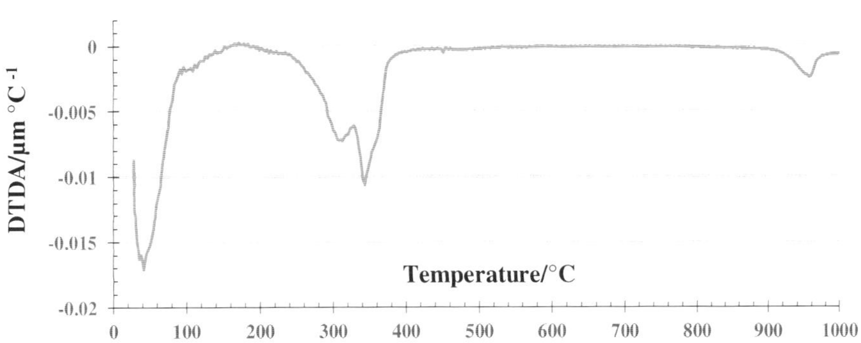

Figure 2. Heating curve (DTDA-diagram) of CA.

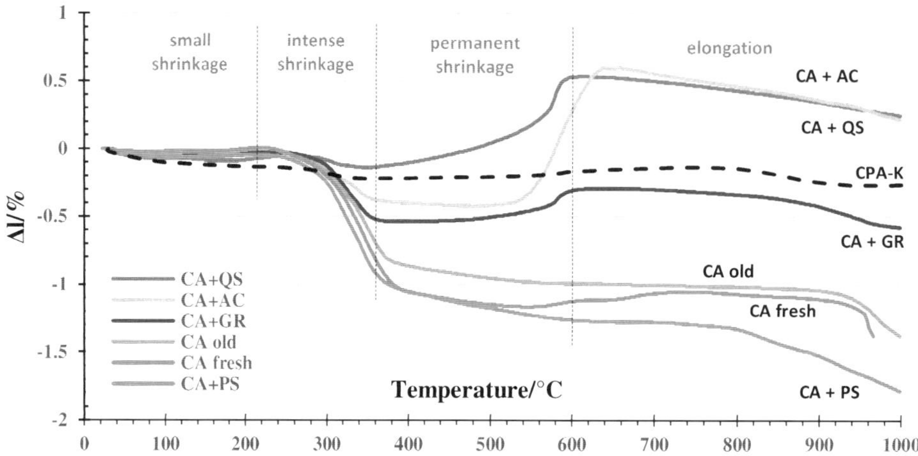

Figure 3. The shrinkage-expansion curves of CA based materials.

3.2. Behavior of CA in the hardening process

It was demonstrated in our preliminary experiments that in the hardening process, a notable increase of the temperature of hardenable specimens takes place, first, with CA just milled.

For this propose, a special study with pure CA with particles size <125 μm was carried out: old (early milled and aged) and fresh (just milled) CA were used. Hardening time was 16 h.

Figure 4a shows that the temperature at the hardening of specimens made from fresh materials increases up to 100 °C whereas with specimens from old CA, the temperature is not significantly increased. It reveals mechanical activation in the high velocity impact milling process. The mechanical activation of materials due to the disintegrator milling was demonstrated in our earlier work [9, 10]. It causes a notable increase in the mechanical properties as well as the bending strength; materials made from fresh milled CA demonstrated up to 2 times higher strength (Figure 4b). It takes about two days to achieve the necessary minimal strength level, e.g., 4 MPa with old CA, while with fresh CA, only 10 h are required. One of the ways to suppress this phenomenon is using ballast materials, i.e., inert fillers. Using fillers noticeable heating of cast mixture was not observed and was similar to old CA.

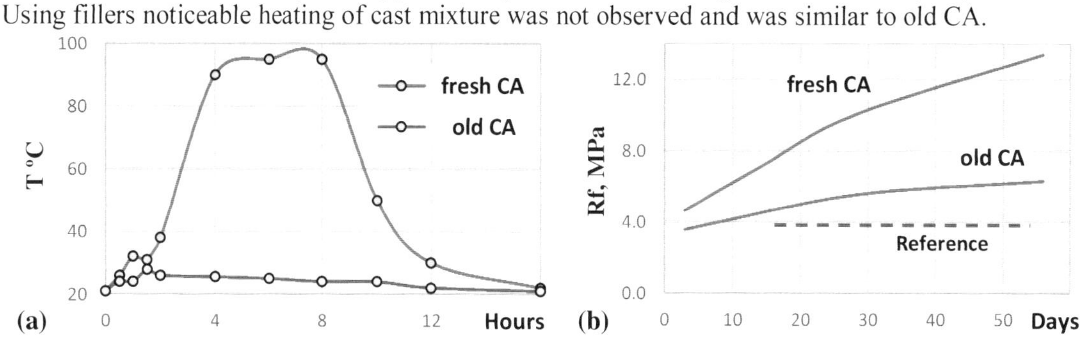

Figure 4. Influence of mechanical activation on the behavior and properties of CA: a – self-heating at the hardening process; b – increase of flexural strength Rf during hardening.

3.3. Optimization of composition

To suppress the high activity of fresh CA, the inert filler materials like granite sand (GS), polarstone sand (PS) and CEN standard sand – quartz sand (QS) were used. The studied mixtures consisted of filler materials from 33 to 66%. Materials were made by the same technology: mechanical mixing of components, casting of mixtures to molds and hardening at room temperature for 24 h, followed by drying at 100 °C for 24 h. Three prismatic specimens were subjected to the bend load at the three-point test. Results of the bending test are given in Figure 5.

At all studied compositions, increase of the filler content causes a decrease in the bending strength. With fillers GS and QS content over 50%, a deep increase in the strength can be seen (Figures 5a and 5b). Up to that point, at the increase of the filler content, flexural strength decreases proportionally. At the same time, the strength of materials increases with the use of fresh immediately disintegrator milled filler materials, being 30% higher (GS and QS) except for PS.

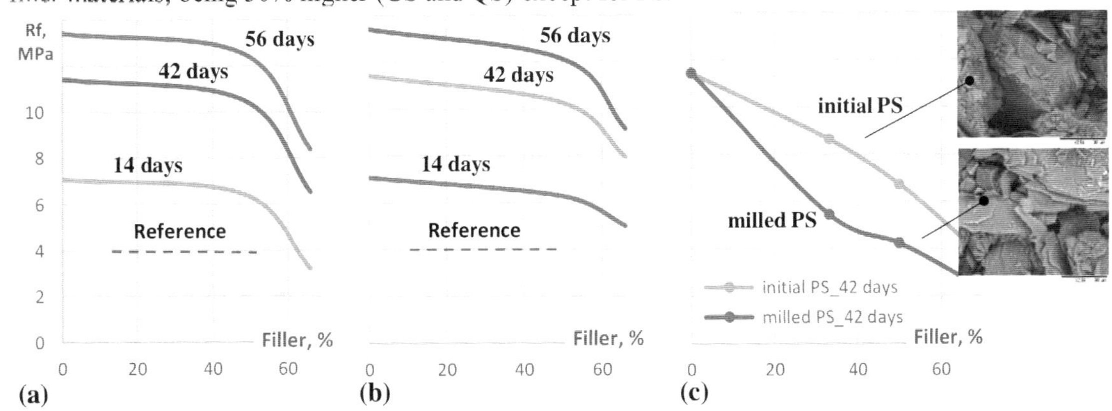

Figure 5. Dependence of flexural strength on the amount of filler in the mixture: a – granite sand (GS), b – quartz sand (QS) and c – polarstone (PS)

Notable influence of hardening time on the strength was observed. Samples gained half of their strength in the first 14 days of hardening. Over time (42 and 56 days), the rate of increase in the flexural strength slowed down.

Adding polarstone (PS) as a filler has a strong effect on to the strength. The dependence of the decrease in strength on the amount of filler has a pronounced linear character (Figure 5c). Better strength was observed when using non-milled PS filler. It can be explained with a change of morphology of PS particles as a result of milling. Due to the lamellar structure of PS particles, we obtained flake powder (Figure 5c).

3.4. Thermal conductivity of CA based materials

Results of the thermal conductivity study are given in Figure 6. The figure shows that the CA+QS material has better conductivity, about 55% higher than the reference material CPA-K, followed by CA+GS material (about 29% higher) and CA+PS (10% higher).

Taking into account the mechanical and thermal properties and the economic aspect, CA-based materials with a quartz sand filler are more prospective.

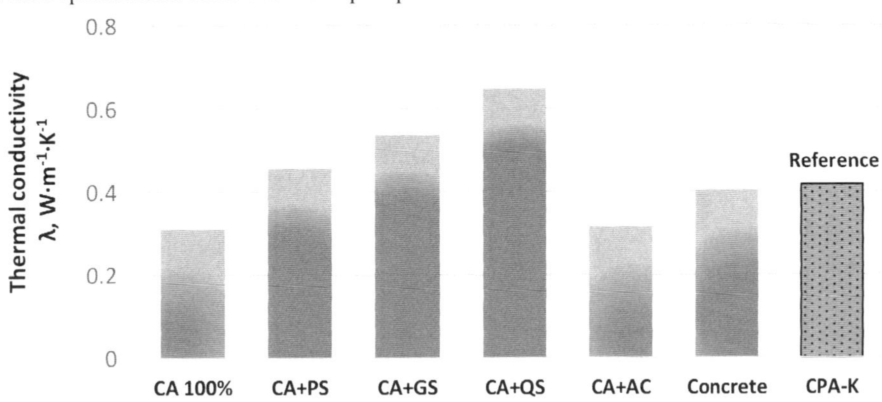

Figure 6. Thermal conductivity of CA based materials.

4. Conclusions

1. Novel Nb-slag based materials were obtained and prospectives of application as heat conductive materials were demonstrated.
2. Based on the study of mechanical and thermal properties of calcium aluminate (CA) materials of different compositions and conditions, we can draw the following conclusions:

- With mechanical milling of CA, activation of ground material is accompanied due to the fresh surfaces formed. As a result, this hardening process is more intensive, accompanied by higher strength (up to 2 times). At the same time, an increase in the temperature accompanied in the hardening process results in the decrease of density and distortion of material. It can be suppressed by optimizing the composition by the use of fillers.
- Optimal CA: filler proportion in fresh CA based materials is 50:50 percentage. Increase of strength using a filler is about 10 – 15% (by GS) and 25 – 30% by QS and PS.
- In view of thermal properties, the composition of CA with a granite and quartz sand fillers is favorable. The thermal conductivity properties is 30-50% higher and strength properties about 3 times higher than that of the reference thermoplates.

Acknowledgements

This work was supported by the Estonian Research Council grants PRG643 and PRG665 and by the Environmental Investment Centre Foundation grant KIK19019.

References

[1] Batalin, B. Harm and benefits of slags landfills, *J. Nature*. 2001, 1/10, 162-165.
[2] Kulu, P., Goljandin, D., Külaviir, J., Hain, T., Kivisto, M. Recycling of Niobium Slag by Disintegrator Milling. In: K. Karjust, J. Kübarsepp, T. Otto, I. Hussainova (Ed.). *Modern Materials and Manufacturing* 2019 Trans Tech Publications Ltd.. (Key Engineering Materials; 799), 2019, (97–102).
[3] Kulu, P., Goljandin, D., Viljus, M., Traksmaa, R., Gregor, A. Heat Conductive Plates from Recycled Niobium Slag. 2020 (in print)
[4] EVS-EN ISO 10081-1:2005. Classification of dense shaped refractory products - Part 1: Alumina-silica.
[5] Cebud. Eu/rh. Fireplace stoves made of Akubet http://cebud.eu/en/
[6] Shishkin, A., Mironov, V., Zemchenkov, V., Antonov, M., & Hussainova, I. Hybrid Syntactic Foams of Metal – Fly Ash Cenosphere – Clay. *Key Engineering Materials*. 2016, 674, 35–40.
[7] Standard CEN - EN 196-1. Methods of testing cement - Part 1: Determination of strength.
[8] Standard EVS-EN 12667 Thermal performance of building materials and products – Determination of thermal resistance by means of guarded hot plate and heat flow meter methods.
[9] Bumanis. G., Bajare, D., Goljandin, D. Performance evaluation of cement mortar and concrete with incorporated micro fillers obtained by collision milling in disintegrator. *Ceramics-Silikáty*. 2017, 61, (3) 231 - 243.
[10] Goljandin, D., Kulu, P. Disintegrators and Disintegrator Treatment of Materials. Germany: *LAP LAMBERT Academic Publishing,* 2015.

IOP Conf. Series: Materials Science and Engineering 1140 (2021) 012029 doi:10.1088/1757-899X/1140/1/012029

Performance of Al_2O_3-cBN materials and perspective of using hyperspectral imaging during cutting tests

M Antonov[1], A Zahavi[1], R Kumar[1], M Tamre[1] and P Klimczyk[2]

[1]Tallinn University of Technology, Estonia

[2]Krakowski Instytut Technologiczny, Poland

E-mail: Maksim.Antonov@taltech.ee

Abstract. The performance of cutting tool materials (CTMs) influences the quality and lifetime of parts produced with their help. Unexpected fracturing or other failures of the tools lead to defects of parts that exaggerate materials fatigue and fracture processes. For the purpose of Industry 4 and future generations of factories it is important to enable *in-situ* monitoring of cutting processes while hyperspectral imaging can serve as a powerful tool. Cubic boron nitride (cBN) has extreme hardness and can provide improved wear resistance if mixed with other CTMs. Besides, such materials can be used without cooling liquids that helps to mitigate workplace health risks. The aim of the current work was to understand how well current hyperspectral imaging technologies can track the changes in performance of CTMs with addition of cBN. The paper presents the results of multiple in-situ (obtained during cutting with real lathe) and static (before or after cutting) tests performed with hyperspectral camera. The wear rate of CTMs and roughness of workpieces were measured with the help of scanning electron microscope and 3D optical profiler respectively. The effect of cBN content and effect of TiN or ZrO_2 additives on performance of alumina-based CTMs produced by spark plasma sintering technique is presented

1. Introduction

Hyperspectral imaging (HSI) is known for processing the information over the electromagnetic spectrum to identify various materials by obtaining the spectrum of each pixel of the image [1][2]. In other words, HSI combines the power of spectroscopy and digital image processing to identify the materials in a scene. While HSI is a fast-growing area in the remote sensing field for earth observation; emerging mobile hyperspectral cameras made it possible to use them for material classification and object recognition in the real-world scale [3].

2. Materials and methods

2.1. Materials

Five alumina-based cutting tool materials (CTMs) reinforced by cubic boron nitride (cBN) without or with TiN or ZrO_2 additives were produced by spark plasma sintering (SPS) technique – see table 1. Their physical and mechanical properties were measured in previous work of P. Klimczyk et al. [4]. Workpiece material was hot rolled normalized unalloyed medium carbon (main constituents: 0.45 % C, 0.2 % Si, 0.7 % Mn, base - Fe) steel C45 according to EN 10083-2.

Table 1. Properties of cutting tool materials investigated [4].

Designation	Composition, %	Relative density, %	Youngs modulus, GPa	Vickers hardness HV30, GPa	Fracture toughness, K_{IC} (indentation method), $MPa \cdot m^{1/2}$
A100	Al_2O_3	>99	398 ± 7	16.3 ± 0.9	3.9 ± 0.6
A10B	Al_2O_3-10cBN	98	421 ± 9	16.4 ± 0.5	5.4 ± 0.5
A20B	Al_2O_3-20cBN	96	442 ± 8	18.7 ± 0.3	5.8 ± 0.4
A20B10T	Al_2O_3-20cBN-10TiN	96	428 ± 11	17.2 ± 0.3	5.5 ± 0.5
A20B15Z	Al_2O_3-20cBN-15ZrO_2	99	409 ± 13	18.8 ± 0.7	5.4 ± 0.5

2.2. Wear rate measurements

The cutting tests were performed on *Sumore* SP2138 CNC lathe. The test sample produced by SPS was cut into required shape and then the edges were ground (figure 1A). Total number of cuts with depth of 0.075 mm and width of 80 mm was 70; feed rate was 0.05 mm per rotation (0.88 mm s^{-1}); average cutting speed was 100 m min^{-1} (1.67 m s^{-1}); total duration of cutting was 105.66 min (6342 s); total length of removed chip was 10566 m, figure 1B. The linear and volumetric wear rates were measured according to scanning electron microscope (*Hitachi* TM1000) images. Volumetric wear was approximated by triangular pyramids, figure 1C. Roughness was measured with the help of *Bruker* ContourGT-K0+ optical 3D profiler.

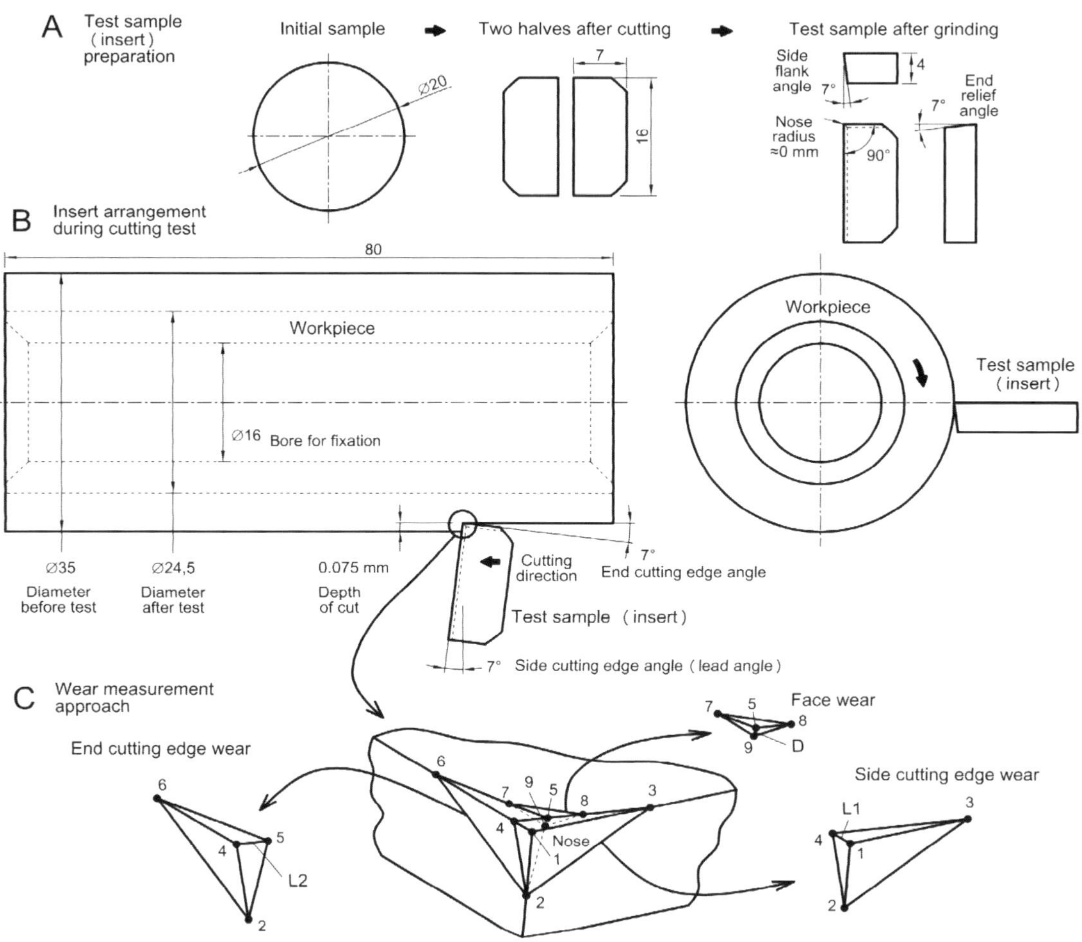

Figure 1. The sample (insert) preparation (A), it's arrangement during cutting test (B) and wear rate measurement approach (C).

2.3. Hyperspectral investigations during and after cutting test

The system was comprised mainly of a hyperspectral camera, lighting source, the test sample fixed in the lathe, and a frame for fixation of components (figure 2). The mobile hyperspectral camera used for conducting these experiments was *SPECIM* IQ (Spectral Imaging LTD, Finland), a VNIR (Visible – Near-infrared) camera which collects the spectrum of each pixel of the image in the range of 400 – 1000 nm [5]. Recording of one image took approximately 5 minutes. Similar to other computer vision applications, the light source used during the image acquisition phase, plays a vital role in the validity of the results. Incandescent lights were applied as the best illumination source for indoor hyperspectral imaging applications [6]. A frame was assembled to install the incandescent light in an appropriate height above FOV (Field of View).

The spectrum of the insert material (before testing) and steel workpiece (after testing) was extracted in static conditions (without movement), while in addition, images of cutting zone were taken *in-situ* during machining to study the effect of feed rate (2.5, 5.0 mm/s), rotational speed of workpiece (500, 1000 rpm / 8.33, 16.66 rps) and depth of cutting (0.1, 0.2 mm).

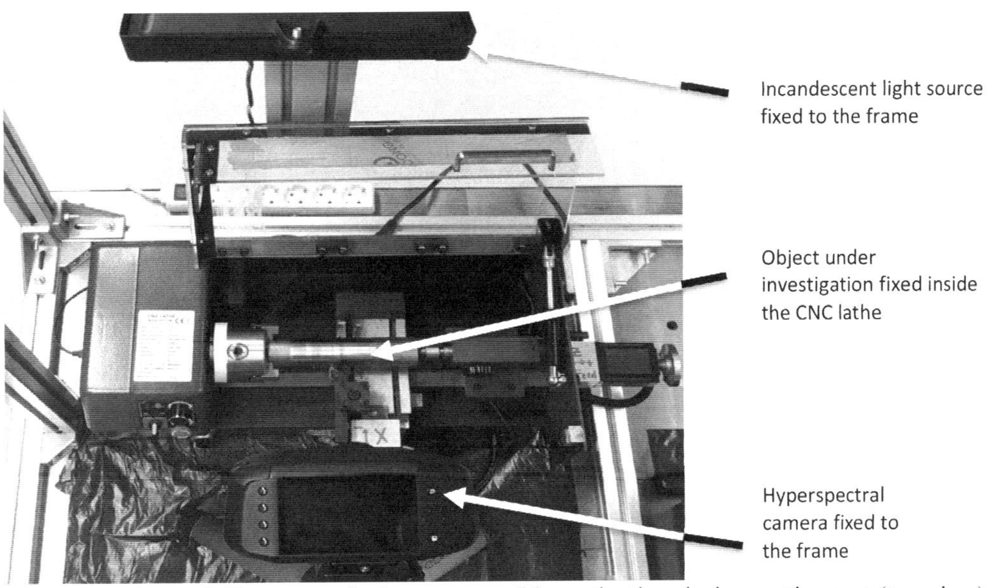

Figure 2. Test setup for hyperspectral investigation during cutting test (top view).

For acquiring the suitable images for analysis, more than 50 images from the scene from different angles with different scenarios were taken. These images were taken to find the best distance from the illumination source to the material, and material to the camera, as well as the proper angle and proper field of view. After taking these images, going through the images one by one, less than ten images have been selected. Also, the best position of the illumination source, camera and material have been found. Therefore, after these experiments, by finding the optimum values of these parameters, further experiments have been done in this optimum situation.

3. Experimental results

3.1. Wear rate measurements

The wear rates of CTMs are presented in figure 3 along with roughness of workpieces after 70 cuts.

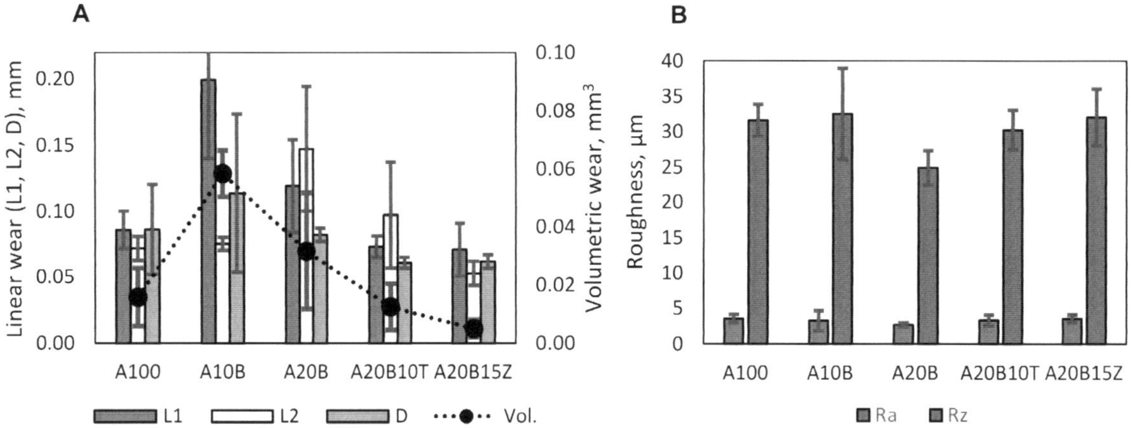

Figure 3. The wear rates of tested samples (A) and roughness of the workpiece after cutting test (B).

It is possible to see that wear resistance of alumina during dry cutting can be significantly (up to 3 times according to volumetric wear) improved by addition of 20 % of cBN and 15 % of ZrO_2 while providing the same roughness of produced parts. According to SEM images, pure alumina (A100) was suffering from chipping of the cutting edge. Reduced performance of materials with 10 and 20 % of cBN could be explained by loss of alumina between grains that facilitates adhesion of steel onto cutting insert and its overheating.

3.2. Hyperspectral investigations during and after cutting test

For minimizing the error that might happen during the image acquisition phase, each cutting tool material was under the experiment five times, and the spectral signatures of ten different pixels of the material were conducted. Then, the average of these 50 spectra (5 multiplied by 10) was calculated to have a final spectrum of the material which may better represent the object or the material. Also, since spectral signature extracted from the hyperspectral image has small fluctuation (less than one percent) over the electromagnetic, for smoothing the graph in the preprocessing stage, Savitzky-Golay filter was used similar to *SPECIM* IQ Studio software; the software for processing the hyperspectral images acquired by *SPECIM* IQ camera [7]. The graphs show the spectral of cutting tool materials (figure 4A), and the spectral signature of steel workpiece after being cut by them (figure 4B).

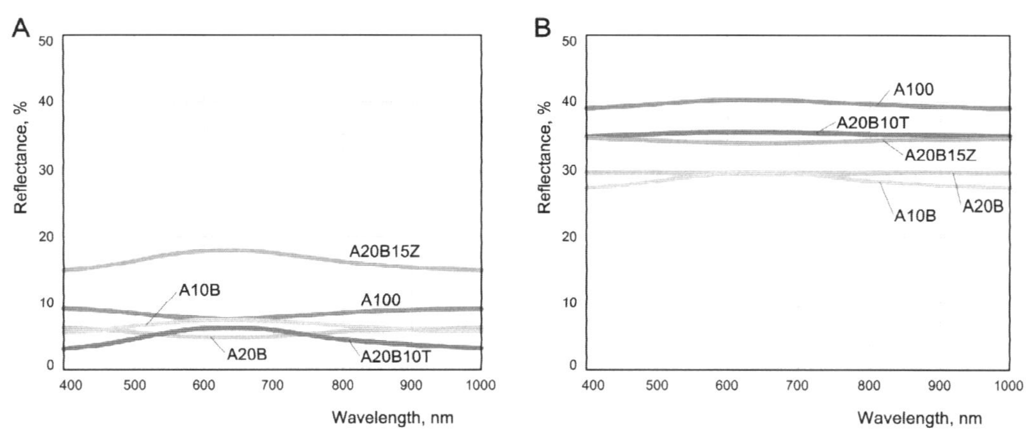

Figure 4. Spectral signature of the cutting tool materials (A) and spectral signature of steel workpiece taken after cutting test (B) in the range of 400 – 1000 nm.

It can be interpreted from figure 4A, that A20B15Z material (blue line) has a brighter colour than others, and it causes to reflect more energy to the camera. Therefore, its spectral signature has a higher percentage of reflectance in comparison with other materials over the spectrum.

However, for identifying the material using Figure 4A, there is a need to observe significant peaks or depths, while in these experiments, the spectral signatures of all materials were almost flat (less than three percent change over the range). In other words, it was not possible to see any significant peaks or depths related to these materials (the characteristic peaks) in the VNIR range. Also, in figure 4B, taken by VNIR camera, the spectrum of workpiece is almost flat over the entire range in all experiments. It shows that the characteristic peaks of steel are outside of the VNIR range.

Spectral signatures taken during *in-situ* testing have shown that doubling of the value of feed rate, rotational speed of workpiece and depth of cutting resulted in max 0.1, 1.0 and 1.0 % change of the workpiece reflectance respectively. Using of near-infrared camera (1000-1700 nm) might lead to more precise identification of materials and control of cutting process.

4. Conclusions

- It was found that wear resistance of alumina during dry cutting can be significantly (up to 3 times according to volumetric wear) improved by addition of 20 % of cBN and 15 % of ZrO_2 while providing the same roughness of produced parts. Such material can be used without cooling liquids to mitigate workplace health risks.
- It was found that unreinforced reference pure alumina exhibit chipping of cutting edge that can lead to defects of parts that exaggerate their fatigue and fracture processes.
- It was possible to establish in-situ monitoring of cutting process by hyperspectral camera.
- Spectral signatures taken during *in-situ* testing has shown that doubling of the value of feed rate, rotational speed of workpiece and depth of cutting resulted in max 0.1, 1.0 and 1.0 % change of the workpiece reflectance respectively that was not enough for tracking of the tool performance. It is advised to use near-infrared camera (1000-1700 nm) to achieve more precise identification.

Acknowledgments

The authors would like to acknowledge the support from Estonian Ministry of Education and Research (M-ERA.NET DURACER ETAG18012, PRG643, SS427).

5. References

[1] Chang C 2003 *Hyperspectral* Imaging: Techniques for Spectral Detection and Classification (Springer Science & Business Media LLC)
[2] Grahn H and Geladi P 2007 *Techniques and Applications of Hyperspectral Image Analysis* (John Wiley & Sons)
[3] Zahavi A 2020 *Deep Learning for Material and Object Identification Using Hyperspectral Imaging* (Tallinn University of Technology)
[4] Klimczyk P, Wyżga P, Cyboroń J, Laszkiewicz-Łukasik J, Podsiadło M, Cygan S, and Jaworska L 2020 Diamond Relat. Mater. 104, 107762
[5] Specim IQ Technical Specifications https://www.specim.fi/iq/tech-specs/ (accessed Nov. 02, 2020)
[6] Zahavi A, Palshin A, Liyanage D and Tamre M 2019 Proc. 20th Int. Conf. Res. & Ed. in Mech. doi: 10.1109/REM.2019.8744086.
[7] Specim IQ Studio - Software for the Specim IQ camera https://www.specim.fi/iq/specim-iq-studio/ (accessed Dec. 09, 2019)

IOP Conf. Series: Materials Science and Engineering 1140 (2021) 012030 doi:10.1088/1757-899X/1140/1/012030

Basic material and technology investigations for material bonded hybrids by continuous hybrid profile fabrication

K Schubert[1], M Gedan-Smolka[1], A Marschner[1], T Rietzschel[1], K Uhlig[1], D Löpitz[2], D Wagner[2], M Knobloch[2]

[1] Leibniz-Institut für Polymerforschung Dresden e.V. (**IPF**), Hohe Straße 6, 01069 Dresden, Germany

[2] Fraunhofer Institute for Machine Tools and Forming Technology (**IWU**), Reichenhainer Straße 88, 09126 Chemnitz, Germany

schubert@ipfdd.de

Abstract. The development of multi-material hybrids by injection molding has been studied very intensively at the IPF in the past. For that, a material bonding between the different substrates was achieved by using a newly developed two-step curing powder coating material as latent reactive adhesive. The aim of the project "Hybrid Pultrusion" was to perform a novel approach for the fabrication of material bonded metal-plastic joints (profiles) in a modified pultrusion process. Therefore, powder pre-coated steel coil is combined with a glass-fiber reinforced epoxy resin matrix. For initial basic studies, the impregnated fiber material has been applied on the pre-coated steel sheets using the Resin Transfer Molding process (RTM-process). It was proved via lap shear tests, that this procedure resulted in very high adhesive strengths up to 35 MPa resulting from the formation of a covalent matrix-steel bonding as well. In addition, the failure mechanism was subsequently studied. Furthermore, by adapting the successful material combination to the pultrusion process it was demonstrated that material bonded hybrids can be achieved even under these continuous processing conditions.

1. Introduction

In many areas, weight reduction plays an important role in saving energy, especially in future electric vehicles. For that reason, multi-material combinations became more and more important during the last years. According to the state of the art, there are several ways to produce multi-material structures [1]. In the past those joints were achieved mostly mechanically, e.g. by form fitting or clamping. Now, material bonding plays an outstanding role for the production of lightweight structures, especially in the automotive and transportation industry, but also for machines and household appliances. In practice, material bonding is usually achieved by applying an adhesive or primer. However, high adhesive strengths in combination with a high durability under application conditions, e.g. for metal-plastic joints, can only be reached by additional cost and time-consuming pre-treatment steps. A direct adhesion works only for less material combinations.

During the last few years a two-step curable powder coating system based on commercially available uretdione (internally blocked isocyanates) cross linkers, OH-functionalized polyester resins and a specific catalyst system was developed at the IPF (Figure 1) [2]. The curing mechanism runs via two crosslinking steps that can be separated by curing time and temperature selectively. At temperatures lower than 160 °C a polyallophanate structure is obtained that shows a high level of post-formability.

At temperatures higher than 180 °C the polyallophanate structure can be transferred to polyurethane within a short time period. Additionally, the post formability of the coating system as well as the surface appearance can be maintained at the high level. For this reason, the second process step can subsequently be used for interlayer reactions and a covalent linkage of different materials. Based on that, the two-step crosslinking for the fabrication of Al-powder coating-plastic hybrids by injection molding was studied very intensively at the IPF. In that manner, 3- or 4-layer metal-plastic hybrids with a very high bond strength were generated [3].
The subject of the current investigation was the adaptation of this concept to the manufacture of metal-plastic joints consisting of powder-coated steel and a glass-fiber-reinforced epoxy matrix. At IPF, powder coating and material development was performed using the discontinuous RTM process. Subsequently, preferred material combinations were transferred to the pultrusion process and a proof of the technological feasibility of such material joints was provided under continuous conditions at IWU.

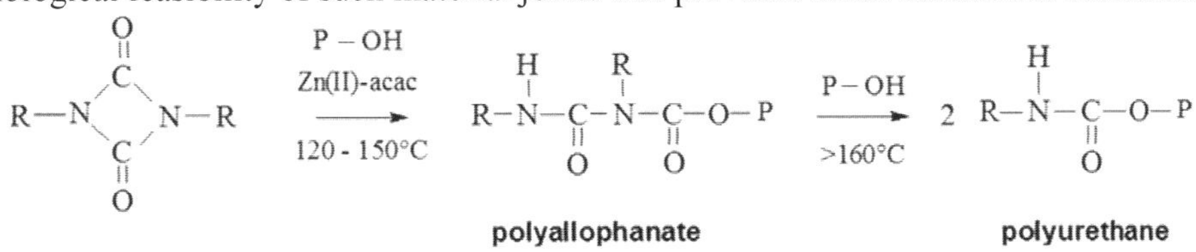

Figure 1: Curing reaction of uretdione-based powder coatings in the presence of a special catalyst [2]

2. Hybrid-materials by using the RTM process

Test plates with a total thickness of 4 mm were fabricated via RTM process in extensive laboratory tests in order to investigate the best process conditions and material combinations.

2.1. Manufacturing of the hybrids

First, the steel sheet substrate (DC01, 2 mm thick) was sandblasted with high-grade corundum and pre-treated with zinc phosphate in an industrial facility, resulting in a fine crystalline layer. In this manner, a good corrosion protection was received for the metal. In addition, previous studies have shown that such treatments are necessary to achieve a high adhesion of the powder coating film in the subsequent hybrid structure [4]. Then, the pretreated steel sheets were coated with various latent reactive adhesives in the form of powder coatings by using corona discharge. The electrostatically applied powder particles on the steel sheet were then cured (150 °C / 15 min). During the crosslinking step, the powder coating, while strongly bonding to the metal, was transformed into a polyallophanate network.
In the next step, the powder coated steel sheets was combined with glass fiber reinforced epoxy resin via RTM process. For that, an epoxy resin system was used (Huntsman, Table 1).

Table 1. Epoxy matrix formulation.

Component	Component name	Supplier
Resin	Araldite LY 3585 CH	Huntsman
Curing agent	Aradur 917-1 CH	Huntsman
Catalyst	Accelerator DY080	Huntsman
Internal Mold Release	IC25	ChemTrend
Filler	ASP 600 (0.6 µm)	BASF

The steel sheet and the fiberglass mats was placed in the RTM plate tool (Figure 2) and heated up to 90 °C under vacuum after closing. Afterwards, the previously mixed epoxy resin system was injected by using a pressure pot. After a curing cycle of 90 °C / 90 min the cross-linked hybrid structure was demolded. Half of hybrid panels (Figure 3) were then thermally treated at 160 °C for 3 h or at 200 °C

for 1 h, respectively, to investigate the effect of post-curing on the resulting bond strength between the powder coating layer and the fiber reinforced epoxy matrix.
It was expected that during the thermal treatment the 2nd reaction step from polyallophanate to polyurethane would occur. As a result, more covalent bonds should be generated in the interface between the powder coating layer and the epoxy resin, leading to higher bond strengths. When investigating the influence of the powder coating formulation, the main focus was on varying the OH number of the polyester resins contained and on adding OH-functionalized dendrimers at 1 or 3 wt% to study their influence on the adhesion strength to the epoxy matrix.

Figure 2. Opened RTM-tool for hybrid preparation.

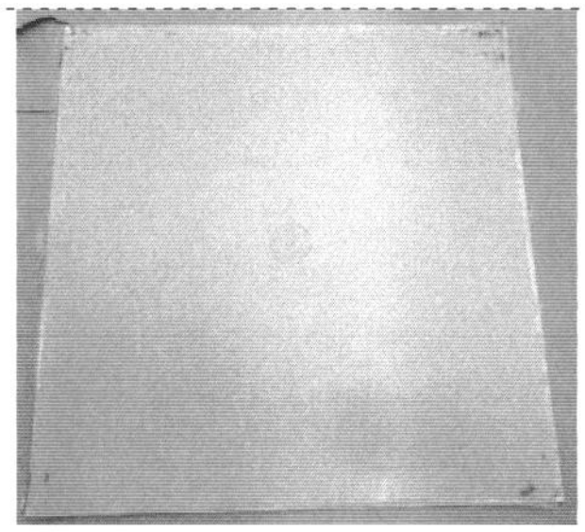

Figure 3. Material bonded steel - thermoset (glass fiber reinforced epoxy) hybrid material plate by RTM.

Figure 4. Standardized tensile test specimen according to DIN 65148 before and after the lap shear test.

2.2. Lap shear tests
To determine the bond strength, a lap shear test was carried out in accordance to DIN 65148 [5]. For this purpose, the hybrid parts were cut to standardized tensile specimen (dimension 250 mm x 25 mm x 4 mm) on a water-cooled separating device. In addition, two notches are milled into the specimen, resulting in a shear area of 25 mm x 12.5 mm, as recommended in DIN 65148 (Figure 4).
After testing, the occurred fractures were analysed by the possible failure patterns sketched in Figure 5 in detail. The percentage was determined from every possible type of breakage.

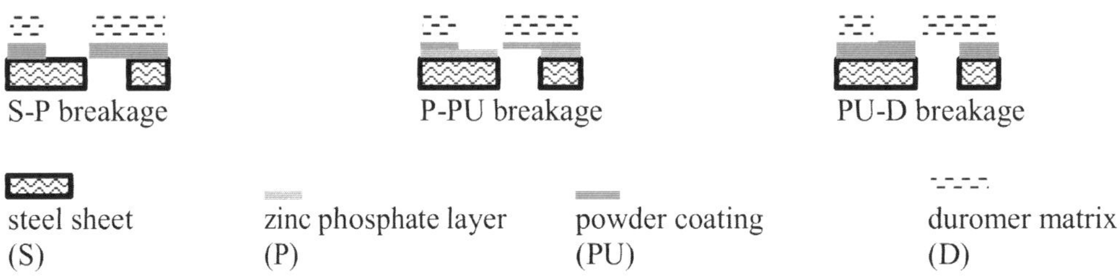

Figure 5: Main failure patterns/failures occurring during the strength test.

Figure 6 shows the influence of the different OH numbers of the polyester resin in the powder coating formulation on the resulting shear strength. Shear strength was expected to increase when larger hydroxyl numbers were used. However, it was found, that the shear strength decreases with the increasing hydroxyl number. It can be assumed that other parameters affect the tensile testing results as well, for example the glass transition temperatures of the powder coatings and resulting thermal stresses during the cooling process after the manufacturing in the RTM plate tool. It was found that the HP-01 powder coating (OH number: 51 mg/g) should be preferred for further testing. Additionally, it was observed, that the failure mainly happens between the zinc phosphate layer and the powder coating (P-

PU) and not the real interesting PU-D interlayer. So only can be concluded that the adhesion between powder coating and the epoxy resin (PU-D) is higher than the determined shear strength values.

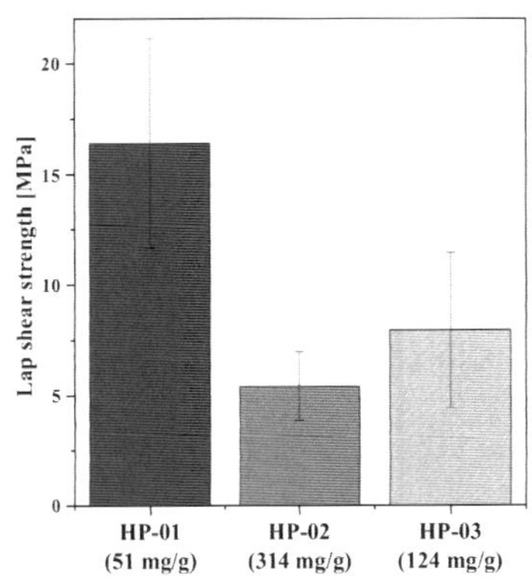

Figure 6: Lap shear strength by varying the OH-group number of the containing polyester resins in the powder coating formulation.

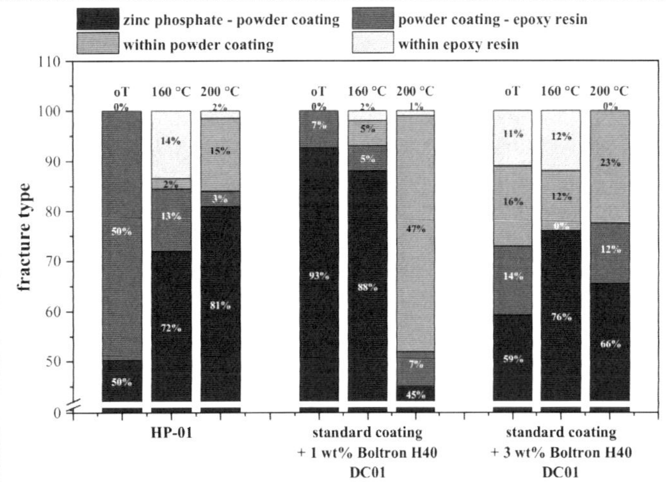

Figure 7: Fracture percentage of the failure modes occurring in the lap shear test using different dendrimer amounts compared to the standard powder coating and influence of post-curing (3 h / 160 °C or 1 h / 200 °C) before the lap shear test.

When considering the influences of the use of dendrimers in the preferred powder coating HP-01 have on the interlayer strength, almost no effect on the resulting shear strength was found in comparison with the standard formulation. Thus, the addition of the expensive dendrimer to the coating is not effective in this case. In contrast, annealed test specimens showed a change of the occurring failure types in the shear test (Figure 7). In the absence of dendrimers in most cases the failure happened between P-PU and PU-D. By using dendrimers the failure percentage between PU-D decreased and therefore the amount of the interfacial failure between P-PU increased. However, also other failure types occurred beside those. The influence of the dendrimers has to be studied more intensively later on.

2.3. Hybrid pultrusion

The preferred results from the RTM process then were adapted to the continuous pultrusion process at IWU (Figure 8). The individual fabrication steps can be described as follows: Semi-finished fiber products (1) are pulled from bobbins by alternately moving pulling devices (4) and pass through a resin bath (2). Afterwards, the impregnated fibers are pulled through a heated die (3), in which the liquid thermoset plastic cures completely within seconds. A saw (5) cuts the profiles to the desired length [6].

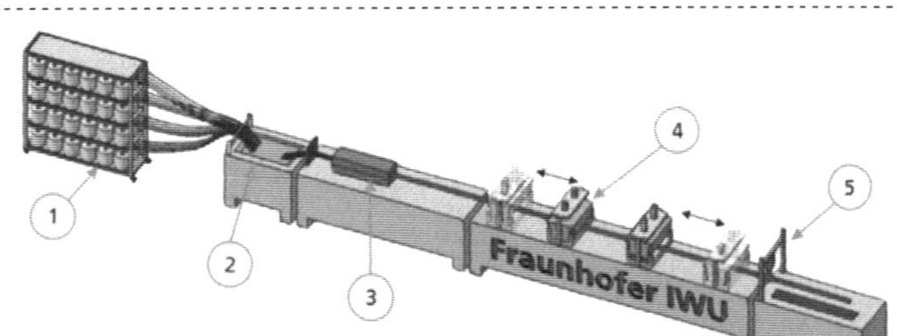

Figure 8: Pultrusion line

With a special manufactured tool and steel sheets, that were powder coated with HP-01, it was possible to produce a side sill demonstrator (Figure 10) via the continuous pultrusion process successfully. The dimension of the 3 chamber-hollow hybrid-demonstrator has a wall thickness of 4 mm, a cross-section area of 3.567 mm^2 and two stands with a wall thickness of 6 mm in the inner for the necessary stiffness (Figure 9 (a)). To further improve the crash properties, a powder-coated steel sheet/coil of 90 mm length was integrated on top of the demonstrator. The pultrusion process was performed with various temperature zones from 150 to 190 °C and a processing speed of 100 mm/min. A simulation (finite element method – FEM) with the program "ANSYS Workbench 2019 R2" and the special program add-on "ANSYS Composites Cure Simulation" of the pultrusion process was realized to identify the best process parameters. The starting point for a valid calculation was formed by a large number of basic trials and corresponding measurements of the process parameters (e.g. temperature, production speed). These pultrusion trials were carried out on strip samples with a profile cross-section of 50x4 mm^2. In addition to different sheet metal thicknesses (0.5; 1.0; 1.5 and 2.0 mm), the influence of the sheet position in the thickness direction was also investigated. The results obtained in this way were incorporated in the construction of a simulation model. Thus, it was possible to calculate not only the different temperatures of the materials but also the degree of cure of the epoxy resin. Coupled with a structural mechanical simulation, it was also possible to determine the distortion due to the different material properties (Figure 9 (b)). This knowledge gained was used for the design of the demonstrator component, so that, for example, the layer structure of the fiber composite profile including the sheet metal insert could be optimized to minimize the resulting distortion. However, the focus of the simulation was on determining relevant process parameters in order to minimize time-consuming and cost-intensive trials. Thus, after only three trials with minor adjustments to the process parameters, a qualitatively appealing demonstrator could be realized.

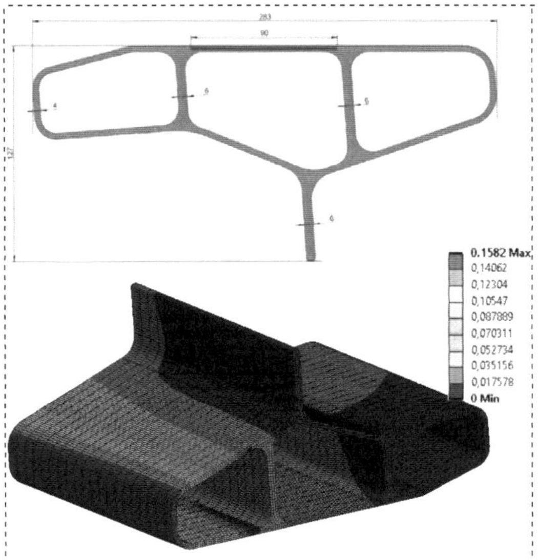

Figure 9:
(a) Cross-section geometry of the side sill demonstrator with a powder coated steel sheet on top.
(b) Simulation result on distortion [mm] due to different shrinkage behavior for the demonstrator

Figure 10: Finished side sill demonstrator manufactured via pultrusion process.

3. Summary

Studies have shown that it is possible to generate material bonded hybrids based on powder coated steel and glass fiber reinforced epoxy resin with a very high bond strength via the RTM process as well as continuous pultrusion by manufacturing a demonstrator for a side sill. The variation of the OH number of the containing polyester resins in the powder coating showed an influence on the resulting adhesive strength. In contrast, the addition of dendrimers to the powder coating formulation showed no clear effect on the bond strength of the manufactured hybrid materials so far.

4. Acknowledgement

This work was carried out within the project "Hybrid Pultrusion", which was funded by the Sächsische Staatsministerium für Wissenschaft, Kultur und Tourismus (SMWK).

European Union

Europe funds Saxony. EFRE European Regional Development Fund

Diese Maßnahme wird mitfinanziert durch Steuermittel auf Grundlage des von den Abgeordneten des Sächsischen Landtags beschlossenen Haushaltes.

References

[1] Erhard G 2004 *Konstruieren mit Kunststoffen 3* p 309

[2] Gedan-Smolka M and Lehmann F and Lehmann D 2001 *New catalysts for the low temperature curing of uretdione powder coatings (Proceedings of the 28'th International Waterborne, High Solids, & Powder Coatings Symposium)* (New Orleans) , p 405–419

[3] Kühnert I and Gedan-Smolka M and Fischer M 2017 *Prefinished metal polymer hybrid parts. - Technologies for Lightweight Structures 1* vol 2 p 89-97

[4] Gedan-Smolka M and Lehmann F and Lehmann D 2003 *Catalysis in uretdione powder coatings enables innovative processing lines* (Amer Chemical Soc) (Washington, DC 20036 USA)

[5] DIN 65148:1986-11 *Prüfung von faserverstärkten Kunststoffen; Bestimmung der interlaminaren Scherfestigkeit im Zugversuch* (Luft- und Raumfahrt)

[6] Starr T F 2000 *Pultrusion for engineers* (Woodhead Publishing Series in Composites Science and Engineering) (Boca Raton: CRC Press)

IOP Conf. Series: Materials Science and Engineering 1140 (2021) 012031 doi:10.1088/1757-899X/1140/1/012031

Nature inspired optimization of jerk limited feedrate profile for NURBS toolpaths in CNC machines

Krystian Erwinski, Rafal Szczepanski and Tomasz Tarczewski

Departament of Automatics and Measurement Systems, Institute of Engineering and Technology, Nicolaus Copernicus University in Torun, Grudziadzka 5/7, 87-100 Torun, Poland

erwin@umk.pl

Abstract. In this paper a novel approach to solving the feedrate optimization problem in Computerized Numerically Controlled (CNC) machines for Non-Uniform Rational B-Spline (NURBS) toolpaths is presented. Particle Swarm Optimization (PSO) gradient-free algorithm is used to determine the optimal shape of the feedrate profile. The shape of the profile is optimized to achieve shortest travel time within the constraints imposed by the machine's axes. Compared to more common approaches the profile is initialized with a quasi-optimal shape determined by a feedrate limit curve and then optimized to the final shape. Simulation results are presented that show the performance and computational effectiveness of the proposed method.

1. Introduction

The problem of optimizing CNC machine velocity tangent to a curved path (feedrate optimization) has been a topic of interest to researchers for some time [1]. Its goal is to minimize the travel time along a curved toolpath under various machine constraints. Initial works focused on generating a trapezoid acceleration limited [2] and s-curve jerk limited [3] feedrate profiles. These did not take into account limitations in the machine's axes. For highly curved paths the axial acceleration and jerk can vary significantly due to their non-linear relationship with the toolpath.

This creates a difficult non-linear optimization problem that requires robust optimization algorithms such as Genetic Algorithms [4] or Particle Swarm Optimization [5]. These methods usually start with multiple random solutions and then explore the solution space to find the best solution based on randomness. These methods are very resistant to local minima but are also slower to converge than traditional gradient based algorithms [6]. In this work, the authors propose using a hybrid approach to enhance the convergence of a nature inspired Particle Swarm Optimization algorithm in the feedrate optimization problem by applying a suitable initial guess.

In modern CNC toolpaths are often expressed as Non-Uniform Rational B-Splines (NURBS) polynomial curves. Position on the curve corresponds to the curve's parameter which is usually a unitless value between 0 and 1 and is computed using DeBoor's algorithm [7]. Shape of the curve is defined by control points which form a control polygon around the curve. Moving each control point only influences the curve's shape locally. The shape is furthermore influenced by weights of each control point which defines the distance between the curve and the control point. The feedrate profile is usually defined as a polynomial function of time or toolpath's parameter. In this work the feedrate profile is defined as a one-dimensional second order B-Spline curve. B-

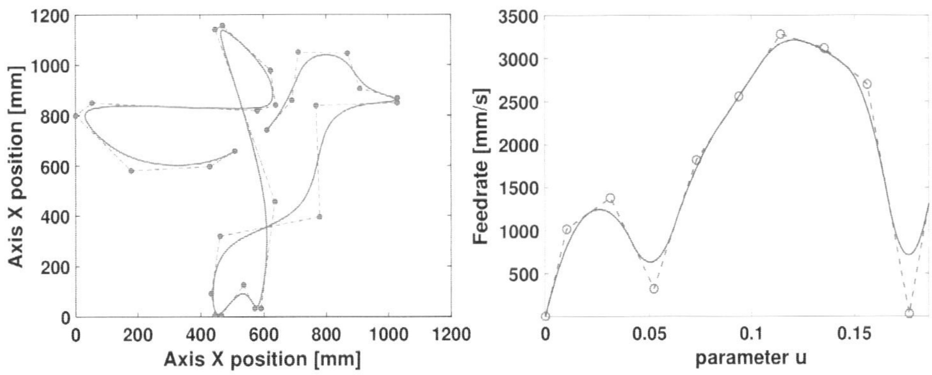

Figure 1: NURBS toolpath (left) and B-Spline feedrate profile (right) with control points

splines are defined the same way as NURBS curve's by using control points but without weight values. An example planar NURBS curve and B-Spline feedrate profile with control points are shown on Fig.1.

2. Feedrate optimization using constrained PSO

The aim of feedrate optimization is to traverse the NURBS toolpath as fast as possible without violating velocity, acceleration and jerk constraints in the machine's axes. This is done by modyfing the shape of the feedrate profile by changing values of its control points. Because acceleration and jerk depend non-linearly on the feedrate and toolpath shape solving the aforementioned problem requires a robust algorithm that can handle complex non-linear objective and constraint functions.

In this work the Particle Swarm Optimization (PSO) algorithm was chosen. This algorithm is a global optimization technique with proven performance. It is much more robust against being trapped in local minima compared to traditional gradient based optimization. The algorithm can also handle highly non-linear and discontinuous objective functions. Compared to more traditional nature-inspired algorithms such as Genetic Algorithms (GA), PSO is simpler to implement and has faster convergence [8–10].

PSO is a swarm algorithm inspired by behavior of fish and birds [11]. Multiple solutions are generated and then updated until they converge to the neighborhood of a minimum. These solutions are called particles. Each particle is described with position in the solution space and velocity. In each algorithm iteration fitness values for each particle is computed and for each particle the position of its best solution is remembered (personal best or pbest). The overall best solution found by the swarm is also stored (global best or gbest). The velocity and position values of each particle are updated according to the following equations[12].

$$v^*_{i,j} = \omega \cdot v_{i,j} + \phi_1 \cdot rand^U_{(0,1)} \cdot (p_{i,j} - x_{i,j}) + \phi_2 \cdot rand^U_{(0,1)} \cdot (g_i - x_{i,j}) \tag{1}$$

$$x^*_{i,j} = x_{i,j} + v^*_{i,j} \tag{2}$$

where: $v^*_{i,j}, v_{i,j}$ - new and previous i-th velocity coordinate of j-th particle, $x^*_{i,j}, x_{i,j}$ - new and previous i-th position coordinate of j-th particle, $rand^U_{(0,1)}$ - pseudorandom number between 0 and 1, ϕ_1,ϕ_2, ω - acceleration and inertia coefficients, $p_{i,j}$ - i-th coordinate of personal best position of j-th particle, g_i - i-th coordinate of global best position.

Coefficients ϕ_1,ϕ_2, ω control the rate of convergence of the swarm and balance between exploration (ability to find new solutions) and exploitation (ability to refine existing solutions).

These can vary widely for different problems but in case of feedrate optimization it was found that default parameters of $\phi_1 = \phi_2 = 1.4962$ and $\omega = 0.7298$ work well for all tested cases so these values were used. The algorithm is terminated when maximum number of iterations is reached which is generally problem based and has to be determined empirically to guarantee convergence of the swarm.

Standard version of the PSO algorithm can only constrain values of the optimization parameters by limiting the positions of the particles. In order to incorporate arbitrary non-linear constraints, necessary for the feedrate optimization problem, an extension of the basic algorithm is required. The most popular constraints handling methods are: Deb's rules and penalty method. The comparison of the above-mentioned constraint handling methods are given in [13]. It was shown that the penalty method provides better convergence and repeatability of obtained solution. It also allows to approach the solution through infeasible solutions which increases convergence speed. Therefore, the penalty methods is applied to the feedrate optimization problem. For each constraint a penalty is added to the objective functions and it's minimization also decreases the violation. This method only modifies the objective function and does not require any modification of the optimization method. In this work an extension of the basic penalty method is used, called Augmented Lagrangian method. The objective function is replaced by the Augmented Lagrangian $L(x)$ defined as [14]:

$$L(\mathbf{x}) = f(\mathbf{x}) + \frac{\rho}{2} \sum_{i=1}^{M} \left(\max \left\{ 0, c_i(\mathbf{x}) + \frac{\lambda_i}{\rho} \right\} \right)^2 \tag{3}$$

where: $f(x)$ - objective function value for solution vector x, ρ - penalty factor, $c_i(x)$ - value of constraint i for solution x, λ_i - Lagrange multiplier for constraint i, M - number of constraints. This replaces a constrained optimization problem with an equivalent unconstrained optimization problem if correct values of ρ and λ_i are used. Optimization is performed in two nested loops. In the inner loop unconstrained minimization of the Augmented Lagrangian objective function is performed using PSO for a predefined number of iterations. In the outer loop the Augmented Lagrangian objective function parameter update is performed after minimization in the inner loop. Compared to a standard quadratic penalty method this two-level approach has the advantage of optimally selecting the penalty factors so that the unconstrained problem is in fact equivalent to the initial constrained problem. The authors use penalty factor and Lagrange multiplier update rules provided in [14].

For the feedrate optimization problem the fitness function is defined as:

$$f(P_i) = \frac{1}{N} \sum_{i=1}^{N} \left(1 - \frac{P_i}{F_{max}} \right)^2 \tag{4}$$

where: F_{max} - maximum feedrate, P_i - value of the i-th control point of the B-Spline feedrate profile, N - number of control points. For a two axis machine the constraints are defined as:

$$c_i = \left\{ \frac{|v_{ax}(u_i)|}{V_{x_{max}}} - 1, \frac{|v_{ay}(u_i)|}{V_{y_{max}}} - 1, \frac{|a_{ax}(u_i)|}{A_{x_{max}}} - 1, \frac{|a_{ay}(u_i)|}{A_{y_{max}}} - 1, \frac{|j_{ax}(u_i)|}{J_{x_{max}}} - 1, \frac{|j_{ay}(u_i)|}{J_{y_{max}}} - 1 \right\} \tag{5}$$

where: $v_{ax}, ..., j_{ay}$ - values of axial velocity, acceleration and jerk, $V_{x_{max}}, ..., J_{y_{max}}$ - maximum values of axial velocity, acceleration and jerk. Both the objective function and constraints are normalized to their specific maximum value and offset by 1 so that violated constraints are positive and constraints without violation are negative.

Dimension of the problem is equal to the number of optimized feedrate profile control points. Large complex toolpaths will also have complex feedrate profiles with many control points. In order to limit the optimization problem dimensionality, a moving window approach is applied. Only a fragment of the profile is optimized at any given time. After optimization is finished the optimization window is shifted by a fixed amount of points and optimization is performed for the next set of feedrate control points. In each window each constraint is evaluated at fixed locations corresponding to equally spaced values of parameter u. Locations of these points are set so that they are spaced on average every 2mm.

3. Generation of initial guess

In this paper in order to present the PSO algorithm with a reasonable initial solution a jerk limited S-Curve profile is first generated [3, 15]. Axial constraints are not considered so axial velocities, accelerations and jerk can vary significantly even for constant feedrate profiles. To indirectly limit the axial values a Feedrate Limit Funcion (FLF) is introduced [16]:

$$FLF = \min\left(\frac{||\mathbf{C}'(u)||F_{max}}{|C'_m(u)|}, \frac{\sqrt{||\mathbf{C}'(u)||^2 A_{m_{max}}}}{|C''_m(u)|}, \frac{\sqrt[3]{||\mathbf{C}'(u)||^3 J_{m_{max}}}}{|C'''_m(u)|}\right) \tag{6}$$

where: F_{max} - maximum feedrate, $C'_m(u), C''_m(u), C'''_m(u)$ - NURBS curve 1st, 2nd and 3rd derivative for axis m, $A_{m_{max}}, J_{m_{max}}$ - maximum acceleration and jerk for axis m.

The FLF linearizes axial constraints aggregating them into one feedrate constraint. Limits for velocity, acceleration and jerk are computed for each axis and the lowest feedrate value is chosen for each u. In order to generate the initial S-curve profile for a given NURBS toolpath the FLF is first scanned and it's minima are located. Then the arc-length between these points is computed and then a S-curve profile is fitted between them. Duration of the acceleration and deceleration phases are computed so that the distance covered by the acceleration-deceleration profile is equal to the arc-length between the FLF minima. The initial S-curve feedrate profile is generated that limits feedrate, tangent acceleration and jerk and passes under the feedrate limit imposed by the FLF.

The initial S-curve, parametrized with time has to be converted to a B-Spline representation parametrized by the toolpath parameter u. A set of feedrate data points on the S-curve are picked and stored with their corresponding parameter value u. A global B-Spline interpolation of these data points is performed and the obtained curve is a close approximation of the initial S-curve profile. The advantage of using a B-Spline representation of the feedrate profile is it's guaranteed continuity of a given degree and the property of only local influence of each parameter on the curve shape. The B-Spline representation also has good numerical properties when interpolating data.

The generated initial feedrate profile is used to initialize the PSO algorithm. One particle is initialized with the feedrate profile's control points and it's position becomes the initial global best. Other particles are initialized by adding a random uniform disturbance to the initial guess position. This disturbance is set to be 10% of the possible value range of the optimization space. Each of these initial particle positions is set to be the initial personal best of this particle. Feedrate optimization is then performed as is described in the previous sections.

4. Feedrate optimization algorithm simulation results

To test the performance of the algorithm the example NURBS toolpath from Fig.1 was used. The algorithm was tested with maximum axial velocity, acceleration and jerk set to $2500mm/s$, $25000mm/s^2$ and $500000mm/s^3$ respectively for each axis and a $250\mu s$ interpolation period - parameters typical for high performance laser cutters with linear motor drives. For these parameters an S-curve initial guess profile was generated and a B-Spline curve with 100 control points was interpolated to obtain an initial guess feedrate profile. The initial feedrate profile was optimized using the algorithm described in the previous sections. Optimization was performed

IOP Conf. Series: Materials Science and Engineering 1140 (2021) 012031 doi:10.1088/1757-899X/1140/1/012031

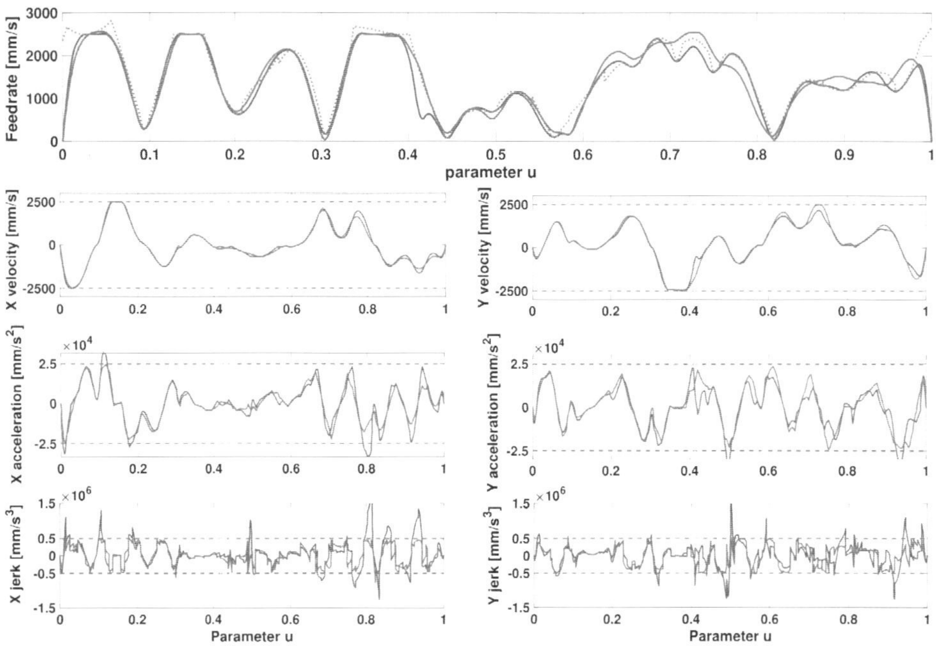

Figure 2: Feedrate initial (blue) and optimized (red) feedrate and axial profiles with Feedrate Limit Function (green)

with windows of 30 control points and an overlap of 10 control points. Constraint evaluation points were spaced approximately 1mm apart. Optimization termination conditions were set to 100 inner iterations and 20 outer iterations for each window. The average optimization time for these settings was around 3 minutes and 50 seconds. The initial and optimized feedrate and axis profiles were shown on Fig.2.

It can be seen that the initial and final feedrate profiles are very similar. Main difference can be seen in the axial plots of acceleration and jerk. The initial profile exhibited some constraint violations of acceleration and jerk around the 0.5 and 0.8 parameter values. These correspond to the curvature extremes of the Bird toolpath which are the "beak" and "tail" regions. The optimization algorithm rectified these violations and the final axial profiles are all within constraints. Optimization was performed several times and each time results were very similar to those presented. For comparison a random initial profile was also generated and optimized using the same algorithm with identical parameters. Final profile obtained from random initialization was very similar but exhibited some jerk constraint violations and the trajectory took longer to execute (4.75s vs. 4.32s). The value of the Augmented Lagrangian objective function for both initialization types in each window was shown in Fig.3

Optimization convergence rate of the randomly initialized profile is slower than the one with a good initial guess. The objective function starts from a higher value and converges to higher values compared to the S-curve initialized case. The objective function value (in fact the value of the Augmented Lagrangian function) changes significantly every 100 iterations due to update of the penalty factors. This is not the case if optimization is initialized with the S-curve initial guess due to much lower constraint violation. It can be seen that for window 2 the objective function value was increased due to constraint violations (increase in penalty factor). Also for window 4 it can be seen that the objective value still decreases after 2000 iteration. The S-curve initial guess indeed improves the convergence rate and quality of the feedrate optimization.

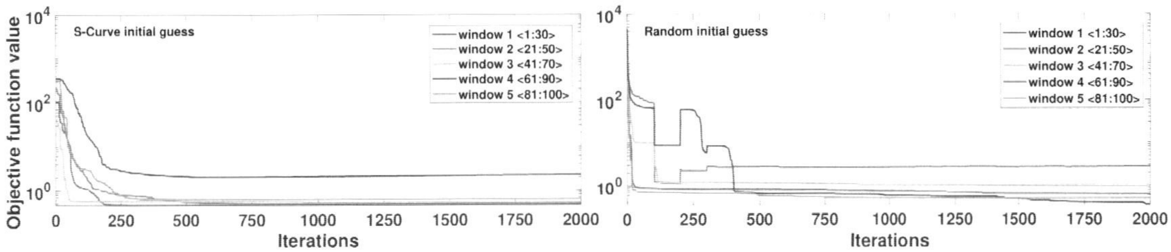

Figure 3: Objective function minimization progression for each optimization window

5. Conclusion

The paper presents a feedrate optimization method based on nature inspired Particle Swarm Optimization algorithm for CNC machine tools. Using polynomial NURBS toolpaths and applying axial constraints on the feedrate optimization makes this a difficult non-linear constrained optimization problem. The algorithm is enhanced with the Augmented Lagrangian constraint handling technique in order to solve the feedrate optimization problem with axial velocity, acceleration and jerk constraints. The authors proposed generating an S-curve feedrate profiles between minima of a Feedrate Limit Function. The S-curve is then converted to a B-Spline profile and used as an initial guess for optimization. It is shown that generating an initial guess based on S-curve jerk limited profiling can significantly improve convergence rate of the algorithm decreasing computation time of this difficult problem. Further work will focus on generalization of the algorithm for machine types with non-trivial kinematics such as serial manipulators, parallel machines (delta, h-bot).

References

[1] Mansour S Z and Seethaler R 2017 *J. Manuf. Sci. Eng.* **139**
[2] Timar S D and Farouki R T 2007 *Robot. Comput. Integr. Manuf.* **23** 563–579
[3] Du X, Huang J and Zhu L M 2015 *J. Comput. Des. Eng.* **2** 206–217
[4] Xu C, Chen J, Zhou H, Yang J, Hu P and Dai W 2019 *Int. J. Adv. Manuf. Technol.* **101** 715–731
[5] Szczepanski R, Erwinski K and Paprocki M 2017 *Proceedings of the 22nd International Conference on Methods and Models in Automation and Robotics (MMAR)* 431–436
[6] Sencer B, Altintas Y and Croft E 2008 *Int. J. Mach. Tools Manuf.* **48** 733–745
[7] Piegl L A and Tiller W 1997 *The NURBS book (2nd ed.)* (New York, NY, USA: Springer-Verlag New York, Inc.) ISBN 3-540-61545-8
[8] Majak J, Pohlak M, Eerme M and Velsker T 2012 *Mechanics* **18** 453–460
[9] Kaganski S, Majak J, Karjust K and Toompalu S 2017 *Procedia CIRP* **63** 283–288
[10] Elbes M, Alzubi S, Kanan T, Al-Fuqaha A and Hawashin B 2019 *Evolutionary Intelligence* 1–17
[11] Kennedy J, Kennedy J and Eberhart R 2001 *Swarm intelligence* (Morgan Kaufmann)
[12] Clerc M and Kennedy J 2002 *IEEE Trans. Evol. Comput.* **6** 58–73
[13] Szczepanski R, Tarczewski T, Erwinski K and Grzesiak L M 2018 *Proceedings of the International Conference on Informatics in Control (ICINCO 2018)* 279–286
[14] Birgin E and Martinez J 2008 *Optim. Methods Softw.* **23** 177–195
[15] Ni H, Yuan J, Ji S, Zhang C and Hu T 2018 *IEEE Access* **6** 66403–66417
[16] Sun Y, Chen M, Jia J, Lee Y S and Guo D 2019 *Sci. China Technol. Sci.* **62** 1067–1081

IOP Conf. Series: Materials Science and Engineering 1140 (2021) 012032 doi:10.1088/1757-899X/1140/1/012032

Safety Assessment and Simulation of Autonomous Vehicle in Urban Environments

Mohsen Malayjerdi, Barış Cem Baykara, Raivo Sell and Ehsan Malayjerdi

Mechanical and Industrial Engineering Department, Tallinn University of Technology, Tallinn, Estonia

momala@taltech.ee

Abstract. The Autonomous Vehicle (AV) industry aims to design strategic plans to ensure the safety of the developed systems before their mass deployment. Real-road testing is shown to be impractical for validating these systems as it requires many years if not decades of testing in different environmental conditions. Therefore, this method must be complemented with simulation for a feasible solution. The primary goal of this research is to develop advanced techniques in the safety validation area by using immersive simulation technologies. This study led to a simulation approach for safety evaluation of an AV shuttle, iseAuto, currently operating at the TalTech campus. First, we create a virtual environment based on the specified path on the university campus including all relevant features. The environment is created by using geospatial data which is collected by a drone, then they are converted to a 3D map applicable for the LGSVL simulator. Next, the developed shuttle 3D model is imported to the simulation environment and with the aim of the LGSVL, sensor data are provided for the Autoware perception algorithms, which is the control software of the shuttle. This system enables us to evaluate AV's decision-making performance and safety in different situations.

1. Introduction

Autonomous vehicles development is one of the top trends in the automotive industry and the technology behind them has been evolving to make them safer. In this way, engineers are facing new challenges especially moving toward the Society of Automotive Engineers (SAE) levels 4 and 5. To put Autonomous Vehicles (AV) on roads and evaluate the reliability of their technologies they have to be driven billions of miles [1], which takes a long time to achieve unless with the help of simulation. Furthermore, due to the past real crash cases of AVs, a high-fidelity simulator has become an efficient and alternative approach to provide different testing scenarios for control of these vehicles, also enabling for safety validation before real road driving [2, 3, 4, 5]. Different high-resolution virtual environments can be developed based on the real world for simulators by using cameras or lidars to simulate the scenarios as close as possible to the real [6]. Also, virtual environment development enables us to customize and create various urban backgrounds for testing the vehicle. Creating a virtual copy of an existing intelligent system is a common approach nowadays called digital twin [7, 8].

In this paper, we focus on the utilization of simulation for an AV shuttle at Tallinn University of Technology(TalTech), Estonia. TalTech AV research group is well-known for its AV shuttle, iseAuto [9] that is operational in the campus for experimental research purposes (figure 1). The vehicle is designed and developed from scratch by implementing previously proposed mechatronic design methodology [10, 11, 12] with a special focus on early design stages. The first prototype development was a joint venture with TalTech and local industry Silberauto [13]. This shuttle is controlled by Autoware [14], a Robotic Operating System (ROS) based platform for self-driving vehicles.

Figure 1. TalTech iseAuto - an AV shuttle

The overall research project is planned to be executed in two stages. First, the virtual environment was built from the campus AV road area, where most of our real experiments take place, to build the simulation framework. We use geospatial images to create the environment as a Unity terrain. Among different modern AVs simulators like CARLA [15], LGSVL [16] and Gazebo we took LGSVL as our simulator due to compatibility with our control software (Autoware) and our terrain generation platform, Unity. Second, creating different scenarios and performing software-in-the-loop (SIL) simulation by connecting Autoware with LGSVL. This enables us to find better sensors configuration and settings in addition to the verification of the decision-making system that leads to safety assessment.

2. Simulator

Simulation has been widely used in vehicles manufacturing, particularly for mechanical behaviour and dynamical analysis. However, AVs need more than that due to their nature. Simulation in various complex environments and scenarios included other road users with different sensors combination and configuration enables us to verify their decision-making algorithms. One of the most popular robotic simulator platforms is Gazebo. It is based on ROS and utilizes physics engines and various sensor modules suitable for autonomous systems. Nevertheless, Gazebo lacks modern game engines features like Unreal and Unity which gives the power to create a complex virtual environment and realistic rendering.

On the other hand, CARLA and LGSVL are modern open-source simulators based on the game engines, Unreal and Unity respectively, which also have good compatibility with our AV stack Autoware. Although, comparing these two is beyond our discussion but we selected the LGSVL as our simulator because of compatibility with our terrain generator which is Unity.

Figure 2 shows a full map of the simulation workflow and relation between Autoware and the simulator. Vehicle 3D Model and the virtual environment, which are created inside the unity, are imported to the simulator. The simulator allows customizing the environment to create different scenarios such as adding/removing other road users, putting traffic systems and adjusting the time of day and the weather of the scene. Then, virtual sensors provide information for the perception of the environment. This information is transferred via a ROS bridge to our control software platform to use in our perception algorithms for localization and detection. Perception results are used in the Autoware planning section which makes the control commands for the vehicle. These control commands are sent back to the simulator via the ROS bridge to navigate the vehicle inside the simulator.

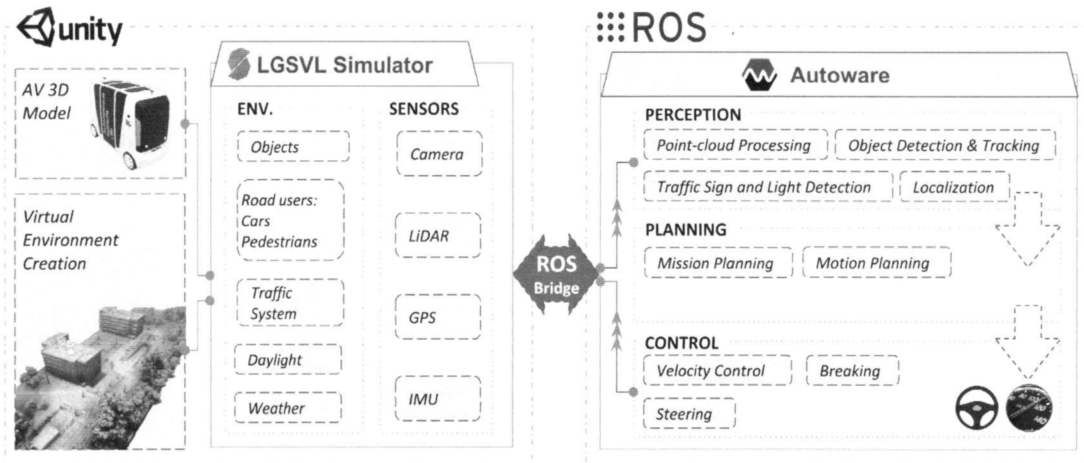

Figure 2. High-level architecture of simulation and the AV system

The iseAuto 3D model and its lidar sensors are illustrated in figure 3. Two Velodyne VLP-16 are installed at the top front and back of the vehicle. Furthermore, two Robosense Bpearl are installed at the sides left and right of the vehicle. This lidar configuration creates a good point-cloud coverage around the vehicle for perception purposes.

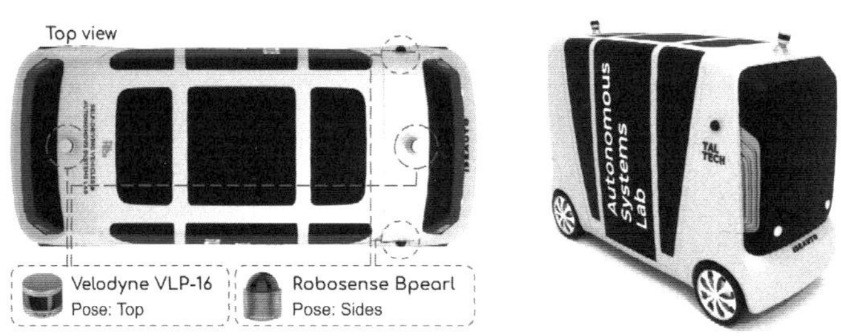

Figure 3. iseAuto simulated Model with different lidars installed

3. Virtual Environment Creation

Nowadays the fierce competition in the gaming industry has brought many features to the table in terms of game engines. These engines can simulate physics and thus can be exploited as simulators aside from game development. LGSVL and others have already taken advantage of these engines and have created a framework for testing autonomous vehicles within these physics simulators. However, even though these simulators provide some basic tools and assets to get started, it is not enough. To make it more realistic, we need real life terrains simulated.

3.1. Data Collection and Processing

Capture of aerial imagery with a drone, over the area to be mapped, has to be conducted. The images are captured at a grid based flight path. This ensures that the captured images contain different sides of a subject. In order to make sure the images have maximum coverage, the flight path is followed three times in different camera angles but at a constant altitude. Taking aerial photo is one of the most important steps in the mapping process as it will significantly affect the outcome of the process and the amount of work to be done to process those images.

There are also external factors that may affect the quality of the pictures taken off the ground. Weather conditions and scene lighting may create artifacts on the pictures that may disturb the photogrammetric process. The images taken are georeferenced by the drone and if necessary a stationary RTK device can be utilized to mitigate errors and shifting on the positioning data stamped on the pictures. The onboard IMU provides the pictures with the orientation so that later they can be stitched together and used for photogrammetric processing. Third party software aligns and creates the dense point-cloud from the pictures that were captured. Once the dense point-cloud is created, the segmentation and classification of the points is needed in order to separate unwanted objects and vegetation from the point-cloud data. However removing is not to be done in the point-cloud as the positional information they provide for their respective objects will aid terrain generation to spawn details. Figure 4 shows the three main steps to generate the Unity train from geospatial data.

Figure 4. Steps for virtual environment generation

3.2. Terrain generation

Digitalization of a real life environment can be used for simulating AVs in countless different scenarios without taking the vehicle out for once. Terrain generation from point-cloud is done right in Unity. In-house developed plugin reads a pre-classified point-cloud file and based on chosen parameters it creates a normal map, a heightmap and a color map to use in conjunction with the unity's terrain engine to create realistic environments.

4. Simulation and Safety Assessments

Based on the simulation architecture illustrated in figure 2, we can run the AV inside the virtual environment. In our collaboration with Florida Polytechnic University and Embry Riddle Aeronautical University, we developed a regime for creating edge-case scenarios for safety validation of the shuttle working on our campus pilot road [17]. Now, by using a high-fidelity simulator we can simulate different scenarios close to real to evaluate the control algorithm performance and safety. In terms of defining these scenarios, LGSVL provides a python API for spawning different objects like cars and pedestrians inside the virtual environment with different motion plans.

Figure 5 shows iseAuto facing a stopped Non-Player Character (NPC) vehicle that is spawned in front of the AV. Picture (a) is inside the LGSVL environment while picture (b) shows the lidar perception of the environment in RViz visualization tool. There is no filtering applied on this point-cloud; therefore, everything is mixed together and it is hard to distinguish objects for later processings. One of the challenging topics of self-driving development is overtaking. The way that the AV should decide for this mission and the risks that it faces is under study. Our experience with the vehicle trying to pass a stopped npc or an object has led us to focus on this topic more. This way, simulations can help us first, to improve our perception and detection system then, the mission and motion planning for a safe overtake. The first steps for detection

Figure 5. LGSVL environment (a) versus RViz point-cloud visualization (b)

are filtering and clustering the point-cloud. Autoware has some predefined features for them. One of the common point-cloud filterings is ground filtering in which some part of the point-cloud that is defined as ground will be separated. Each lidar point-cloud can be filtered separately or once after the concatenation with other lidars. Filtering parameters have an intensive effect on the detection result. Sometimes losing 10 to 20 points due to the improper filtering will result in the object not to be detected.

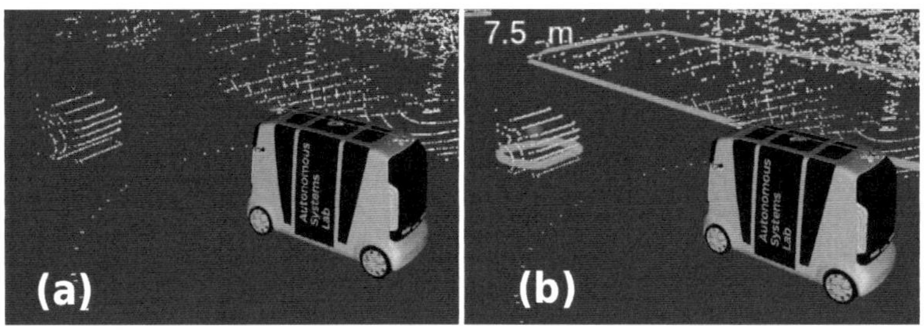

Figure 6. The ground filtering of the point-cloud (a) and applying euclidean clustering (b)

Figure 7. Maximum distance for detect a stopped npc after filtration with two filtering parameters

Filtering and clustering are illustrated in figure 6. Filtering is applied to the picture (a) and the ground, which was detected by lidars in figure 5 (b), is almost removed from the point-cloud. However, the npc points remained and they were clustered as an object in the picture (b). Filtering accuracy results in high performance object detection and safe decision making [18]. For example, Figure 7 shows how different ground filtering parameters can change maximum distance for detecting a stopped npc in front of the AV shuttle, although, both cases have similar clustering parameters. Picture (b) shows that the npc is detected by the AV shuttle from 32 meters distance but picture (c) shows that the maximum distance that it is able to detect an object decreased to 18 meters. The more distance we have for detection, the more time we have for making a smooth control decision. In AVs with multiple lidars, filtering accuracy can be improved by performing it before point-clouds concatenation.

5. Conclusion

Safety validation is crucial for most of the AVs developments and deployments. The simulation as a validation approach in this paper offers a practical and effective way to evaluate the safety in different levels. This paper presents the simulation architecture of the iseAuto with SiL testing that shows how the virtual environment and vehicle model are used in combination with Autoware to simulate different scenarios. The development and utilization of this testing scheme will enable the safety improvement and autonomous vehicles performance development.

Acknowledgments

The research is supported by the EC H2020 project Finest Twins (grant No. 856602).

References

[1] Kalra N and Paddock S M 2016 Driving to safety: How many miles of driving would it take to demonstrate autonomous vehicle reliability? *Transportation Research Part A: Policy and Practice* **94** 182–193
[2] Chao Q, Jin X, Huang H W, Foong S, Yu L F and Yeung S K 2019 Force-based heterogeneous traffic simulation for autonomous vehicle testing *2019 International Conference on Robotics and Automation (ICRA)* (IEEE) pp 8298–8304
[3] Aeberhard M, Rauch S, Bahram M, Tanzmeister G, Thomas J, Pilat Y, Homm F, Huber W and Kaempchen N 2015 Experience, results and lessons learned from automated driving on germany's highways *IEEE Intelligent transportation systems magazine* **7** 42–57
[4] Anderson S J, Peters S C, Pilutti T E and Iagnemma K 2011 *Robotics Research* (Springer) pp 39–54
[5] Razdan R, Lumina J, Balachandran A, Cheng C, Sreenivas S, Fernando X, Taiber J, Kalia A, Keel N, Zuby D, Krishnan K, Langer D and Sell R 2019 Unsettled technology areas in autonomous vehicle test and validation Tech. rep. SAE Technical Paper
[6] Malayjerdi M, Kuts V, Sell R, Otto T and Baykara B C 2020 Virtual simulations environment development for autonomous vehicles interaction *IMECE2020* (ASME)
[7] Kuts V, Modoni G E, Otto T, Sacco M, Tähemaa T, Bondarenko Y and Wang R 2019 Synchronizing physical factory and its digital twin through an iiot middleware: a case study. *Proceedings of the Estonian Academy of Sciences* **68**
[8] Kuts V, Cherezova N, Sarkans M and Otto T 2020 Digital twin: industrial robot kinematic model integration to the virtual reality environment *Journal of Machine Engineering* **20**
[9] Sell R, Leier M, Rassõlkin A and Ernits J P 2018 Self-driving car iseauto for research and education *2018 19th International Conference on Research and Education in Mechatronics (REM)* (IEEE) pp 111–116
[10] Christophe F, Sell R, Bernard A and Coatanéa E 2009 Opas: Ontology processing for assisted synthesis of conceptual design solutions *International Design Engineering Technical Conferences and Computers and Information in Engineering Conference* vol 49026 pp 249–260
[11] Sell R, Coatanéa E and Christophe F 2008 Important aspects of early design in mechatronic *6th International Conference of DAAAM Baltic Industrial Engineering, Tallinn*
[12] Wang R, Sell R, Rassõlkin A, Otto T and Malayjerdi E 2020 Intelligent functions development on autonomous electric vehicle platform *Journal of Machine Engineering* **20**
[13] Rassõlkin A, Sell R and Leier M 2018 Development case study of the first estonian self-driving car, iseauto *Electrical, Control and Communication Engineering* **14** 81–88
[14] Kato S, Tokunaga S, Maruyama Y, Maeda S, Hirabayashi M, Kitsukawa Y, Monrroy A, Ando T, Fujii Y and Azumi T 2018 Autoware on board: Enabling autonomous vehicles with embedded systems *2018 ACM/IEEE 9th International Conference on Cyber-Physical Systems (ICCPS)* (IEEE) pp 287–296
[15] Dosovitskiy A, Ros G, Codevilla F, Lopez A and Koltun V 2017 Carla: An open urban driving simulator *arXiv preprint arXiv:1711.03938*
[16] Rong G, Shin B H, Tabatabaee H, Lu Q, Lemke S, Možeiko M, Boise E, Uhm G, Gerow M, Mehta S *et al.* 2020 Lgsvl simulator: A high fidelity simulator for autonomous driving *arXiv preprint arXiv:2005.03778*
[17] Medrano-Berumen C, Malayjerdi M, Akbaş M İ, Sell R and Razdan R 2020 Development of a validation regime for an autonomous campus shuttle *2020 SoutheastCon* (IEEE) pp 1–8
[18] Habermann D, Hata A, Wolf D and Osório F S 2013 3d point clouds segmentation for autonomous ground vehicle *2013 III Brazilian Symposium on Computing Systems Engineering* (IEEE) pp 143–148

Variations of ice friction regimes in relation to surface topography and applied operating parameters

Igor Velkavrh[1]*, Joël Voyer[1], Thomas Wright[1], Jānis Lungevičs[2], Ernests Jansons[2], Irina Boiko[2]

[1]V-Research GmbH, Stadtstrasse 33, 6850 Dornbirn, Austria

[2]Riga Technical University, Department of Mechanical Engineering and Mechatronic, Kipsalas str. 6b, Riga, 1048, Latvia

* igor.velkavrh@v-research.at

Abstract. To evaluate the effects of surface topography, ice and ambient conditions on the sliding behaviour of steel samples on ice, in the present study friction and velocity measurements were performed with steel samples having different surface roughness values and distinct surface structures. It was shown that the influence of surface roughness on friction and sliding velocity is strongly dependant on ice conditions and the applied experimental parameters due to the formation of different friction regimes.

1. Introduction

In ice friction, depending on the contact conditions and operating parameters such as temperature, sliding velocity and contact pressure, different processes and mechanisms prevail, which define different friction regimes: boundary, mixed or hydrodynamic friction [1]. One of the main parameters defining the ice friction regimes is the thickness of the lubricating liquid-like layer (LLL) on ice with respect to the roughness of the ice and the slider [2,3]. Generally, in different friction regimes different friction levels prevail, thus understanding of boundaries and transitions between different friction regimes is essential for the control of ice friction.

In winter sports, high velocities are desired and therefore the drag between the slider and the ice/snow surface should be minimal. Minimal drag can typically be correlated with the optimal LLL thickness which is dependent on a myriad of parameters and boundary conditions and is anything but straightforward. In a previous study [4], it was shown that depending on the tribometer type, different effects of surface roughness on the coefficient of friction and sliding velocity were observed. On an inclined ice track tribometer, the samples with higher roughness reached lower velocities than the samples with lower roughness, while on oscillating tribometer samples with higher roughness provided lower friction than the samples with lower roughness.

To further evaluate the observed effects, in this study, ice friction measurements were performed with steel samples having different surface roughness values and distinct surface structures. An oscillating tribometer measuring the coefficient of friction and an inclined ice plane measuring the sliding velocity of the steel samples were used to determine the sliding ability.

2. Experimental

2.1. Samples

All samples were manufactured from the Uddeholm Ramax HH steel, which is a chromium alloyed, corrosion-resistant holder steel supplied in a high pre-hardened condition. All samples were cut to dimensions of 35 mm x 18 mm x 14 mm and polished to mirror polish (Ra < 0.01 μm). Steel samples having two distinct surface textures were prepared: one by polishing and abrading the samples in sliding direction and another one by sandblasting and polishing [5]. In Figure 1, sample topographies and roughness values Ra of the tested samples are presented. For sample denotation, G stands for "grooved samples", while SP stands for "sandblasted and polished samples". Sample G1 is the control sample, while samples G2 and G3 were further treated by roughening with different grade sandpapers to achieve parallel grooves in the direction of sliding, having two different depths respectively. Samples SP240, SP150 and SP30 were sandblasted and polished to achieve different surface structures and roughness values.

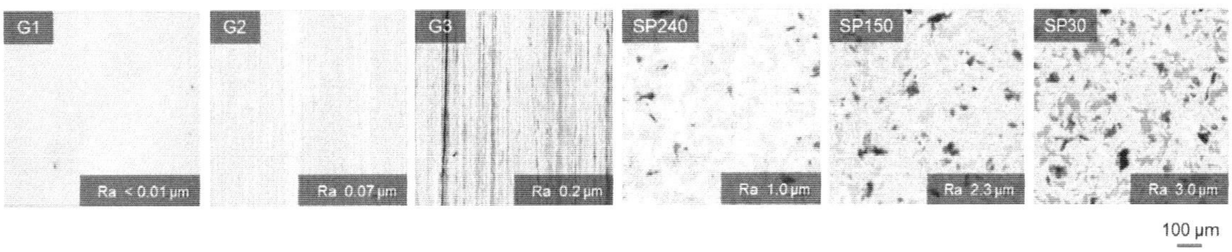

Figure 1. Surface topographies and roughness values Ra of the tested samples.

2.2. Ice preparation

In tribological tests on the oscillating tribometer, four different test conditions were applied within 2 different test series (Table 1). As can be seen from Table 1, the main difference between the test series 1 and 2 lies in the steel sample temperature: in the test series 1, before the test, the steel sample was kept in a freezer at around -18°C, while in the test series 2, the steel sample was kept in the cooling chamber and had a temperature close to that of the ambient atmosphere (around 5°C).

Before each test series, a new ice surface was prepared. The ice rink was 20 mm wide, 80 mm long and 5 mm deep. For ice formation, 18 ml of distilled water was used to which 0.5 ml of tap water was added to accelerate ice crystallization. To achieve a homogeneous temperature and crystallization rate within the ice volume, the sample was cooled for 40 min. Due to the expansion of the water volume under sub-zero temperatures, when ice was formed, it had a curved surface, therefore, the surface was smoothened with an aluminium plate having surface dimensions of 45 mm x 28 mm. The smoothening of the ice surface was performed at a normal load of 692 N at a sliding velocity of 0.08 m/s until the height difference between the left and the right side of the ice track was lower than 100 μm. Usually, this was achieved within 10-20 min of sliding the aluminium plate back and forth.

Table 1: Ambient and ice conditions for different test setups on the oscillating tribometer.

Test series	Test setup	Ambient temperature (°C)	Ambient RH (%)	Ice surface temperature (°C)	Ice bulk temperature (°C)	Sample temperature (°C)
1	1-1A	7.7±0.8	42±7	ca. -8.0	ca. -9.0	ca. -18.0
2	2-1A	2.7±0.3	15±1	-8.6±0.5	-9.1±0.0	ca. 5.0
	2-2A	3.0±0.5	20±3	-8.7±0.3	-8.7±0.3	ca. 5.0
	2-2B	6.7±0.3	53±4	-8.3±0.4	-8.7±0.2	ca. 5.0

Tests on the inclined plane tribometer were conducted at different ice and air temperatures to see how ambient conditions influence the results. All known ambient conditions for inclined plane tests are summarized in Table 2. Test 1 was conducted at lower ambient temperatures than Test 2. The ice was frozen identically for all the test setups. The temperature was -10°C while the ice was frozen layer by layer (5 layers in total). Each new layer was poured on the previous one and warm water was used to ensure a better mixture of the layers. Before experiments, the ice surface was levelled with a specifically developed planer that also creates a small groove on the ice surface. The groove guides freely sliding samples in a straight line while they are sliding down the inclined plane. Small ice debris was removed from the planed ice with a moist sponge to ensure a smooth ice surface. The same ice was used in all experiments which were conducted on the same experiment day. The sample groups were tested on different days (two experiment days in total).

Table 2: Ambient and ice conditions for different tests on the inclined plane tribometer.

Samples	Test setup	Ambient temperature (°C)	Ambient RH (%)	Ice surface temperature (°C)	Sample temperature
G	1	-1.6±0.2	63±2	-9.1±0.2	ca. -1.6°C
	2	-0.8±0.3	64±3	-9.0±0.2	ca. -0.8°C
SP	1	-3.5±0.5	64±3	-7.0±0.2	ca. -3.5°C
	2	-0.4±0.3	75±3	-2.0±0.3	ca. -0.4°C

2.3. *Friction measurements*

A universal modular designed tribometer (RVM1000, Werner Stehr Tribologie GmbH, Germany) was used in oscillating mode. Tests were performed at a constant normal load of 52 N. In each friction test, initially, a running-in phase was employed for 60 s at 0.1 m/s resulting in 120 cycles. Afterwards, 7 increasing velocity steps (0.02, 0.05, 0.10, 0.14, 0.19, 0.29 and 0.38 m/s) were employed. In each velocity step at least 10 cycles were performed and at the same time, for each velocity step, a minimum of 3 s duration was employed. Further details on the experimental setup can be found elsewhere [4].

2.4. *Velocity measurements*

Steel samples were slid down a 3 m long ice path tilted at an angle of 16 ± 0.5°, which was sufficient to promote sliding of steel samples without stacking in the start position. Samples always started the movement from a steady-state (using a start gate) and accelerated freely sliding down the ice surface. To measure the sliding time, optical sensors allowing time measurements with 0.01 s resolution were used. From the time measurements, the average sliding speed of the samples was calculated. The total distance between the first and the last optical sensor was 2850 mm. Each sample was tested 40 times. From 40 measurements, 3 fastest and 3 slowest results were eliminated. The results shown in the graphs herein represent the average of the selected 34 measurements. Further details on the experimental setup (inclined plane tribometer) can be found elsewhere [4].

3. Results

In Figure 2, the coefficient of friction of sandblasted and polished samples at two different ice and atmospheric conditions are presented. From Figure 2a it can be seen that for test setup 1-1A coefficient of friction decreased with sliding velocity and slightly increased with decreasing surface roughness. Higher friction of the smoother samples could be correlated with their large contact area which resulted in higher adhesive forces than for the rougher samples with a smaller contact area. From Figure 2b, for test setup 2-2A, coefficients of friction significantly decreased as compared to test setup 1-1A. This is most probably correlated with a higher sample temperature resulting in the formation of a LLL. At the same time, in Figure 2b the influence of surface roughness is not as pronounced; however, it can be observed that it has an inverse effect on friction as in test setup 1-1A. Namely, for test setup 2-2A, coefficient of friction with rougher samples was higher than for the smoother samples, which was especially pronounced at low velocities.

In Figure 3, results with grooved samples are presented for different ice and atmosphere conditions. From Figure 3a, it can be observed that the coefficient of friction increased with sliding velocity and at the same time, the coefficient of friction increased with surface roughness. From Figure 3b, at higher ambient temperature and humidity and/or on a run-in ice surface (in test 2B, ice surface was additionally run-in) coefficient of friction decreased, the influence of roughness remained unchanged, while the influence of velocity was less pronounced than in test setup 2-1A.

IOP Conf. Series: Materials Science and Engineering 1140 (2021) 012033 doi:10.1088/1757-899X/1140/1/012033

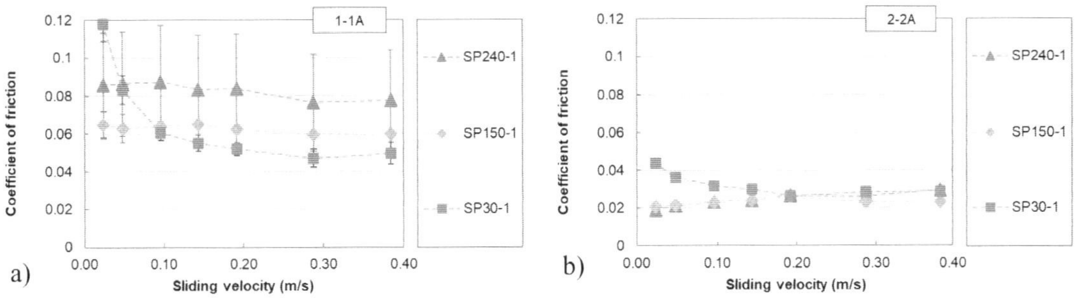

Figure 2. Coefficient of friction of sandblasted and polished samples at two different ice and atmosphere conditions: (a) test setup 1-1A, (b) test setup 2-2A.

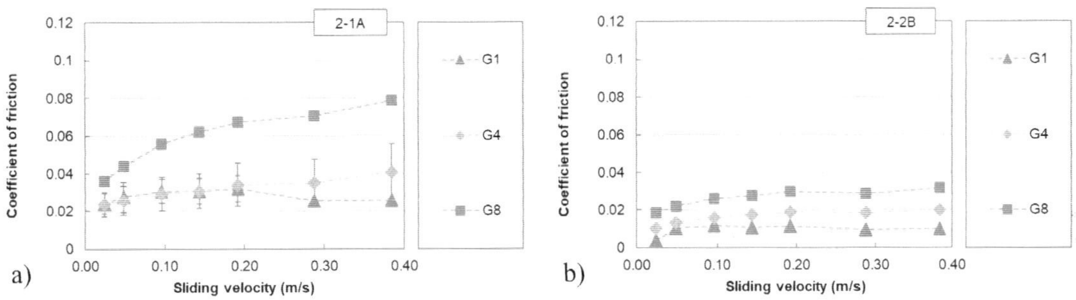

Figure 3. Coefficient of friction of grooved samples at different ice and atmosphere conditions: (a) test setup 2-1A (b) test setup 2-2B.

In Figure 4, test results from the inclined plane tribometer are shown. As shown in Table 2, test 1 was conducted at lower ambient temperatures thus providing a "dryer" ice surface while test 2 provided "wetter" ice surface. For both sample types, a trend of decreased sliding velocity for rougher surfaces can be observed. With grooved samples, at higher ambient temperature (test 2) higher sliding velocities were achieved, while with sandblasted and polished samples an opposite trend was observed. The decrease of sliding velocity at higher temperatures could be caused by a thicker LLL resulting in increased hydrodynamic drag for such isotropic surfaces at low normal loads. Figure 4 indicates a pronounced influence of the ambient conditions on the sliding velocity and suggests that changes in the ambient conditions might have a larger effect on sliding performance of the steel samples than their surface texture modifications.

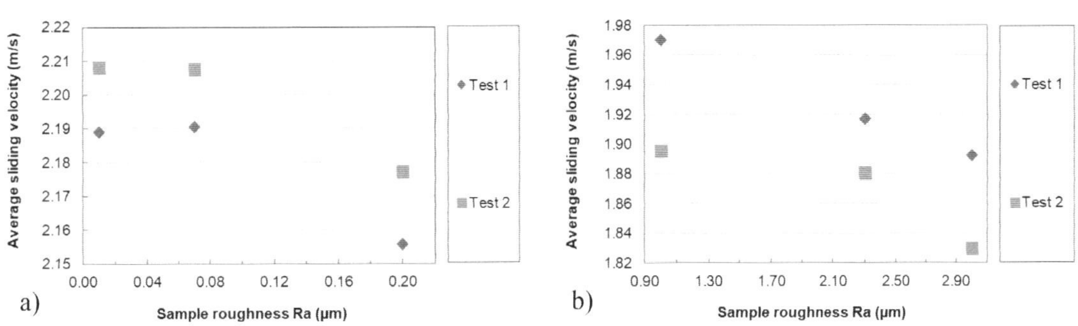

Figure 4. Average sliding velocity on the inclined plane tribometer for (a) grooved and (b) sandblasted and polished samples. Test 1 was performed at lower air and ice temperatures than test 2. Each sample group was tested on a different day.

4. Discussion

In the present study, an inverse roughness-friction correlation was observed for the sandblasted and polished samples tested at different boundary conditions on an oscillating tribometer. Most probably, the inverse roughness-friction correlation occurred due to the difference in the steel sample temperature. In the test 1-1A, where samples with higher roughness provided lower friction values, the initial sample temperature was approximately -18°C, thus in the contact areas between the steel sample and ice, LLL viscosity was lower or even freezing of the LLL could have occurred on a small scale. Since the smoother samples have a larger contact area with the ice surface, more "freezing" points could have formed which resulted in higher friction values. In test series 2, where warmer samples were used, freezing of the LLL did not occur. On the contrary, steel samples melted the ice surface due to heat transfer from the relatively warm samples. Due to a larger contact area with the smoother samples, LLL formed more effectively, resulting in lower friction values. For the grooved samples, the influence of surface roughness on friction was similar as for the sandblasted and polished samples in test series 2, which most probably is correlated with the similar sample temperature in both tests and thus more efficient LLL formation for the smoother steel samples.

Inclined ice plane tests showed a deteriorated sliding ability at higher ambient temperatures for the sandblasted and polished samples, while for the grooved samples an opposite effect was observed. Different influence of the ambient temperature indicates that with different surface structures, thickness of the LLL and the corresponding hydrodynamic drag can vary significantly.

5. Conclusions

In the present study it was observed that:

1. At very low sample temperature, the influence of surface roughness had an inversely different effect on friction compared to tests which were performed with higher sample temperatures. This effect was correlated with different heat transfer effects between the sample and the ice for different sample temperatures.
2. In the inclined ice plane tests sliding velocity decreased with the roughness of the steel samples regardless of their texture type.
3. For similar sample conditions, results on the oscillating tribometer and the inclined ice plane are in good correlation.

Acknowledgements

Parts of this work were funded by the Austrian COMET Program (Project InTribology, no. 872176) and carried out at the "Excellence Centre of Tribology" (AC2T research GmbH) in cooperation with V-Research GmbH and Riga Technical University. Financial support of Austrian Cooperative Research (ACR) is gratefully acknowledged.

References

[1] Kietzig A-M, Hatzikiriakos S G and Englezos P 2010 Physics of ice friction *Journal of Applied Physics* **107** 081101
[2] Kietzig A-M, Hatzikiriakos S G and Englezos P 2009 Ice friction: The effects of surface roughness, structure, and hydrophobicity *Journal of Applied Physics* **106** 024303
[3] Marmo B A, Blackford J R and Jeffree C E 2005 Ice friction, wear features and their dependence on sliding velocity and temperature *J. Glaciol.* **51** 391–8
[4] Velkavrh I, Lungevičs J, Jansons E, Klien S, Voyer J and Ausserer F 2019 The Influence of Isotropic Surface Roughness of Steel Sliders on Ice Friction Under Different Testing Conditions *Lubricants* **7** 106
[5] Jansons E, Lungevics J and Gross K A 2016 Surface roughness measure that best correlates to ease of sliding *Proceedings of 15th International Scientific Conference Engineering for Rural Development* Engineering for Rural Development (Jegava, Latvia: Latvia University of Agriculture, Latvia) pp 687–95

Modern Robot Integrated Manufacturing Cell According to the Needs of Industry 4.0

Madis Moor[1,2]; Kristo Vaher[1,2]; Jüri Riives [1]; Tavo Kangru [2]; Tauno Otto[1]

[1] Tallinn University of Technology (TalTech), Estonia
[2] TTK University of Applied Sciences, Tallinn, Estonia

madis@tktk.ee

Abstract. The continuous need to develop Industry 4.0 branches has led to a position, where highly sophisticated and multi-layer smart robotic systems are conducting the way in future manufacturing. This study aims to build a connectivity and system intelligent layer on top of a Co-bot integrated CNC-based Manufacturing cell. The connectivity layer is used to bypass all the data from machines to the upper intelligent layer vice versa. When raw data is arriving in the intelligent layer it is converted to information and again to knowledge for reflection back to the cell. Machine to Machine Communication and Digital Twin process for optimization is used for data conversions. This study is a down-scale example of the CPS for further development of existing robot cells.

1. Introduction

The combination of technologies, such as Smart Sensors, Internet of Things (IoT) [1], Machine to Machine Communication (M2M) [2], and AI-based solutions have fundamental importance in the process of further development of existing manufacturing cells. The next phase for production units in the future is to implement data collecting level as criteria for decision-making possibilities for existing robot cells. This can be achieved by using virtual environments for data processing and management of the actual robot cell. In this article, possible solutions and methods have been investigated for upgrading existing production cells to the level of automation and intelligence needed for Industry 4.0 (I4.0).

The goal is to define possibilities how to develop existing production system to meet the needs of I4.0 principles by applying Smart Sensor technology. The information is collected, processed and controlled by Manufacturing Execution System (MES) [3] to ensure continuous workflow in every production unit as a whole system.

2. Ontology of Modern Manufacturing

In a knowledge-based economy and production which is characterized by continuously shortening product life cycles, but continuously increasing demands towards products' functionality, quality, and other customers' requirements. The orientation towards development and improvement of the production system and its efficient use is extremely important. A company is an entire system that operates in a certain location and customer-oriented field of activity. A company may belong to a group or network, whereby its belonging to the network may be abstract or the company may have certain connections or functions with the network [4].

The production system has certain resources, processes, and strategies (Fig. 1). The production system is characterized by the physical environment (number, type, layout, and location) and functional environment that is expressed by technological possibilities of machine tools. Machine tools have mutual logistical relations inside the system as well as relations with the external environment. Technological possibilities of a company's production system evolve mainly based on machinery (CNC tools, IRs and Co-bots, presses, welding equipment, etc). Technological possibilities can be defined as a set of characteristics of the current device for producing a specified workpiece or performing a certain technological task.

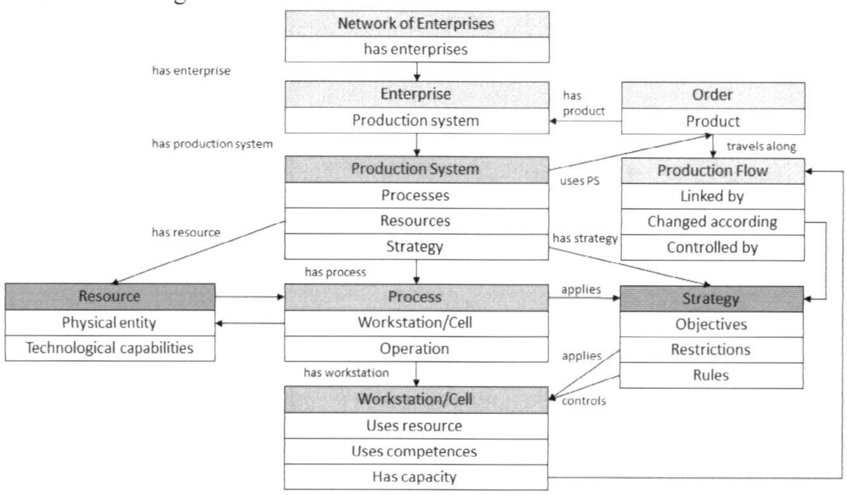

Fig. 1 Ontology in manufacturing [4]

The Industrial Production of the future will be characterized by the strong individualization of products under the conditions of highly flexible production, the extensive integration of customers and business partners in value-added processes. Future manufacturing combines technology, knowledge, information, and human ingenuity to develop and apply manufacturing intelligence. It comprises the smart use of networked information for demand-dynamic economics such as: integrated enterprise and supply chain and broad-based workforce engagement; industrial robots that work safely with people in shared spaces; and metal-based additive manufacturing. Driven by the software connected to the Internet, the real and virtual worlds are growing closer and closer together to form the IoT.

In this article, we focus mainly on the workplace level, which we mean a robotic workplace, and look at it as an integration of different hardware and software tools forming a Cyber-Physical System (CPS) (Fig. 2). The CPS is based on a multi-dimensional complex space that generates and evolves diverse subspaces to contain different types of individuals interacting with, reflecting, or influencing each other directly or through the cyber physical subspace. "CPS are engineered systems that are built from and depend upon the synergy of computational and physical components."

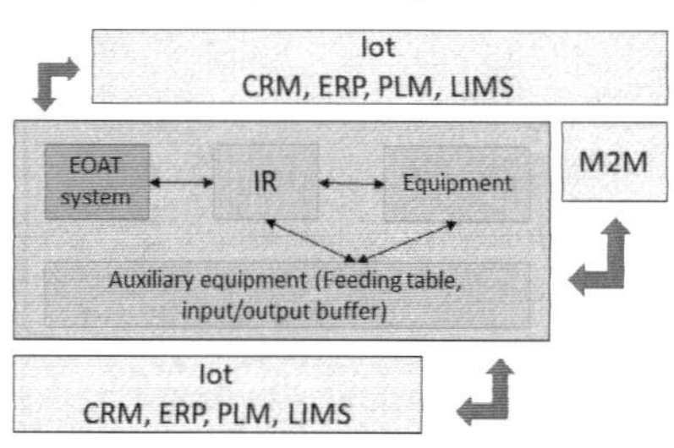

Fig. 2 Robot Cell with CPS functionality

Three main functionalities for modern robot-cells are:

- Internet of Things (IoT),
- Machine to Machine communication (M2M),
- Big Data.

IoT gives the possibility to connect more devices to a central controller. This set-up allows manufacturers to gather more data and streamline and digitize their processes more efficiently. With the M2M the success of data exchange and autonomous automation would be achieved, and it depends on the machines' ability to communicate and their real-time responses with one another. The role of data and information is increasing currently in manufacturing. Extended supply chain, but also horizontal and vertical value stream are based on data exchange and information processing. Thanks to IoT, data transfer possibilities, data analytics, different decision-making algorithms, it will come possible to use AI possibilities in the workplace for intelligent manufacturing with adaptive control, agile planning and real-time management.

3. Integration and Connectivity

The new manufacturing architecture follows the need to integrate vertical and horizontal value chains. As the production is going more intensive, more digital solutions (ERP, PLM, MES, CAD/CAM/CAQ, etc) in both value chains are implemented to the general computing architecture trends. Over time smaller computers and PCs distributed functions over networks, and distributed computing has been refined in many ways with Service Oriented Architecture, HTTP, Remote Procedure Call (RPC), and other functions.

This integration and connectivity are formalized with the ISA 95.00-xx standard [5]. Traditionally, ISA95 has been implemented in a 5-layer system network architecture, implying that each level only communicates through adjacent levels (Fig. 3). This implementation was built around the technology available at the time it was conceived, which has changed significantly with high power computers, high-performance networking, and embedded edge computing, all supported by more sophisticated and refined software that has been developed for general computing [6].

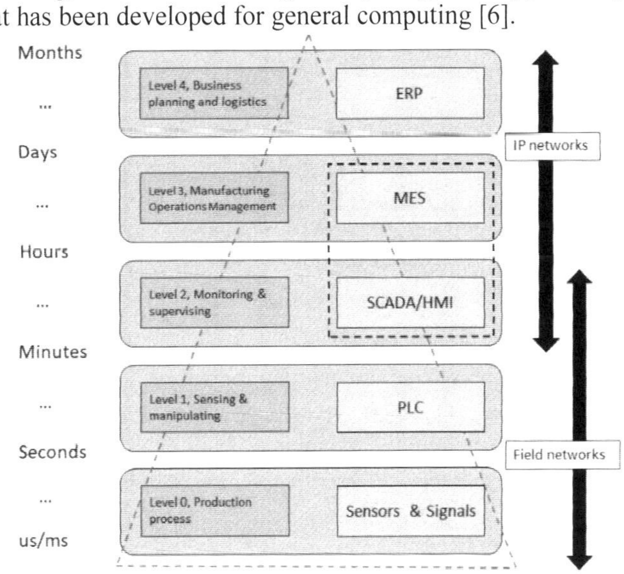

Fig. 3 Integration and connectivity in manufacturing according ISA 95.00 [6]

The technical innovation of Industry 4.0 towards smart manufacturing is characterized by the integration of manufacturing systems, using digital twin principles in the different value chain positions, virtual reality and augmented reality in designing and execution stages, big data, and data analytics for continuous improvement.

4. Integrated Manufacturing Implementation

In this article, we focused on implementation of miniature production system which have been under development at TTK University of Applied Sciences. The equipment and component description have been used in [7], and the multi-layer CPS uses similar ontology of manufacturing stated above (Fig. 1).

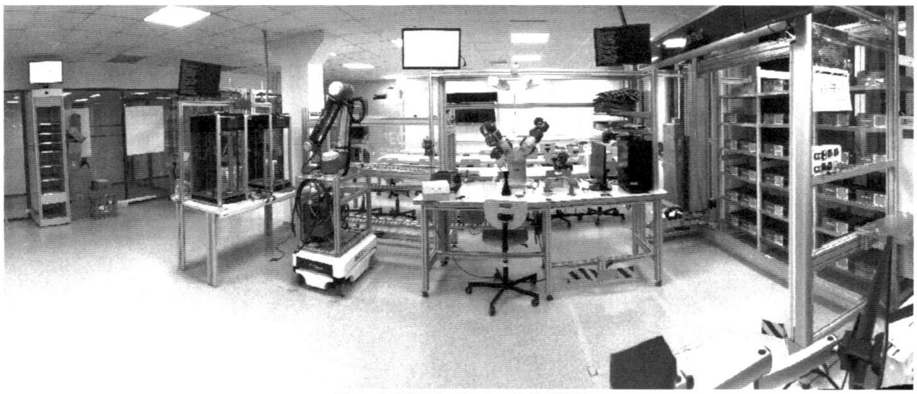

Fig. 4 TTK Industry 4.0 CPS

The entire system works autonomously, where incoming orders are planned in the ERP system, M2M communication and the sequence of production operations is managed by a program created at the MES level. All devices in the system are connected to MES via LAN or Wi-Fi but are independent of each other and do not require constant communication with a higher-level program (Fig. 5). All the units are programmed for each operation, and the MES program distributes only start commands and waits for execution confirmation to distribute new commands. Error reports from devices are forwarded to MES where necessary decisions are made for further operations. According to the number of details needed to produce, MES is capable of evaluating unused workstations and is capable to decide how to produce parallelly similar items or details.

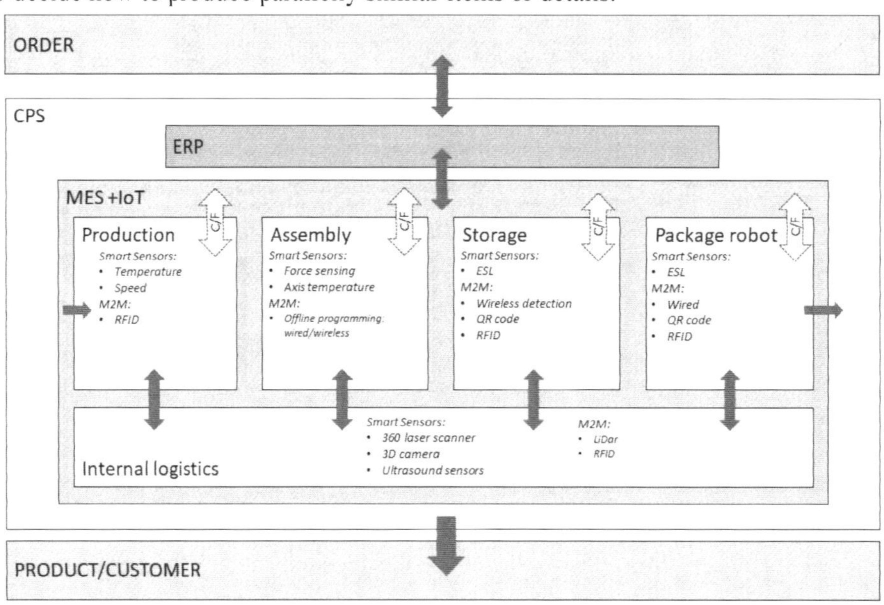

Fig. 5 Developed system schematic diagram

The system design is flexible and easy to reconfigure according to I4.0 principles, with the possibility to add new production equipment to the entire system without the need to make major changes to the control system. The mobile robot arm plays a major role in this factory, which makes it possible to apply a similar method in the existing factory without making major adjustments for the robots. Each unit is equipped with photoelectric sensors (PNP output) [8]to monitor and give feedback about the current situation at the workplace, first of all, to PLC and then to MES for management of other units. Another important part of CPS is RFID technology [9] with the RFID chip [10], -reader [11], -communication module, and power module. Each production plate and storage box are equipped with an RFID chip and every unit and also MES has live information.

5. Conclusion

Modern Manufacturing has been changing rapidly in the last decade. The sophisticated and multi-layered robot integrated manufacturing is not far in the future. Our entrepreneurs have the machinery and equipment to manufacture products according to the customer needs, but continuously shortening product life cycle and product's increasing complexity forces us to raise the efficiency of the existing system. In this paper, we have reviewed the possibilities to implement I4.0 principles to existing downscaled IR and Co-bot based manufacturing system, by using Smart Sensors and M2M connectivity combined with IoT based manufacturing software. The study we have conducted indicates that the usage of I4.0 technologies is strictly based on the needs and the purpose is to raise the efficiency of the certain robot-based manufacturing cell.

References

[1] R. S. Batth, A. Nayyar and A. Nagpal, "Internet of Robotic Things: Driving Intelligent Robotics of Future - Concept, Architecture, Applications and Technologies," 2018 4th International Conference on Computing Sciences (ICCS), Jalandhar, 2018, pp. 151-160, doi: 10.1109/ICCS.2018.00033.

[2] I. Stojmenovic, "Machine-to-Machine Communications With In-Network Data Aggregation, Processing, and Actuation for Large-Scale Cyber-Physical Systems," in IEEE Internet of Things Journal, vol. 1, no. 2, pp. 122-128, April 2014, doi: 10.1109/JIOT.2014.2311693.

[3] B. Saenz de Ugarte, A. Artiba & R. Pellerin (2009) Manufacturing execution system – a literature review, Production Planning & Control, 20:6, 525-539, DOI: 10.1080/09537280902938613

[4] Riives, J.; Lõun, K.; & Otto, T. Framework for Extended Use of Technological Resources in the Network of Enterprises. 7th International DAAAM Baltic Conference "Industrial Engineering" 22-24 April 2010, Tallinn, Estonia.

[5] ANSI / ISA-95.00-xx. Enterprise Control System Integration

[6] ISA 95 in the IoT and Digitalization Era. By Bill Lydon, Contributing Editor, Automation.com. January 06, 2020

[7] Vaher, K.; Vainola, V.; Otto, T. (2019). Industry 4.0 Laboratory. Industry 4.0, Technological Basis of "Industy 4.0", 1/5: IV International Scientific Conference, Industry 4.0. Summer session, 24-27.06.2019, Burgas, Bulgaria. Industry 4.0 Proceedings: 108 Rakovski str; 1000 Sofia, 52−53

[8] Omron Corporation (2021, January 24), Photoelectric sensor E3F1-DP22, Retrieved from https://industrial.omron.eu/en/products/E3F1-DP22

[9] R. Want, "An introduction to RFID technology," in IEEE Pervasive Computing, vol. 5, no. 1, pp. 25-33, Jan.-March 2006, doi: 10.1109/MPRV.2006.2.

[10] Siemens Corporation (2021, January 24), RFID chip MDS D124, Retrieved from https://mall.industry.siemens.com/mall/en/WW/Catalog/Products/10047932

[11] Siemens Corporation (2021, January 24), RFID reader RF340R (GEN2), Retrieved from https://support.industry.siemens.com/cs/pd/467302?pdti=pi&dl=en&lc=en-GB

On the Selection and Application of a Convenient Energy Management Software for Industrial Purposes

Larissa Wurster[1]**, Gregor Thiele**[2]**, Clemens Briese**[3] **and Jörg Krüger**[4]

[1]Plant Development Infrastructure / Assembly Trucks (TG/MAD), Daimler Truck AG
[2]Department for Process automation and Robotics, Fraunhofer Institute for Production Systems and Design Technology
[3]Department for Machine Vision, Fraunhofer Institute for Production Systems and Design Technology
[4]Department for Industrial automation at the Institute for Machine Tools and Factory Management, Technical University of Berlin

E-mail: larissa.wurster@daimler.com

Abstract. The profitability of companies, especially in Germany, is affected by rising energy prices and increasingly stringent energy legislation. Certification according to ISO 50.001, an international standard for energy management systems, can help to meet legal requirements and increase energy efficiency. This article offers assistance in the form of a methodology that provides a structure for the selection process of an energy management system (EnMS). Considering properties of the individual company, it helps to identify an appropriate software solution. It is applied to a globally operating metalworking company in an example study. Furthermore, the measurement infrastructure in particular is a basic requirement that should not be overlooked in addition to the process of choosing an appropriate software solution. In this context, suitable metering points or criteria must be defined as well as sufficient attention paid to the anticipated costs for acquiring and integrating new measuring transducer and energy meters. The type and number of relevant measuring points must be considered in particular. Discussing these aspects, this article facilitates the process of planning and implementing an energy monitoring.

1. Introduction

The subject of energy management is becoming increasingly important. This is attributable to, among other things, stricter and stricter regulations being imposed on companies to meet energy targets and ensure that production changes towards carbon neutrality. Therefore, the certification in the complicated area of energy management according to ISO 50.001 is a major milestone for many companies. The purpose of this article is to illustrate a methodology for the selection of an appropriate energy management system (EnMS) solution. We will define the related terms and state reasons that support the introduction of an EnMS (Sec. 1). Accordingly, we will explain how to deduce the general conditions and requirements for the implementation in Sec. 2. The individual aspects, including the possible costs of implementation, will be presented in Sec. 3. Furthermore, we will examine the selection of suitable meter operators and energy metering equipment as well as the metering infrastructure in Sec. 4. Sec. 5 summarizes our contribution.

1.1. The terms "Energy Management" and "Energy Management System"
The term energy management is defined in the VDI 4602 standard as "the forward-looking, organized and systematic coordination of procuring, distributing and using energy to meet requirements while considering ecological and economic objectives" [1, p. 2]. In this paper, energy management is limited to the interaction of energy controlling and energy monitoring. It comprises energy data collection and energy data analysis.

The standard ISO 50.001 defines an EnMS as a "management system to establish an energy policy, objectives, energy targets, action plans and processes to achieve the objectives and energy targets" [2, p. 15]. The organizational and informational structure plays a crucial role for the EnMS implementation. Special attention should be applied into the selection of appropriate software and hardware. [1, p. 5]

1.2. Motivation for Introducing an EnMS
It is becoming increasingly important for companies, especially in Germany, to obtain certification according to ISO 50.001. In addition to financial benefits, such as the EEG levy, this offers further advantages like an increase in general energy efficiency, especially a lower electricity consumption, and transparent insights in the CO_2 output. [3, p. 3]

Many production units lack transparency into how much energy they consume, especially in their infrastructure and production processes. Moreover, their system of energy reporting is relatively complicated and time consuming. An EnMS can help overcome these challenges. [4, p. 5] A key aspect is increasing production efficiency. By reducing, the amount of material and energy consumed and raising the quantities produced while at the same time reducing the time needed to produce them, it strengthens a company's competitiveness. Another key aspect is compliance with statutory regulations and guidelines, for example ISO 50.001, which requires a continuous improvement process (CIP). [2, p. 17]

2. General Conditions
The essential tasks of an EnMS include the systematic collection and communication of energy flow as well as automatic control of facilities and production equipment for improving energy efficiency [5]. It is useful to collect the most significant energy and media consumptions within the production unit. At the very least, the consumption on the reviewed production hall as well as the major individual consumers should be monitored. The more detailed the breakdown with respect to consumption is, the higher the density of the data collected will be. But accumulating in a too fine granularity will cause an unnecessary amount of data. It appears more beneficial to concentrate on the essential consumers. One option is to install additional energy meters as time goes by in order to obtain higher data density. Key factors in the production environment are also the monitoring of grid quality and the integration of the company's internal databases – for example, the production calendar [4, p. 11].

3. Methodology for Selecting an EnMS
The basis for selecting EnMS software is an organized procedure devised by Treutlein and Sontow [6], originally created for choosing Enterprise-Resource-Planning (ERP) software solutions. It is used to narrow down the EnMS solutions examined as part of a market analysis (see Fig. 1). [6]

3.1. Best Fit Principle
First available providers are narrowed down by conducting a comparative market analysis. The focus laid on the basic features of the various solutions. [6] One possible method of performing the screening is to create a matrix table and use it to evaluate the different EnMS solutions based on specific criteria. Essential criteria are presented in Fig. 2, which shows an example of a matrix table generated in accordance with the best-fit approach. The energy management solutions can be pre-screened using the matrix table. The following prerequisite must be established for doing so: The companies wishing to implement an appropriate software solution needs to have examined in advance the criteria essential to it and demands placed on a suitable EnMS solution.

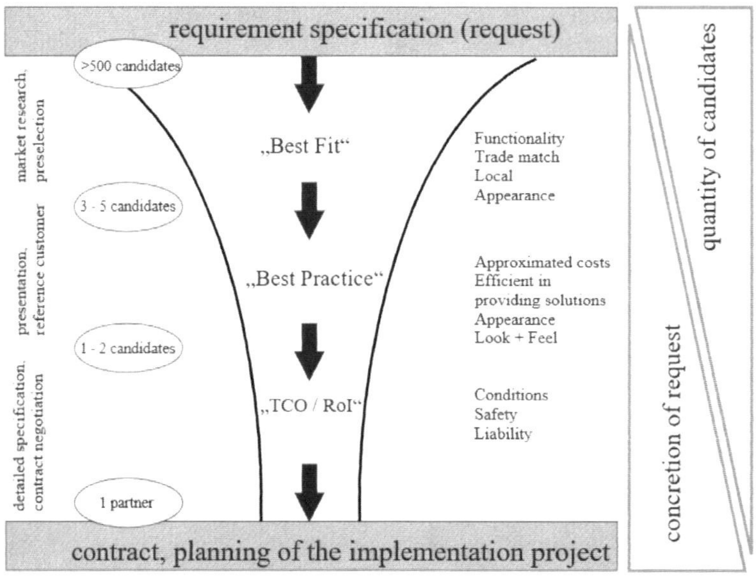

Fig. 1. Selecting ERP software [6].

software		criteria					results of the evaluation				
		1	2	3	4	5	1	2	3	4	
manufacturer	software name	EnMS functionality	attractiveness (dashbord, diagrams)	interfaces, compatibility with existing systems	reference installation in similar branch	reference installation in the company	criteria 1+2	criteria 3	criteria 4	criteria 5	notes
manufacturer A	software A	✓	✗	-	-	-	✗	✗	✗	✗	
manufacturer B	software B	✓	✓	✓	✗	-	✓	✓	✗	✗	
manufacturer C	software C	✓	✓	✓	✓	?	✓	✓	✓	✗	unsure if there's an installation at the company

Fig. 2. Section of a matrix table.

3.2. Best Practice

As a next step we analyze the costs incurred when implementing an EnMS [6]. The comparability of costs and licensing models for the different vendors pose a special challenge. To illustrate the procedure, we use an example of a large company in the metalworking industry with global operations and that, thus far, has no EnMS.

First, the individual production plants of this sample company are divided into different categories based on size. That establishes three separate categories of plant sizes that differ based on assumptions of how many metering points they wish to install. Tab. 1 shows these assumptions. The next step is inserting the costs determined for the software solution in Tab. 2.

Tab. 1. Plant sizes and metering points.

		quantity	counters / plant (total)	existing counters / plant	server quantity / plant	user quantity / plant
plant size	S	2	1.000	200	1	10
	M	5	2.000	400	2	25
	L	1	2.500	800	3	50
	total	8				

Tab. 2. Sample calculations of software cost.

		manufacturer x													
		license		update	support	server connection		installation data points		training		project management		total costs [€]	
		cost [€]													
		one-off payment	annual payment	cost [€]	cost [€]	quantity	cost [€]	quantity	cost [€]	piece [piece]	cost [€]	days [d]	cost [€]	one-off payment	annual payment
basic license															
plant size	S														
	M														
	L														

The one-time and annual costs for each vendor are first extracted from Tab. 2 and determined for each plant size.

The one-time costs consist of the following expenses:

Total one-time costs [€] = basic license [€] + one-time license fee [€]
+ (number of servers [ea.] x cost of server connection [€])
+ (number of data points [ea.] x cost of connecting the data points [€])
+ (number of training sessions [ea.] x cost per training session [€])
+ (number of project management days [d] x cost per project management day [€]) (1)

The annual costs consist of the following expenses:

Total annual costs [€] = annual license fees [€] + cost of updates [€] + support fees [€] (2)

The next step is to take the one-time and annual costs calculated per vendor and determine the amount for each location of the example factory. The results are shown in Tab. 3.

Tab. 3. Excerpt from comparison of EnMS costs by plant size.

			manufacturer A		manufacturer B		manufacturer C	
			cost [€]		cost [€]		cost [€]	
		quantity	one-off payment	annual payment	one-off payment	annual payment	one-off payment	annual payment
plant size	S	2	96.000	17.000	115.000	14.000	53.000	11.000
	M	5	340.000	42.500	412.500	65.000	220.000	55.000
	L	1	94.000	9.000	122.500	29.000	57.000	13.000
	total	**8**	**529.000**	**68.500**	**650.000**	**108.000**	**330.000**	**79.000**

The costs are calculated for each vendor using this formula:

One-time costs [€] = number of plants per size category x total one-time costs per vendor [€] (3)

Annual costs [€] = number of plants per size category x total annual costs per vendor [€] (4)

An extrapolation of the individual results based on a useful life of five years, including an annual increase of 2% in running costs is recommendable.

The EnMS implementation costs can depend on the complexity of the system, the needed customizations and also the company size plays a crucial role. [7]

4. Meters and Infrastructure

4.1. Selecting Appropriate Meter Operators and Energy Recording Equipment

Preparing a careful metering strategy is crucial to determining the number of metering points and designing their layout. The operational reality reveals the applicability of the Pareto principle, also known as the 80/20 Rule. In the area of energy management, this means that approximately 20% of the energy loads are responsible for 80% of energy consumption. [8], [9] For that, it is a good idea to start energy monitoring considering the most extensive loads first [9].

In selecting the energy metering equipment, attention must be paid to ensure that they are appropriate for the respective metering task. They should guarantee high levels of data quality and operational reliability. When implementing additional metering points and equipment, the components should be added ordered by the consumption. This requires to scout all production processes with their respective energy consumption and detailed consultations with plant representatives and process owners. [9]

4.2. Infrastructure Costs

The infrastructure costs are the foundation of an EnMS. For the calculation of them let us remember Tab. 1. Based on that, we get the infrastructure costs for each plant size and, finally, the total cost of implementation on the infrastructure end as found in Tab. 4.

Tab. 4. Excerpt of infrastructure costs.

infrastructure																
plant		per plant														
		counter								server			user			
			new					amount								
				initial cost		connection cost					cost			cost		
plant size	quantity	existing counter / plant	quantity	per counter [€/pc.]	total [€]	per counter [€/pc.]	total [€]	total quantity	cost [€]	quantity	per server [€/pc.]	total [€]	quantity	per user [€/pc.]	total [€]	total cost [€]
S	2	200	800	350	280.000	600	480.000	1.000	760.000	1	5.000	5.000	10	100	1.000	766.000
M	5	400	1.600	350	560.000	600	960.000	2.000	1.520.000	2	5.000	10.000	25	100	2.500	1.532.500
L	1	800	1.700	350	595.000	600	1.020.000	2.500	1.615.000	3	5.000	15.000	50	100	5.000	1.635.000
all plants		3.200	11.300					14.500	10.735.000	15		75.000	195		19.500	10.829.500

5. Summary and Outlook

The introduction of an EnMS is an essential requirement for obtaining certification according to ISO 50.001. Over the medium term, every manufacturing enterprise should strive to implement such a system. Within this paper a detailed method for the selection and application of an EnMS has been presented. The terms energy management and energy management system (EnMS) has been introduced and defined. A model for the calculation of the implementation costs has been explained.

Key steps in this are the installation and the daily work with an EnMS, plus expansion of the metering infrastructure down to the machine level. When procuring new machinery and equipment, companies should factor account energy data collection equipment into the scope of their orders. Last but not least, a holistic view and optimization of energy and materials cycles within production will be indispensable measures for achieving energy policy targets and, thus, making factories energy-efficient in the future.

References

[1] VDI-Richtlinie 4602-1 2016 *Energy Management - Terms, Definitions*

[2] DIN EN ISO 50.001:2018-12, Energy management systems – Requirements with guidance for use

[3] Schulze M, Nehler H, Ottosson M and Thollander P 2016 *Energy management in industry – a systematic review of previous findings and an integrative conceptual framework*. Journal of Cleaner Production 112, p. 3692–3708. DOI: 10.1016/j.jclepro.2015.06.060.

[4] SIEMENS AG 2019 Energy Efficiency begins with Transparency: SIMATIC Energy Manager V7.2. - *System description (ID: 109748841)*.

[5] Federal Ministry for the Environment, Nature Conservation and Nuclear Safety (BMU) 2012 *Energy Management Systems in Practice. ISO 50001: A Guide for Companies and Organisations.*

[6] Trovarit AG 2021 *Aufgaben der ERP-Auswahl* URL https://www.it-matchmaker.com/de/software-suchen/software-kategorien/erp/erp-auswahl-aufgaben/.

[7] Mey, J 2011 *How can we facilitate the introduction of energy management systems (EnMS)?* EGEE 2011 Summer study, p. 759–766

[8] Cambridge Business English Dictionary 2020 *The Pareto principle. Definition.* (Cambridge University Press) URL https://dictionary.cambridge.org/de/worterbuch/englisch/pareto-principle.

[9] WEKA MEDIA GMBH & CO. KG 2020 *Energy lexicon. Energy Data.* URL https://www.energiemanagement-und-energieeffizienz.de/energie-lexikon/energiedaten/.

IOP Conf. Series: Materials Science and Engineering 1140 (2021) 012036 doi:10.1088/1757-899X/1140/1/012036

Production environment of toMorrow (ProMo): Partially automated repair process of small tool moulds, forming tools, injection moulding tools and sand casting tools

Jan Kuschan, Vinzenz Müller, Stephan Mönchinger, Oliver Heimann, Carsten Niebuhr, Oday Kabha

Fraunhofer Institute for Production Systems and Design Technology IPK, Germany

E-mail: `jan.kuschan@ipk.fraunhofer.de`

Abstract. Small defects in the grain or major damage to a moulded part or tool can bring production to a standstill. SMEs in particular have neither the personnel nor the equipment to repair such damage on their own, so they send it to specialised contractors. The repair process is carried out manually, depending on the accuracy requirements, and is usually completed by a finishing process. This work requires qualified personnel and, at the same time, requires a lot of time in case of larger damages. In this paper we present a way to map the Maintenance, Repair and Operations (MRO) process chain in a partially automated manner.

The symbiosis of individual technologies results in a significantly increased efficiency of the MRO process chain, which continues to focus on people and their process knowledge.

While Directed Energy Deposition (DED) for the MRO of moulded parts is used widely, usually a high manual effort in measuring the component geometries and teaching of the machine tool paths is necessary. However, there are clear advantages compared to the manufacture of new parts or manual laser welding repair. At the same time, the resource and energy requirements can often be significantly reduced compared to new part production. ProMo focuses on automating the time-consuming machine programming by reducing the number of necessary work steps in CAD/CAM-based program creation.

Based on a subsequent robot-guided scan, a digital actual 3D model is generated. Due to intelligent path planning algorithms, no manual programming of the robot is necessary and at the same time it is possible to detect components of different sizes, shapes and covers in this system with a minimum of effort. In addition, the operator passes on elementary information, such as the approach path of the milling head, to the subsequent processes by means of finger gestures and can thus significantly reduce tedious CAM programming steps. Now, the scanned component is transferred to a 3D-CAD model and a target/actual comparison is created for the damaged areas. Those are milled out in a defined manner and then restored using DED.

1. Introduction

Turbine blades, like in stationary gas turbines or aircraft engines are exposed to extreme conditions. High temperature, pressure and depending on the use case also foreign objects reduces the life time of the blades. Since, a broken blade leads to a reduced turbine efficiency or even to a destruction of other parts, the Maintenance, Repair and Operations (MRO) has a major impact on the operator of turbines. The main defects are related to blade tip erosion or cracking defects, but in most cases, damaged parts are discarded and replaced by new parts, resulting in the loss of resources and high costs. [1]

To reduce industrial greenhouse gas emissions with its approximately 30% of the total greenhouse gas emissions, is one of the major challenges of our time [2]. One of the main reason for industrial greenhouse gas originate manly from material processing, i.e. from the conversion of natural resources into materials to convert into manufacturing and construction goods to produce the final product [3]. The use of additive manufacturing (AM) could improve both, an efficient material/resource and energy management. An important technology in AM for the MRO of metal components is laser-based DED which allows a material and cost efficient repair process with high deposition rates. It is widely used in the industry to repair complex gas turbine or moulding parts. [4]

If nowadays, blades are repaired this results in high engineering and machine time. One main disadvantage is the reconstruction and CAD programming of the free-form object. For the engineer this results in a very complex and detailed tool path planning of the DED machine and also the CNC milling machine. This is time consuming not only for the engineer but also in case of machine time.

2. Contribution

Within this paper we present a novel approach for a semi-automated damage detection and repair process, that we verified on a turbine blade. Therefore, we present an intuitive human robot interaction specifically designed for manufacturing in workshop environments. For the local workshop employee, the system provides a task oriented programming interface through a projection based augmented reality. The system generates most of the program automatically based on sensor input. This reduces the user input to the critical information required to solve ambiguities.

For maintenance personal or an off-side process engineer, the system offers a virtual reality interface. This allows to immediately bring in a remote expert when a critical problem arises. The system not only monitors the machines state but also reconstruct any object, such as the work piece or a human arm, within the working area. This gives the remote user a unique insight into the current problem and helps him provide the necessary support without an expensive and time-consuming trip to the facility. Within this paper we present and tested a holistic process chain for the MRO of a gas-turbine blade.

3. Related Work

Generally, automated Reverse Engineering, based on optical 3D-data is characterized by six stages in succession. First of all, we need to capture geometrical data of the virtualized object. These are then being post-processed and segmented. Then the generated data can be classified and used as input for model creation. The result is a 3D CAD model [5]. Reverse engineering, as a non-automated process, involves numerous of manual, repetitive and highly know-how-dependent activities. Automated data acquisition systems often require high investments, are not yet available in an industrial context, or are currently only being developed for specific applications [6, 7]. Post-processing of the scan data is often manually been done. By tessellating the scan data, a representation of the surface is created. For further use of the data, we have to transfer it into mathematically described models. This is commonly done by manual

Figure 1. Repair process chain of the defective component compare [13]

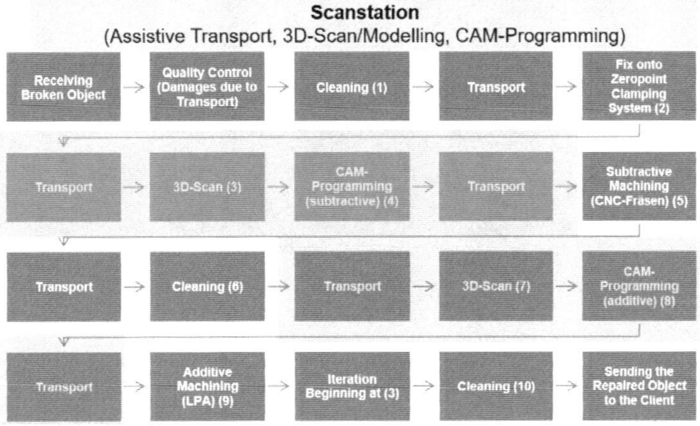

reconstruction of a CAD model based on the tessellated surface model. Cycle time and quality varies depending on the level of knowledge of the executor. Therefore, we work on the automation of these process steps.

Having an automated reconstruction process is highly depending on a well performed digitization of the blade or form. The scanning process itself is combined with a robot. Therefore, Heimann and Guhl [8] classifies robot programming methods into three sections: online, offline and hybrid, each with different advantages and disadvantages. Online programming methods like lead through, walk through or by demonstration need direct interaction with the robot, while offline programming methods like text based, graphic interfaces or simulations are done through comprehensive software tools without the physical hardware. Using a hybrid approach, it is possible to define the robots work space and routing offline and adapt the path to the specific problem.

Current work about human robot interaction splits in two topics. One is focusing the recognition of human action while the other focus point is the interface design. Due to current restriction in recognition of human by the computer the design of the interfaces is still narrowed. [9] gives an overview on the diverse ways to detect hand gestures, starting from traditional computer vision up to today's machine learning based methods. [10] describes a current system using traditional computer vision for accurate hand pointing in a robot environment. To bring the user interface to the robot working place augmented reality is a common way. Projectors and AR glasses are used to display UI elements and state information directly in the working area. In [11] a augmented reality system is introduced, where the robot shows its planed action. Touch interaction on a projective interface is shown in [12].

4. Experimental Setup

In our use case, we project a whole repair chain from receiving the broken part via repairing and finally the transfer back to the costumer (Fig. 1). Therefore, we designed a scanstation (Fig. 2) equipped with a zero-point clamping system mounted on a workstation table and a 3D-hand-scanner attached to a collaborative robot looking upside down. To control the process directly at the station, a stereo-camera for finger tracking and hand-gesture detection and a projector to visualize a GUI was mounted to the scanstation. An a wearable sensor vest giving a haptic real-time feedback in ergonomic critical movements.

Figure 2. Scanstation with overhead mounted robot, scanner and turbine blade.

5. Process Model

The repair process is described with a specific process model, which includes all processing steps that should be done in sequence from the beginning until the end of the process. The process model represents the interface between the user, and the industrial machine used in this project. The connection between the process model, and the machine is established using a REST-Interface. The process model provides several GUI objects, each one has a specific color, and represents a specific functionalities. Each GUI Object has several subclasses, in order to classify the functions in separate groups. Each group is dealing with the same subject. The most important GUI object is the resource, which is the real function that will be performed in a processing step. The resource has subclasses, which represents REST functions, OPC-UA functions, GUI functions, and others.
In this project the process model will be loaded into the AR-Application. Using a C# script the process model will be processed step by step starting with the START node and ending by the END node.

6. Augmented Reality

Augmented reality (AR) provides several options, in order to make the use of industrial machines easier for new employees. Due to the industrial machines complexity, and multiple functionalities; a training course should be given to the employees, in order to enable them to deal with the machine without making mistakes. These trainings are too expensive, and take a long time, while during the training time the machine will be out of work. Augmented reality helps to make training time shorter, and provides more options to explain the functionalities of the machine. One of the most important benefits of integrating augmented reality (AR) in industrial machines is that we need no additional hardware in the machine itself like screens, and computers for showing texts, and communicating with external servers. In this project the AR-Glass HoloLens 2 from Microsoft is used as an info visualization tool to display the processing state of the machine during the repair process. The HoloLens 2 also provides a GUI, which enables the user to change some settings, and to intervene during the processing, when it is necessary. To complement the HMD Hololens 2 interface, a projective AR interface was implemented with a DLP projector. This interface augments the key 3D working area and components (plate with the objects) and simple 2D buttons from above onto the scanstation table. Users interact with this interface via finger gestures, e.g. activate virtual buttons by pointing at them. These buttons are the means by which the user controls and triggers the process flow. Additionally users can define 3D positions with their fingers and by that mark the identified damages on the workpiece. To provide this functionality and correctly project user feedback, the workpiece and its pose are detected via cameras and a CAD model of the object.

7. Realtime Erognomics Feedback

Since workers often handles heavy objects, we are also presenting a solution to reduce the risks of injuries. Therefore, we designed a textile sensor vest for haptic ergomonic assessment feedback called ErgoJack, containing of three separable different layers, an outer, middle and inner layer. The outer layer is a robust textile that prevents the electronics from physical impacts. The middle layer contains all electronic components, a computing unit, five Inertial Measurement Units (IMUs), a vibration unit and a textile on/off button all connected trough textile wires. The inner layer, which has body contact,contains of breathable spacer fiber material which is comparable to the contact surfaces of backpacks. All layers are connectable through buttons and zippers. With this strategy it is possible to wash the material which is either exposed to environmental influences, like dust and fluids, or is close to the body and is therefore exposed to sweat, without destroying the electronics.

The ErgoJack uses the IMUs to detect the pose and movement of the worker and uses the vibration unit to give a feedback in real-time when the wearer performs potentially body harming actions. Details on the algorithm are published by Kuschan et al. [14]. Using this strategy, the wearer has a long-term learning effect, but is not restricted in his movements. By using the vibration unit the feedback is also only recognized by the wearer him- or herself.

8. Natural human-machine interface

Non-intrusive, contact free hand gesture detection can be beneficial for many human-computer interactions [15] and lately has also been implemented in human-robot interaction scenarios [15, 16]. In our use case we apply a simple 3D finger position to allow easy to use position or path definitions and manipulations of a projected interface directly within the working area. Thus no additional devices which may easily be destroyed in an robot assisted production space are needed. The user can naturally interact with the robot and show him where to perform which milling process. GUIs are still the well-established solution to interact with a machine and especially suitable to ease the transfer from a monitor to an augmented reality interface in a workplace. Instead of using a controller we allow users to naturally and efficiently use his/her fingers for 3D pointing interactions. Selection method is dwell time based and supported with projected visual feedback directly at the fingertip. The hand/finger detection is implemented as a mixture of traditional computer vision and machine learning approaches. Gesture recognition is based on the volumetric scene reconstruction data (see section 10) and a 3D separation of scene and the users hand. Parallel to that a hand and fingertip detection is processed based on a Yolov5 neuronal network. After recognizing hand and fingertip region, the exact position of the fingertip is computed using a hough transformation specially adapted to the finger geometry. The gesture recognition process is running on CPU not interfering with the GPU intensive volumetric scene reconstruction. In addition to GUI based human-machine interactions, the operator pass on elementary information, such as the approach path of the milling head, to the subsequent processes by means of finger gestures and can thus significantly reduce tedious CAM programming steps.

9. Robot Integration

To improve and automatize the scan, we mounted an UR5 on top of the scan-station facing downwards with the camera integrated at the end of the robot arm. The working area of the robot is predefined to prevent collisions. Trough a rough 3d reconstruction of the blade or form recorded by the finger tracking cameras the robot path is adapted to the ideal working distance between scanner and object.

10. Scanning and Processing

In a two step process the object on the plate is detected and for this purpose the previously mentioned volumetric scene reconstruction is used. First, the plate itself is detected in the scene by simply fitting a plane around the origin of the work environment in the point cloud. If the plate is successfully detected, the objects on the plate can be detected and their poses estimated. The object pose detection is of importance for the AR projected scene and for the interaction with the scene (e.g. for the gestures) and scanning. The pose estimation by means of point cloud registration is based on the approach described in [17]. The point cloud is segmented based on the plane representing the plate and all voluminous objects over the plate are extracted. By using the CAD models of the known objects, the objects can be registered to the segmented point clouds and their poses estimated. The limitations of our approach is based on the reconstructed objects and/or scene. If the objects are too small or if objects occlude each other it may result in an unsuccessful pose detection. Since the plate can be fixated in four different orientations (shifted in 90° steps) a global reference frame is introduced by means of a QR code (containing the plate identification), which can be detected in a 2D or 3D scene. The status of detection of the plate and object as well as object's pose is stored in variables on the OPC UA server for other software modules to access and to trigger the process flow.

The workflow for generating the difference model is divided into five steps: scanning, filtering the point cloud, registering the target and actual model, cutting out the region of interest (ROI) and generating the difference volume. This is followed by an interface to the CAM software, which includes an automated globally oriented positioning in the virtual installation space. The process of data acquisition has already been described in detail above. Therefore we start here at the process step of filtering. Due to error detections and unnecessary data points, such as the base plate, the data set must first be repaired and filtered. This process is necessary to create a continuous even waterproof and also reduced actual model. Therefore, we integrated PCL based functions to reduce the model and to detect and filter out the baseplate. In the following, the simplified actual model and the corresponding target model are first roughly aligned with each other by means of RANSAC registration (Random Sample Consensus) and then finely aligned with each other by means of ICP registration (Iterative Closest Point). Correspondences of points between the registered actual and target models are also calculated. Points that have no corresponding point in the other model are extracted. From this a correspondence point cloud is generated, which can be divided into many clusters based on Euclidean distance. Many clusters are created at this point because there are production-related deviations between the target and actual model. Since these are irrelevant for the repair of the defect area, it is necessary to determine the cluster which covers the relevant defect area. With the coordinates of the Point of Interest (POI), which is read by the OPC-UA server, the next cluster to the POI is selected as the point cloud of the Region of Interest (ROI). An oriented bounding box is then calculated, which we use as a tool for cutting the ROI mesh. After cutting the ROI mesh, existing disturbances, e.g. overlapping models, are removed. From this, a difference volume between actual and target model is calculated quickly and cleanly with the Boolean difference in FreeCAD Software. However, the difference volume cannot be processed directly by the CAM software because of its irregular base area. Therefore, a common volume of the cut ROI of the target model and the oriented Bounding Box of the ROI is created using a Boolean operation to create a flat and rational base. This allows the toolpaths to be built in the CAM software. The interface between the data processing and the CAM software consists of a Python script that opens the CAM software and a macro that calls the CAM software functions. The parameters for additive manufacturing are then entered manually to generate the tool path and the corresponding NC code.

11. Additive and substractive Manufacturing

Based on the calculated tool paths from the previous process step and defined process parameters such as laser power, feed rate, track widths or track overlap, the defective areas of the component are prepared in CAM software and then welded on using the DED process. A zero-point clamping system is used to clamp the component. To repair the defective volumes, a metallic powder is fed through the nozzle of the DED processing head to the surface of the component, where it is melted by a laser. By moving the head, weld beads an 3D structures are formed, resulting in a near-net-shape geometry. This geometry is specifically manufactured with material allowance which results in an oversized geometry for post processing.

For components with large tolerances, the accuracy of the DED system may already be sufficient, but for the repair of a turbine blade presented here, accuracies in the lower µm range are required. Fig. 3 shows the turbine blade machined by DED. In the subsequent machining post-processing, the repair-welded component is also picked up using a zero-point clamping system, thus ensuring that the component is referenced. The repaired defects are specifically reworked, whereby the oversize is removed from the DED process to produce the target geometry.

Figure 3. Blade tip before (left) and after (right) repair process

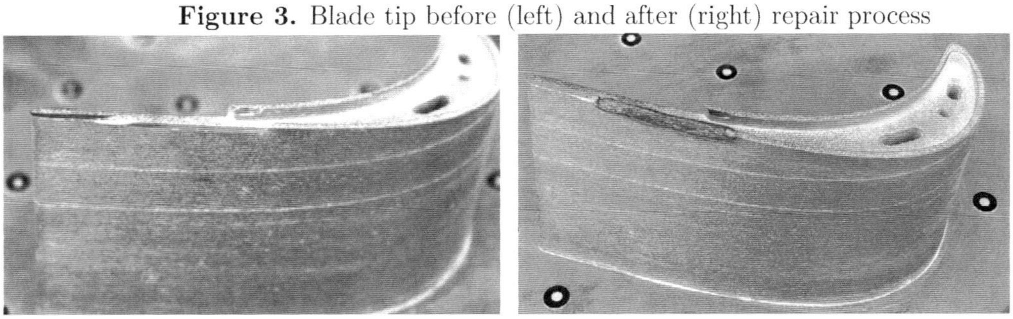

12. Conclusion

The presented system combines different technologies and shows the gained of value of combining them by an interdisciplinary team. The use-case of repairing the tip of a blade was reduced from three hours by programming the DED machine by hand to one hour preparation time and 15 minutes machine time. Integrating all hard- and software components into one workshop environment saves unnecessary transit times and enhances efficiency of the process. Our specially adjusted augmented reality interface with three dimensional hand-interaction provides better user experience and options for on the job training, as well as operation of the system by non expert workers. Such body-gesture controlled, projected or glasses based AR interfaces are applicable to many industrial applications. They are beneficial in spatial tasks and compared to buttons and keyboards less breakable in heavy duty tasks. Currently, we have not included a milling machine, which will be the next step to finish the processing. Also, this strategy is not restricted to blades. Many objects, especially forms, where the CAD is available can be repaired using this process chain.

Acknowledgement

The research described in this article was carried out as part of the "Berlin Center for Digital Transformation" project (project numbers: EFRE 1.8/20). The authors would like to thank the

Berlin Senate and the European Union for funding. Without this funding, it would not have been possible to carry out the work.

References

[1] I. Alfred, M. Nicolaus, J. Hermsdorf, S. Kaierle, K. Möhwald, H.-J. Maier, and V. Wesling, "Advanced high pressure turbine blade repair technologies," *Procedia CIRP*, vol. 74, pp. 214–217, 2018.
[2] M. Leino, J. Pekkarinen, and R. Soukka, "The role of laser additive manufacturing methods of metals in repair, refurbishment and remanufacturing–enabling circular economy," *Physics Procedia*, vol. 83, pp. 752–760, 2016.
[3] M. Fischedick, J. Roy, A. Acquaye, J. Allwood, J.-P. Ceron, Y. Geng, H. Kheshgi, A. Lanza, D. Perczyk, L. Price *et al.*, "Industry in: Climate change 2014: Mitigation of climate change. contribution of working group iii to the fifth assessment report of the intergovernmental panel on climate change. technical report." 2014.
[4] B. Graf, S. Ammer, A. Gumenyuk, and M. Rethmeier, "Design of experiments for laser metal deposition in maintenance, repair and overhaul applications," in *Procedia CIRP*, 2013, pp. 245—-248.
[5] Z. Geng and B. Bidanda, "Review of reverse engineering systems–current state of the art," *Virtual and Physical Prototyping*, vol. 12, no. 2, pp. 161–172, 2017.
[6] F. Buonamici, M. Carfagni, R. Furferi, L. Governi, A. Lapini, and Y. Volpe, "Reverse engineering modeling methods and tools: a survey," *Computer-Aided Design and Applications*, vol. 15, no. 3, pp. 443–464, 2018.
[7] S. Mönchinger, M. M. Schmidt, S. Dreßen, P. Wissmann, and R. Stark, "Automatized 3d-scanning application for the virtualization of large components," in *Gas Turbine India Conference*, vol. 83532. American Society of Mechanical Engineers, 2019, p. V002T11A002.
[8] O. Heimann and J. Guhl, "Industrial robot programming methods: A scoping review," in *2020 25th IEEE International Conference on Emerging Technologies and Factory Automation (ETFA)*, vol. 1, 2020, pp. 696–703.
[9] Z. Xia, Q. Lei, Y. Yang, H. Zhang, Y. He, W. Wang, and M. Huang, "Vision-based hand gesture recognition for human-robot collaboration: A survey," in *2019 5th International Conference on Control, Automation and Robotics (ICCAR)*, 2019, pp. 198–205.
[10] L. Yang, Z. Li, Q. Lei, J. Xu, Y. Deng, and Y. Zhong, "Teaching a robot to draw: hand gesture demonstration based on human-robot interaction," in *International Conference on Machine Vision*, 2020.
[11] G. Bolano, C. Juelg, A. Roennau, and R. Dillmann, "Transparent robot behavior using augmented reality in close human-robot interaction," in *2019 28th IEEE International Conference on Robot and Human Interactive Communication (RO-MAN)*, 2019, pp. 1–7.
[12] Y. Gao and C.-M. Huang, "Pati: A projection-based augmented table-top interface for robot programming," in *Proceedings of the 24th International Conference on Intelligent User Interfaces*, ser. IUI '19. New York, NY, USA: Association for Computing Machinery, 2019, p. 345–355. [Online]. Available: https://doi.org/10.1145/3301275.3302326
[13] J. Kuschan, A. Beleke, O. Heimann, V. Müller, S. Mönchinger, C. Niebuhr, and J. Krüger, "Mehrwert der digitalen vernetzung für den reparaturprozess von freiformbauteilen / modellerstellung für additive und spanende verfahren," *wt Werkstattstechnik online*, vol. 11/12, 2020.
[14] J. Kuschan, H. Schmidt, and J. Krueger, "Improved ergonomics via an intelligent movement and gesture detection jacket," in *Proceedings of ISR 2016: 47st International Symposium on Robotics*. VDE, 2016, pp. 1–6.
[15] R. de la Barré, P. Chojecki, U. Leiner, L. Mühlbach, and D. Ruschin, "Touchless interaction-novel chances and challenges," in *International Conference on Human-Computer Interaction*. Springer, 2009, pp. 161–169.
[16] P. Tsarouchi, A. Athanasatos, S. Makris, X. Chatzigeorgiou, and G. Chryssolouris, "High level robot programming using body and hand gestures," *Procedia CIRP*, vol. 55, pp. 1–5, 2016.
[17] D. Holz, A. E. Ichim, F. Tombari, R. B. Rusu, and S. Behnke, "Registration with the point cloud library: A modular framework for aligning in 3-d," *IEEE Robotics & Automation Magazine*, vol. 22, no. 4, pp. 110–124, 2015.

Properties of technical aluminum under the effect of dynamic alloying

Y Usherenko[1], V Mironovs[1], S Usherenko[2], O Reut[2], V Lapkovskis[1]

[1]Riga Technical University, Scientific Laboratory of Powder Materials, 6B Kipsalas str., Lab 110, Riga, LV-1048, Latvia
[2]Belarusian National Technical University, Nezavisimosti 65, Minsk, 220013, Belarus

E-mail : Osher_yu@mail.ru

Abstract Dynamic alloying in the mode of super-deep penetration (SDP) allows us to evaluate the effects that occur in the materials of the spacecraft when interacting with high-speed streams of cosmic dust. The results of the experiments are presented, confirming that dynamic alloying in the SDP mode creates anisotropy in a solid body and leads to a change in physical properties (electrical resistance, electron work function). The dynamic alloying increased the anisotropy of electrical resistance in technical aluminum 99.7% Al in 2.05 times (105%), led to a decrease of the electrical resistance in the longitudinal direction by 16% and increase in the transverse direction by 41%.

1. Introduction

Failure of control systems during long flights significantly reduces reliability and, accordingly, increases the cost of using satellites and space stations. Therefore, the reasons for such failures are continually being searched. An increase in the flight time of satellites and space stations in near-earth space leads to an increase in the likelihood of collisions with space debris [1,2,3]. Dodging macro-objects (space debris) can be accomplished by maneuvering. The detection of streams of micro objects (cosmic dust) is currently unrealistic.

For a long time, the collisions of streams of dust particles with metal modules of spacecraft were considered only from the position of erosion of the outer surface. The researchers have now concluded that such collisions lead to the appearance of plasma and damage to electronic control systems [4,5,6].

Loss of tightness by modules is traditionally used as the main criterion for registering shell penetration. As a rule, no depressurization is observed upon collision with dusty objects. Currently, there are research data in the field of dynamic alloying, which shows that the process of penetration of micro-objects occurs by the collapse of the channel-crater. Therefore, there is no depressurization of metal modules [7]. In the case of implementation of dynamic alloying in the super-deep penetration (SDP) mode, the barrier limitation on penetration is 10^2- 10^4 calibers of the striker. Accordingly, the possible penetration depth of dust microparticles can reach tens and hundreds of millimeters [8,9]. The change in mechanical properties under dynamic loading [7,10,11] indirectly points a change in the physical properties of the material, and the peculiarities of the dynamic alloying process in the SDP mode should increase the anisotropy of the material being processed.

Lightweight constructions, metals and their alloys are used as the materials for the shell of aircraft and spacecraft. The necessity for high performance and low-cost materials caused the researchers worldwide to switch the focus from monolithic to composite materials. Also, a significant effort has been made to create numerous combinations of metal matrix composites. Intensive labor is dedicated to creation of the lightweight constructions: reinforced Al composites [12,13], hybrid lightweight composites design [14], production of the special lightweight fillers [15] and applications of it in Al- and Mg- lightweight composites [16]. Among the metal matrix composites, aluminum-based composites are treated as the most promising structural materials due to their high corrosion, wear resistance, specific modulus, and weight for automobile and aerospace applications [12].

Super-deep penetration (SDP) is a complex physical phenomenon, when in a split-second stream of powder particles with a fraction less than 100 microns, accelerated to speeds of 700-3000 ms^{-1}, penetrates the solid metal body at depth in tens, hundreds mm. At the same time, the high and ultra-high pressure (0.2-20 GPa), intensive deformation, local heating, friction is occurred [17]. Dynamic alloying in the SDP mode is characterized by the simultaneous action of various physical factors on the material (high pressure, significant pressure gradients inside a solid metal body, zones of intense tension and compression, temperature, radiation, etc.). Such a complex effect occurs in the time interval of 10^{-9}-10^{4} s, which significantly changes the conditions of heat and mass transfer. The intensity of loading is so high that destruction or irreversible changes are possible in the body on which it acts. Unlike classical alloying, SDP provides the introduction of alloying elements into an already solid body, and to a greater depth than when using surface hardening and ion implantation methods [7-9].

The SDP process, in a simplified form, is the formation of extended high-pressure zones with different levels of stress state in mutually perpendicular directions. These zones are reinforcing a metal solid. In particular, the SDP process involves the creation of alternating zones of tensile and compressive stresses and, accordingly, extrusion into the channel regions of the interaction products of the introduced and matrix material [7,17].

The anisotropic nature of the flow of energy and mass transfer processes during SDP suggests, due to the freezing of defects during unloading onto a metal matrix at the stage of termination of penetration, a formation of a residual anisotropic structure of the composite material. The processes of multidirectional intense deformation and freezing of defects are most clearly reflected in such physical properties as electrical conductivity and thermal conductivity. Therefore, the verification of the hypothesis, that the anisotropy in the nature of the course of SDP processes should lead to a strong anisotropy of physical properties, should be carried out based on experiments with electrical conductivity in aluminum and the distribution of the electron work function (EWF) over the surface.

The electron work function (EWF) is one of the fundamental characteristics of matter in a condensed state. The EWF value is influenced by such factors as the chemical composition of the surface, crystal lattice defects, adsorption of substances from the surrounding atmosphere, etc. [18]. The high sensitivity of this parameter to the state of the surface determines the interest in using ERW for characterizing the metal surface after various types of processing.

According to the existing theoretical models, local deviations of the surface EWF values are mainly associated with the deformed state of the metal and the adsorption of atoms or molecules on its surface. The effect of the deformed state of the metal on the EWF of the surface is determined by the change in the length and angle of interatomic bonds due to changes in the volume and density of atomic clusters under load [19].

Experimental measurements of the EWF of a metal surface and its changes caused by mechanical stresses have also been published [20-24]. In particular, it has been shown that alternating loading of aluminum and titanium specimens leads to a decrease in the EWF in the stress concentration region. It can be assumed that the observed changes in the EWF are associated with both an increase in the number of dislocations and the formation of new surfaces in the process of sample destruction associated with the development of a crack. Various researchers independently showed that elastic tensile deformations lead to a decrease in the recorded values of the EWF, while elastic compression deformations, on the contrary, to their growth [22]. Changes in the contact potential difference can also be associated with the transfer of matter to the metal surface. The adsorption of atoms or molecules on the surface is accompanied by charge transfer between the adsorbate and the substrate, which leads to the modification of the surface double layer. The consequence of this is a change in the surface potential nor a corresponding change in the external EWF [23,24.] The treatment of a metal surface by electrochemical or high-energy (in particular, plasma) impact also leads to a change in its contact potential difference [25,26]. In this case, both a change in the EWF of the surface and a change in its electrochemical potential take place.

A significant number of materials show different mechanical behavior under dynamic loads compared to quasi-static. Therefore, a comprehensive study of material dynamic behavior is essential for applications in which dynamic loads are dominant [27,28]. Let's consider results of such processes using technical aluminum as an example.

2. Experimental Methods

Dynamic processing in SDP mode was carried out in the following conditions [17]: average particles speed 1000 ms^{-1}, exposure time - 400 microseconds, the material of billets: casting technical aluminum A7 (99.7% Al, interstate standard GOST 11069-2001, DIN 3.0275), the material of powder particles (strikers) – SiC powder (60%), 63- 70 microns and Pb powder (40%), 10-90 microns.

The samples preparation technique includes cutting, polishing and etching in 4% nitric acid.

To measure the electrical resistance plates were cut out of the target material (∅40 mm, length 150 mm) using electro-spark processing. 4-5 plates were cut from each sample in the transverse and longitudinal directions. Electrical resistance was determined as the average of the measurements. The electrical resistance measured on the treated sample by two-probe method was compared with measurements on the initial material.

Besides, the distribution of the electron work function over the surface of the plates in the longitudinal and transverse directions was studied. The topology of the electron work function of the plate surface was registered by the scanning Kelvin probe method [23,24].

3. Results and Discussion

To determine the anisotropy of electrical resistance measurements was carried out in mutually perpendicular directions: longitudinal and transverse.

Measurement of electrical resistance in the longitudinal direction are shown in Table 1:

Table 1. The electrical resistance of plates of technical aluminum A7, longitudinal direction

Sample	Average electrical resistance ρ_{av}, 10^{-8} Ohm·m
1.Initial	5.2792
2. SDP treated	4.4192
3. SDP treated	4.4118

In the treated aluminum $\rho_{avSDP} = (\rho_{avSDP1} + \rho_{avSDP2})/2 = 4.4155 \cdot 10^{-8}$ Ohm·m

Thus $\rho_{avSDP}/\rho_{avInitial}$=0.8364 and therefore, it has been established that electrical resistance in the longitudinal direction after treatment decreased by 16.36%.

Measurement of electrical resistance in the transverse direction are shown in Table 2:

Table 2. The electrical resistance of plates of technical aluminum A7, transverse direction

Sample	Depth of sample, m	Average electrical resistance ρ_{av}, 10^{-8} Ohm·m
1. Initial	-	6.3785
2. Initial	-	6.4756
1.SDP treated	0.043	9.3896
2. SDP treated	0.093	8.7530
3. SDP treated	0.097	8.9770
4. SDP treated	0.136	9.2366

Therefore, in the initial aluminum A7 $\rho_{av\ initial} = (\rho_{av2} + \rho_{av3})/2 = 6.4270 \cdot 10^{-8}$ Ohm·m.

In the treated aluminum $\rho_{av\ SDP} = (\rho_{avSDP1} + \rho_{avSDP2} + \rho_{avSDP3} + \rho_{avSDP4})/4 = 9.08905 \cdot 10^{-8}$ Ohm·m

Then $\rho_{av\ SDP}/\rho_{av\ initial} = 1.4126$, i.e. the electrical resistance after the treatment in the SDP mode increased by 41.26%.

If we consider the changes in the electrical resistance after processing from the upper part (the surface of the introduction of the SiC (60%) + Pb (40%) stream, then the sample 1 is cut at a depth of 43 mm ($\rho_{avSDP1}=9.3896 \cdot 10^{-8}$ Ohm·m) $\rho_{avSDP1}/\rho_{avInitial}= 1.4610$), sample 2 at a depth of 93 mm ($\rho_{avSDP2}=8.7530 \cdot 10^{-8}$ Ohm·m) ($\rho_{avSDP2}/\rho_{avInitial}=1.3619$), sample 3 at a depth of 97 mm ($\rho_{avSDP3}= 8.9770 \cdot 10^{-8}$ Ohm·m) ($\rho_{avSDP3}/\rho_{avInitial}=1.3960$), and sample 4 at a depth of 136 mm ($\rho_{avSDP4}=9.2366 \cdot 10^{-8}$ Ohm·m) ($\rho_{avSDP44}/\rho_{avInitial} = 1.4370$). Thus, in the transverse direction, the relative electrical resistance in processed technical aluminum changes markedly up to a depth of 136 mm (Figure 1).

The obtained data are summarized in Table 3. A production of an ordinary aluminum billet by casting leads to the anisotropy of electrical resistance in the longitudinal and transverse directions by 21% (ρ_{i1}/ρ_{i2}). Additional high-energy processing of such billet, as expected, increased the anisotropy of electrical resistance up to 2.05 times (ρ_{SDP1}/ρ_{SDP2}).

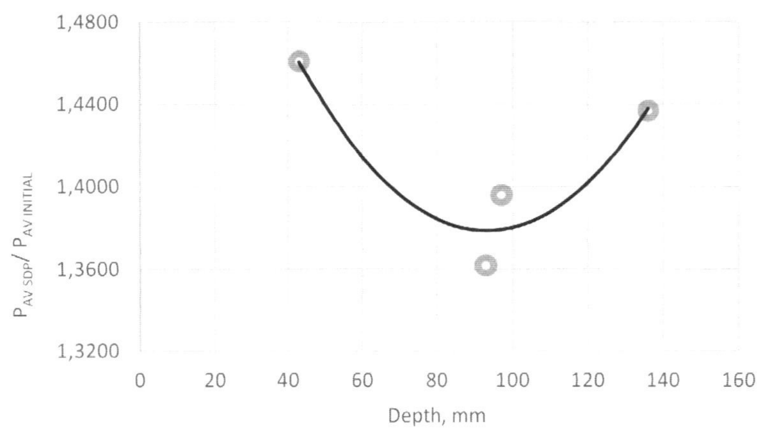

Figure 1. Change in the relative electrical resistance over the depth of the A7 technical aluminum sample

Table 3. Electrical resistance of technical aluminum A7 in mutually perpendicular directions, depending on the nature of the processing

Measurement direction	Initial electrical resistance ρ_i, 10^{-8} Omh·m	Electrical resistance after processing ρ_{SDP},Omh·m	$\frac{\rho_{i1}}{\rho_{i2}}$	$\frac{\rho_{SDP1}}{\rho_{i1}}$	$\frac{\rho_{SDP2}}{\rho_{i2}}$	$\frac{\rho_{SDP1}}{\rho_{SDP2}}$
1-transverse	6.427	9.089	1.217	1.413	0.836	2.058
2-longitudinal	5.279	4.416				

The Kelvin probe method measures the contact potential difference arising between the measured surface and the surface of the reference sample. In this case the measured surface and the reference sample form a flat capacitor and do not contact each other. The value of the contact potential difference will be determined by the difference between the EWF of the reference sample and the studied one [20]. Due to its high sensitivity, the method is used in physicochemical studies, in the study of friction processes, as well as in the study of deformation and destruction of metallic solids 23,24]. It should be noted that the EWF [25] is the energy that must be spent to remove an electron from a solid or liquid substance into a vacuum. Hence, it follows that the EWF depends both on the state of the volume of metal and on the state of its surface. This second component depends on the crystallographic orientation of the surface, adsorbed atoms, the presence of defects on the surface, surface microroughness, etc.

Figure 2 shows the topology of EWF of the sample of A7 technical aluminum. The distribution of the EWF over the surface is nonuniform. The upper part of the longitudinal section is the crater zone, the area of impact on the sample of a high-energy stream of powder particles. The upper and the central zone, which are marked in red, correspond to the lowest EWF. With an increase in the distance from the crater, the region of low EWF, on the one hand, is localized in the center, on the other hand, at a depth of about 90 mm, it dissipates with the formation of several regions with a reduced EWF. This distribution of the zone of the low value of the EWF shows the distribution of the impact of the high-energy stream of powder particles over the sample.

The distribution of EWF over the transverse sections of the sample depends on the depth of the section obtained. The transverse sections of the top and bottom of the sample, as expected, demonstrate a greater uniformity of the EWF. At the same time, the section at a depth of 50 mm from the top of the sample demonstrates a pronounced area with a reduced EWF.

This non-uniformity of the distribution of the EWF in the longitudinal and transverse sections of the sample correlates with the recorded change in the electrical resistance in the longitudinal and transverse directions (Table 3).

The observed changes in the EWF along the depth of the sample coincide with the data on the change in hardness along the depth of the sample (Brinell) obtained earlier [29]. It was found out that the effect of processing by dynamic alloying in the SDP mode on the hardness of technical aluminum A7 was significant up to a depth of 100 mm. The study of the change in the EWF, in turn, showed that the zone of low EWF along the depth of the sample also extends to approximately the same depth.

Measurement of EWF with a Kelvin probe method confirms that dynamic alloying in the SDP mode provides alternating loading of samples, material transfer and leads to a creation of many structural defects of various levels, located unevenly throughout the sample volume. An increase in the density of defects, in turn, provides anisotropy of the mechanical and physical properties of the material.

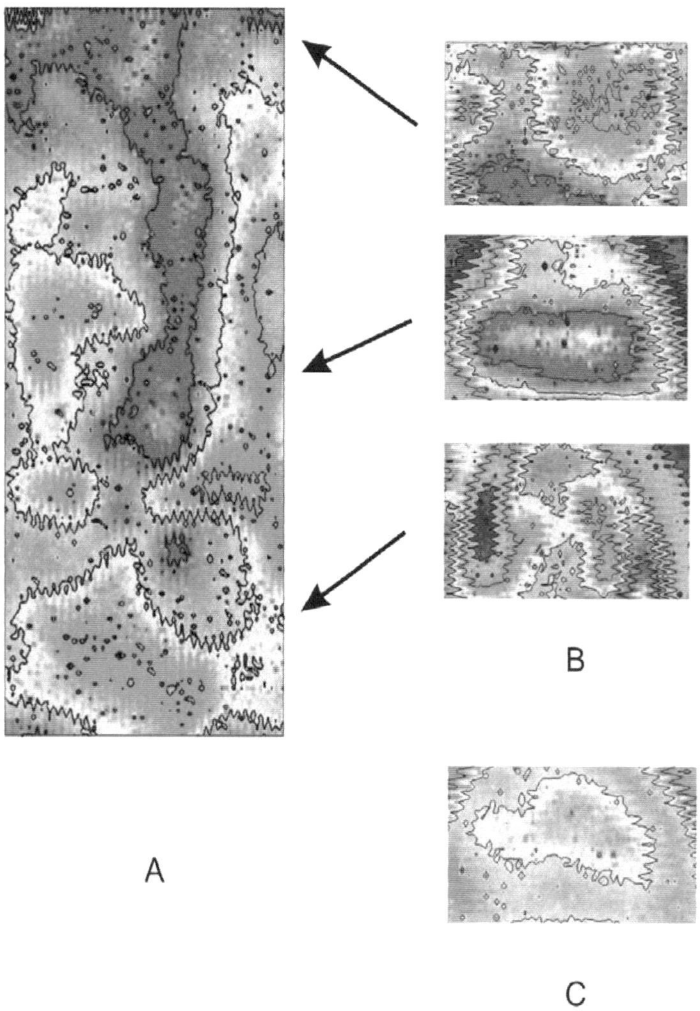

Figure 2. The topology of the electron work function of A7 technical aluminum plates after treatment with SDP.A - longitudinal section, B - transverse section, C - transverse section of initial A7 technical aluminum

4. Conclusions

Based on the analysis of literature sources describing the behaviour of a solid metal body exposed to a high-energy stream of powder particles (in the SDP mode) and experimental data, the following main conclusions can be drawn:

Dynamic alloying in the SDP mode leads not only to a change in mechanical properties [22] but also in physical properties (electrical resistance, EWF);

A casting of an aluminum billet leads to the anisotropy of electrical resistance in the transverse and longitudinal directions in 1.21 times (21%). Additional high-energy processing in the SDP mode increased the anisotropy of electrical resistance in 2.05 times (105%).

Dynamic alloying in the SDP mode led to a decrease of the electrical resistance in the longitudinal direction by 16% and to increase in the transverse direction by 41%.

Measurement of EWF by a Kelvin probe confirms that dynamic alloying in the SDP mode provides a creation of many structural defects of various levels, located unevenly throughout the sample volume. An increase in the density of defects, in turn, provides anisotropy of the mechanical and physical properties of the material.

The hypothesis that the anisotropic nature of deformation processes and redistribution of the energy of pressure fields during dynamic treatment in the SDP mode leads to the anisotropy of the structure and physical properties in a metal solid after exposure is confirmed.

Acknowledgements

This work has been supported by the European Regional Development Fund within the Activity 1.1.1.2 "Post-doctoral Research Aid" of the Specific Aid Objective 1.1.1 "To increase the research and innovative capacity of scientific institutions of Latvia and the ability to attract external financing, investing in human resources and infrastructure" of the Operational Programme "Growth and Employment" (No. 1.1.1.2/VIAA/2/18/335).

PostDoc Latvia
NACIONĀLAIS ATTĪSTĪBAS PLĀNS 2020
EIROPAS SAVIENĪBA
Eiropas Reģionālās attīstības fonds
IEGULDĪJUMS TAVĀ NĀKOTNĒ

References

[1] Belous A I, Soloduha V A, Shvedov S V 2019 *High Velocity Microparticles in Space: Influence Mechanisms and Mitigating Effects of Electromagnetic Irradiation* (Springer) p 309
[2] Carrillo-Sánchez J D, Nesvorný D, Pokorný P, Janches D and Plane J M C 2016, *Geophys. Res. Lett.* 43 11979– 986
[3] Wozniakiewicz P J, Burchell M J 2019 *Astronomy and Geophysics* 60 (3) 38 -42
[4] Fletcher A C, Close S 2017 *Physics of Plasmas* 24 (5) 053102-1-7
[5] Hew Y M, Goel A, Close S, Lee N 2018 *Int. J. Impact Eng* 121 1-11.
[6] Nuttall A, Close S 2020 *Int. J. Impact Eng.* 144 103645
[7] Usherenko Y, Usherenko S,Yazdani J 2016 *Key Eng. Mater.*7 21290–4
[8] Usherenko Y, Usherenko S,Yazdani J 2017 *Procedia Eng.* 172 1198-1203
[9] Aleksentseva S E, Krivchenko A L 2013 *Vestnik of Samara State Technical University. Technical Sciences Series* 21(2) 71-78 (In Russian)
[10] Petrov E V, Trofimov V S, Kopytskiy V O 2020 *J. Phys. Conf. Ser.* 1431(2) 012044
[11] Petrov E V, Saikov I V, Saikova G R,Trofimov V S 2020 *Russ. J. Non-ferrous Metals* 61 549–553
[12] Priyaranjan S, Pandu R V, Arabinda M, Manas M. M 2020 *J. Manuf. Process.* 59 2020 131-152
[13] Shishkin A, Mironov V, Goljandin D, Lapkovsky V 2012 *Key Eng. Mater.* 527143–147
[14] Shishkin A, V. Mironov, Zemchenkov V, Antonov M, Hussainova I 2016 *Key Eng. Mater.* 674 35-40
[15] Shishkin A, Drozdova M, Kozlov V, Hussainova I, Lehmhus D 2017 *Metals* 7, 16S
[16] Sankaranarayanan S, Nguyen QB, Shabadi R, Almajid A H, Gupta M 2016 *Powder Metall* 59(3) 188-196
[17] Usherenko Y, Mironovs V, Lapkovskis V, Usherenko S, Gluschenkov V 2019 *Agronomy Research* 17(6) 2445–2454
[18] Henrich V E, Cox P A 1994 *The surface science of metal oxides* (Cambridge:Cambridge

university press) p 464
[19] Nazarov A, Prosek T, Thierry D *Electrochimica acta* 2008 53(25) 7531-7538
[20] Pogosov V V, Shtepa O M 2002 *Ukr. J. Phys* 47(11) 1065-1071
[21] Levitin V V, Loskutov S V, Pravda M I, Serpetzky B A 1994 Solid State Commun. 92(12) 973-976
[22] Loskutov C B, Levitin V V, Gordienko V N 2002 *J. Frict. Wear* 23 (2) 176-180 (in Russian)
[23] Zharin A 2010 *Scanning Probe Microscopy in Nanoscience and Nanotechnology. NanoScience and Technology* ed B Bhushan (Berlin Heidelberg:Springer) Chapter 19 687-720
[24] Pantsialeyeu K U, Mikitsevich U A, Zharin A L 2016 *Devices and Methods of Measurements* 7(1) 7-15 (In Russian)
[25] Tyavlovsky A K 2015 *Instruments and methods of measurement*, (1), 30-36 (In Russian),
[26] DeVecchio D, Bhushan B 1998 *Rev. Sci. Instrum.* 69(10) 3618-3624
[27] Rahmat M 2019 *Mech Time-Depend Mater* 23, 385–405
[28] Zhang P, Song A Fang Y, Yue X, Wang Y, Yu X 2020 *Vacuum* 173 109119.
[29] Usherenko Y, Mironovs V, Usherenko S 2021 *Solid State Phenomena* 320 8-13

IOP Conf. Series: Materials Science and Engineering 1140 (2021) 012038 doi:10.1088/1757-899X/1140/1/012038

Numerical study on the effect of geometry on mechanical behavior of triply periodic minimal surfaces

Mansoureh Rezapourian[1], Nikhil Kamboj[1], Irina Hussainova[1]

[1] Mechanical and Industrial Engineering Department, Tallinn University of Technology, Tallinn, Estonia

Email: Mareza@taltech.ee

Abstract. Metallic cellular structures with triply periodic minimal surfaces (TPMS) is a great approach to enhance multifunctionality for several engineering disciplines. In the present study, we aimed to generate metal-based Neovious, Schoen I-WP and Schwarz Primitive structures, leveraging on TPMS based design methods. Neovious, Schoen I-WP and Schwarz Primitive structures are mathematically formulated. Geometric factors like wall thickness, number of cells and the length of unit cell can affect the mechanical response namely elastic modulus and compressive strength. Herein, Neovious, I-WP and Primitive based materials are fabricated through Selective Laser Melting (SLM) to evaluate the difference between the aforementioned geometrical factors both experimentally and through simulations. The Taguchi method was implemented to explore the effect of geometric factors on structural response. These results were further extended to the structures to optimize the elastic modulus and compressive strength.

1. Introduction

High performance of cellular structures related to excellent energy absorption, lightweight, high strength-weight ratio and excellent thermal, acoustic, mechanical and electrical properties are gaining increasing interest in their specific design. Cellular materials have a wide range of applications in biomedicine, energy absorption and aerospace [1,2]. Recently, many researches have been focused on controllable topology and design of shell-based structures of minimal surfaces which are known as Triply Periodic Minimal Surfaces (TPMS) [2-7]. Triply periodic demonstrates a repeated structure in 3D which is devoid of any self-intersections. TPMS are specified as surfaces with a zero mean curvature at each point on the surface [8]. Nowadays, the fabrication of complex architectures has been perceived due to development of Additive Manufacturing (AM) approaches. Conventional ways to produce cellular materials have many limitations about using material and simplicity in design which can be addressed by application of AM technologies to mitigate the challenges. Selective laser melting (SLM) as a kind of AM process has a great potential to fabricate complex structures [9]. Traditional structures such as a strut-based lattice fabricated by SLM with a lack of continuous transition on their surface and sharp edges demonstrated a poor manufacturability in the AM process [10,11]. It is notable that the absence of nodes and other stress concentration regions in TPMS make them valuable structures with improved strength and toughness [12-15]. The absence of joints and struts in TPMS lattice structure (in comparison with strut-based architectures) is a considerable benefit in terms of processability [16]. SLM as an AM method is known an appropriate technique to printing metallic structures. Variety of metal powders, such as Ti6Al4V, 316 L stainless steel, Fe–Mn and Zn can be produced with design in TPMS lattice architecture with the help of SLM techniques [17, 18].

The geometric parameters such as wall thickness, sample size, number of surface periods, and the associated iso-value can affect the mechanical response of gyroid structure fabricated of Ti6Al4V by SLM [19]. To quantify the influence of mentioned geometric factors on structural response, OFAT (One Factor at a Time) and Taguchi methods were used and it was shown that an increase in the number of cells and wall thickness strongly effect the elastic modulus and compressive strength whilst a decrease in the sample size and iso-value can lead to improved mechanical response.

This research focuses on finding optimum design and studying the effect of wall thickness and number of cells on a structural response of Neovius (Neo), Schoen I-WP (IWP) and Schwarz Primitive (Pri) structures by implementing Taguchi method. To identify the effect of mentioned factors on mechanical properties, numerical and experimental procedures are required. Validity has been verified by the experimental data of compression test performed on the SLM printed structures of Ti-6Al-4V.

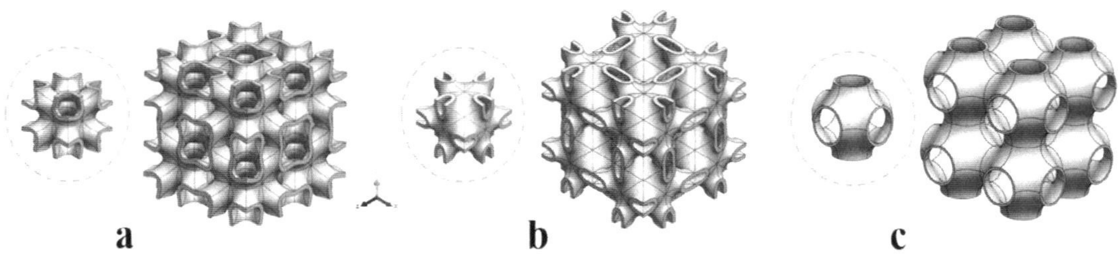

Figure 1. TPMS unit cells and designed specimens of a) IWP, b) Neovius and c) Primitive

2. Material and method

2.1. Modeling and computer-aided design of cellular lattice

In x-y-z spatial coordinates, the TPMS can be approximated using implicit methods and introduced by a level-set function $\varphi = c$, which is defined for Neovius (Neo), Schoen I-WP (IWP) and Schwarz Primitive(P) surfaces, respectively in equation (1) [20]:

$$\varphi_{Neo} = 3(\cos kx + \cos kz + \cos ky) + 4kx \cos kz \cos ky = c$$

$$\varphi_{IWP} = 2(\cos(kx)\cos(ky) + \cos(ky)\cos(kz) + \cos(kz)\cos Kx) - (\cos 2kx + \cos 2ky + \cos 2kz) = c \quad (1)$$

$$\varphi_P = \cos kx + \cos ky + \cos kz = c$$

Where $k = 2\pi q/l$. q defines the number of cells in each direction, and l is the length of the unit cell. The value of c controls the shape of surfaces and is known as iso-value parameter. In this research, the value of c is assumed to be zero. MATLAB software was used to generate the initial surface of TPMS and SolidWorks followed by a CAD software which was used to thicken and create 3D models of TPMS. Labeling of structures has been shown in figure 1. For example, the label of "$Neo050 - 2$" describes the Neovius structure with wall thickness of 0.5 mm and 2 repeated cells at each edge. Designed models are shown in figure 1.

2.2. Material, Machine, and Methods

The Ti6Al4V powder of particle size 20 – 63 μm was printed by Realizer SLM50 metal printer. The associated laser parameters were applied for fabrication with laser power of 84 W, layer thickness of 30 μm, hatch distance of 120 μm, at a scanning speed of 750 mm/s, rotate hatch at 60 degree and 35 degree with y axis was selected to print structure. According to these parameters, the calculated laser energy density is 31.11 J/mm^3.

2.3. Uniaxial compression test and finite element analysis

Pri030-2 specimen with the length size of 16.67 mm in x, y and z direction was chosen for experimental study, which was performed by a servo-hydraulic model 8500 universal testing machine (Instron ltd., UK). A computational finite element (FE) analysis was performed to evaluate mechanical performance with experimental compression test (figure 3). For static compression at a room temperature, a strain rate of 0.1 mm /s was applied with compressive load and displacement recorded at each 0.1 s intervals during testing. It is notable that in all simulations from now on, the coefficient of the friction was defined as 0.1 to prevent sliding during applied load. The ANSYS 2019 software was used to simulate compression tests.

2.4. Taguchi method

To investigate the effect of individual geometric factors on mechanical properties, statistical Taguchi method was implemented to improve the quality of the outcomes. According to full factorial approach, with three significant factors considering three level of each TPMS surface, the designed experiments could reach up to 81 which henceforth is very time-consuming. Therefore, in order to decrease the number of tests without any loss in accuracy, it is recommended to implement Taguchi approach. In total 9 test were designed, which were assigned to three factors such as type of structures; surface wall thickness and number of cells (L/l). For designing, a cube, as a bulk sample with constant length 25 mm in x, y and z direction was assumed. Bulk size can be defined by multiple of cell size to the number of cells on each direction. It is noted that with changing cell number, cell size can be altered. This study represents the importance of above-mentioned factors on mechanical properties with Taguchi method. All designed topologies and calculated modulus of elasticity with maximum compressive strength at 4% strain was based on Taguchi orthogonal L9 matrix henceforth represented in Table 1.

Table 1. *Taguchi L9 Matrix for study of factor influence*

Taguchi L9 Matrix	Factors			Results		Porosity (%)
Topology	Structures	Thickness (mm)	Number of cells (L/l)	Elastic Modulus (GPa)	Max. Stress (MPa)	
IWP050-2	I-WP	0.5	2	2.804	34.336	85.87
IWP040-3	I-WP	0.4	3	3.730	45.718	82.93
IWP030-4	I-WP	0.3	4	3.839	44.084	83.09
Neo050-3	Neovius	0.5	3	6.839	70.121	79.12
Neo040-4	Neovius	0.4	4	7.561	73.060	77.71
Neo030-2	Neovius	0.3	2	2.109	22.318	91.50
Pri050-4	Primitive	0.5	4	2.557	39.633	81.33
Pri040-2	Primitive	0.4	2	0.536	10.030	92.50
Pri030-3	Primitive	0.3	3	0.744	14.108	91.56

3. Results

3.1. Evaluation compression test result

FE model was performed to validate the result with experiment compression test of Ti6Al4V SLM printed Pri030-2 with 91% porosity and figure 2 represent force-displacement result of both FE analysis and experimental data. The outcomes of elastic modulus and maximum stress indicate 3% and 1.9% deviation in the result of the numerical and experiment which is acceptable and shows strong agreement with the physically obtained result.

3.2. SEM Analysis

The fabricated lattice sample Pri030-2 with its SEM analysis figure 3 which is done for top surface of printed structure to investigate dimensional variations between the designed and 3D-printed structures. The as-fabricated lattice structure clearly followed the CAD model. Residual particles were found attached to the fabricated structures as shown figure 3(a-c). Residual particles can cause the thickness variations.

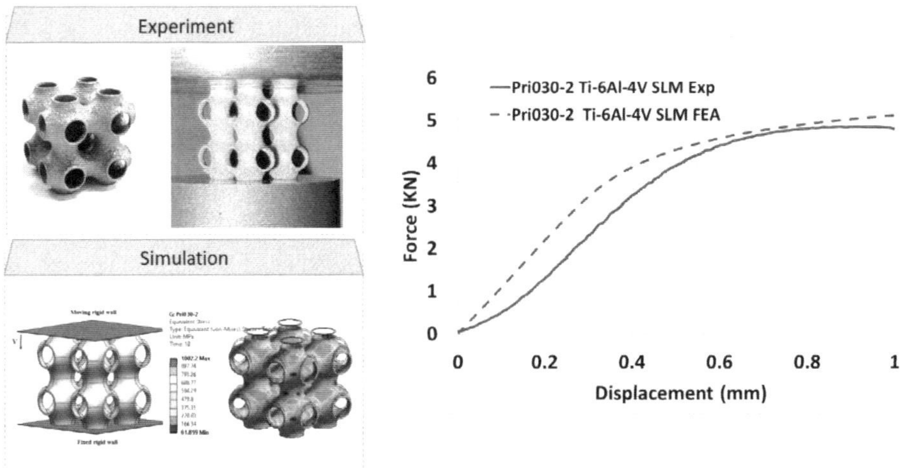

Figure 2. Compression test of Pri030-2 at 6% strain at t=10 s in experiment and simulation, contour of equivalent (Von-Mises) stress with undeformed wireframe, force-displacement curve

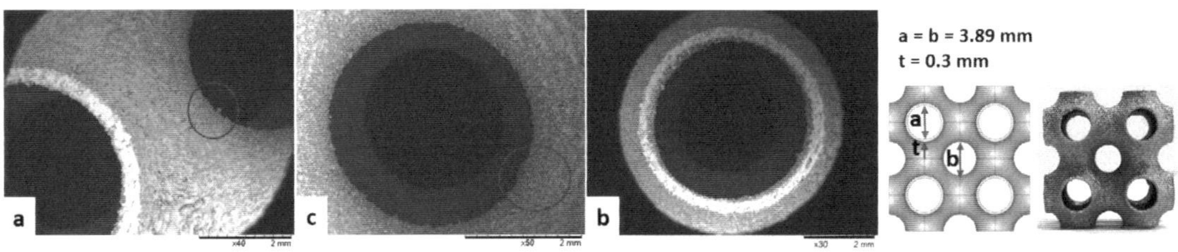

Figure 3. SEM image of top surface of printed Pri030-2

3.3. Taguchi Method

Based on Taguchi method, a signal to noise ratio (SNR) defines the ratio of calculated data to the error. The accuracy of measured value and calculated data for each test were determined by SNR. In statistical analysis, main effect is the influence of different levels of each factor on outcomes when they are contrary. If SNR and main effect shows the same trend, the obtained data and designed data are correct.

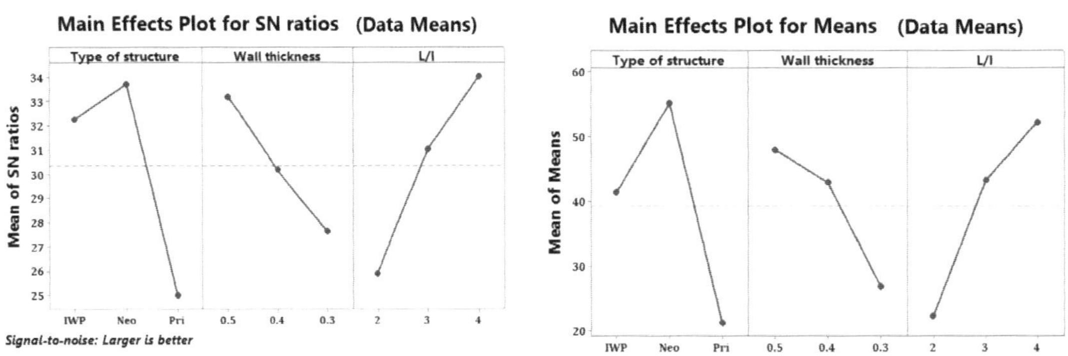

Figure 4. SNR and mean values for compressive stress

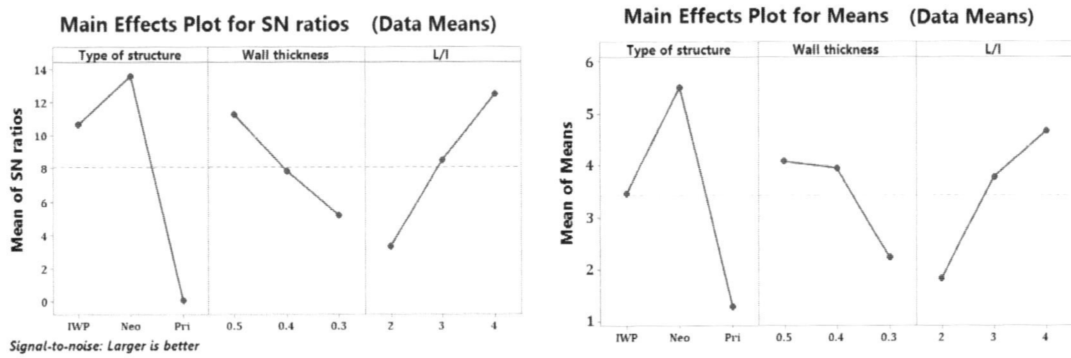

Figure 5. SNR and mean values for elastic modulus

The criteria 'large is better' was chosen to maximize the modulus of elasticity and compressive strength. Figure 4 and figure 5 illustrate the main effect and SNR for both modulus of elasticity and compressive strength. As mentioned above, they have the same trend indicating the results to be correct. Table 2 shows the ranking of three factors for compressive strength based on 'large is better' criteria. First of all, the average response for all parameters is determined. Then the delta (the difference between minimum and maximum average) is calculated. Finally, all factors are categorized from largest to smallest. The result for the means reveals the type of structure as the best type for maximizing compressive strength with the first rank. The factor number of cells (L/l) and wall thickness are identified as the second and third effective factors.

Table 2. *SNR and mean values for compressive strength, 'Larger is better'*

Response Table for Signal to Noise Ratios				Response Table for Means			
Level	Type of structure	Wall thickness	L/l	Level	Type of structure	Wall thickness	L/l
1	32.27	33.20	25.90	1	41.38	48.03	22.23
2	33.72	30.17	31.04	2	55.17	42.94	43.32
3	24.99	27.62	34.04	3	21.26	26.84	52.26
Delta	8.73	5.58	8.14	Delta	33.91	21.19	30.03
Rank	1	3	2	Rank	1	3	2

Table 3 shows the ranking of three factors for elastic modulus based on 'large is better' method. Similar to compressive strength according to the means result, the type of structures is the first effective factor to maximize elastic modulus, with number of cell (L/l) is in the second ranking. Based on the main effects, the wall thickness of 0.5 and 0.4 mm shows that there is no significant difference for maximizing elastic modulus. Also, the ranking of the SNR represents the same result for all the three factors.

Table 3. *SNR and mean values for elastic modulus, 'Larger is better'*

Response Table for Signal to Noise Ratios				Response Table for Means			
Level	Type of structure	Wall thickness	L/l	Level	Type of structure	Wall thickness	L/l
1	10.6929	11.2709	3.3448	1	3.458	4.067	1.817
2	13.5856	7.8666	8.5252	2	5.504	3.943	3.771
3	0.0626	5.2035	12.4711	3	1.279	2.231	4.653
Delta	13.5229	6.0674	9.1263	Delta	4.224	1.836	2.836
Rank	1	3	2	Rank	1	3	2

These results revealed that the Neovious structure has the most influential effect on maximizing both elastic modulus and compressive strength. Increasing number of cells and wall thickness can decrease the volume fraction which means less densification with less porosity that results heavier structure and higher elastic modulus and compressive strength.

4. Case study

Regarding the Taguchi results, among different levels of selected factors, the optimum values were chosen. As result represents, the most influential effect on mechanical behavior is related to the type of structure, maximum cell number and wall thickness. A custom designed Neo050-4 with defined parameter length of 25 mm, with Neovius structure, wall thickness of 0.5 mm and 4 unit cells in each direction and porosity of 72% was developed using Ti6Al4V material. The outcome represents a simulated yield stress of 93.01 MPa and ultimate strength of 140 MPa and elastic modulus of 9.7555 GPa.

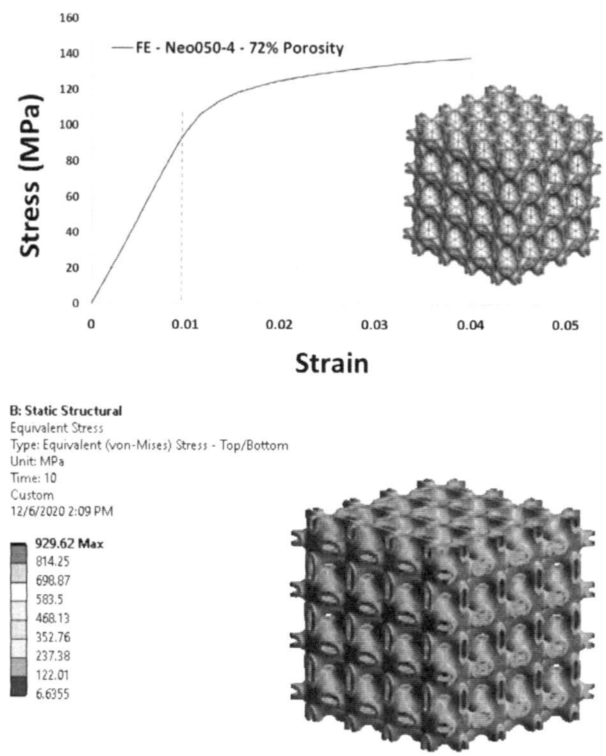

Figure. 6 Mechanical response of Neo050-4 – FE analysis

5. Conclusions

This study has demonstrated the effect of geometric parameters on mechanical response of Neovius (Neo), Schoen I-WP (IWP) and Schwarz Primitive (P) complex structures based on a statistical approach. Triply Periodic Minimal Surfaces (TPMS) were produced from Ti6Al4V alloy using SLM. A numerical model has been defined to validate with experimental data which shows a close agreement. Based on the results, Neovius (Neo) has shown to outperform other structures in terms of strength and stiffness, which can lead to significant effect on elastic modulus and compressive strength. The number of cells and wall thickness are the second and third important factors which can affect structural response directly. A candidate structure of Neovius structure with wall thickness of 0.5 mm and cell number of 4 with the porosity of 72% was developed to validate the result and FE analysis of structural response indicated high elastic modulus and compressive yield strength respectively (9.75 GPa) and (93.01 MPa).

6. Acknowledgements.

This work was supported by the Estonian Research Council grant PRG643 (I. Hussainova).

7. References

[1] L I Gibson M F Ashby 1997 Cellular solids: structure and properties. New York: Cambridge university press

[2] M Helou, S Kara 2018, Design, Analysis and manufacturing of lattice structures: An overview *Int. J. Comput. Integr. Manuf.* **31** 243–261

[3] I Maskery, L Sturm, A O Aremu, A Panesar, C B Williams, C J Tuck, R D Wildman, I A Ashcroft, R J M Hague 2018 Insights into the mechanical properties of several triply periodic minimal surface lattice structures made by polymer additive manufacturing, *Polymer* **152** 62–71

[4] O Al-ketan, R Rezgui, R Rowshan, H Du, N X Fang 2018 Micro architected stretching-dominated mechanical metamaterials with minimal surface, *Topologies* **20** 1–15

[5] I Maskery, N T Aboulkhair, A O Aremu, C J Tuck, I A Ashcroft 2017 Compressive failure modes and energy absorption in additively manufactured double gyroid lattices *Addit. Manuf.* **16** 24-9

[6] L Bai, C Gong, X Chen, Y Sun, J Zhang, L Cai, S Zhu, S Q Xie 2019 Additive manufacturing of customized metallic orthopedic implants: Materials, structures, and surface modifications *Metals (Basel)* **9** 1004

[7] A Ataee, Y Li, M Brandt, C Wen 2018 Ultrahigh-strength titanium gyroid scaffolds manufactured by selective laser melting (SLM) for bone implant applications *Acta Mater.* **158** 354–68

[8] L Han, S Che 2018 An Overview of Materials with Triply Periodic Minimal Surfaces and Related Geometry: From Biological Structures to Self-Assembled Systems *Adv. Mater.* **30** 1705708

[9] A Nazir, K M Abate, A Kumar, J Jeng-Ywan 2019 A state-of-the-art review on types, design, optimization, and additive manufacturing of cellular structures *Int J Adv Manuf. Technol* **104** 3489–3510

[10] M Mazur, M Leary, S Sun, M Vcelka, D Shidid, M Brandt 2016 Deformation and failure behavior of Ti–6Al–4V lattice structures manufactured by selective laser melting (SLM) *Int J Adv Manuf. Technol* **84** 1391–411

[11] L Liu, T Minasyan, N Kamboj, S Aydinyan, I Hussainova, 2020 Bio-inspired TiB_2-TiB-TiN lattices by selective laser melting *Mater. Lett.* **277** 128337

[12] M-T Hsieh, B Endo, Y Zhang, J Bauer, L Valdevit. 2019 The mechanical response of cellular materials with spinodal topologies *J. Mech. Phys. Solids* **125** 401–19

[13] R Schwaiger, L R Meza, X Li 2019 The extreme mechanics of micro- and Nano-architected materials *MRS Bull.* **44** 758–65

[14] S C Han, J W Lee, K Kang, 2015 A new type of low-density material: Shellular *Adv. Mater.* **27** 5506–11

[15] A E Garcia, C S Wang, R N Sanderson, K M McDevitt, Y Zhang, L Valdevit, D R Mumm, A Mohraz, R Ragan 2019 Scalable synthesis of gyroid-inspired freestanding three-dimensional graphene architectures *Nanoscale Adv.* **350** 1508–13

[16] K A Kamran, and K R Abu Al-Rub 2017 Time dependent response of architectured Neovius foams *Int. J. Mech. Sci.* **126** 106–19

[17] C Y Yap, C K Chua, Z L Dong, Z H Liu, D Q Zhang, L E Loh, and S L Sing, 2015 Review of selective laser melting: materials and applications *Appl. Phys. Rev.* **2** 041101

[18] M Antonov, R Ivanov, Y Holovenko, D Goljandin, R Rahmaniahranjani, L Kollo, I Hussainova, 2019 3D printing of plain and gradient cermets with efficient use of raw materials *Key Eng. Mater.* **799** 239-245

[19] E Yang, L Leary, B Lozanovski, D Downing, M Mazur, A Sarker, A Khorasani, A Jones, T Maconachie, S Bateman, M Easton, MQian, P Choong and M Brandt 2019 Effect of geometry on the mechanical properties of Ti-6Al-4V Gyroid structures fabricated via SLM: A numerical study *Mater. Des.* **184** 0264-1275

[20] M T Hsieh and L Valdevit 2020 Minisurf – A minimal surface generator for finite element modeling and additive manufacturing *Software Impacts* **6** 100026.

Development of Virtual Learning Factory Toolkit for Production Engineering Education

Kashif Mahmood[1*], Tauno Otto[1], Vladimir Kuts[1], Walter Terkaj[2], Marcello Urgo[3], Geza Haidegger[4]

[1]Department of Mechanical and Industrial Engineering, Tallinn University of Technology, Tallinn, Estonia.
[2]STIIMA-CNR, Institute of Intelligent Industrial Technologies and Systems for Advanced Manufacturing, National Research Council, Milan, Italy.
[3]Manufacturing and Production Systems Division, Department of Mechanical Engineering, Politecnico di Milano, Milan, Italy.
[4]SZTAKI – Institute for Computer Science and Control, Budapest, Hungary.

*E-mail: kashif.mahmood@taltech.ee

Abstract. The advent of industry 4.0 and the continuous digitalization of production ask for the enhancement of human skills and competences in the field of information and communication technology (ICT). Therefore, higher education has to keep pace with the global market needs for the necessary ICT skills and the overall understanding of the complexity of industries in 21st century. This paper focuses on the development and integration of Virtual Learning Factory (VLF) tools that can be used in production management and engineering education. The digital tools integrated in VLF toolkit can help students to exploit enabling technologies like simulation and virtual reality in their manufacturing studies and practical projects with industrial companies.

1. Introduction

There is no doubt, the main wealth generating activity for many nations is the manufacturing. Manufacturing has a significance contribution in the Gross Domestic Product (GDP) of Europe. Although there is recent development and growth in service sector, international markets are still persuading the importance of manufacturing in Europe and manufacturing remains a dominating factor in the total global trade with the contribution of 70% [1]. The current and prospect importance of manufacturing can be observed in many fora, a comprehensive report on “the future of manufacturing in Europe” is one of the great example of future-oriented study about manufacturing [2], where the technological development, its exploration and adaptation perceived as a key-game changer across the Europe.

Since manufacturing faces the rapid advancement in production related technologies, tools and techniques. Therefore, in this new era of manufacturing, industrial workers and engineers need to rely on the new learning schemes to keep up with those advancements. Higher education plays a vital role and a major driver to build a skill and knowledge-based workforce of new generation in manufacturing [3]. Furthermore, the latest advances and incorporation of Information and Communication Technology (ICT) in manufacturing leads the digitalization of manufacturing, and similar trend like ‘industry 4.0’ demand the training and professional development of workforce in this field, which is a challenge for the educational institutions. The concept of learning factories can be an effective enabler and contribute to establish the relevant theoretical and practical knowledge about the new advancements [3, 4]. There are advanced learning labs, FESTO didactics and competence centres, adopted by many educational institutions to cope up with the challenges of digitalization in manufacturing and to learn new technological skills, knowledge and expertise in manufacturing [5, 6].

On the other hand, the learning can be organized virtually by using digital tools to design, analyse, and visualize the manufacturing processes and systems. In many production engineering and management related curricula, learning via creating the virtual environment of a real production system or even a whole production factory has been practiced due to its prevalent benefits. Moreover, this kind of learning is quite appealing and justified in the current pandemic of COVID-19 and its impact on educational institutions. There are developments in the virtual learning approaches with the help of simulation based digital tools and many studies are currently carried out in the domain of virtual and digital factory [7, 8]. Nowadays, several technical institutions and research centres are developing their own virtual factory tools or they are using and enhancing existing software platforms and systems to develop optimization and control algorithms for factory simulation and visualizations. Nevertheless, the developed digital tools are actually employed by scientists and practitioners involved in research activities and these software tools developed during the research are not yet widely available for teachers and students.

The purpose of this paper is to present a Virtual Learning Factory Toolkit (VLFT) framework that contains digital tools and their integration. The study provides a systematic approach about the activities and digital tools used to design, evaluate and visualize manufacturing processes and systems. Moreover, it gives an overview – how the learning process should be conducted, which is validated through a case scenario of manufacturing system configuration. The case was tested in the form of joint learning lab among different students, where the digital tools are used and improvements suggested. The paper is connected to the ERASMUS+ VLFT project that aims to integrate the existing digital tools into a usable toolkit for students.

2. VLFT Framework and Workflow

The VLFT is a set of existing digital tools to support advanced engineering education in manufacturing. Modelling, analysis, virtual and augmented reality, as well as the role of the human workers in the factories have been among the main building blocks of VLFT framework. The functionalities of VFLT framework are production system design, exploiting virtual and augmented reality in manufacturing, and human modelling in manufacturing with the help of digital tools. An integrated user interface and a guideline to support the adoption of the tools in an engineering curricula are included in the framework. The framework is shown in Figure 1.

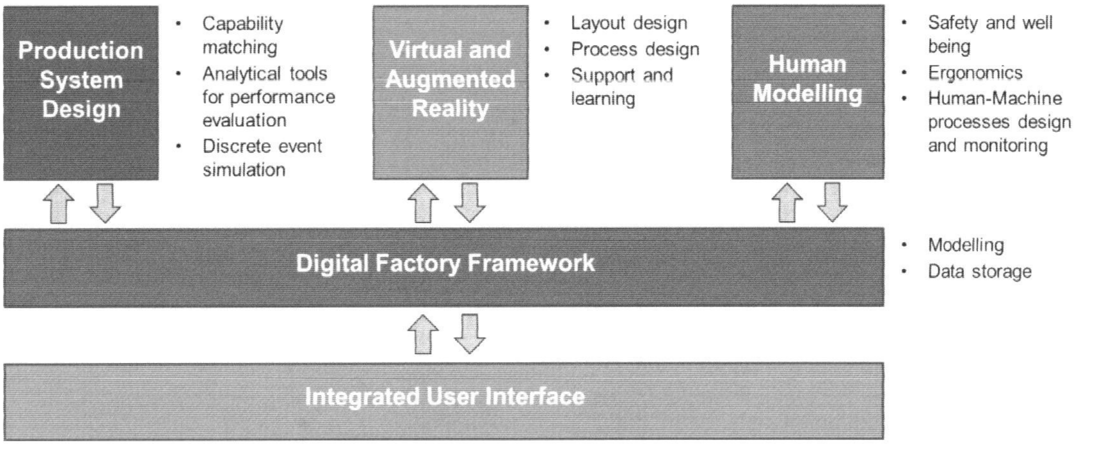

Figure 1: VLFT Framework

2.1. Digital tools and related activities

For production systems design, evaluation and visualization, the following tools are included in VLFT and are being used for the case study with their corresponding activities.

- Graphical User Interface (GUI) Tool i.e., *OntoGUI* [9] – for the definition of production systems configuration. It includes part types, process plans, operations, capabilities of the production resources like machine tools, transporters, storage systems.

- *Java Modelling Tools (JMT)* [10] and ARENA – Rockwell Automation [11] – for performance evaluation via discrete event simulation of production systems and respective Key Performance Indicators (KPIs) analysis i.e., throughput, utilization, lead time, etc.

- Unity3D [12], libraries of *Babylon.js [13]* and ApertusVR [14] – for the 3D visualization, animation and interaction with production systems and resources via virtual and augmented reality.

The use of these digital tools at different levels of manufacturing helps not only the industrial users to execute a series of analysis for system design and performance evaluation of resources. It also helps in a teaching context, the adoption of digital tools supports the application of complex engineering approaches such as simulation and analytical performance evaluation, which can assist to reach a set of learning objectives and to improve the students' learning effectiveness. More information about the VLFT tools can be found in [15].

2.2. Workflow for practicing digital tools

The set of digital tools for the Virtual Learning Factory Toolkit are aiming at improving the learning effectiveness of engineering students. To examine the digital tools, we identified a series of case studies covering different manufacturing topics representing both real engineering application and learning use cases. Four different case studies are taken into consideration addressing manufacturing at different levels of detail and from different points of view. They are as follows and consider the manufacturing *process* to be executed, the *equipment*, the *human workers* and the *monitoring*:

1. Manufacturing system (re-)configuration
2. Process-System design and analysis
3. Humans in manufacturing
4. Manufacturing system monitoring

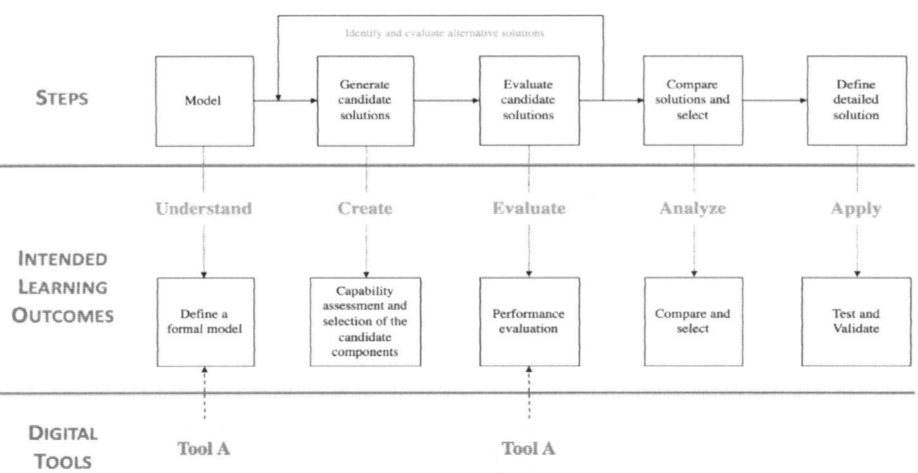

Figure 2: Structure of a workflow

For each case study, a workflow can be defined to represent the logical steps to solve it and identifying the best solution. The standard three-layer workflow structure is illustrated in Figure 2, acting as the definition of a common skeleton to be adapted and tailored for the different application cases. The first layer describes a sequence of steps, starting from the formalization and modelling of the problem under study, to the definition of the final solution, passing through design and analysis steps. For each step, the Intended Learning Outcomes (ILOs) are declared and linked to proper digital tools whose utilization aims at improving the quality of the ILOs and the pace to reach them. Moreover, ILOs describing what students should know and be able to do at the end of the learning activity that they do not know before.

3. Case Study of Manufacturing System Configuration

The configuration of a manufacturing system is a key activity since the decisions to be taken have a long-term impact on the lifecycle of the system under study. Starting from a set of requirements in terms of capability and capacity, the set of pieces of equipment and their arrangement must be analysed to find the best configuration solution, for example – minimizing the equipment cost (under performance constraints). In order to do this and specifically for reconfigurable manufacturing systems, a very large number of alternative configurations should be evaluated and analysed. Thus, digital tools can support the management of real industrial cases even in teaching engineering approaches in Bachelor or in Master courses.

3.1. Workflow for the case study

The configuration and re-configuration process can be described in terms of a sequence of steps, starting from the modelling of a hypothetical system or the existing one (in case of re-configuration) and ending in the virtual commissioning where the selected best solution is validated and tested.

The first step entails the modelling of the system under study. In case of green-field design, these modelling entails some basic decisions, for example, the adopted architecture (job shop, flow shop, etc.), the class of transportation schemes and routes, etc. In case of a re-configuration, this modelling phase is devoted to provide a formal representation of the existing system and the associated constraints.

Once the modelling phase has been completed, the configuration workflow proceed to the selection of the pieces of equipment to be in the system (Select equipment) and, then, a set of candidate solutions are generated by composing the pieces of equipment in different alternative combinations. These two tasks have to take as input the characteristics of the products to be produced, the associated production process, the derived capability associated to the process steps and the requirements in terms of volumes, flexibility, etc. The final output is the formal representation of the candidate system configurations to be further analysed. During the execution of these steps, backtracking actions can be operated to modify/improve the selected candidate configuration grounding on the assessment of the associated capability.

These candidate configurations are evaluated in terms of production performance. First, analytical approaches are used to quickly evaluate and possibly slim down the set of candidate configurations (Analytical evaluation of the candidate configurations). Then, Discrete Event Simulation (DES) approaches are used to provide a more detailed evaluation (DES evaluation of the candidate configurations).

After these two performance evaluation steps. The candidate solutions are transformed in more detailed layouts to assess them in a more realistic way and, in the end, the best solution(s) is (are) selected. The realistic environment of a manufacturing/production system can be created through 3D modelling and visualization tool.

3.2. Testing of tools in joint learning lab of students

The developed workflow for manufacturing system configuration and the identified digital tools in VLFT were tested in the joint learning lab of students. It was the first joint student workshop, held under the climax of ERASMUS VLFT project. The purpose of the joint student workshop was to develop and test the set of identified digital tools, and improve these tools through students' feedback. In the workshop, 11 different students from 3 different institutes were involved in developing and testing the digital tools. Tutors and evaluator from partner institutes were also participated in the workshop.

The development, learning and testing of the digital tools were practiced within two phases. During the first phase, students worked independently at their home institutes. During the second phase they were involved in a joint workshop and were asked to work together by exchanging information, improving their skills and learning from each other. The students were divided into two mixed groups considering the technical skills and the activities they already addressed. Each group addressed a different use-case and following three groups of activities: a, b and c specifically tested with the use-cases. Students submitted their work, which included: the formalization of the use-cases using OntoGui, simulation models using both Arena and JMT, a representation in virtual reality of the use-cases using Unity3D.

a) Formal modelling and defining production (assembly) processes of use-cases by using OntoGui (see Figure 3)

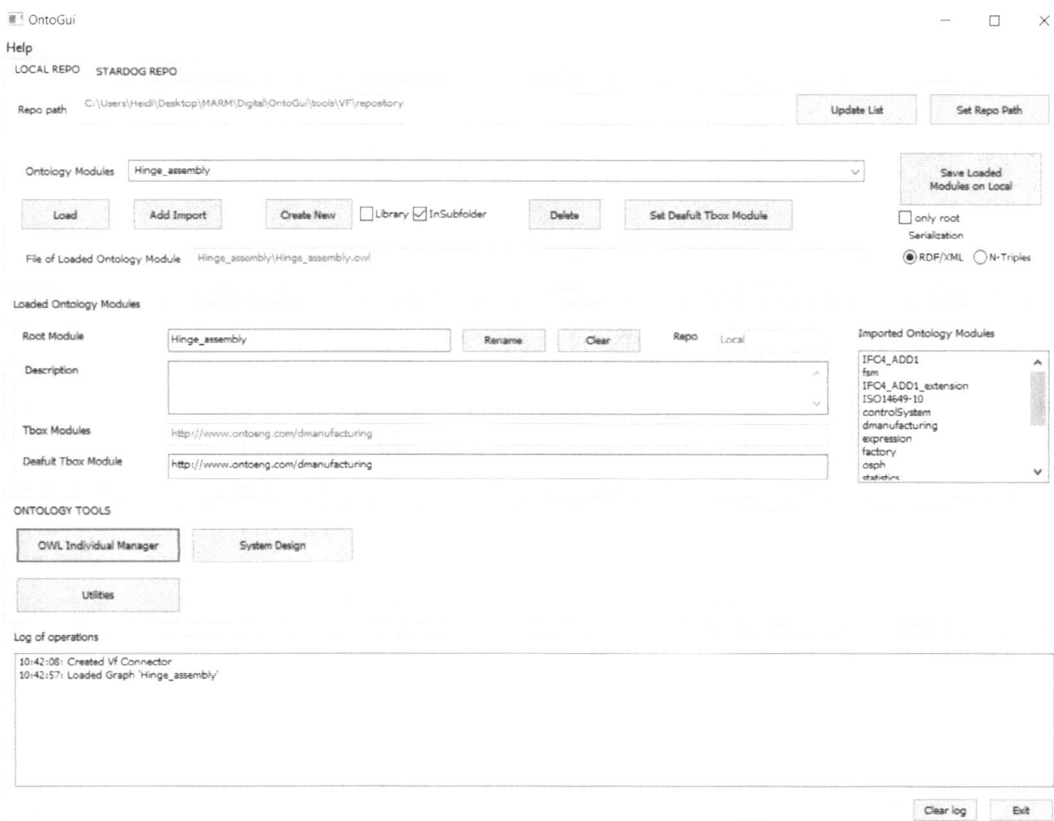

Figure 3: Use-case example of OntoGui

b) DES for performance evaluation by using JMT and ARENA. Prepared a procedure for the automatic generation of a DES model using JMT starting from the OntoGui repository for both use-case (see Figure 4).

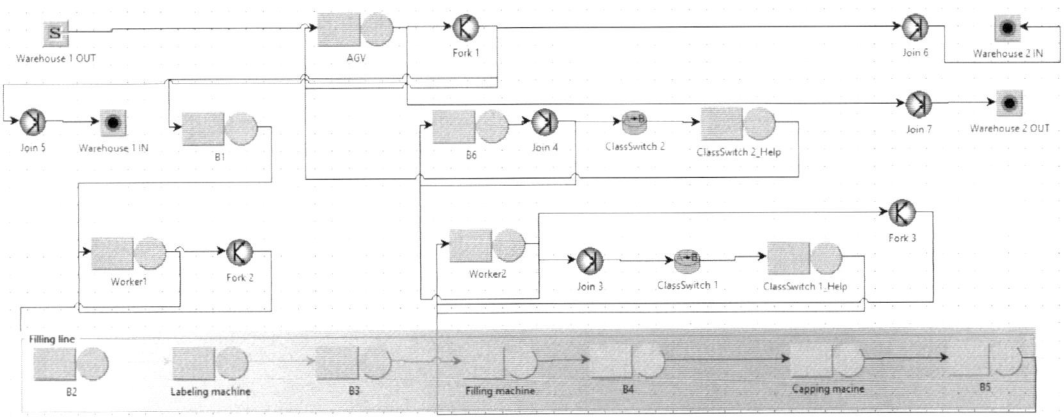

Figure 4: Use-case example of JMT

c) 3D modelling, visualization and animation for products and systems associated to the use-cases. 3D scenes were generated via Unity3D (see Figure 5) and CAD models were created through solid modelling software and open source libraries. The animated output was visualized by means of VR.

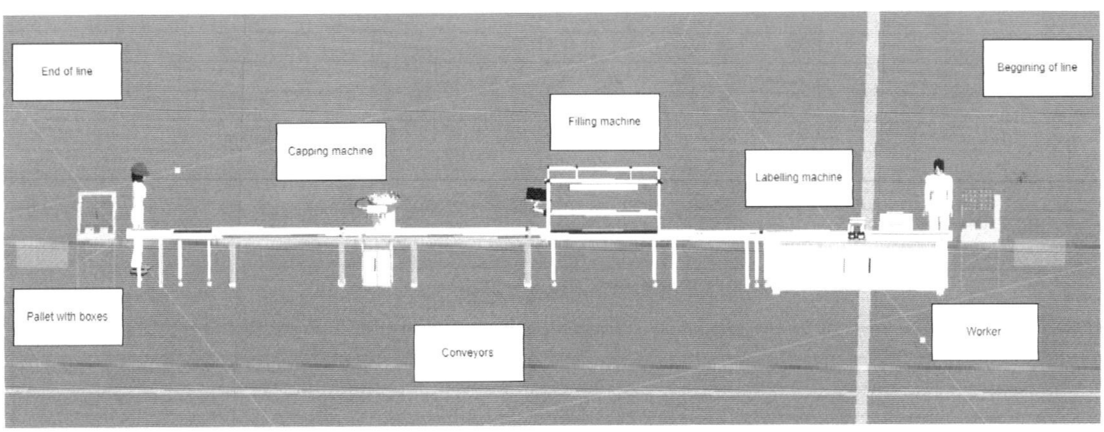

Figure 5: Use-case example of Unity3D

Some concluding observations of tutors about the students' work: *(pros)* – the virtual representation of both the use-cases was in line with the expectation accordingly to the available time. *(cons)* – the formal representation of the assembly use-case was not in line with the expectations, the quality of the simulation models was low, the integration between digital tools was not addressed sufficiently.

3.3. Student feedback

After the workshop, students' feedback was collected through a survey. The survey consists of the questions related to the following aspects (few points are included in this study), the corresponding graph of responses for each aspect is portrayed in the Figure 6.

a. General appreciation for the workshop
b. Learning ability
c. Learning level reached
d. Self-evaluation

The satisfaction of the students is expressed in a 1-5 scale: *1 = Very dissatisfied, 2 = Dissatisfied, 3 = OK, 4 = Satisfied, 5 = Very satisfied*

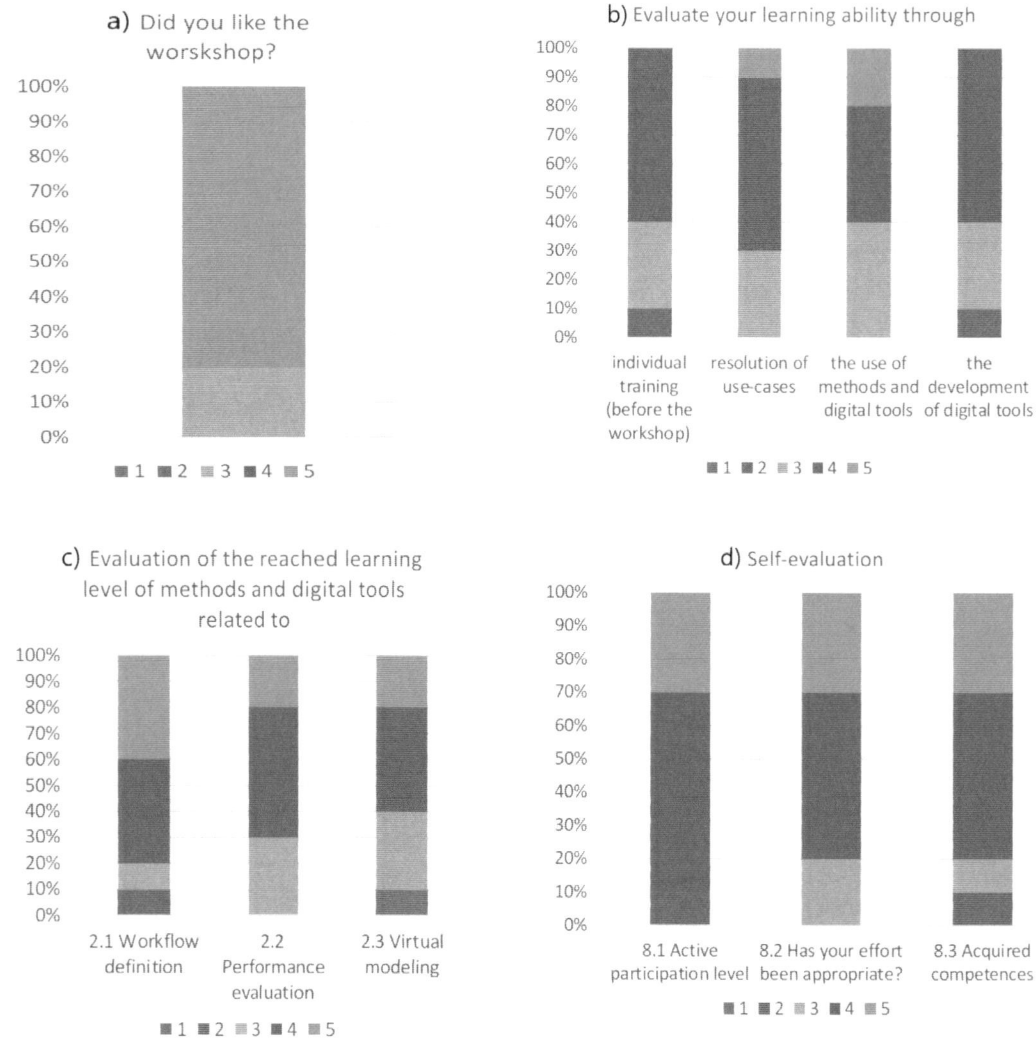

Figure 6: Graphical representation of students' feedback about the workshop

From the survey outcomes, it concludes that the student workshop was satisfactory, students have learned, contributed and tested the digital tools and methods. However, there were few shortcomings like in the learning outcomes of the digital tools used with respect to workflow definition, performance evaluation and virtual modelling. It was also observed, there were some missing competences related to programming and the integration of digital tools. Moreover, communication and cooperation skills were found out as very necessary aspects in the team work.

4. Conclusion

The development of virtual learning factory toolkit is a contribution to the modern learning in manufacturing engineering education. This work presents the framework for the VLFT, the suitable digital tools were identified and included into the toolkit, learning activities were defined, and workflows were designed and tested that help in the study of production systems and enhanced the ICT skills in manufacturing curricula. The selected and tested digital tools facilitates to define, evaluate the performance, visualize the production systems and related processes in the virtual environment based on the 3D simulations, virtual and augmented reality concepts. Testing and using the new VLF toolkit, improve the ICT skills of teachers and students, promote better understanding of the actual processes and production management, including developing of virtual factories. The tools and related learning activities were experimented in the form of joint students' workshop, which resulted in a positive experience and a way to improve the VLFT. The integration of digital tools was not sufficiently addressed in this study. Within the last segment of VLFT project, it will be enhanced and disseminated in the future workshops.

5. Acknowledgements

This study is a part of the ERASMUS+ VLFT project and it has been supported by ERASMUS+ program. Some part of this work was (partially) supported by the EC through the H2020 project EPIC under grant No. 739592. [16].Authors are thankful to the students and mentors from TALTECH, POLIMI, CNR-STIIMA and SZTAKI, who participated in the workshop and provided a valuable feedback.

6. References

[1] World Trade Organization (WTO), 2018, *World Trade Statistical Review 2018*, Geneva. [https://www.wto.org/english/res_e/statis_e/statis_e.htm]
[2] Eurofound, 2019, *The future of manufacturing in Europe*, Publications Office of the European Union, Luxembourg. [http://eurofound.link/fomeef18002]
[3] Abele E, *et al*, 2017, Learning factories for future oriented research and education in manufacturing, *CIRP Annals – Manufacturing Technology*, vol. 66, pp. 803–826
[4] Wagner U, AlGeddawy T, ElMaraghy H and Müller E, 2012, The state-of-the-art and prospects of learning factories, *Procedia CIRP*, vol. 3, pp. 109 – 114
[5] Baena F, Guarin A, Mora J, Sauza J and Retat S, 2017, Learning Factory: The Path to Industry 4.0, *7th Conference on Learning Factories, Procedia Manufacturing*, vol. 9, pp. 73 – 80
[6] Tvenge N and Ogorodnyk O, 2018, Development of evaluation tools for learning factories in manufacturing education, *8th Conference on Learning Factories, Procedia Manufacturing,* vol. 23, pp. 33-88
[7] Terkaj W, Tolio T and Urgo M, 2015, A virtual factory approach for in situ simulation to support production and maintenance planning, *CIRP Annals – Manufacturing Technology*, vol. 64, pp. 451–454
[8] Terkaj W, Gaboardi P, Trevisan C, Tolio T, Urgo M (2019) A digital factory platform for the design of roll shop plants. CIRP Journal of Manufacturing Science and Technology, 26:88-93
[9] Terkaj W, 2017, OntoGui: A Graphical User Interface for Rapid Instantiation of OWL Ontologies, *CEUR Workshop Proceedings, 2050*
[10] Bertoli M, Casale G and Serazzi G, 2009, *JMT:* performance engineering tools for system modelling. *ACM SIGMETRICS Performance Evaluation Review*, Vol. 36 (4), 10-15, New York, US, ACM press
[11] Arena – simulation software, 2021, Rockwell Automation, [https://www.arenasimulation.com/academic/students]
[12] Unity – real time development platform, 3D, 2D, VR & AR, 2021, Unity Technologies, [https://unity.com/]
[13] Babylon.js – 2021, [https://www.babylonjs.com/]
[14] ApertusVR – 2021, [http://apertusvr.org/]
[15] Virtual Learning Factory Toolkit (VLFT), 2021, TalTech 2020 [https://vlft.eu/tools.html]
[16] Centre of Excellence in Production Informatics and Control – EPIC [https://www.centre-epic.eu/]

Risk Assessment and Calibration of Risk Matrices Aspects

A Sivitski, P Põdra
Mechanical and Industrial Engineering Department, Tallinn University of Technology, Tallinn, Estonia

E-mail: alina.sivitski@taltech.ee

Abstract. Modern quality management systems, including ISO 9001 standard based, are widely used by companies. Availability of ISO 9001 certificate and quality management system, that runs like well-lubricated machine, gives the company the tool to concentrate on customer needs, to keep the high quality standard of all processes in the company, to collaborate with reliable partners and as a result to offer the high quality products and services. There is a tendency that management systems are becoming more risk thinking based and refer to risk management standard ISO 31000. Risk management standard stands for managing risks with consequences related to safety, economic performance, environmental and other outcomes thus support company continuous improvement.

For machine industry risk assessment is crucial as this is a part of CE marking documentation. According to EU legislation, to sell the machinery within European Economic Area, CE marking of machines showing the machine correspondance to Machinery directive requirements is mandatory. Risk assessment in machine industry is applied for evaluation of the risks associated with corresponding hazards to guarantee the safety of design, production process and machinery produced. Risk assessment is conducted to evaluate machine design, exploitation or manufacturing process risks. For production process risk estimation the risk levels of subprocesses should be considered.

There are no available risk assessment industry standards. Risk matrices developed by companies are used to analyse and estimate risks for correct decision making and risk reduction. Unfortunately faulty risk matrices cause difficulties during risk estimation process and lead to inappropriate safety measures application. Risk matrices use numerical values and consider minimally hazardous event severity and probability of hazardous event levels. Numerical values of risk levels are listed in the risk matrices during risk assessment. Unfortunately calibration information with defined numerical risk tolerance criteria is often missing. Miscalibrated risk matrices lead to critically wrong decisions on risk elimination issues. This paper describes the aspects of risk matrices calibration and highlights the importance of risk tolerance definition applied during risk assessment.

1. Introduction

Most companies, that stand for high quality products and international collaboration, implement quality management systems (QMS) including ISO 9001 standard based QMS. Risk based thinking is one of the main issues of the latest versions of QMS and it is applied to provide high standards of safety and quality of processes and products of the company. Therefore, risk management standards including ISO 31000 are getting high priority. According to risk management standards all kind of risks related to safety, economic performance, environmental and other outcomes should be managed to guarantee company continuous improvement. For risk levels estimation and elimination risk assessment different techniques listed in risk management standard ISO 31000 can be used. However often risk and hazard definitions are mixed up. Hazard is an issue that can cause any harm, for example health loss, social and economic disruption or environmental degradation. Risk definition is more complex as it defines the probability that hazard will lead to some consequence. As the consequence can be positive and negative, risks also can have positive and negative influence on the process or issue being managed, investigated. During risk estimation and risk assessment hazards or hazardous scenarios are listed. Hazard probability of occurrence and severity of consequence levels are considered for risk level estimation. Risk elimination methods and safety measures are offered. Risk assessment can be divided by stages. Final risk assessment is performed after risk elimination techniques are applied and should reveal low risk levels for most issues.

For company context and economic performance risk estimation preliminary hazard analysis (PHA) is often used (see Table 1). Risk matrices usually consider only three risk levels low, medium and high. Probability of occurrence of hazard or issue is considered in percentage. Severity of consequence is presented in form of the economic impact. It should be also stated in the matrices to whom the risk is assigned, expected results, uncertainty, opportunity to eliminate the risk and also cost of risk elimination. The hazards or issues listed are divided onto internal issues related to company environment and external issues that depend on political, social, ethical, technological and economic environment the company is operating in. For example, external issues include changes in the legislation, certification and standardization (see Table 1).

Table 1. Example of PHA risk assessment matrices

Hazards/ Issues (Internal / External)	**To whom risk are assigned**	**Results to expect**	**Uncertainty or reasons of hazard**	**Proba-bility of occur-rence**	**Conse-quence or economic impact**	**Risk level (low, medium, high)**	**Opportunity or risk elimination**	**Cost of risk elimina-tion**
Lack of multi-skilled workforce (Internal)	Company	Competent workforce	Company all present workforce is not skilled enough	30% - 60%	in €	Medium	Creation of multi-skilled teams	in €
Noncon-formance to standards within industry (External)	Company, Clients, State and EU	Well informed about changes in standardi-sation	Standards are being updated and changed frequently	0% - 30%	in €	Low	Designers participate in conferences on standards update	in €

In machine industry risk assessment is performed by companies for safety of manufacturing processes and safety of produced machines. Risk assessment is one of the main parts of machine design process and it is the first stage of design process. Risk assessment report is included in CE marking technical file documentation. Within European Economic Area CE marking of machines is mandatory and shows the machine correspondence to Machinery directive requirements. There are no standards for risk assessment procedure. Companies mostly use self-composed risk assessment matrices with references to ISO 12100:2010 Safety of machinery - General principles for design - Risk assessment and risk reduction standard and ISO/TR 14121-2:2012 Safety of machinery - Risk assessment - Part 2: Practical guidance and examples of methods (see Fig. 1).

Severity (Se) levels from 1 to 4 starting from light and ending with irreversible injury are defined. Likelihood parameter is divided by three: frequency, probability and avoidance. Frequency (Fe) levels vary from 1 to 5 and correspond to the hazard occurrence duration time starting from interval more than 1 year and ending with the interval 1 hour or less. Probability (Pr) of occurrence varies from 1 to 5 and corresponds to negligible up to very high probability of hazard occurrence. Avoidance (Av) parameter vary from 1 - likely, 3 - possible and to 5 - impossible level. Risk level (RC) is the sum of all likelihood parameters. Combining severity (Se) and risk level (RC) parameter values the risk index (RI) level is determined (low, medium, high risk). Numerical values of risk along with risk levels are presented in the risk assessment risk matrices (see Fig. 1). Numerical values of risk give more detailed overview of extent of risk connected with specific hazard.

Hazard Number	**Severity**	**Likelihood**				**Risk Index RI**
	Se	**Frequency Fr**	**Probability Pr**	**Avoidance Av**	**Risk Level RC**	RC = Fr + Pr + Av and Se level combination
1.	4	3	2	1	6	High risk
2.	2	2	2	1	5	Low risk
3.	2	4	2	3	9	Medium risk

Risk Index RI:		
	≤ 64	Low risk
	98 ... 512	Medium risk
	686 ≤	High risk

	Risk level RC:	4	5	6	7	8	9	10
Severity Se	4	343	686	1372	2744	5488	10976	21952
	3	49	98	196	392	784	1568	3136
	2	7	14	28	56	112	224	448
	1	1	2	4	8	16	32	64

Figure 1. Example of Initial risk estimation and evaluation; Method: Hybrid ISO/TR 14121-2:2012.

To provide reliable risk assessment data, risk matrices should be calibrated. Decision of the risk acceptability is made based on risk matrices results. Majority of companies do not consider risk matrices calibration and do not define numerical risk tolerance criteria. Without calibration criteria wrong decisions on risk elimination issues are made. For successful risk matrices calibration process the risk assessment techniques should be applied in the early design stage before whole system structure is designed.

2. General aspects of risk matrices calibration

Risk matrices are widely used as a method for risk assessment procedures. However, risk matrices methodology has its limitations in risk management decision making [1]. The example of the simplest risk assessment matrix is shown in Table 2. In the risk matrix the boundary conditions value for probability and consequence are eighter 0 and 1. Three levels of risk considered: low, medium and high for different probability and consequence quantitative values combinations.

Table 2. A simple 2 × 2 Risk Matrix [1].

Consequence/ Probability	Low	High
High	Medium (1,0)	High (1,1)
Low	Low (0,0)	Medium (0,1)

Risk matrices appear to be reliable for risk assessment if probability and consequence values are positively correlated, both probability and consequence have quantitative values either (0,0) and (1,1) correspondingly. (0,0) corresponds to minimal or zero probability and consequence and low risk level and (1,1) to maximum both parameters value and high risk level. On the contrary if probability and consequence values are negatively correlated and are placed along diagonal (0, 1) to (1, 0), then no information about greater of lower risk level is available as for both risks "Medium" risk level is assigned and decision error probability increases up to 50% [1]. Such pairs as high probability and low consequence and low probability and high consequence are commonly presented in the risk assessment.

Risk matrix calibrating allows companies to estimate their risks comparing them to risk reference value, for example, risk considered acceptable by society [2]. The calibration method of risk matrix requires the application of calibration cell in the matrix. The calibration cell corresponds to the highest consequence level (depth, lots of injuries, big monetary loss etc.) and the lowest probability of occurrence. As an example of the calibration cell risk value, risk of death in a motor vehicle accident on the way from and to work over a 25-year working period is assigned [2]. Then the highest board value of risk level is selected, other risk zones can be defined.

In process risk assessment research [3] the concept of matrix allocation is highlighted. According to allocation method the number of hazardous scenarios possible are defined and risk tolerance criteria is divided by the number defined. Furthermore, general facility tolerable risk can be allocated to individual hazardous scenario. If each individual scenario risk criteria has not been exceeded, then general facility risk tolerance criteria will be not exceeded. The number of hazardous scenarios estimation method depends on the process complexity. If process contains of subprocesses general facility risk tolerance criteria allocation should consider total number of hazardous scenarios not only individual process ones [3].

The importance to define difference between individual and group risk is also important. If individual risk represents the probability of occurrence of hazardous situation harm any individual may experience, then group risk is considered as the relationship between probability of occurrence and the number of experiencing a level of harm from hazardous scenario [3].

Failure design analysis along with statistical analysis for efficient risk values determination can be performed. To compose risk matrices having allowable residual risks and presenting optimal quantitative distributions of the risk, the optimal design approach methods should be integrated with risk assessment techniques.

3. Calibration aspects of machine safety risk assessment matrices

For successful machine manufacturing risk assessment of design and production stages along with quality control should be performed. Due to absence or lack of statistical data on machines accidents more subjective and qualitative not quantitative methods are often used in risk assessment matrices by companies. The worst-case scenarios or the highest severity hazards are often supposed to have low probability. Integrating the optimal design and manufacturing methods, failure analysis methods with risk assessment tools in the early stages of design and manufacturing process will guarantee allowable residual risks along with effective quantitative distribution of the risk. Design for Manufacture and Assembly (DFMA), Failure design method such as Failure Mode and Effects Analysis (FMEA) and Failure Modes, Effects and Criticality Analysis (FMECA) with aid of statistical tools like decision trees, vector quantization and rough sets ensure risk matrices contain optimal quantitative distribution of risk.

DFMA process is successfully used for optimizing the manufacturing and assembly process. Integrating DFMA process flow charts, that divide manufacturing and assembly processes into steps, with risk assessment matrix along with red, yellow and green colors or numerical risk estimation values for risk acceptability control is the first step to risk matrix calibration and tolerable risk evaluation. The most critical safety aspects can be spotted on the very first design stage of manufacturing and assembly process. Each manufacturing and assembly concept stage can be evaluated with response to the complexity, cost and risk level. As a result, most simple, cheaper and low risk concepts are defined.

For machine or technical system potential failures identification and elimination FMEA and FMECA techniques are used [4]. FMEA was initiated by SAE reliability engineers and is widely used for mechanical systems reliability control [2]. Current standards are available that provide guide for failure and effect analysis: SAE ARP 5580 Recommended failure modes and effects analysis practices for non-automobile applications; SAE J1739 Potential Failure Mode and Effects Analysis in Design and Potential Failure Mode and Effects Analysis in Manufacturing and Assembly Processes; IEC 60812 Procedures for failure mode and effect analysis. FMEA and FMECA differ by criticality aspects for reliability estimates and criticality is estimated in conjunction with risk assessment. Bottom-up approach initiates the part failure mode and its failure effect from lower level to higher level in the system. The top-down approach in the contrary investigates the main system functions and functional fails and is performed in the beginning of design process. Failure mode, failure effect and risk assessment severity, probability and risk level parameters are considered during failure mode analysis. Based on the failure analysis results more effective and optimal risk evaluation criteria and risk tolerance criteria can be obtained.

4. Conclusions

However, risk matrices are popular tools among companies and are often used for design, manufacturing processes, products, company economic performance hazardous scenarios probability of occurrence, consequences and risk evaluation and estimation, there are some limitations and uncertainties that can arise during that risk assessment method application. Calibration of risk matrices requires additional analysis tools application. Integrating efficient design, manufacturing process tools and failure mode analysis approaches with risk assessment procedures lead to more optimal quantitative distribution of the risk levels and more reliable risk assessment results.

References

[1] Cox L A Jr 2008 What's Wrong with Risk Matrices? *Risk Analysis* **28**, No. 2, 497-512
[2] Clements P 1996 Sverdrup System Safety Course Notes
[3] Baybutt P 2015 Calibration of risk matrices for proc. safety *J. Loss Prev. Process. Ind.* **38**, 163-8
[4] Rausand M 2011 Risk Assessment Theory, Methods, and Applications Hoboken NJ Wiley

Supply Chain Digitalization Framework for Service/Product Satisfaction

Lea Murumaa[1,2], Eduard Shevtshenko[1,2], Tatjana Karaulova[1], Kashif Mahmood[1], Janek Popell[2].

[1]TalTech Ehitajate tee 5, 19086, Estonia, Tallinn
[2]TTK UAS Pärnu mnt 62A, 10135, Estonia, Tallinn

Corresponding author e-mail lea.murumaa@gmail.com

Abstract. It is essential for every enterprise existence to satisfy customer and consumer. The main idea of the current research is to apply customer satisfaction level Key Performance Indicators (KPIs) for supply chain reliability improvement. The Supply Chain Operations Reference (SCOR) model based KPI metrics enables to increase the quality of product/service by monitoring, visualization and further digitalization of directly involved processes. In the long run, the solution will ultimately help to reduce/eliminate the number of customer reclamations in the supply chain.
The paper includes a customer satisfaction improvement feasibility case study for the approval of findings, where the garment field business processes are connected to KPI-s of product and service models. In addition, this work suggests modelling tools for faster product/service and business process modelling and evaluation (assessment) based on the selected strategic goal for each management level. The authors defend that the current research helps to create new digital solutions to increase reliability by moving toward the customer satisfaction.
Nevertheless, the proposed approach is adaptable to other fields also, whereas the small and medium enterprise (SMEs) focal player or project owner defines the processes that have the biggest impact on the product/service reliability.

1. Introduction.

The service sector is the largest and fastest growing sector in the world economy, accounting largest share in total output and employment in most developed countries [1]. Service enterprises have a lot of problems that they face every day, and the main of them is the Quality management problem. At the same time, it is essential for every enterprise existence to satisfy customer and consumer.

The main problems that small and medium garment producing enterprises have been facing are 100% order fulfillers expectation by customers and agile deliveries at minimized cost level while the order amounts and ordered products vary significantly. Garments are sold mostly via internet shops nowadays and customers are placing orders as restocking causing high level variety at small ordered amounts. Working at 95% quality level is not enough to respond to the increasing customer needs and companies are looking for a new solution, which helps to fulfil today's customer's expectations.

One of the main directions of the formation of strategic competitive advantages is the provision of services of higher quality in comparison with competitors [3]. There are different methodologies available to achieve that. For example, the SCOR reference model can be applied to define process architecture in a way that aligns with key business functions and goals [4, 5]. At the same time, the fast development of digitalisation, Industry 4.0 tools and working practices have had a significant impact on the performance of Supply Chain in recent years, and this will continue in the future. The motivation of current research is to provide companies with the Supply Chain digitalization framework to satisfy current customer needs and at the same time to fulfil the strategic goals of the companies.

This article aims to define supply chain reliability metrics at various levels of process modelling which supports the Supply chain movement toward digitalisation. Introduced Supply Chain digitalisation framework defines business processes that directly influence the level of customer satisfaction and support the decision-making process to select the most efficient tools to improve them. In the current research, authors introduce how to connect SCOR model third level KPI to reclamations and quality data collected from the companies, and how to apply BBN methodology for supply chain reliability improvement.

The empirical part of the current research is limited to garment sector companies, however, the results can be successfully applied to different fields.

Supply Chain digitalisation framework analyses the current reliability level of the companies based on reclamations and processes quality data provided and points out the improvement needed business processes. The business process KPI-s are developed based on the SCOR model, and the efficiency of digitalised business processes is predicted by BBN tool. Authors recommend applying the framework to improve the reliability of the companies, by adjustment of current business processes accordingly to the customer needs.

The second chapter provides the reader with an important background, in the third chapter authors introduce the framework for supply chain digitalisation. The practical implementation is given in the forth chapter followed by conclusions and discussion section.

2. Background

2.1. Service sector

The service sector is the largest and fasted growing sector in the world economy, accounting largest share in total output and employment in most developed countries. The share of the services sector in total GDP is 47 percent in low income countries, 53 percent in middle income countries and 73 percent in high income countries [1].

Service sector is the lifeline for the social economic growth of a country. It is today the largest and fastest growing sector globally contributing more to the global output and employing more people than any other sector. For most countries around the world, services are the largest part of their economy [2]. Almost all service enterprises belong to Micro, Small and medium-size enterprises.

Service enterprises have a lot of problems that they face every day, and the main of them is the Quality management problem. Dissatisfaction with the service leads, as a rule, to large losses in the market share. That is why the service provider must as accurately as possible identify the needs and expectations of its target customers. Unfortunately, the quality of the service is harder to judge and even harder to determine. This is the main reason for our paper to considerate the quality and reliability aspects in supply chain definition.

2.2. Supply chain reliability metrics

The SCOR model supports the operational evaluation metrics at three levels. The metrics of level one provides an approach to the supply chain to assess management, and the metrics of levels two and three include more specific and detailed criteria regarding the categories and elements of the processes. The metrics of level one are systematically divided into five operational criteria, three of which, reliability, flexibility, and responsiveness, are customer-facing attributes, and the other two, costs and assets, are internal-facing ones. Each of these metrics is further divided into minor metrics at the lower levels [4]

The purpose of the SCOR reference model, or business process framework, is to define process architecture in a way that aligns with key business functions and goals. The architecture here references how processes interact and perform, how the SCOR processes are configured [5] (see Figure 1).

The SCOR reference model consists of 4 major sections [5]:

• Performance: Standard metrics to describe process performance and define strategic goals
• Processes: Standard descriptions of management processes and process relationships
• Practices: Management practices that produce significantly better process performance
• People: Standard definitions for skills required to perform supply chain processes.

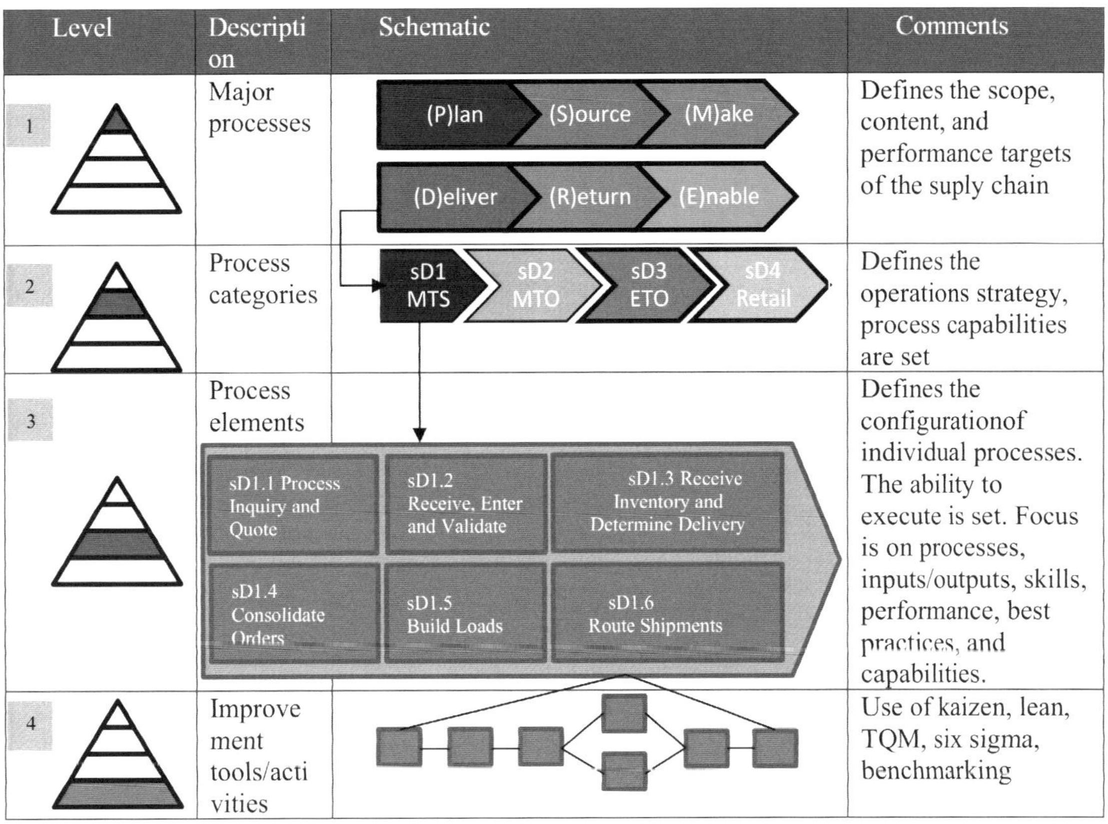

Figure 1. SCOR - hierarchical process model processes [5].

"A metric is a standard for measurement of the performance of a supply chain or process. SCOR recognizes three levels of pre-defined metrics:

• Level-1 metrics are diagnostics for the overall health of the supply chain. These metrics are also known as strategic metrics and key performance indicators (KPI). First level metrics help to establish realistic targets to support the strategic directions.

• Level-2 metrics serve as diagnostics for the level-1 metrics. The diagnostic relationship helps to identify the root cause or causes of a performance gap for a level-1 metric.

• Level-3 metrics serve as diagnostics for level-2 metrics." [5].

The analysis of the performance of metrics from level-1 through 3 is a first step in identifying the processes that need further investigation. (Processes are linked to level-1, level-2 and level-3 metrics).

IOP Conf. Series: Materials Science and Engineering 1140 (2021) 012041 doi:10.1088/1757-899X/1140/1/012041

2.3. Description of Perfect Order Fulfillment (POF) KPI

"The reliability factor is also one of the most effective criteria which mean the probability of the intact and flawless performance of the system for a definite and pre-scheduled period of time" [6]. The scope and boundaries of cooperation among the constituents of the supply chain include various activities from material ordering to final product control.

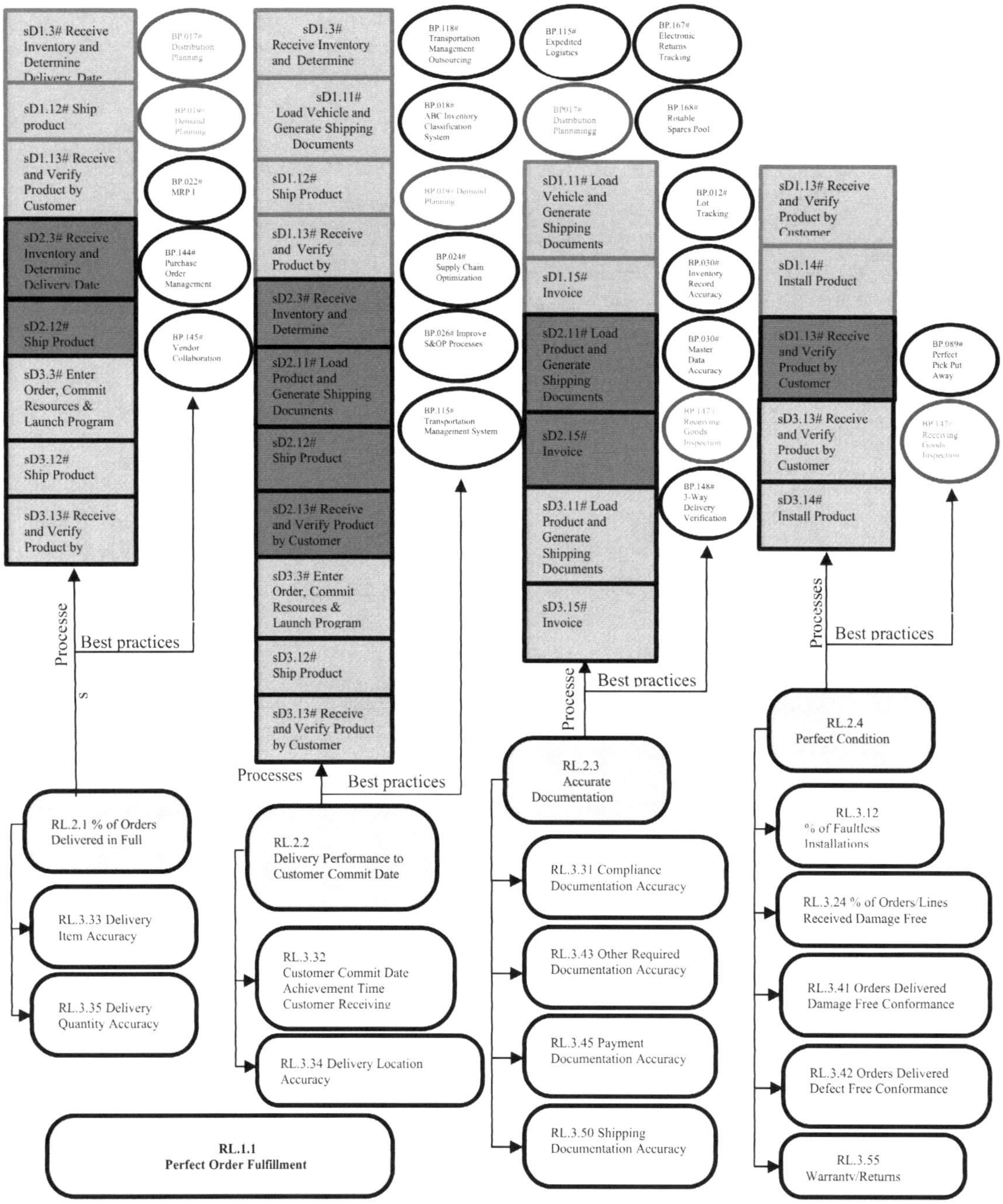

Figure 2. SCOR- POF framework for "Perfect Order Fulfilment" strategic metrics, its processes, and best practises.

"The reliability factor is also one of the most effective crit-eria which mean the probability of the intact and flawless performance of the system for a definite and pre-scheduled period of time" [6]. The scope and boundaries of cooperation among the constituents of the supply chain include various activities from material ordering to final product control. Defect in one part of the system creates problems in other parts [4, 7]. To establish a successful new relationship in the supply chain, assessing the reliability of the relationship is among the crucial factors in this field [8].

"The rate of reliability, which is the operational criterion discussed in this article, is also assessed and measured at level one of the supply chain based on SCOR model through the metrics of perfect order fulfilment; at level two through the metrics of perfect order fulfilment, delivery performance to customer commit date, accurate documentation, and perfect condition of the order.

Level three of the supply chain understudy also has minor and more detailed metrics for the assessment of the above-mentioned metrics." [9]. In the Figure 2 are shown the structure of reliability metrics in the SCOR model and codification of metrics at three levels and their relationships with processes and processes elements. Best practices for the second level of metrics is also introduced in the framework. SCOR reliability metrics interconnection will allow as more correctly select the effective metrics and process elements for various levels of supply chain processes. Since we are dealing with network metrics structure, to calculate their values different network calculation methods may be used, such as Bayesian Belief Network (BBN) [10, 11, 12], TOPSIS [13,14], Petri nets [15, 16] (see Table 1).

Table 1. Comparison of different network calculation models.

Network calculation models	Type	Description	Advantages	Parameters for calculation	Relationship type
Bayesian Belief Network (BNN), [10, 11, 12],	Graphic probabilistic model	Dependency model	Structure is the same as the SCOR structure; Successors of statistical approaches, AI and Data Mining	From ERP system	Causal
Technique for Order of Preference by Similarity to Ideal Solution (TOPSIS),[13,14]	Multi-criteria Analysis Method	Utility-based method, compares each alternative directly depending on data in the evaluation matrixes and weights.	Simple computation process;	From ERP system	Concept of distance measures
Petri nets, [15, 16]	Basic model of parallel and distributed systems.	Graph model for the control behavior of systems exhibiting concurrency in their operation	For modeling systems containing concurrent processes	From ERP system	Factor-process model with two types of components

For our research, we introduce to use the Bayesian Belief Network (BBN), because the structure of the BBN is the same as the structure of the SCOR metrics (see Figure 3). Parameters for calculation is possible to obtain from ERP system.The Bayesian networks are natural successors of statistical approaches, Artificial Intelligence and Data Mining. Particularly suited to considering uncertainty, they can be easily described manually by experts in the field. Bayesian Belief Network (BBN) is a graphic probabilistic model through which one can acquire, capitalize on and exploit knowledge. It consists of a set of interconnected nodes, where each node represents a variable in the dependency model and the connecting arcs represent the causal relationships between these variables [10].

3. Main idea of the research

Authors introduce a novel approach for Supply Chain digitalisation based on Service/Product Satisfaction. The developed framework enables Supply Chain dynamically to respond to the returns, reclaims, and quality related issues in the way of definition and digitalisation of involved business processes. BBN model enables to select the digital tool, which will give the highest impact on Perfect Order Fulfilment KPI and predict the expected result after tool implementation.

The closed loop of a suggested framework (see Figure 3) consists of ten steps, that guarantee the continuous improvement of Supply Chain Business Processes and increase the level of customer satisfaction.

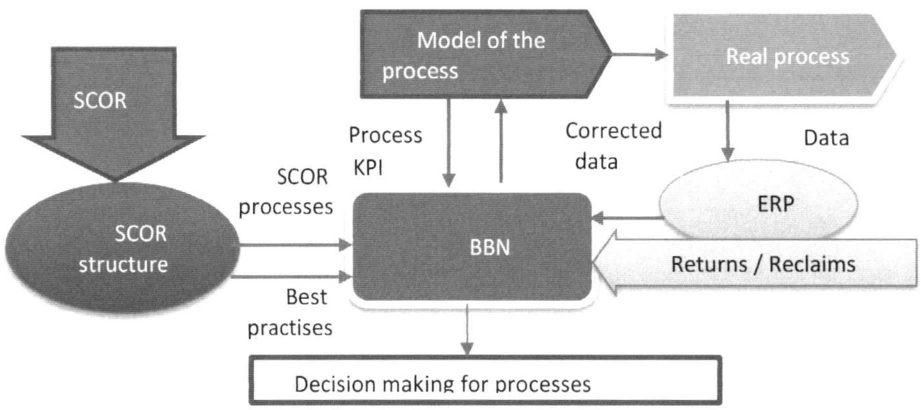

Figure 3. The main idea of the research: Service/Product Satisfaction Based Supply Chain Digitalisation Framework

Step 1. To develop the SCOR standard based BBN model to connect third level KPI RL.1.1 - Perfect Order Fulfilment POF KPI-s to appropriate AS-IS business processes.
Step 2. To select the Service/Product KPI-s data used to measure quality and customer satisfaction used in Supply Chain and to connect them to appropriate third level KPI-s
Step 3. To transfer the Returns and Reclaims data together with Service/Product KPI data from the ERP system to the BBN model.
Step 4. Assess the current POF level calculated by the BBN model.
Step 5. Define the AS-IS business processes to be improved, which are connected to Returns/ Reclaims received.
Step 6. To select the best practice SCOR digital tools for current AS-IS business processes improvement and to assess the predicted POF KPI.
Step 7. To decide on tools to be implemented in physical business processes and create a TO-BE business process model
Step 8. Simulate the TO-BE business process and assess the result achieved.
Step 9. Apply change management process to current AS-IS business process.
Step 10. Monitor the improved business process and proceed with step a to close the continuous process improvement loop

3.1. SCOR-POF structure

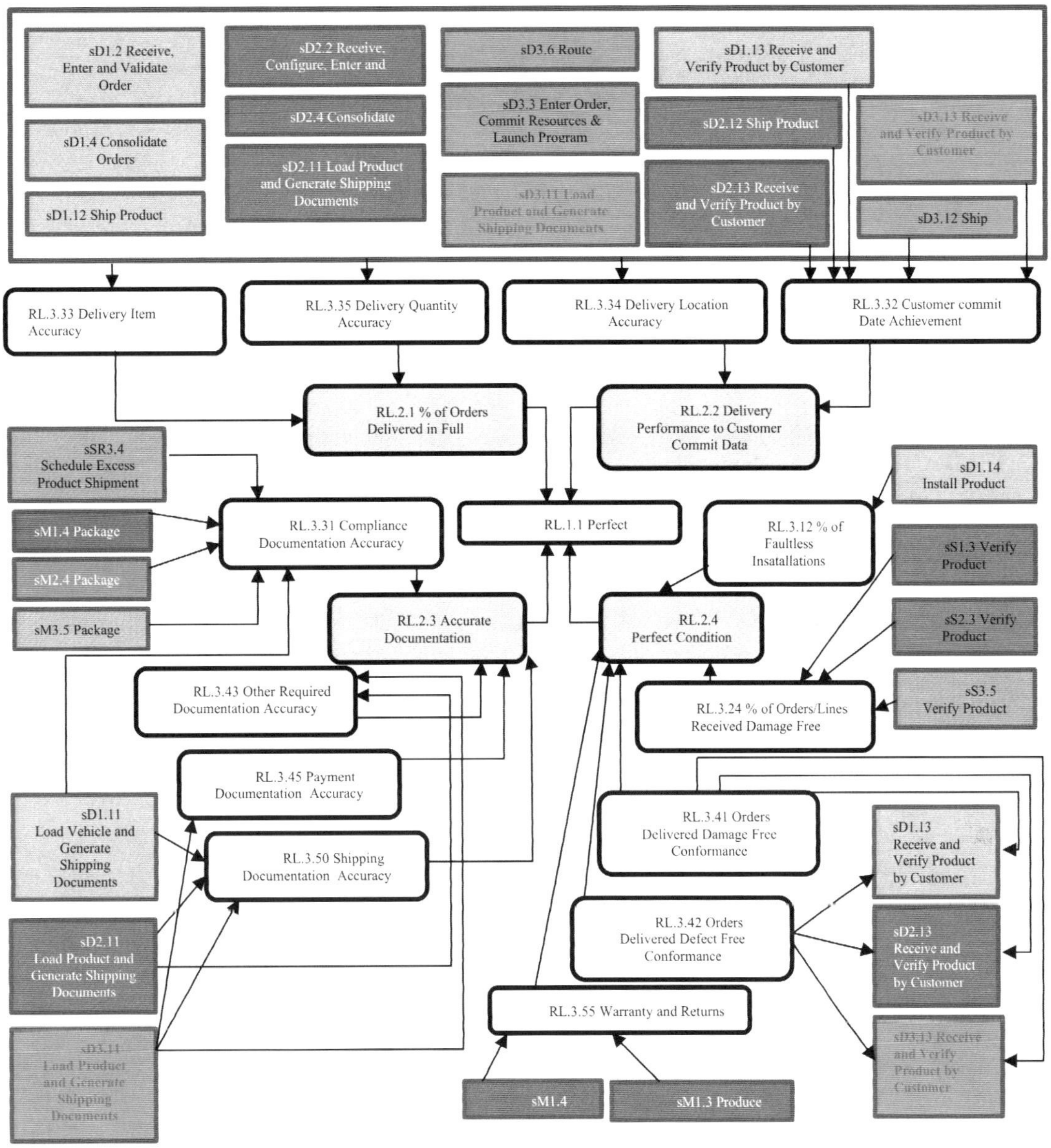

Figure 4. Processes for 3-level of metrics. A common scheme for BBN creation.

NOTES
sS1 - Source Stocked Product; sM1 - Make-to-Stock; sD1 - Deliver Stocked Product;
sS2 - Source Make-to-Order Product; sM2 - Make-to-Order; sD2 - Deliver Make-to-Order Product;
sS3 - Source Engineer- to-Order Product; sM3 - Engineer-to-Order; sD3 - Deliver Engineer-to-Order Product.

3.2. SCOR - BBN integration

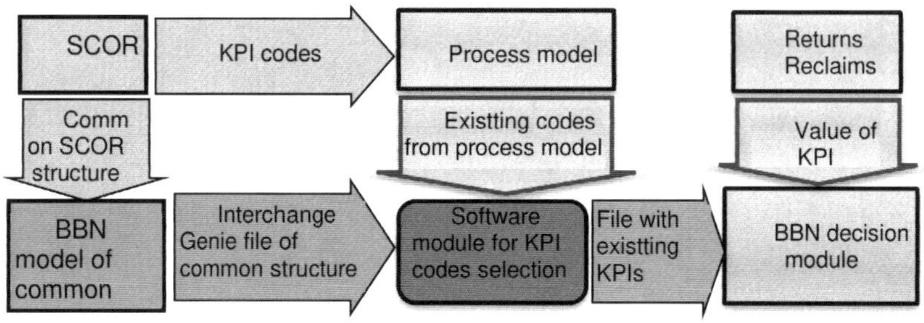

Figure 5. SCOR - BBN integration framework

The *Module for KPI codes selection* is a software which helps to select from common SCOR structure needed KPI-s for definite types of production. It is possible by using of *Interchange* GeNIe file.

The integration module receives information about the probabilities of failures from the model of the process data (ERP) and reclamations related empirical data gathered from companies and start scanning the generated code of the template to find and leave active KPI code.

3.3. BBN common scheme for SCOR- POF

The template based on the common classifier must be created in the GeNIe (BBN tool) and this operation must be done once. The developed template based on the SCOR- POF structure given in Figure 6.

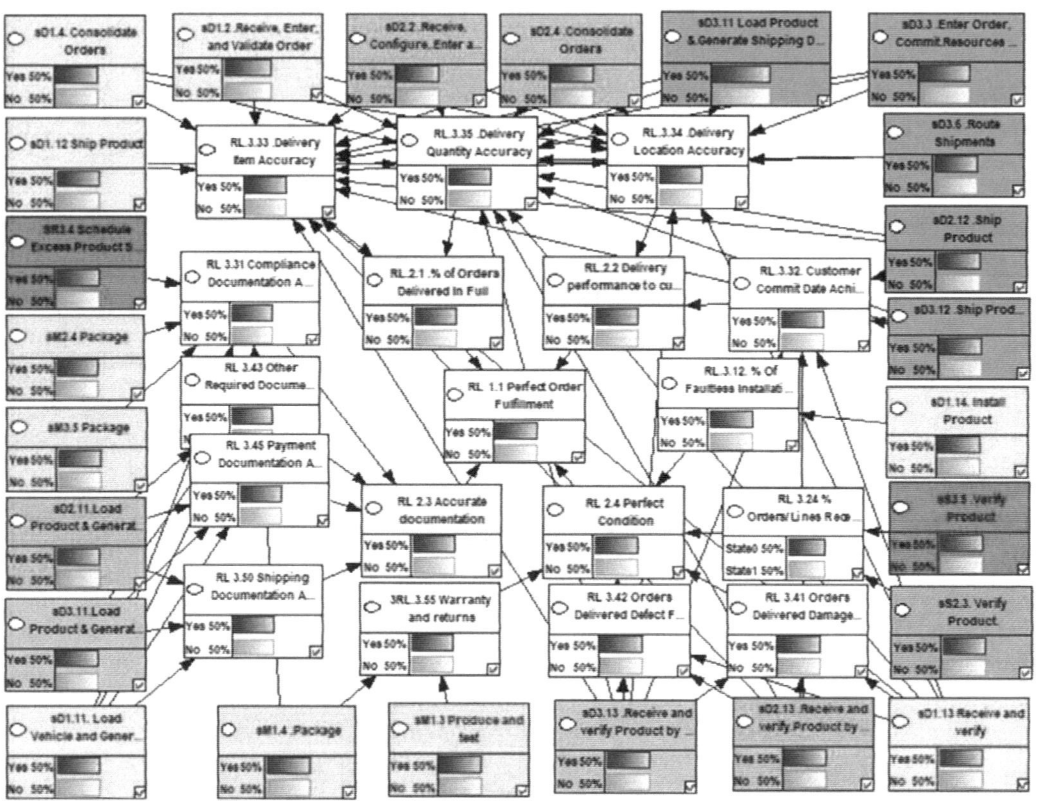

Figure 6. BBN network based on SCOR- POF structure.

4. Case study

The main problems that small and medium garment producing enterprises have been facing are 100% order fulfillers expectation by customers and agile deliveries at minimized cost level while the order amounts and ordered products vary significantly. Companies were chosen from garment production area of activities because this sector is facing severe transformation. Garments are sold mostly via internet shops nowadays and customers are placing orders as restocking causing high level variety at small ordered amounts. Working at 95% quality level is not enough to respond to the increasing customer needs.

To achieve high quality level expected by customers every garment is checked fully before delivery, defects found should be corrected and details with material faults replaced. Any deviation in quality causes a delay in payment or price reduction in the worst cases. Companies divide defects as internally found and externally found defects. All those defects that can be corrected before delivering the items are influencing companies results without influencing on customer relationship if the corrections are done within agreed delivery time and orders delivered in full amount. Some defects can be discovered too late then the orders are delivered without those items. Due to this the order is not perfectly fulfilled for the customer. Considering reduced ordered amounts even one item can be perceptually high part from the ordered amount.

External defects can be found in some cases by the end customer, it can be a defect that occurs when wrong care procedures are used, or physical damage happens. Technological defects are avoided by quality checking in the production process and before delivery.

Defects can be divided by cause to technological, finishing and information quality. Technological defect includes an unsuitable solution for the material chosen for the order. Those defects usually are discovered already in process and corrected to avoid appearing at final checking. Some defects can occur at the final check, the technological solution often proposed by the customer is not always considering material peculiarities. Finishing defects can happen due to material singularities, the fibres used may have different finishing and temperature regulations. Sometimes the customers can choose materials that are not supporting their ideas in the best way and remain to them hence the producer is suggesting changes. Information quality is poor order placement quality, all product characteristics wished by the customer are not clearly clarified, in those cases the result of the process depends from knowing the customer preferences. Materials used have also impact on product´s fitting due to its peculiarities.

Table 2. Empirical data received from companies.

	Returns / Reclaims														
	Internal			External				Returns %	Prepayment %	Warranty	OTIF	DIFOT	% of warranty returns	% of materials rejected	Ontime payments
Enterprise	Finishing, ironing, checking	Technological	Information quality, CAD-CAM	Finishing - ironing, checking	Technological		Information quality, CAD-CAM			10 days after received					
A	30%	20%	50%	49%	13%	62%	38%	4..5 %	50.. 100%		93%	93%	7%	9%	81%
B	39%	45%	16%	44%	32%	76%	24%	4..5 %			90%	90%	10%	8%	74%
C	51%	28%	21%	63%	35%	98%	2%	4..5 %			95%	95%	5%	5%	70%
D	71%	22%	7%	31%	46%	77%	23%	4..5 %			92%	92%	8%	6%	89%

NOTES: OTIF – On time, in full; DIFOT – Delivered in full, on time

The interviews conducted with four small garment producing companies gave qualitative data about defects occurred both internally and externally. Amount of defects is not similar, defects are presented as a percentage of total amount (see Table 2.) Based on data third level KPIs relevant to processes of interviewed enterprises were chosen and calculated, results are given in Table 3.

Table 3. Calculation of third level SCOR KPI-s

Faultless Installations	RL 3.12=100- %of internal rejected%of materials rejected of %Warrancy returns/100 = 0,70
Orders/lines received damaged free	RL 3.24 = OTIF=0,93
Warranty and Returns	RL 3.55=%Warrancy returns/100=0,07
Orders Delivered Defect free conformation	RL 3.42 =78%x7%
Another required documentation accuracy	RL 3.43=22%x%7=0.15
Payment documentation accuracy	RL 3.45 = 1%=0.01

Third level SCOR KPIs probabilities calculated based on qualitative data gathered via interviewing companies were entered into BBN model (see Figure 7). Probability of processes having an impact on third level KPIs in interviewed companies is not known due to lack of analysis in companies. This is a matter of future discussion and thorough analysis. Until the impact of processes is cleared their probability is given in model as 50% / 50%.

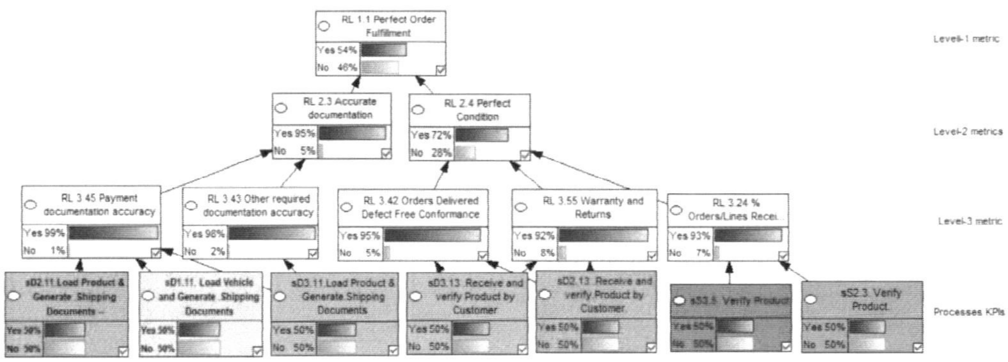

Figure 7. Definition of processes KPI

However it is possible to analyse which third level KPI have the highest influence on Second level KPI and Perfect order fulfilment Strategic Target to select best practice, which helps to achive the target.

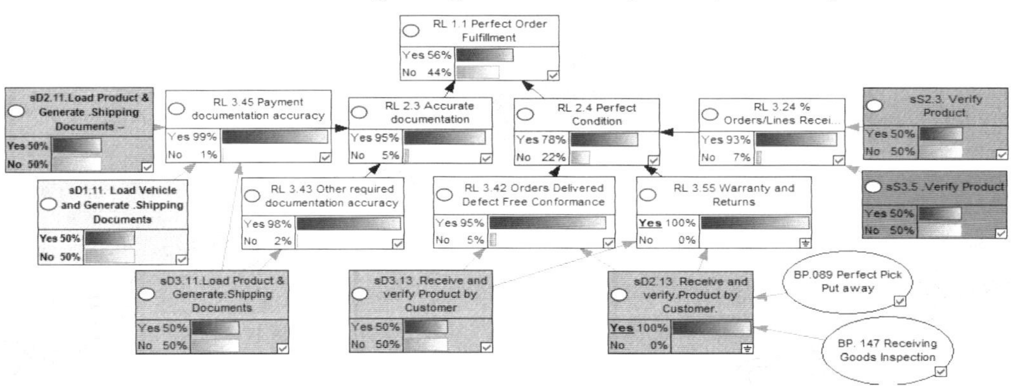

Figure 8. Evaluation of processes KPI BBN network after best practice tools implementation.

For current case study the highest impact on second level KPI have the third level KPI “RL 3.35 Warranty and Returns”. Score is recommending to apply the “BP.089 Perfect Pick Put away” and “BP.147 Receiving Goods Inspection” best practices and after implementation of those practices it is expected to eliminate the problems related to current returns (see Figure 8).

Conclusion

A novel approach for Supply Chain digitalisation based on Service/Product Satisfaction introduced and tested on a selection of small garment producing companies working as service providers for different customers shows possibilities to solve digitally quality problems companies are facing due to rising consumer expectations. The developed framework enables Supply Chain dynamically to respond to the returns, reclaims, and quality related issues in the way of definition and digitalisation of involved business processes. BBN model enables to select the digital tool, which gives the highest impact on Perfect Order Fulfilment KPI and predict the expected result after tool implementation.

The preliminary results shown in BBN model created based on data gathered to determine the content of the future work to develop the approach and digital tool useful for service providers to dynamically respond the growing customer expectations (figure 7). For making usage of developed tool wider the processes and third level KPIs of other fields of activities should be studied and analysed for clearing out similarities and specificities. Also, the selection of companies in a field should be bigger and data should be ERP-based for enabling the use of quantitative analysis.

Authors are aimed to carry on the work at the developed framework in aim to offer the digital tool enabling solutions to quality issues. For current case study it is recommended to implement the digital solution that guarantee the Pefect Pick and Put Away and Receiving Goods Inspection. Based on empirical data authors proved that those tools have the highest influence on Second level KPI and Strategic Target fulfillment (see Figure 8).

References

[1] Home Care Services And Development Perspective – A Study On Customer Perception And Acceptability In The Urban Parts Of India. The article is available for download from http://hdl.handle.net/10603/3077
[2] Service businesses https://www.referenceforbusiness.com/small/Qu-Sm/Service-Businesses.html#ixzz5ZCZHYABw
[3] Medini K., Bourey J.P. (2012) SCOR-based enterprise architecture methodology, International Journal ofComputer Integrated Manufacturing, 25:7, 594-607, DOI: 10.1080/0951192X.2011.646312
[4] Taghizadeh and Hafezi (2012) The investigation of supply chain's reliability measure: a case study. Journal of Industrial Engineering International 2012 8:22. doi:10.1186/2251-712X-8-22
[5] Supply Chain Operations Reference Model SCOR, Version 12.0, 2017 APICS
[6] Haj Shirmohammadi A (2002) Programming maintenance and repair (Technical management in industry), 8th edn. Ghazal Publishers, Esfahan
[7] Bozarth C, Handfield RB (2007) Introduction to operations and supply chain management, 2nd edn. Prentice Hall, New Jersey
[8] Xujie L (2009) Modeling and analyzing supply chain reliability by different effects of failure nodes. 2009 International Conference on Information Management, Innovation Management and Industrial Engineering 4:396–400
[9] Stephan J, Badr Y (2007) A quantitative and qualitative approach to manage risks in the supply chain operations reference. 2nd International Conference on Digital Information Management 1:410–417
[10] Neapolitan, R. E., *Learning Bayesian Networks*, Prentice Hall, (2003)
[11] Karaulova, T., Kostina, M., Sahno, J. (2012). *Framework of Reliability Estimation for Manufacturing Processes.* SCIENTIFIC JOURNAL "MECHANIKA". ISSN: 2029-6983
[12] Shevtshenko, E.; Wang, Y., *Decision support under uncertainties based on robust Bayesian networks*, International Journal of Computer Applications in Technology, Vol.36, pp.247 – 258, (2009)
[13] Sachdeva, A., Kumar, P. & Kumar, D., *Maintenance criticality analysis using TOPSIS*, Proceedings of the

2009 IEEE IEEM, 978-1-4244-4870-8/09, pp.199-203, (2009)

[14] Shevtshenko, E.; Polyantchikov, I.; Mahmood, K.; Kangilaski, T.; Norta, A.; Karaulova, T.; Perm, A. (2019). Collaborative Project Management Framework for Partner Network Initiation in Machinery Domain. Proceedings of the 18th Online World Conference on Soft-Computing in Industrial Applications (215−223). Springer. (Advances in Intelligent Systems and Computing; 864).10.1007/978-3-030-00612-9_19

[15] Bertolini, M., Bevilacqua, M. & Mason, G., *Reliability design of industrial plants using Petri nets*, Journal of Quality in Maintenance Engineering, Vol. 12, Iss: 4, pp. 397 – 411, (2006)

[16] Maja Bärring (2020) Digital Twin for Smart Manufacturing: The Practitioner's Perspective. IMECE2020. ASME

IOP Conf. Series: Materials Science and Engineering 1140 (2021) 012042 doi:10.1088/1757-899X/1140/1/012042

Investigation of technological factors influencing the strength of bonded Al – alloy

Vainius Reneckis [1], Almontas Vilutis [2], Vytenis Jankauskas [3]

[1]JSC "Kaunas Aircraft Maintenance Services", Oro uosto str. 4, Karmelava, LT–54460, Kaunas, Lithuania

[2]JSC "Poliplastas", Engineering Department, Geliu str. 97, LT–53330 Kaunas, Lithuania

[3]Vytautas Magnus University, K. Donelaičio str. 58, 44248, Kaunas, Lithuania

E-mail: vreneckis@gmail.com

Abstract. Adhesive bonding makes up a significant part of rigid joints of machine elements. The aim of this study was to investigate the contamination effect of moisture and oil on the bond strength of aluminium alloy 2024 T3 bonded by film and liquid adhesives using mechanical and chemical surface preparation methods. Two types of joints were used: homogeneous (Al-Al) and heterogeneous (Al-GFRP) ones. Surface preparation highly increased surface roughness Ra. The surfaces were moistened with deionized water and mineral oil and glued after a constrained time delay (outlife). Tensile shear strength tests of the joints showed higher strength of the chemically prepared surfaces by gluing with film adhesive. Contrarily mechanically prepared surfaces were stronger with liquid adhesive. Film adhesive seemed less sensitive to surface contamination in general.

1. Introduction

Aluminium alloys are one of the most important construction materials [1]. The global aluminium consumption volume of nearly 68.6 metric tones attained in 2019 and it is projected to reach almost 85 mln. metric tons by 2025 [2]. Aluminium alloys have wide industrial use (automotive, aviation etc.). Lighter vehicles lower fuel consumption (reduce CO_2 emissions) and more than 90% recyclability makes aluminium an ideal material for circular economy [1].

Adhesive bonding of aluminium alloys started in 1940's in England and since then has become more popular among other traditional connections (riveting, welding, etc) [3]. Bonding is simple and effective type of connection not reducing the strength of constructional elements [4]. Adhesive bonding is the most suitable method of joining both for metallic and non– metallic structures where strength, stiffness and fatigue life must be maximized at a minimum weight [5]. Mechanical interaction and degradation of bonds are complex, strength of the connection is influenced by various factors – materials used, geometrical parameters, etc. Degradation in a bonded joint begins with a mixed load, both the load and the operating environment must be considered, as the joint is affected by temperature, humidity, lubricants, etc. [6].

Surface preparation is particularly important for the joint strength and durability. Its influence on surface preparation of glued joints was investigated [7]. Aluminium is special in that its surface is coated with an oxide layer, so various surface preparation methods are used the most effective of which are combined ones: degreasing, etching, anodizing, etc. [8, 9]. The strength of the connection in operation is influenced by the environment, the most important factor of which is humidity. Water enters through surfaces, cracks and other defects in adhesive and adherend surfaces, weakening the bond and accelerating loss of adhesion [10].

The aim of the research is to investigate the influence of technological factors on the strength of the bonded aluminium alloy joint by evaluating the influence of surface contamination with water and mineral oil.

2. Materials and methods

2.1. Materials

There were 126 specimens prepared for testing all of the same dimensions in length and width being 60x15 mm (gluing area 15x15=225 mm^2) and 6×3 test plan was selected with 6 groups of materials and 3 subgroups in each. The plan is presented in table 1.

Table 1. Design of experiment (DOE) matrix

Code	Adher.	Adhes.	Surface prepar.	Surface contamination	Constrained time delay	Strength of 126 samples	Elongation at break of 126
1AK	Al-Al	film	mech	uncontaminated	0	n=3	n=3
1A1	Al-Al	film	mech	water	5, 15, 30 min	n=9(3x3)	n=9
1A2	Al-Al	film	mech	min. oil	1, 12, 24 h	n=9	n=9
2AK	Al-Al	film	chem	uncontaminated	0	n=3	n=3
2A1	Al-Al	film	chem	water	5, 15, 30 min	n=9	n=9
2A2	Al-Al	film	chem	min. oil	1, 12, 24 h	n=9	n=9
3AK	Al-Al	liquid	mech	uncontaminated	0	n=3	n=3
3A1	Al-Al	liquid	mech	water	5, 15, 30 min	n=9(3x3)	n=9
3A2	Al-Al	liquid	mech	min. oil	1, 12, 24 h	n=9	n=9
4AK	Al-Al	liquid	chem	uncontaminated	0	n=3	n=3
4A1	Al-Al	liquid	chem	water	5, 15, 30 min	n=9	n=9
4A2	Al-Al	liquid	chem	min. oil	1, 12, 24 h	n=9	n=9
5AGK	Al-GFRP	liquid	mech	uncontaminated	0	n=3	n=3
5A1	Al-GFRP	liquid	mech	water	5, 15, 30 min	n=9(3x3)	n=9
5A2	Al-GFRP	liquid	mech	min. oil	1, 12, 24 h	n=9	n=9
6AGK	Al-GFRP	film	mech	uncontaminated	0	n=3	n=3
6A1	Al-GFRP	film	mech	water	5, 15, 30 min	n=9	n=9
6A2	Al-GFRP	film	mech	min. oil	1, 12, 24 h	n=9	n=9
					Total:	n=126	n=126

4 groups consisted of homogeneous adherends (Al-Al), the remaining groups consisted of heterogeneous (Al – GFRP) ones. Two adherend materials were used: aluminium alloy 2024 T3 and glass fiber Gilliner 1076C Type59 (E – glass, polyester resin system). Aluminium alloy chemical composition is presented in table 2. The mechanical properties of fiberglass are given in table 3. The properties in the warp and weft directions are calculated and given averaged.

There were 3 subgroups in each group: uncontaminated (control joints) and contaminated (with deionized water and mineral oil). Low viscosity mineral oil AeroShell Fluid3 was used.

Epoxy adhesive types used: film AF163–2K and liquid EC 2216 B/A Gray (two component). Both are 3M (USA) materials. The technical data of the adhesives are given in table 4. Thickness of dried film epoxy adhesive is 0.25 mm, liquid – 0.1 mm, GFRP – 1.45 mm.

Table 2. Material properties of Al 2024-T3 alloy [11]

Main chemical composition Al 2024-T3 alloy [wt, %]							
Al	**Cu**	**Mg**	**Mn**	**Fe**	**Zn**	**Ti**	**Cr**
90.7-94.7	3.8-4.9	1.2-1.8	0.3-0.9	≤ 0.5	≤ 0.25	≤ 0.15	≤ 0.1

Mechanical properties							
Density, [g/cm³]	**Hardness, HB**	**Tensile strength, [MPa]**	**Young modulus, [GPa]**	**Poisson's ratio**	**Shear modulus, [GPa]**	**Shear strength, [MPa]**	**Elongation at break, [%]**
2.78	120	483	73.1	0.33	28	283	18

Table 3. Material properties of GFRP Gilliner 1076 Type 59 [12]

Thickness, [mm]	**Density, [g/cm³]**	**Bond strength, kN**	**Avg. tensile strength, MPa**	**Avg. flexural strength, [MPa]**	**Avg. flexural modulus, [MPa]**
1.45	1.91	3.86	499.5	396	17.55

Table 4. Material properties of epoxy adhesives used [11]

Adhesive material	**Type**	**Density, [g/cm³]**	**Young modu-lus, [MPa]**	**Shear modu-lus, [MPa]**	**Poisson's ratio**	**Max strength, [MPa]**	**Elong. at break, [%]**
AFK163–2K	film	1.21	1.10	413.7	0.34	48.26	13
2216 B/A Gray	liquid	1.30	0.56	192.4	0.47	18.2	4.1

2.2. Surface preparation and testing

Before gluing surfaces were prepared mechanically (mech) and chemically (chem). During mechanical preparation surfaces were abraded with P180 emery paper in 3 angular directions (90^0, 180^0 , 45^0), cleaned with acetone and after water film test coated with the chromate conversion coating Henkel BODERITE M–CR 1200S RTU AERO. During chemical preparation, surfaces were cleaned with NaOH solution and afterwards kept in the sulfuric acid– dichromate (FPL (Forest Product Laboratory) is a long used surface treatment for Al in the aerospace industry) solution for 60 s. Surface roughness in μm Ra was measured with MarSurf PS10 (Mahr, Germany) device. Gluing was done with a special gluing station HC9200B (HEATCON, USA), connections heated and joined by vacuum (0.88 bar). The samples with film and liquid adhesives were heated and cooled at the same rate of 16.1 C^0/min and all were cooled down till 25 C^0. The samples with film adhesive were heated till 121.1 C^0 and hold for 90 min, while with liquid adhesive were heated till 93.3 C^0 and hold for 120 min.

Surface contamination was created by shortly immersing planned part of samples to be bonded into water, the other ones – into oil. The test samples were kept in a special holder to remove excess of liquid and then after a set time were bonded at 22±1 °C temperature and 40±5 % humidity. Samples taken from water were bonded after 5, 15 and 30 min, from oil – after 1, 12 and 24 hours. Tensile shear strength tests made with INSTRON 5965, tension speed 20 mm/min, one end of the samples were immovable. Recorded were max force and elongation at break according to ISO 4587 standard [13], calculations done by the equation: $\tau = F_{max}/(LW)$. Here τ – tensile shear strenght [MPa], F_{max} – maximum force [N]; L – lapping length [mm]; W – lapping width [mm].

The fracture structure was described according to EN ISO 10365:1995 standard [14]: SF (substrate failure), CSF (cohesive substrate failure), DF– delamination failure, CF (cohesive failure), SCF (special cohesion failure), AF (adhesion failure), ACFP (adhesion and cohesion failure with peel). The number next to the letter indicates the amount of disintegration as a percentage.

Software packages IBM SPSS Statistics 2020, MS Excel add-in Data Analysis and add-on XLSTAT were used for data entry and analysis.

3. Experimental results and discussion

Surface roughness of the aluminium alloy 2024 T3 before preparation was R_a 0.22 ±0.032 (the standard error SE= 0.018); ground in 3 directions – 0.95 ±0.080 (SE= 0.046); ground and covered with the

conversion coating – 0.93 ±0.049 (SE=0.028); cleaned with NaOH solution – 0.83 ±0.075 (SE= 0.043); etched in the FPL acid solution – 0.85 ±0.027 (SE= 0.015).
Close to one Eta=0.986 value received after Crosstabs analysis show high degree of association between variables (surface preparation and roughness). Generalized linear model allows to state that 97.2% of the scatter of roughness values are influenced by the surface preparation methods used (the coef. of determ. R^2=0.972, p=0.05).

Based on the tensile and elongation at break test data mathematical bond failure model was developed and 50% failure probability estimated, taking tensile and elongation at break separately as failure criteria (Figure 1). Sample size of each group n=21 (minimum bound ratio requirements of n=20 are being met for this Weibull analysis).

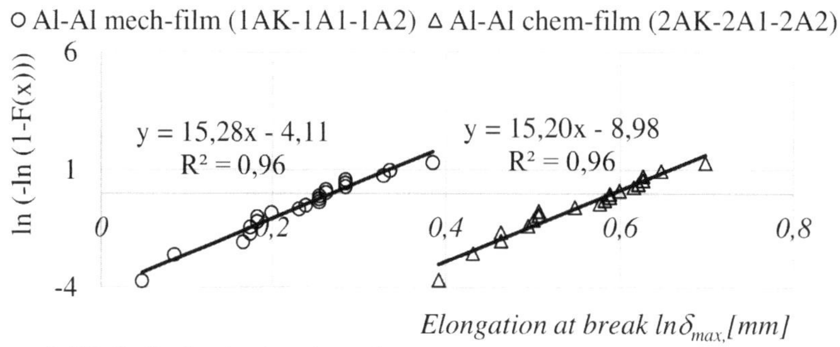

Figure 1. Weibull plot (estimation of parameters by using log-linear regression)

Data are presented in table 5. Higher values of Weibull modulus m are observed of the strongest bonds (it shows lower standard deviation and lower spread of failure criteria values from the mean).

Table 5. Weibull distribution data of bond failure criteria (significance level α = 0.05)

Group	Bond description	Regression equation	Failure probability 50% (F(0.5))		
			Elongat. at break δ_{max}, [mm]	Modulus m	Scale param. σ_0
Bond failure criterion – elongation at break:					
1AK-1A1-1A2	Al-Al mech-film	y=15.28x-4.11, R^2=0.962	1.28	15.28	1.31
2AK-2A1-2A2	Al-Al chem-film	y=15.20x-8.98, R^2=0.957	1.76	15.20	1.81
3AK-3A1-3A2	Al-Al mech-liquid	y=9.18x-0.52, R^2=0.911	1.02	9.18	1.06
4AK-4A1-4A2	Al-Al chem-liquid	y=4.34x+0.04, R^2=0.833	0.91	4.34	1.01
5AK-5A1-5A2	Al-GFRP mech-liq.	y=7.91x+0.03, R^2=0.960	0.95	7.91	1.00
6AK-6A1-6A2	Al-GFRP mech-film	y=3.96x-0.22, R^2=0.914	0.96	3.96	1.06
Bond failure criterion – shear tensile strength:			τ, [MPa]		
1AK-1A1-1A2	Al-Al mech-film	y=9.99x-28.64, R^2=0.987	16.97	9.99	17.60
2AK-2A1-2A2	Al-Al chem-film	y=29.59x-92.16, R^2=0.987	22.24	29.59	22.52
3AK-3A1-3A2	Al-Al mech-liquid	y=7.58x-19.91, R^2=0.958	13.20	7.58	13.85
4AK-4A1-4A2	Al-Al chem-liquid	y=6.83x-16.86, R^2=0.946	11.20	6.83	11.82
5AK-5A1-5A2	Al-GFRP mech-liq.	y=5.30x-12.49, R^2=0.979	9.85	5.30	10.56
6AK-6A1-6A2	Al-GFRP mech-film	y=6.74x-16.55, R^2=0.977	11.02	6.74	11.64

Figure 2 shows contaminated bonds strength with varied time between contamination and bonding. An average strength of uncontaminated surfaces is given at the 0 time or before the test. For statistical evaluation of results XLSTAT mixed linear model was used with 6 groups of materials and 7 constrained time delay intervals (a–0 min (control), b–c–d (5–15–30 min H_2O) and e–f–g (1–12–24 h min. oil) were selected and bond strength evaluated.

IOP Conf. Series: Materials Science and Engineering 1140 (2021) 012042 doi:10.1088/1757-899X/1140/1/012042

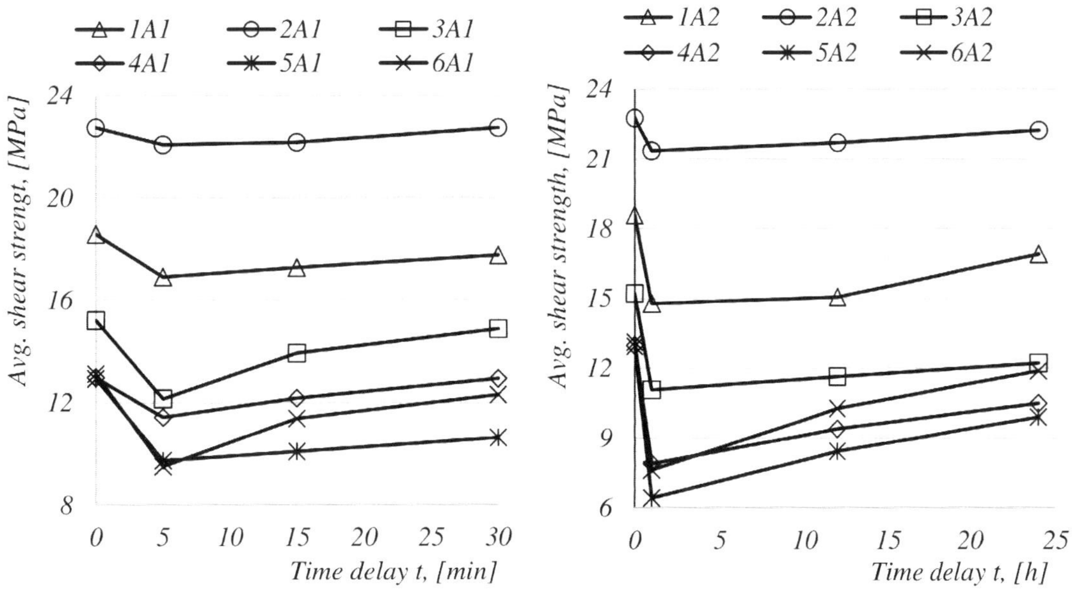

Figure 2. Impact of time delay on the avg. change of strength of contaminated adhesive bonds with water (a) and oil (b)

It has been determined that time delay interval after contamination has a great influence on the model (error covariance $p<0.0001$) and doesn't have on covariance among material groups, but is on the limit ($p=0.058$). Statistically the most significant are uncontaminated samples ($p<0.0001$) when compared with contaminated ones and time interval until the first measurements of the strength of the contaminated bonds (the less time elapsed since the start of contamination, the weakest are the bonds).

Time delay after contamination with water is insignificant from the period of 5 min ($p=0.493<0.05$) till 15 min ($p=0.191<0.05$), slow change of strength is observed, however, after 30 min elapsed it started to show higher impact ($p=0.006$). Time delay after contamination with oil has the highest impact on strength after 1 ($p<0.001$) and 12 ($p<0.009$) hours elapsed, significant change of strength is observed.

Statistically significant are only two groups: 2 nd ($p=0.21\text{x}10^{-3}$) – contamination with water or oil has the lowest influence on the average strength and was the highest in average strength values, 5th ($p=0.04$) – the lowest average strength (table 6).

Table 6. Impact of contamination of bonds with water and oil by the bond groups

Contamination	**Mechanical treatment**				**Chemical treatment**		
Water	**1A1 film**	**3A1 liquid**	**5A1 liquid**	**6A1 film**	**2A1 film**	**4A1 liquid**	**Avg.**
Uncont. (t_0) avg. strength, [MPa]	18.58	15.21	12.94	13.12	22.75	12.98	**15.93**
Reduction of avg. strength [MPa] (t_0–t_{30} min)	0.81	0.33	2.31	0.82	0.01	0.05	**0.72**
Fracture denomination of contaminated bond:	AF100	AF100	DF10 AF90	DF80 AF20	CF100	AF100 start, AF30CF70 end	
Mineral oil	**1A2 film**	**3A2 liquid**	**5A2 liquid**	**6A2 film**	**2A2 film**	**4A2 liquid**	**Avg.**
Uncont. (t_0) avg. strength, [MPa]	18.58	15.21	12.94	13.12	22.75	12.98	**15.93**
Reduction of avg. strength [MPa] (t_0–t_{30} min)	1.68	3.0	3.04	1.24	0.53	2.49	**2.00**
Fracture denomination of contaminated bond:	AF100	AF100	DF85 AF15	DF70 AF30	AF25 CF75	AF100	

Images of the strongest and weakest disintegrated adhesive bonds are presented in figure 3.

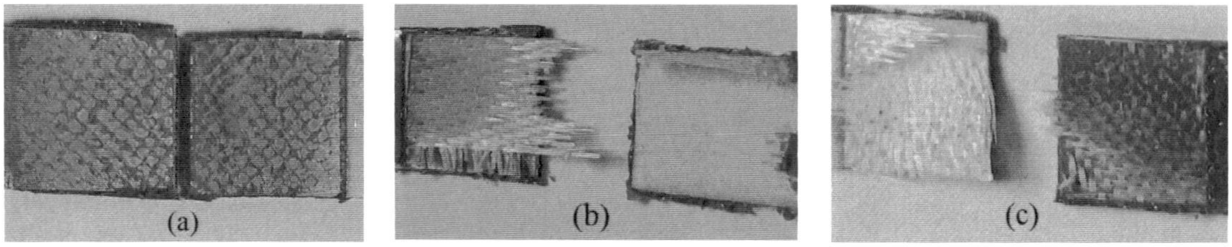

Figure 3. Images of fracture surfaces: a) the strongest Al – Al (group 2), the weakest groups: b) Al – GFRP (group 5) and c) Al – GFRP (group 6)

As can be seen from the image (a), the strongest bonds failed in cohesion. These are chemically prepared surfaces and thermal gluing of film adhesive in vacuum greatly increased adhesion. In general film adhesive is nearly twice higher in technological thickness compared to liquid ones and have got better mechanical properties. Meanwhile the weakest bonds are those with dissimilar materials (Al-GFRP) which surfaces were prepared mechanically. Despite they are liquid (b) or film (c) ones, it can be explicitely seen, that they mainly failed due to delamination of fibers with the most severe damages and delamination observed on the glass laminate surfaces. It reveals weaker interlaminar strength of the glass fiber in tensile shear. In order to increase the bond strength between metal and glass fiber, stronger then polyester fibers and resins should be used, special structural primer applied, etc. Besides, it is know that in case of differences in the thickness and type of adherend materials, due to eccentricity such single lap bonds are impacted by higher forces and moments and their failure seems more complex compared to similar ones [6,7].

4. Conclusions

- Surface roughness and scatter of its values is 97.2% affected by the surface preparation method used. The most uniform distribution of roughness Ra values achieved with FPL etch solution–the lowest standard deviation and error (SD = ±0.027, SE= 0.015).
- Contamination with water is less dangerous than with mineral oil (strength recovers faster). Bonds contaminated with water have lower strength by 5% in average, type of the adhesive does not influence the strength of the bond. The lowest strength decrease has been observed of bonds with homogeneous materials (Al-Al) and both by chemical (0.02%) and mechanical (2.2%) surface preparation methods.
- Contamination with oil reduces the strength by 14% in average. The hybrid material (Al-GFRP) bonds mechanically prepared and glued with liquid glue has the highest reduction of strength (~24%).
- Mechanically prepared AL-GFRP and contaminated with water and oil bonds showed reduction in strength in 10-20% (film-liquid), prevailing mixed disintegration DF+AF.

References

[1] Information on https://www.statistica.com/ (access 01.12.2020)
[2] Information on https://www.expertmarketresearch.com/ (access 01.12.2020)
[3] Miroslav M Analysis of production parameters of single-lap bonds adhesive bonded with composites based on aluminium filler *Agricultural Engineering*, Vol. **63**, 2017 (1) p 36–44
[4] H S da Costa Mattos, A H Monteiro, R Palazzetti Failure of adhesively bonded joints in composite materials. *Materials and Design* **33** (2012) 242–247

[5] A Baldan Adhesively-bonded joints and repairs in metallic alloys, polymers and composite materials: Adhesives, adhesion theories and surface pretreatment *Journal of Material Science* **39** (2004) 1–49
[6] J Benias, P Jacubczak Interlaminar Shear Strength and Failure Analysis of Aluminium-Carbon Laminates with a Glass Fiber Interlayer after Moisture Absorption *Journal of Materials*, 2020,**13**, 2999.
[7] A Rudawska Comparison of the adhesive joints' strength of the similar and dissimilar systems of metal alloy/polymer composite *Journal of Applied Adhesion Science* **7** (2019)
[8] S G Prolongo, A Urena Effect of surface pre-treatment on the adhesive strength of epoxy-aluminium joint *International Journal of Adhesion and Adhesives* **29** (2009) 23–31
[9] F Cavezza, M Boehm A Review on Adhesively Bonded Aluminium Joints in the Automotive Industry. *Journal of Metals* 2020,**10**,730.
[10] C Bellini, G Parodo Effects of operating temperature on aged single lap bonded joints *Journal of Defense Technology*, https://doi.org/10.1016/j.dt.2019.05.015
[11] Information on https://www.matweb.com/ (access 10.11.2020)
[12] Information on https://www.thegillcorp.com/ (access 10.11.2020)
[13] ISO 4587:2003 Adhesives–Determination of tensile lap shear strength of rigid–to–rigid bonded assemblies.
[14] EN ISO 10365:1995 Adhesives–Designation of main failure patterns.

IOP Conf. Series: Materials Science and Engineering 1140 (2021) 012043 doi:10.1088/1757-899X/1140/1/012043

Alloying of TiC-FeCr cermet in manganese vapor

M Kolnes[1], M Tarraste[1], J Kübarsepp[1], K Juhani[1], M Viljus[1]

[1] Tallinn University of Technology, Ehitajate tee 5, Estonia

E-mail: mart.kolnes1@taltech.ee

Abstract. Conventional vacuum sintering is not suitable for producing manganese containing cermets because of the physical-chemical properties of manganese, high vapor pressure, combined with high sintering temperature of cermets (1400-1600 °C). Sintering in Mn-rich *microatmosphere* does not only prevent Mn loss, but also enables additional *in-situ* alloying of the binder phase during sintering. We studied alloying of TiC-based cermet bonded with high chromium steel during sintering in Mn-rich atmosphere. Sintering in manganese vapor was found to increase sinterability of the TiC-FeCr cermet, resulting in the formation of ~1 mm thick Mn-rich surface layer with homogeneous microstructure while the core region of the material remained unaffected. This core region exhibited highly increased Mn content and competitive mechanical properties – hardness of ~1200 HV30 and indentation fracture toughness of ~13 MPa*m$^{1/2}$. In addition, an unusual ε-type martensitic phase was observed in the surface layer.

1. Introduction

Nickel and cobalt are toxic for humans. Therefore, efforts have been made to find nontoxic alternatives for a binder in ceramic-metallic composites (cermets). Iron-based binder systems, in particular FeCr-alloys, are promising [1, 2] since iron is known as a nontoxic, accessible and cost-effective metal.

Manganese is a chemical element present in all commercial steel grades; moreover, it is a low-priced alloying element. Characterized as an active deoxidizer that improves the machinability, it enhances mechanical properties. Besides interstitial elements (carbon and nitrogen), manganese is also an effective γ-Fe (austenite) stabilizer. Our suggestion is that alloying iron-bonded TiC cermet with manganese can improve the mechanical properties of cermet. Manganese belongs to the group of crystalline materials, which, based on their physical–chemical properties, sublimate at the barometric pressure at temperatures below their melting point. The manifestation of the high vapor pressure during sintering is sublimation of elements from the powder or bulk material and its rate is dependent on the sintering temperature. These temperatures embrace practically the main range of sintering temperatures [3].

Alloying of powder compacts by the manganese gas phase was addressed by Salak et al. [4]. The study shows uniformly alloyed surface layers on iron particles in the compacts formed as a result of condensation of manganese vapor combined with high rate surface diffusion. Manganese vapor treatment for alloying is also used for platinum group metals [5].

In this study we investigated the alloying effect of the manganese vapor during the sintering of the TiC-FeCr cermet. The main focus was on the changes in the chemical composition and its effect on the microstructure.

2. Experimental Details

Nickel- and cobalt-free TiC-FeCr cermet powder mixture with 30 wt% of binder (initial, calculated value) was prepared using a common powder metallurgy process for ceramic-metallic composite powders: ball milling for 72 h in ethanol by WC-Co balls with ball-to-powder weight ratio of 10:1. Chemical composition, characteristics and producer of starting powders are presented in Table 1.

Table 1. Description of starting powders

Powder	Chemical composition, wt%		Particle size, μm	Producer
	Basic components	Impurities		
TiC	C_{comb} - 19.12; Ti - bal.	O - 0.30; N - 0.02	2-4	PPM Ltd.
AISI430L (X2Cr17)	Cr - 16.8; Mn - 0.69; Si - 0.64; Fe - bal.	C - 0.02; P - 0.01; S - 0.01	10-45	Sandvik Osprey Ltd.
Mn	Mn - 99.84	O - rest	7.95	PPM Ltd.

Paraffin wax was added to the powder mixture as compacting aid and green samples were prepared employing uniaxial pressing. *In-situ* alloying sintering was carried out in a vacuum furnace (Red Devil) using a specially designed chamber for the generation of manganese vapor *microatmosphere* [6]. Heating rate, maximum sintering temperature and holding time were 10 °C·min^{-1}, 1400 °C and 30 min, respectively. The cross-sections of sintered specimens were prepared (ground and polished) for characterization.

Scanning electron microscope (SEM) Zeiss EVO MA15 equipped with the X-ray spectroscopy (EDS) system INCA was used for microstructural and elemental composition analysis. Phase composition was analyzed with Rigaku Smartlab using Cu Kα radiation and the step size of 0.02°. Vickers hardness and microhardness were measured with Indetec 5030 KV (load 294.2 N) and Micromet 2001 (load 1 N), respectively. The indentation fracture toughness (K_{IC}) was measured using the Palmqvist method and the equation described in [7] (Eq. 8):

$$K_{IC} = 0.16 \times \left(\frac{c}{a}\right)^{-1.5} \times \left(H \times a^{\frac{1}{2}}\right),$$

where c is the average length of the cracks emanating from the tips of the Vickers indentations (μm); a is the half length of the diagonal of Vickers indentations (μm); H is Vickers hardness (MPa).

3. Results and Discussion

Table 2 demonstrates the calculated chemical composition before milling and after vacuum and alloying sintering (EDS elemental analysis). EDS analysis was performed near the surface and from the core region of the sintered samples. Chromium and manganese content in the surface and in the core of vacuum sintered TiC-FeCr cermet decreased considerably. The vacuum level during the vacuum sintering was approximately 10^{-2} mbar, which causes reduction of chromium and intense decrease of manganese content due to the sublimation and evaporation of these elements. On the other hand, during alloying sintering in Mn-rich atmosphere, the manganese content increases drastically in the surface layer of the TiC-FeCr cermet. It correlates with the change in weight – the mass decrease of vacuum sintered cermet was 3.1 wt% while the mass increase of alloying sintered cermet was 8.3 wt%. Nevertheless, the manganese content of the cermet core remained at the same level as the calculated chemical composition. This demonstrates that Mn losses can be prevented by creating a manganese vapor *microatmosphere* and manganese vapor in the sintering device has an alloying effect, however only in the surface layer of the cermet. The sublimation of manganese and formation of manganese vapor starts already at temperatures around 900-1000 °C [3]. The major densification of TiC-FeCr cermet does not occur earlier than the formation of the liquid phase at sintering temperatures around 1300 °C. Despite the fact that open porosity and manganese vapor coexist, alloying of specimen core does not occur during solid-state sintering (see Table 2.). The small fraction of tungsten in sintered specimens is caused by the wear of WC-Co milling balls.

Table 2. Chemical composition and weight change of TiC-FeCr cermet

			Chemical composition, wt%							**Weight change****
			Ti*	**C***	**Fe**	**Cr**	**Mn**	**Si**	**W***	
Calculated		Total	56.63	13.37	24.56	5.04	0.21	0.19	-	-
		Binder	-	-	81.87	16.80	0.69	0.64	-	
Vacuum sintering	Surface layer	Total	63.06	15.02	14.22	3.09	0.12	-	4.49	-3.1 %
		Binder	-	-	91.96	6.34	0.50	1.20	-	
	Core	Total	59.26	14.44	17.91	4.20	0.11	-	4.08	
		Binder	-	-	92.97	5.41	0.65	0.98	-	
Alloying sintering	Surface layer	Total	40.56	12.87	26.67	5.58	**10.97**	-	3.35	+8.3 %
		Binder	-	-	67.58	6.22	**25.47**	0.73	-	
	Core	Total	58.84	13.96	18.42	4.35	0.19	-	4.24	
		Binder	-	-	91.95	6.18	0.91	0.96	-	

*Although small Ti, C and W signals were detected in the binder as well, an assumption was made that the majority of them come from nearby TiC grains and not from the binder. Thus, the binder composition was recalculated excluding Ti, C and W.
**Weight change that occurs due to the debinding (removal of paraffin wax during sintering) is excluded.

Different sintering conditions have dissimilar influence on the microstructure of TiC-FeCr cermets sintered at 1400 °C. The SEM images of visually two-phase microstructures of TiC-FeCr cermets (darker TiC grains embedded in a brighter binder) after vacuum sintering and alloying sintering in Mn-rich atmosphere are presented in Figure 1. Microstructures of the surface layer and the core region are presented. As can be seen, the whole cross-section of the vacuum sintered cermet and the core of the

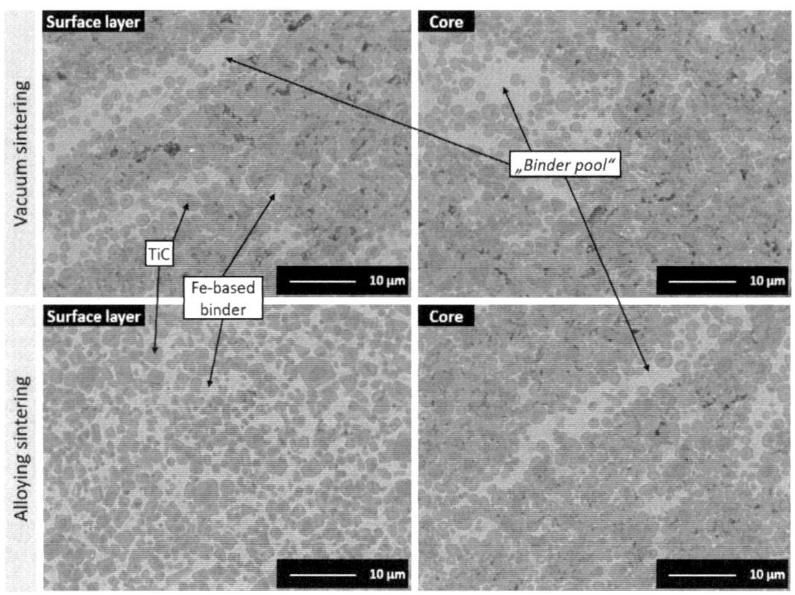

Figure 1. The SEM images of microstructures of TiC-FeCr cermet after vacuum sintering (top) and alloying sintering in Mn *microatmosphere* (bottom). Microstructures of the surface layer (left) and the core region (right) are presented.

alloying sintered cermet is characterized by an uneven distribution of the metallic binder phase (*binder pools*) and residual porosity. The inferior sinterability is caused by the insufficient reduction of oxides and limited wettability. In case of vacuum sintering, the TiC-FeCr cermet requires temperatures above 1475 °C for developing a uniform microstructure with low porosity [8]. However, the distribution of the binder on the surface of the sintered cermet is significantly improved during alloying sintering in manganese *microatmosphere* – the microstructure is homogeneous and the porosity has been decreased.

It is the result of high deoxidation ability of manganese influencing the sinterability – it improves wetting conditions between the solid carbide grains and the liquid metallic binder phase. Figure 2 presents the optical microscope (OM) images of the surface layer and the core of the TiC-FeCr cermet after alloying sintering and EDS line scan as an indication of the Mn content. The thickness of the Mn-rich surface layer that exhibits uniform microstructure and low porosity is roughly 1 mm. The hardness of the surface layer is around 1200 HV30 and as the Mn content decreases, the hardness of the cermet increases. Such result was expected since the fraction of a more ductile and plastic metallic binder is considerably higher in the surface layer. However, the hardness near the core of the sample has very high uncertainty due to the highly heterogeneous microstructure. Due to the uniformity of the microstructure, the indentation fracture toughness was measured only from the surface layer and the result, 12.7±0.4 MPa*m$^{1/2}$, is comparable with the research reported in [1].

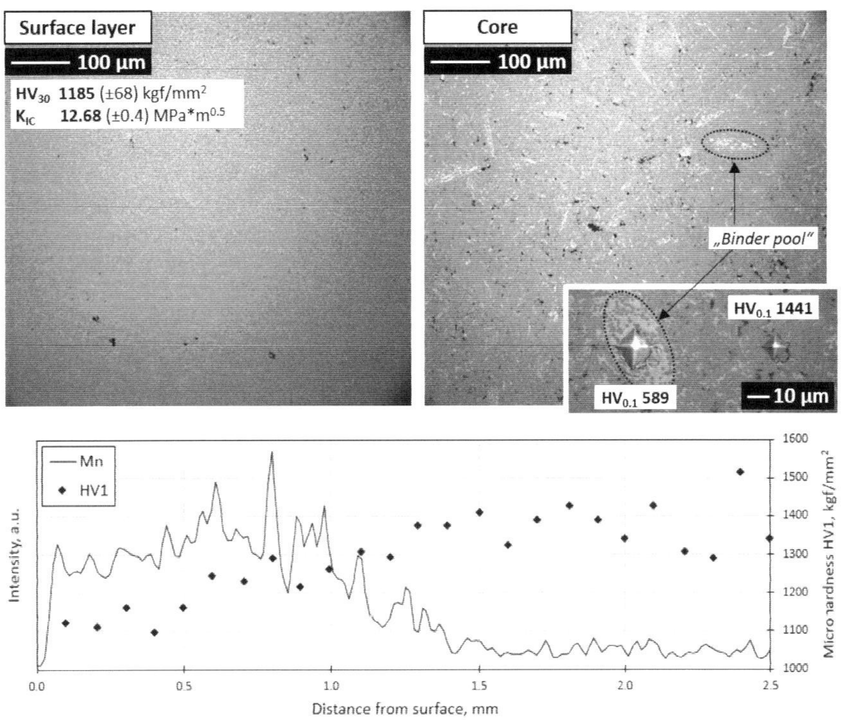

Figure 2. The OM images of TiC-FeCr cermet after alloying sintering in Mn *microatmosphere* (top) and EDS Mn line scan of the cross-section (bottom). Macrohardness and fracture toughness of the surface layer and microhardness of the cross-section and the core are shown.

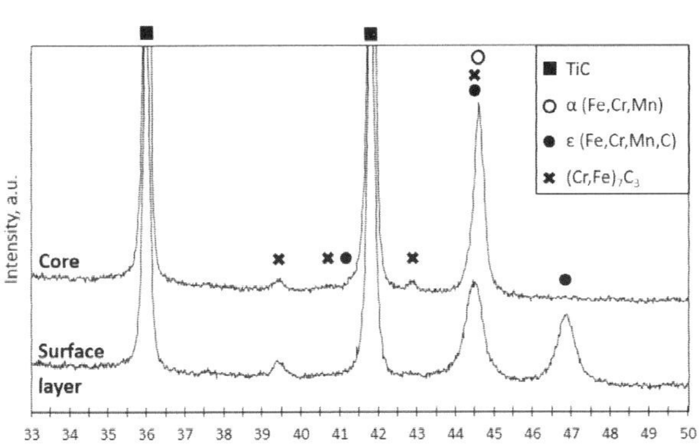

Figure 3. X-ray diffractograms of the core region (top) and the surface layer (bottom) of TiC-FeCr sintered in Mn *microatmospehere*.

XRD phase analysis showed that after the alloying sintering, the bulk of the TiC-FeCr cermet is composed of three phases: TiC, ferrite (α) and chromium-based double carbide $(Cr, Fe)_7C_3$. While the ferrite is expected in the binder phase, double carbide has formed during sintering due to the high affinity of chromium for carbon [9]. Interestingly, an additional phase, ε-type hexagonal martensite, is present in the surface layer. While equilibrium state of the Mn-rich binder is austenite (γ), XRD results indicate that during cooling $\gamma \rightarrow \varepsilon$ transformation took place, resulting in the ferritic-martensitic binder. To the authors' knowledge, ε-type martensite has not been reported in TiC cermets with the Fe-based binder before.

4. Summary

Titanium carbide cermet bonded with ferritic high-chromium steel was alloyed with manganese employing sintering in Mn-rich *microatmosphere*. The following conclusions can be drawn:

- Mn vapor *microatmosphere* in the sintering device prevents Mn losses. However, regardless of open porosity before the formation of the liquid phase, the alloying effect of the bulk of TiC-FeCr cermet body was not achieved. As a result, composition and non-uniform microstructure of the samples sintered using conventional vacuum sintering and alloying sintering remained similar.
- Sintering in Mn *microatmosphere* allows considerable increase in Mn content only in the surface layer of the TiC-FeCr specimens with a thickness of about 1 mm. The metallic binder of the Mn rich surface layer is composed of ε-type martensite and α-type ferrite. This surface layer is characterized by structural homogeneity and high mechanical properties – hardness of ~1200 HV30 and indentation fracture toughness of ~13 MPa*m$^{1/2}$.

Acknowledgements

The research was supported by Estonian Research Council grants PRG665 and PRG1145.

References

[1] Kübarsepp J and Juhani K 2020 Cermets with Fe-alloy binder: A review *International Journal of Refractory Metals and Hard Materials* **92** 105290

[2] Norgren S, García J, Blomqvist A and Yin L 2015 Trends in the P/M hard metal industry *International Journal of Refractory Metals and Hard Materials* **48** 31–45

[3] Hryha E and Dudrova E 2011 *Application of Thermodynamics to Biological and Materials Science*, ed Mizutani T (London: IntechOpen) pp 573–602

[4] Šalak A and Selecká M 2012 *Manganese in Powder Metallurgy Steels* (New York City: Springer Publishing) pp 39-71

[5] Taninouchi Y and Okabe T H 2017 *Rare Metal Technology* (New York City: Springer Publishing) pp 119–27

[6] Schubert W D, Fugger M, Wittman B and Useldinger R 2015 Aspects of sintering of cemented carbides with Fe-based binders *International Journal of Refractory Metals and Hard Materials* **49** 110-123

[7] Evans A G and Charles E A 1976 Fracture toughness determinations by indentation *Journal of the American Ceramic Society* **59** 371-372

[8] Kolnes M, Mere A, Kübarsepp J, Viljus M, Maaten B and Tarraste M 2018 Microstructure evolution of TiC cermets with ferritic AISI 430L steel binder *Powder Metallurgy* **61** 197–209

IOP Conf. Series: Materials Science and Engineering 1140 (2021) 012044 doi:10.1088/1757-899X/1140/1/012044

Quality of Photovoltaic Modules, Experimental Evaluation and Mathematical Modelling

P Tšukrejev*, **K Karjust and J Majak**

Department of Mechanical and Industrial Engineering. Tallinn University of Technology. Ehitajate tee 5, Tallinn, Estonia

*E-mail: pavel.tsukrejev@taltech.ee

Abstract. Over the world rapid growth of demand for photovoltaic systems installations brings forward magnificent increase in production numbers in manufacturing facilities of PV systems. Production companies are facing challenges in providing the best quality simultaneously with rising manufacturing quantities. Due to technology behind not all the quality decisions can be done in real time. This study is focused on the development of experimental study and mathematical modelling of the PV modules quality control parameters, which could only be tested during chemical processes and could not be monitored constantly by operators at the production line.

1. Introduction

Current trends in smart manufacturing show the direction to stay competitive on the market and to deliver the maximum return on assets for production related companies. To achieve this, companies have to continuously search for innovative ways to improve their production and quality control processes, to optimize manufacturing processes using new I4.0 based technologies and perform work in a faster and better way [1-2]. Production processes should be effectively monitored and controlled to avoid malfunction and unplanned downtime.

Quality is becoming an increasingly important function for the company due to the increased customer demands and product quality requirements. Manufacturing companies apply modern quality control techniques to improve the production line and its processes quality. A range of techniques are available to control product or process quality [3]. These include seven statistical process control (SPC) tools, acceptance sampling, quality function deployment (QFD), failure mode and effects analysis (FMEA), six sigma, and design of experiments (DoE). Quality Control (QC) and Quality Assurance (QA) can be defined as fulfilling specification or customer requirements, without any defect. A product is said to be high in quality if it is functioning as expected and is reliable. Quality control refers to activities to ensure that produced items are fulfilling the highest possible quality.

Photovoltaic (PV) modules installations are growing annually, global Compound Annual Growth Rate of cumulative photovoltaic installations between years 2010 and 2019 was as much as 35% [4]. Photovoltaic modules are utilizing the effect that generates flow of electrons inside the materials under the light. There are different possibilities for the materials to be used. According to [4] 95% of production based on silicon-based solar cells which is presented in Figure 1.

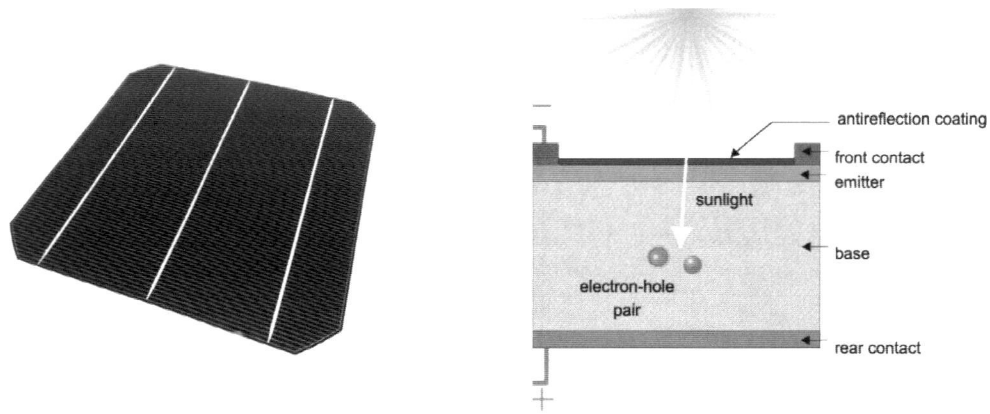

Figure 1. c-Si silicon solar cell and its schematic cross-section [5].

To build a PV module there are also other materials used in order to ensure maximization of light gathering, structural health as well as electric and climate insulation. The structure of PV module considered includes [6]:

- Frontsheet – usually glass or some other transparent material for light transparency and climate and mechanical protection;
- Photovoltaic cells – for current generation;
- Ribbon connections – for electrical circuit;
- Backsheet – for electrical and climate insulation;
- Encapsulant – for laminating everything all together, protection from moisture and air as well as being transparent for light.

Current study aims to gather experimental data and based on these data to build mathematical model(s) for prediction the quality of the encapsulant gel content. The obtained results will allow manufacturers to predict the crosslinking level instantly at place on the basis of real-measured parameters.

2. Experimental evaluation of the quality of encapsulant

Quality of lamination is a general focus of the series of papers and emerging problem for solar companies. Encapsulant under study is Ethylene/Vinyl-Acetate (EVA), as it is mainly used by the partner PV manufacturer of this study.

In terms of particular research assessment of lamination success could be divided into two main branches, represented in Figure 2:

1. Visual component – all possible visual fault that leading to bigger issues in the future;
2. Quality of encapsulant (crosslinking level) – Gel content of the EVA material, should be defined during time consuming process [7].

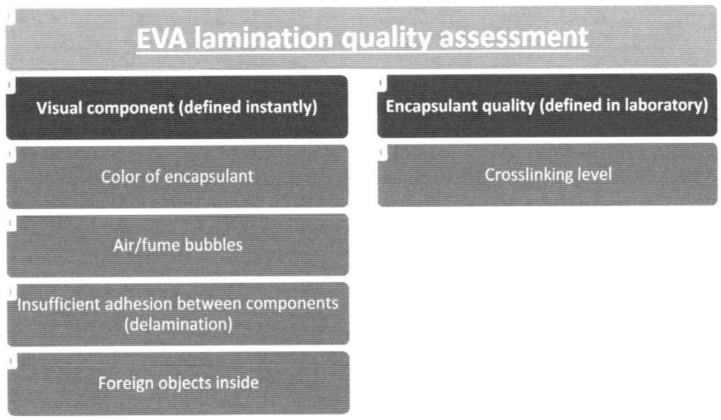

Figure 2. Quality assessment of cured Ethylene/Vinyl-Acetate.

Ensuring quality of encapsulant is a challenging due to lack of possibilities to assess and evaluate quality of lamination on chemical composition level just after the lamination cycle is done. In order to define the crosslinking level laboratory tests are needed. Good cross-linking level is considered to be 65% [8]. Supplier of EVA suggesting target value for PV modules to be between 70% and 80%. Sample gathering is something that is making a PV module not usable anymore.

There are number of inputs that are impacting the quality of the lamination process [9]: temperature, duration, pressure/vacuum time. As temperature and duration of the process are considered by the authors to make the biggest impact on the quality of encapsulation it was decided to measure the temperature from the edge of the module during the real manufacturing lamination cycle. Previous experience showed that measuring from the surface of the module is damaging backsheet and module is becoming visually defected and not usable.

External equipment was employed in order to measure the temperature in real time with the possibility to trace everything via online cloud-based graphical user interface. During the experimental phase of measuring temperature by external equipment research group had faced the fact that there is a difference in real measured temperature from module and the temperature shown by lamination machine which is represented in Figure 3. Also, there is dependence of temperature difference from the time lamination occurred: first laminations after startup, continuous numerous laminations or lamination after long pause. This is the point of interest to the PV manufacturer as the need for tune the receipts used in production appeared.

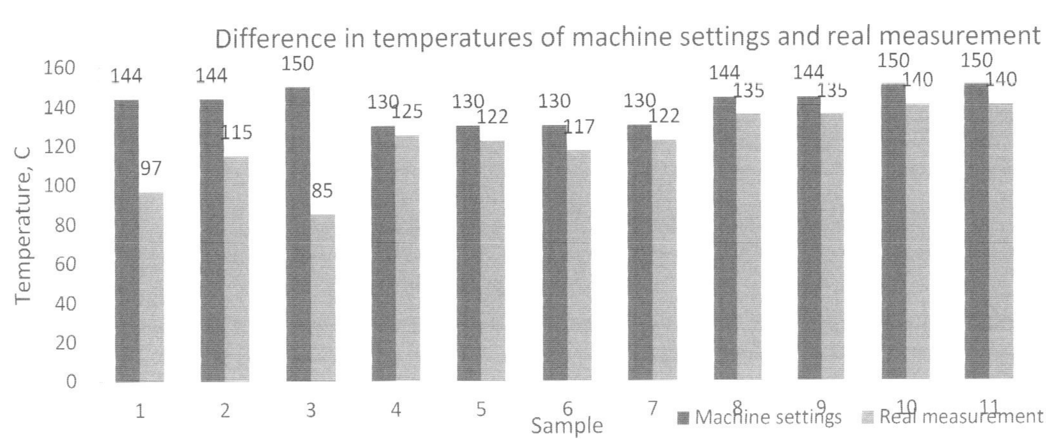

Figure 3. Difference between machine measured temperature against real measurement from module edge.

Total of 11 samples were sent to the laboratory testing. Unfortunately, only four of those are having trustworthy results (see Table 1).

Table 1. Gel content. Experimental data

Temperature by external sensor (°C)	135	135	140	140
Processing time, (sec)	1320	870	870	1320
Gel content (%)	67,9	74,9	62,2	82,5

Other experiments cannot be considered due to the fact that gel content percentage was too low (less than 50%). Obviously, additional gel content tests are needed.

3. Mathematical modelling of the quality of encapsulant

The workgroup has long time experience on adaption of AI tools for wide class of engineering problems [10-11]. Herein, the feedforward artificial neural network (ANN) model with one hidden layer was adapted for modelling gel content. Such an approach provides required accuracy if dataset is trustable and big enough. However, due to limited trustable dataset available from experiments at current time, the final tuning of the ANN is not yet performed (determining optimal number of neurons in hidden layer, adjusting weights). ANN has hierarchical structures and is powerful tool for modelling various problems. However, due to fact that it is based on random generation of initial weights, its application is complicated in the case of limited dataset available. The full factorial design of experiment is performed using at least four levels for both variables. Corresponding test are planned, but trustworthy results not guaranteed, due to complex measurements required.

For this reason, the authors introduce also one new and interesting alternate mathematical model – Haar wavelet based approximation [12]. This model is deterministic, does not include uncertainty and can be utilized in the case of limited dataset. The Haar wavelet expansion based 2D mathematical model is introduced as

$$f(x,y) = \sum_{i=1}^{2M} \sum_{j=1}^{2M} a_{ij} h_i(x) h_j(y) \qquad (1)$$

where the function $f(x,y)$ stand for the gel content, x and y for the temperature and processing time, respectively. The a_{ij} are unknown coefficients, h_i (also h_j) are the Haar functions defined as

$$h_i(x) = \begin{cases} 1 & for \quad x \in [\xi_1(i), \xi_2(i)) \\ -1 & for \quad x \in [\xi_2(i), \xi_3(i)) \\ 0 & elsewhere \end{cases} \qquad (2)$$

where $i = m + k + 1$, $m = 2^j$ is the maximum number of square waves deployed in interval $[A, B]$ and the parameter k indicates the location of the particular square wave,

$$\xi_1(i) = A + 2k\mu\Delta x,\ \xi_2(i) = A + (2k+1)\mu\Delta x,\ \xi_3(i) = A + 2(k+1)\mu\Delta x,\ \mu = M/m,\ \Delta x = (B-A)/(2M),\ M = 2^J \qquad (3)$$

In (3) j=0,1,...,J and k= 0,1,...,m−1 stand for dilatation and translations parameters, respectively. According to higher order Haar wavelet method in (1), the function $f(x,y)$ is replaced with its n-th order derivative, where n=1,2... Latter approach is based on higher order Haar wavelet method introduced recently by workgroup and provide higher accuracy/convergence rate [12], but require extra n test points for determining complementary integration constants.

Both mathematical models, described above, can be utilized for prediction as well as further optimization of the gel content value utilizing traditional gradient based and global optimization methods [13-18]. In the case of limited dataset the wavelet based approximation can be preferred since ANN approach uses random and may lead to different results in different runs is dataset is not satisfactory. The Haar wavelet approximations are commonly treated for uniform mesh. In the case of experimental study not all results may be available for applying uniform mesh. This means that widely used Haar matrices and its integrals derived for uniform mesh cannot by utilized. However, the Haar functions can be evaluated in any points based on simple formula (2). Thus, the increase of complexity is not significant.

4. Summary

The external measurement equipment has been elaborated for measuring temperature in real time. Furthermore, it has been observed that real measured temperature from module and the temperature shown by lamination machine differs. The temperature and process duration are considered for modelling quality of the gel content. The two mathematical models, feedforward ANN and higher order Haar wavelet model, are developed. In order to refine and validate these models, an additional test data should be acquired.

In further study it is planned to measure the pressure/vacuum conditions directly from the lamination chamber with no relying on machine data, to embed a wireless sensor inside the PV module [16].

Acknowledgement
Innovation Framework for Challenge Oriented Intelligent Manufacturing (INforM) founded by the INTERREG programme; and the Estonian Centre of Excellence in Zero Energy and Resource Efficient Smart Buildings and Districts, ZEBE, grant TK146 funded by the European Regional Development Fund.

References

[1] Sell R, Otto T 2008. Remotely controlled multi robot environment. *Proc. of 19th EAEEIE Annual Conf.*: 19th EAEEIE Annual Conference, Tallinn, Estonia, June 29 - July 2, 2008. Tallinn, pp 20–25. DOI: 10.1109/EAEEIE.2008.4610152.
[2] Kuts V, Otto T, Tähemaa T, Bukhari K, Pataraia T 2018. Adaptive industrial robots using machine vision. ASME 2018 International Mechanical Engineering Congress and Exposition, 2: IMECE2018, November 9-15, Pittsburgh, PA, USA. ASME. DOI: 10.1115/IMECE2018-86720.
[3] Judi HM, Jenal R, Genasan D 2011 Applications and Experiences of Quality Control. *Quality Control Implementation in Manufacturing Companies: Motivating Factors and Challenges* (InTech) chapter 25 pp 495-508. DOI: 10.5772/15997.
[4] Fraunhofer Institute for Solar Energy Systems, ISE, Photovoltaics report (Sept. 16, 2020).
[5] Honsberg C, Bowden S 2014 " Solar Cell Structure" [Online]. Available: https://www.pveducation.org/pvcdrom/solar-cell-operation/solar-cell-structure. [Accessed 03.12.2020].
[6] Satpathy R, Pamuru V 2021 *Solar PV Power* ed R Satpathy and V Pauru (Academic Press) chapter 5 pp 135–241.
[7] Jaunich M, Böhning M, Braun U, Teteris G, Stark W 2016 Investigation of the curing state of ethylene/vinyl acetate copolymer (EVA) for photovoltaic applications by gel content determination, rheology, DSC and FTIR. *Polymer Testing*, 52 pp 133-40.
[8] Chinnadurai T, Nalajam P, Vendan AS 2018 Analysis of mechanical and thermal behaviors for cross linked ethylene vinyl acetate (EVA) protective film employed for PV cells. *Materials Today: Proceedings*, 5, 11, Part 2 pp 23369-74.
[9] Karjust K, Kruuser K, Tšukrejev P, 2019 Production monitoring system development for manufacturing processes of photovoltaic modules. *Proc. of the Estonian Academy of Sciences*, 68 (4), pp 401–6. DOI: 10.3176/proc.2019.4.09.
[10] Kaganski S, Majak J, Karjust K 2018 Fuzzy AHP as a tool for prioritization of key performance indicators. In: *Procedia CIRP* pp 1227–32. 1st CIRP Conference on

Manufacturing Systems (CIRP CMS 2018). Stockholm, Sweden. DOI: 10.1016/j.procir.2018.03.097.
[11] Paavel M, Karjust K, Majak J 2017 PLM Maturity model development and implementation in SME. *Procedia CIRP,* 63, The 50th CIRP Conference on Manufacturing Systems, Taiwan, *3-5 May.* Ed. Tseng, M. Elsevier , 651–657. DOI: 10.1016/j.procir.2017.03.144.
[12] Majak J, Shvartsman B, Ratas M, Bassir D, Pohlak M, Karjust K, Eerme M 2020 Higher-order Haar wavelet method for vibration analysis of nanobeams. *Materials Today Communications*, 25, #101290. DOI: 10.1016/j.mtcomm.2020.101290.
[13] Zhu J, Zhang W, Xia L, Zhang Q, Bassir D 2012 Optimal packing configuration design with finite-circle method. *Journal of Intelligent and Robotic Systems: Theory and Applications* 67, 3-4 pp 185-199.
[14] Zhang WH, Domaszewski M, Bassir H 1999, Developments of sizing sensitivity analysis with the ABAQUS code. *Structural Optimization* 17, 2 pp 219-225.
[15] Guessasma S, Bassir D, Hedjazi L 2015 Influence of interphase properties on the effective behaviour of a starch-hemp composite. *Materials and Design* 65, pp 1053-1063.
[16] Nahas M, Alzahrani M 2020 Optimal Stochastic Distribution of CNTS in a Cantilever Polymer Microbeam Using Artificial Neural Networks. *Mechanics of Composite Materials* 56, pp 665–672.
[17] Cui D, Li DK 2019 Optimization of Extension-Shear Coupled Laminates Based on the Differential Evolution Algorithm. *Mechanics of Composite Materials* 54, pp 799–814.
[18] Plotnikova SV, Kulikov GM, Shape Control of Composite Plates with Distributed Piezoelectric Actuators in a Three-Dimensional Formulation. *Mechanics of Composite Materials* 56 pp 557–572.

Detection of cracks in green products of powder metallurgy by means of laser vibrometry

A Tatarinov, V Kurtenoks and V Mironovs

Laboratory of Powder Materials, Riga Technical University, 6B Kipsalas Street, LV-1048, Riga, Latvia

E-mail of corresonding author: aleksejs.tatarinovs@rtu.lv

Abstract. Cracking of green parts in powder metallurgy (PM) is a big concern in PM industry. Detection of cracks in green parts before sintering can prevent recycling of already sintered parts or production of defective ones. As opposed to sintered parts, where the presence of cracks is well revealed by impact spectroscopy, it is challenging for green parts. The reason is an extremely low mechanical quality factor of these parts composed of compacted but non-sintered metal powder particles. Low mechanical integrity of particles causes the absence of pronounced resonances of the parts. The aim of the study was to test a possibility to discerning healthy and cracked green parts using spectral characteristics of signals obtained by laser vibromery. Four similar green PM gear wheels made of steel powder were obtained from a PM manufacturing company, two of them were healthy and two with simulated cracks that appeared in the area of stress concentration between the hub and the disc. The experimental setup included a Polytec PSV-500 scanning laser vibrometer and a Bruel&Kjer vibration stand type 4824. Integrated spectrograms in the frequency range 50 – 3500 Hz were recorded upon a 95-point network on the gear surface at different vibration intensities proportionally graded in the range from 1 to 5. Comparison of the 3D (frequency-intensity-amplitude) vibration spectrograms showed that the defected specimens with cracks differed from the healthy ones in the following features: resonance frequency shift and /or splitting of resonances when the load changes. The study confirmed the principal possibility of cracks detection in green PM parts using laser vibrometry regardless the crack location in the part. The aim of further studies is to determine the sensitivity limit in terms of crack size. Keywords: non-destructive testing, powder metallurgy, green parts, cracks, laser vibrometry.

1. Introduction

Powder metallurgy (PM) makes it possible to obtain products with unique physical and mechanical properties, increasing the utilization rate of metal and reducing the mechanical processing of products to a minimum [1]. PM products are used to operate in responsible unites of machines and structures under a wide range of extreme mechanical loads, thus, a strong defect-free requirement applies to these parts. One of the most dangerous causes of destruction is the formation of cracks that can appear at the stage of pressing the product, i.e. at the green or unsintered state, followed by their growth during sintering and finishing operations. The main causes of green-state cracks can be summarized into four basic categories: improper material composition such as a lack or excess of lubricant and the presence of impurities; particles side shifting with an improper compacting technology; improper elastic strain release in the presence of plastic deformations; and high tensile/shear stresses imposed by external or internal factors during manufacturing [2]. It is desirable to detect cracks even at the early green state that can prevent the production of defective parts. In this case, there is no need of labor-intensive and energy-intensive operations of re-processing of defective sintered products. Although the problem of cracking diagnostics of sintered parts has been more or less solved by impact resonance spectroscopy, including in-line commercial installations [3], cracks detection in green parts still remains challenging. This is due to two reasons: firstly, the inapplicability of impact to green parts that can damage them, and secondly, an extremely low mechanical quality factor of these parts composed of compacted but non-sintered metal powder particles with weak bonds between them. Despite considerable efforts, traditional non-destructive testing (NDT) methods have been largely unsuccessful at defects detecting in green PM parts or showed a limited applicability in real production environment. Electrical resistivity method was realized by the voltage distribution measurements on the surface by an array of spring-loaded needle contacts [4]. The surface wave mediator technique used point contact to induce and receive Rayleigh acoustic surface waves in a megahertz frequency range that was sensitive to surface breaking cracks [5]. Another tested option was realized in the same work by ultrasonic electromagnetic acoustic transducers inducing shear horizontal waves and torsional waves. Increased attenuation of high frequency acoustic waves in green PM parts forced to switch to lower frequencies than traditionally used in ultrasonic NDT of metal objects [6]. Enhanced image processing techniques applied to digital radiographic images allowed recognition of a crack in a green PM compact that was virtually invisible in the original digital x-ray picture [7]. Summarizing the said approaches, it can be concluded that despite the demonstrated positive effects, NTD of green PM parts is still complicated by one or a few of the following reasons: a lack of sensitivity; measurement only in a local accessible plot; impermissible impact or contact action; huge and slow operating equipment and/or inability to adapt the equipment to in-line testing.

The purpose of this study was to assess the possibility of detecting cracks in green PM products by the spectral characteristics of signals obtained by laser vibrometry. The laser Doppler vibrometry [8] was chosen taking into account such its advantages as remote sensing that avoids the necessity of a direct contact with a PM part and a gentle vibration effect on the test object. Besides, it is adaptable to objects of different size and complicated shape. Despite the relatively high cost of laser vibrometry equipment, it can be quite well built into in-line testing and can pay off in bulk mass production.

2. Materials and methods

2.1. Examined PM parts.

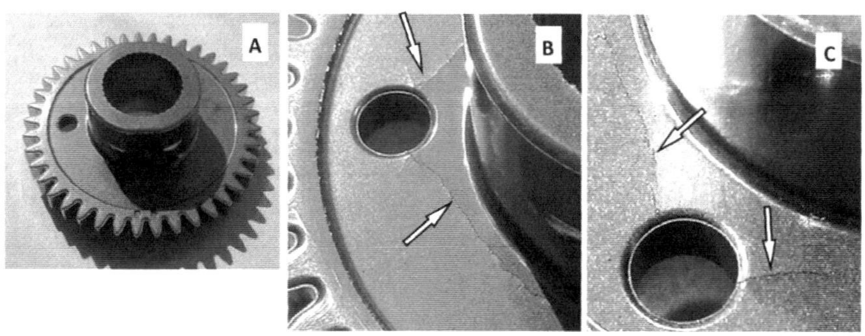

Figure 1. Examined PM gears: A – product general view; B, C – simulated cracks in gears' bodies denoted by arrows.

The examined PM parts were four similar green PM gear wheels made of steel powder with a diameter of 80 mm. The specimens were obtained from a large PM manufacturing company, Miba Sinter Slovakia, S.R.O. Two specimens were healthy and other two ones were artificially damaged with simulated cracks. The cracks were induced by a moderately light blow and appeared in the area of stress concentration on the disc by the trajectory connecting a hole between the hub and the disc (Figure 1).

2.2. Laser vibrometry.

The experimental laser vibrometry equipment consisted of a single-channel laser vibrometer and a vibrometer stand depicted in Figure 2.

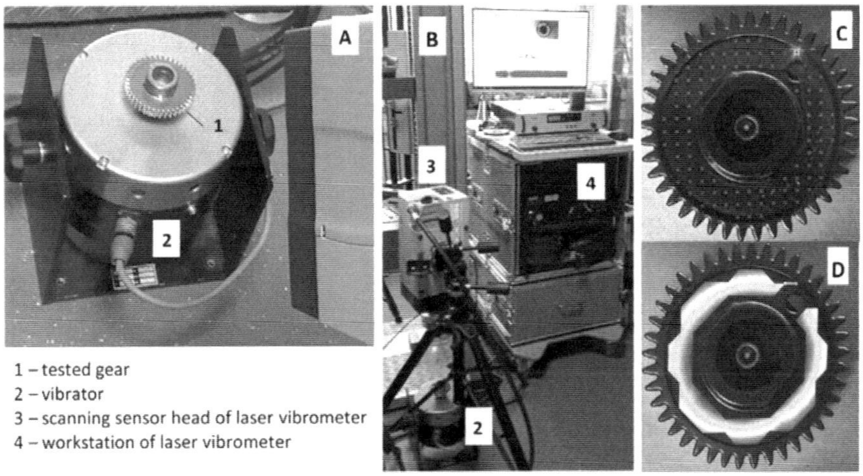

Figure 2. Illustrations of experiment: A – tested gear on the vibrator's platform; B – general view of equipment, C – gear with imposed grid of measurement points; D – gear with imposed virtual oscillation pattern at a resonance.

The laser vibrometer was a Polytec Scanning Vibrometer PSV-500 in 1D optical configuration providing fully automated vibration measurements in the range from DC to 100 kHz. The Polytec vibrometer system is a laser Doppler vibrometer determining the vibration velocity and displacement at a point by sensing the frequency shift and phase variation of back scattered laser light from a moving surface. The scanning head with a high precision scanner and HD video camera helped recording of vibration spectra upon a 95-point network on the gear surface (Figure 2C). and presenting the integral spectrum upon all 95 points as a result. The vibrometer was a Bruel&Kjer electrodynamic Modal Exciter Type 4824 (Figure 2A) intended for modal testing in structural dynamics investigations. It operates in a frequency range from DC to 5 kHz providing sine force rating up to 100N. The selected frequency range was from DC to 4000 Hz since the resonances of natural vibrations of the tested specimens were exactly in this range. The frequency step of the spectra acquisition was 1 Hz. As a result, an integral spectrum was presented for all 95 points of the measurement network on the gear surface. To assess the dependence of the spectral characteristics on the intensity of vibration, the spectra were taken at 5 power levels set by the gradually increased excitation voltage of the vibrator from 1 to 5 V with a step of 1 V.

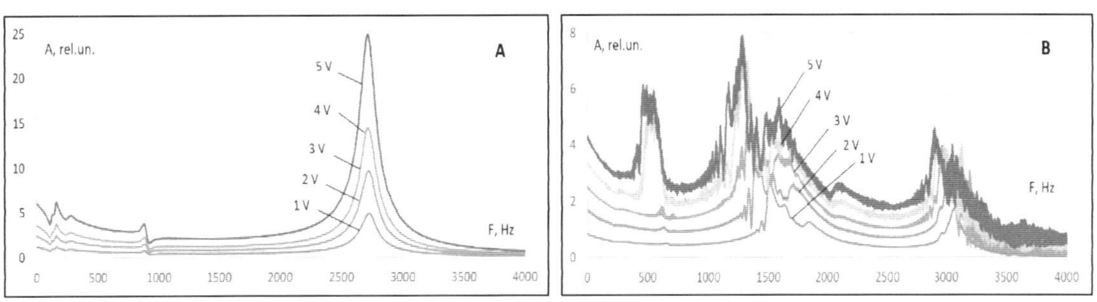

Figure 3. Vibration spectra of empty platform (A) and the platform loaded by a PM gear (B) at 5 levels of vibration intensity controlled by excitation grades from 1 V to 5 V.

Figure 3 represents examples of integral vibration spectra of the unloaded vibrator with empty platform (A) and the same platform loaded with a PM gear specimen from the group. The amplitudes of obtained spectra are proportional to 5 gradually increasing levels of vibration intensity.

3. Results and discussion

The main results are visually presented in Figure 4. It shows the integral spectra of the empty platform and the platform loaded by healthy (H1, H2) and damaged (D1, D2) specimens in the frequency range from 500 to 3500 Hz. The spectra are presented as B-scans, where brightness decodes the spectrum amplitude. One line corresponds to the step of vibration intensity, at which the spectrum was obtained. The spectra are normalized by the peak amplitude. Areas denoted as quadrangles A and B in two-dimensional frequency-intensity patterns outline regions-of-interest, where differences between healthy and damaged specimens are the most expressed.

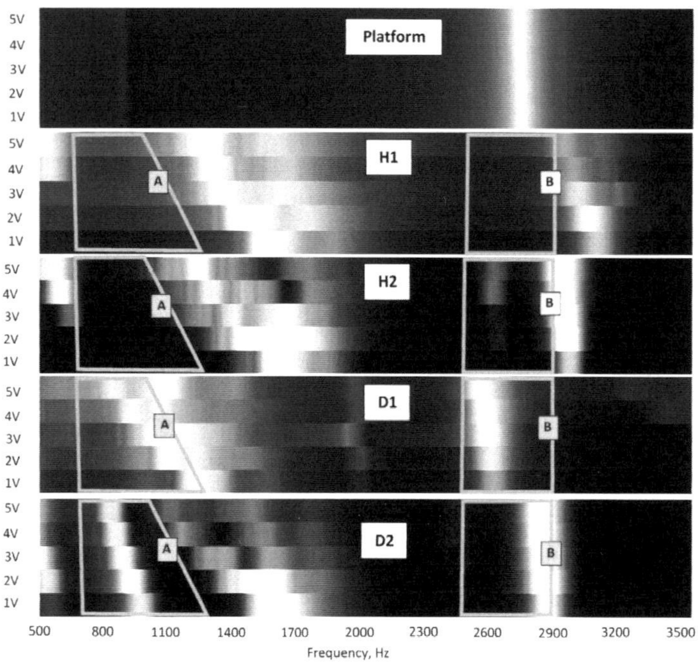

Figure 4. Vibration spectra of empty platform and the platform loaded by healthy (H1 and H2) and damaged (D1 and D2) PM gear specimens at 5 levels of vibration intensity from 1V to 5 V. The spectra are presented as B-mode scans with normalized amplitudes within corresponding vibration intensity levels. Regions-of-interest in frequency-intensity field are denoted by quadrangles A and B.

The differences between healthy (H1, H2) and damaged (D1, D2) specimens are clearly identified in the presented frequency-intensity patterns in Figure 4. There are two main manifestations that separate groups H and D. The first one is the frequency shift towards lower frequency values in damaged specimens with cracks that is shown by appearance of bright plots in quadrangles A and B, the most expressed in the case of D1. The second one is the splitting of spectral peaks and appearance of additional resonances as shown in the case D2, quadrangle A.

In both cases of healthy and damaged samples, changes in the spectra are observed depending on the vibration intensity. An exception is the case with the empty platform, which is stiffly connected with the vibrator. It has a strong resonance at about 2750 Hz, which is invariant with respect to the vibration intensity. Obviously, the main source of nonlinearity in our experiment is the absence of a rigid coupling between the PM specimen and the platform. Nevertheless, this circumstance does not interfere with the identification of differences between healthy and damaged specimens. At the same time, the frequency shift as a function of vibration intensity imposes demands on taking this factor into account when planning testing.

Two-dimensional frequency-intensity patterns formed by vibration spectra at gradually increased levels of vibration intensity demonstrated a good opportunity to use it as a source material for diagnosing the state of PM of parts using the principles of pattern recognition and machine learning. Regions-of-interest in such patterns can be predetermined using a database for a specific type of PM product, obtained in notoriously known healthy and damaged specimens.

With regard to the applicability of the method in production, the key success criterion is the speed of testing products and the rejection of defective ones causing minimal impact. Despite the fact that a complete scanning of the object surface in the current experiment on a multipoint grid took a few minutes, the vibrometry test using the integral spectrum of the object can be reduced to a few (5-6) seconds allowing in-line quality assurance in mass production. Further industrial studies are necessary to determine the ability of the method to detect small inner cracks at different location and orientation within the object on the background of random deviation of spectra in a series of similar healthy objects of different size and shape.

4. Conclusion

The study confirmed the principal possibility of cracks detection in green PM parts using laser vibrometry regardless the crack location in the part and using integral vibration spectra upon a PM part as a function of vibration intensity. Two-dimensional frequency-intensity patterns formed by vibration spectra at gradually increased levels of vibration intensity can serve as a source material for diagnosing the state of PM of parts using pattern recognition. The aim of further studies is to determine the sensitivity limit of the approach in terms of crack size and location, as well as the ability of the method to test objects of complex structure.

References

[1] German R M 1994 Powder Metallurgy Science (Princeton, NJ: Metal Powder Industries Federation)
[2] Zenger D C, Cai H, McNeil J and Ludwig R 1997 Classification strategy to identify and classify common cracks in green-state powdered metallurgy compacts *Review of Progress in Quantitative Nondestructive Evaluation* vol 16B (New York: Plenum Press) pp 1443-1450
[3] Miller P 2020 Next-level defect detection with advanced Acoustic Resonance Inspection *Powder Metallurgy Review* vol 9 (3) Autumn/Fall pp 55-62
[4] Bogdanov G, Ludwig R and Michalson W R 2000 A new apparatus for non-destructive evaluation of green-state powder metal compacts using the electrical-resistivity method *Measurement Science and Technology* vol 11(2) pp 157-166
[5] Zhu Y, Hauck E T and Rose J L 2006 Novel methods for crack detection in green and sintered parts *AIP Conference Proceedings* vol 820 pp 1507 – 1514
[6] Tatarinov A, Mironov V, Rybak D, Stankevich P. 2017 Non-destructive testing of joints of antifriction parts crimped by pulsed magnetic deformation. *Solid State Phenomena* vol 267 pp 248-252
[7] Ponomarev MG, Selcuk C, Gan T-H, Amos M, Nicholson I, Iovea M, Neagu M, Stefanescu B and Mateiasi G 2014 Defect detection and classification system for automatic analysis of digital radiography images of PM parts *Powder Metallurgy* vol 57(1) pp 17-20
[8] Rothberg S J, Allen M S, Castellini P, Di Maio D, Dirckx J J, Ewins D J, Halkon B J, Muyshondt P, Paone N, Ryan T, Steger H, Tomasini E P, Vanlanduit S and Vignola J F 2017 An international review of laser Doppler vibrometry: Making light work of vibration measurement. *Optics and Lasers in Engineering* vol 99 pp 11–22

Acknowledgement

The authors express their gratitude to representatives of Miba Sinter Slovakia S.R.O. for providing green PM specimens for testing and Mr. Eduards Skukis, a researcher at the Institute of Materials and Structures of the Riga Technical University, for his kind help in carrying out experiments using laser vibrometry.

The effect of UV-C radiation on the durability of 3D printed plastic parts in disinfectant devices

T Tähemaa[1*], M Sarkans[1], I Sarand[1], M Pohlak[1], A Niidas[1] and M Saarna[1]

[1] Department of Mechanical and Industrial Engineering, Tallinn University of Technology, Estonia, Harju, Tallinn, Ehitajate str. 5, 19086

*toivo.tahemaa@taltech.ee

Abstract. The usage of far or hard ultraviolet radiation in wavelength 200 – 280 nm or shortly UV-C disinfectors are becoming more and more popular due to their universal ability to fight against bacteria and viruses. It is well known that according to DIN 5031-7, both UV-A and UV-B radiation from sunlight have clear influence on the mechanical and visual properties of plastic parts, but the influence of intensive artificial UV-C is not so widely investigated. As the UV-C is in the shorter end of the UV radiation wavelength spectrum and is completely absorbed by the ozone layer and atmosphere, it is not possible to see the influence of it in everyday life. But in the equipment where the artificial UV-C radiation wavelength is used as a disinfectant, and probably in the open space, the influence of UV-C could be remarkable. Recently, there has been increasing interest in the food industry to avoid contamination in the food packing process. The global COVID-19 coronavirus pandemic has amplified this interest and has forced authors to run a set of tests with printed plastic parts. This paper analyses the outcome of those tests including two different printing directions and two different surface paints. The study shows that there is considerable influence of UV-C radiation on mechanical and aesthetic properties of 3D SLS printed PA12 parts. However, the influence can be diminished by use of appropriate paints.

1. Introduction

3D printing has made continuous growth already throughout many decades [1]. The need for high quality printed parts is still growing and many different areas of use have been implemented, as printed parts are used in final products, not just prototypes. In the era of the COVID-19 pandemic, UV-C disinfection devices have become increasingly popular. However, if 3D printing is also increasingly used in UV-C disinfection devices, new challenges are faced. Different parts like lamp holders, filter housings, shutters etc. can be made of polymers printed with different 3D printing technologies. Studies show that for equivalent incident dosages, UV-C degrades human stratum corneum (the most superficial skin layer) the most [2], so those devices are classified as sources of danger regarding ISO 15858:2016. Therefore, designers of the devices should pay attention and follow all mechanical properties of those parts and materials. And one should follow it during the whole lifecycle of the device, even if the parts are used in prototype making stage. It is understandable that inside the sterilization device near the UV radiation source UV-C radiation level can be very high. As the authors of the current paper are involved in UV-C disinfection cabinet design group the question arose how much we can trust those printed parts in critical places. As first 8 hours in small and light UV-C cabinet did change remarkable the color of high-quality 3D SLS printed part reasonable suspicion occurred, that mechanical properties of printed parts may change as well.

First, main topic of current paper is 3D printing as production method. Two most popular 3D printing technologies are introduced and compared here shortly. We should consider that models for mock-ups and product visualization purpose only are made with simplest printing solutions like Fused Filament Fabrication (FFF)/ Fused Deposition Modeling (FDM). Mechanical properties of FFF technology are weak and surface quality is poor. On the other hand, printing device itself has significantly lower cost and it has a way smaller size as well. Another positive side of FFF/FDM printing technology is that different colors can be combined in order to create colorful parts. The method also uses an exact amount of plastic filament required for the needed detail, meanwhile in Selective Laser Sintering (SLS) methods where plastic powder waste amount is larger. [3] But if we compare from the component quality side then all parameters are better in case of SLS technology. To avoid the temptation to use only FFF printing the Plastic Selective Laser Sintering (PSLS) printing method as a more exact, finer, and more suitable for UV device parts, should be deeply investigated. In current tests only SLS method is used for specimen printing.

Second topic of current paper is UV radiation and its feature to influence polymers. Most polymer parts tend to be affected by UV rays (100…400 nm range of el. magnetic wave), and after a certain time it is noticeable that their surface becomes chalky, brittle and its color changes.

In terms of the components more likely to be at risk of UV damage, all outside parts are endangered. As it was mentioned before, a change of the material's surface layer takes place. But in some cases, if the object is damaged by UV, it may also cause complete component breakage.

If we look more deeply into UV types, one can acknowledge that based on the most investigated material- skin – UV-A (wavelengths 320 to 400 nm) causes aging and tanning of the complexion, while UV-B (wavelengths 280 to 320 nm) may mean it burns. The common effect of UV-C (wavelengths 200 to 280 nm) is germicidal, which kills or inactivates microorganisms by destroying nucleic acids and disrupts their DNA [4].

Our hypothesis is that if we can prove that UV-C influence on mechanical properties of SLS printed parts (with or without paint layer) is not remarkable then those parts have the advantage of working properly and lasting the whole lifecycle of the UV device.

A lot of different research has been carried out with different UV types and for different purposes [5] [6] [7] [8]. However, the data concerning combination of 3D printed plastic parts and UV-C radiation is missing or has not been published yet.

2. Materials and methods, test specimen production for UV-C tests

1.1. The research question is: how does UV-C radiation affect mechanical, tribological and the color properties of SLS printed parts? SLS 3D printer FORMIGA P100 (EOS GmbH) was used to produce all test pieces for this study. The main principle of SLS process is depicted in Figure 1.

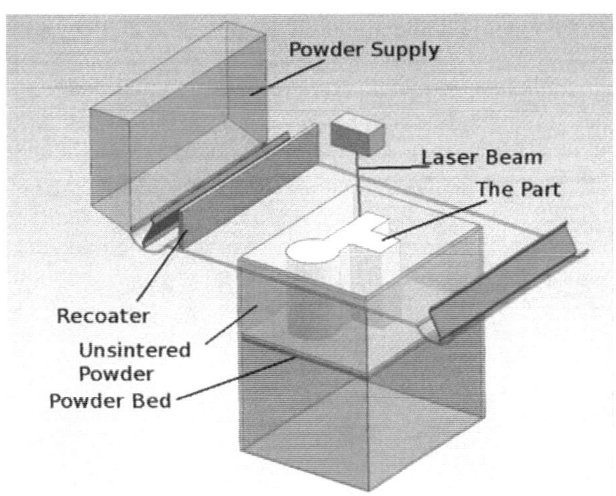

Figure 1. Selective Laser Sintering process

Manufacturer of the powder is EOS GmbH. Powder average grain size: 56 μm (ISO 13320-11). Preheating temperature during the process is 171 C. The layer thickness 0.1 mm and laser power 30 W (CO2 laser) was used. Diameter of focused beam: <0,5 mm. In the process, 3D object is built by sintering powder with laser beam, section by section. The Polyamide 12 (PA2200) powder in the build platform is maintained at an elevated temperature just below the melting point and/or glass transition temperature of the powdered material. The part is supported in the process by unsintered powder. There are four different fusion mechanisms which are present in such AM processes [9]. These include solid-state sintering, chemically induced binding, liquid-phase sintering, and full melting. Most commercial processes utilize primarily liquid-phase sintering and melting [9]. Properties of the parts along layers and perpendicular to layers should be similar but in reality, they can differ considerably.

As experience shows, UV-C light changes the properties of plastic parts, including parts produced with SLS. Polyamide 12 (PA220) as the most universal material for Formiga SLS printer has proven its durability in many applications. Inside of the UV-C devices under radiations influence all material properties may change. The first and most noticeable change is in its color if parts are not covered with paint layer. However, it is not fully known how the radiation coming from UV-C disinfection systems affects the properties of the printed parts. It is also not known how the properties change depending on build position of the parts.

To study this, several tests have been made. Firstly, tensile tests to study global strength properties. Test results are discussed in chapter 3. Secondly, an impact test to focus on possible brittleness effects of UVC processing. Test results are discussed in chapter 3.1. Thirdly, surface color evaluation is discussed in chapter 3.2. Some of the tests were made on parts printed in different directions (the layer position) but vertically positioned specimens measurement uncertainty was too large [10], so more extensive study is needed, and therefore the discussion concerning those tests is skipped here.

The impact test pieces were printed accordingly to standard EN-ISO 179-1:2010 (80x10x4 mm) and tensile test pieces were printed accordingly to standard ISO 527-1:2019 (150x20x4 mm). A total of 5 test specimens were printed for each test type. Temperature on the working table during the tests was 23 °C.

2.1. Test specimen surface processing

The two following paints were used to cover the surface of SLS parts: craft spray for plastics "Chrome effect" (Motip) and Dicco Color Black (Tikkurila OY). The paints were carried onto the SLS specimens by spraying ("Chrome effect") and dipping (Dicco Color Black). First, widely used "Chrome effect" was used as one layer of spray paint. The porous surface of the specimen was very suitable (Figure 2. a) for that kind of vertically positioned and dried spray-painting technology. The painted surface looked excellent as its layer covered firmly and evenly all surfaces of the test specimen.

Figure 2. Cross section of destructed specimen of impact test with
a, chrome effect - b, black stain paint

Second paint, solvent based stain, produced by Tikkurila OY contains lightfast dyes and its viscosity is very low. The porous SLS printed specimen was dipped for 5 seconds into the black stain. Drying took place in a horizontal position on a sheet of office paper. Absorbed layer of stain is shown in Figure 2. b.

Even though it is meant for wooden materials its black matte finish looks also perfect and carefree for most PSLS details.

Finally, as known from the literature [11] that polyamide absorbs water, and the absorbed water tends to act like a plasticizer and can have a significant effect on the plastics properties, it was decided to test the mechanical properties of water dipped specimens as well. In addition, a batch of specimens were dipped into the water for 300 hours and tested without drying.

3. Results and discussion

Tensile test results.

Tensile tests have been carried out on Tinius Olsen H10KT tension-compression testing device in combination with extensometer Epsilon 3542-025M-050-ST. Preload was 0 N and speed 5 mm/min. Extensometer was used to observe precise deformation of specimen. The results are shown in Figure 3.

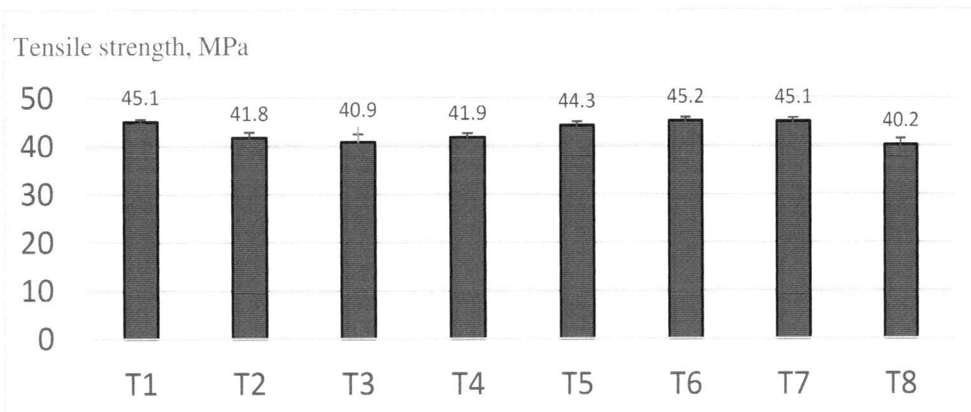

Figure 3 Comparison of different specimen groups.

The coding of specimens shown in Figure 3. is as follows:

T1 Reference test specimen, not UV-C treated, not painted
T2 Medium (200 h) UV-C treated specimen, not painted
T3 Long (300 h) UV-C treated specimen, not painted
T4 Reference specimen for painted tests, not UV-C treated, chrome effect spray paint
T5 Reference specimen for painted tests, not UV-C treated, black dipped stain
T6 Long (300 h) UV-C treated, chrome effect spray paint
T7 Long (300 h) UV-C treated, black dipped stain
T8 Water treated specimen, not UV-C treated and not painted

In first group (T1 to T3) small 10 % change in tensile strength could be noticed. Second group, painted specimens from group T4 and T5 are approximately 5% weaker than the reference specimen. Absorbed paint shows, however, that the reason for those discords could be small cracks on the surface of specimens, Figure 2. b.

It is clearly seen (Figure 2) that objects shown in the figures above have been printed horizontally. It is especially visible in the black stain, Figure 2. b.

Third group, T6 and T7, long UV-C and painted surface has almost the same values of tensile strength as the reference test specimen had. Based on that fact we can assert that painted surface has influence to the tensile strength, paint protects the 3D printed part against UV-C rays.

And lastly, the water treated specimen shows that in wet conditions we lose 10 % of the part's tensile strength under UV-C radiation.

3.1. Impact test results

Impact test has been carried out on ZWICK Charpy impact test pendulum device. Distance between specimen supports was 62 mm and 4 different impact energy pendulums (4, 2, 1 and 0,5 J) were used regarding wide variety of needed impact energy. Results are presented below in the Figure 4.

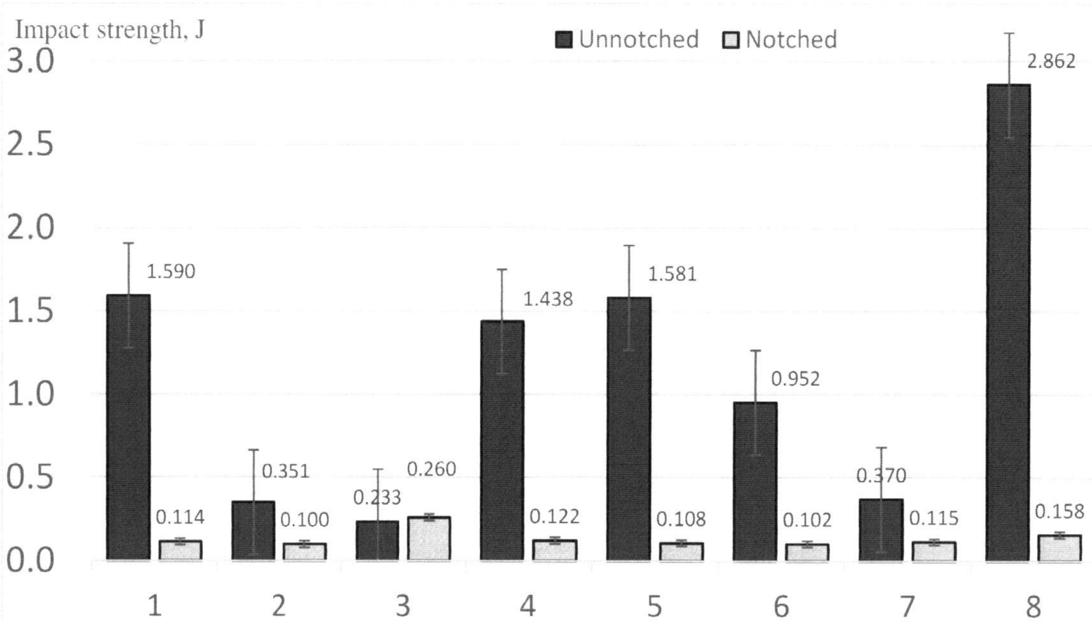

Figure 4. Impact test results, notched specimens are placed horizontally, unnotched vertically.

The coding of specimens shown in Figure 4 is as follows:

1 Reference test specimen, no UV-C, not painted
2 Short (100 h) UV-C treated specimen, not painted
3 Medium (200 h) UV-C treated specimen, not painted
4 Reference specimens for painted test, not UV-C treated, chrome effect spray paint
5 Reference specimens for painted test, not UV-C treated, black dipped stain
6 Medium (200 h) UV-C treated, chrome effect spray paint
7 Medium (200 h) UV-C treated, black dipped stain
8 Water treated specimen, not UV-C treated, not painted

We can see same trends in first group (1, 2, 3) where UV-C treatment has good correlation with impact test results, but the difference is much bigger than in tensile tests, so UV-C treated parts are 5 to 6 times more fragile. The paint layer barely affects the results, as groups 4 and 5 have similar results with reference group no. 1. But in impact test we can see that the paint layer does not protect plastic properties as much as it protected in tensile tests. Chrome effect spray paint gives quite good results, but the black dipped stain has almost zero protection against UV-C radiation.

3.2. Surface colour evaluation

Printed details have normally white matte color. After short, 8-hour UV-C treatment they turned yellow. Longer treatment did not change color remarkably. "Chrome effect" spray paint layer gave a good protection to the surface color. Black matte stain and dipped painting technology had less value in surface protection as the stain did not form protective layer.

4. Conclusion

After 120 valid tests we can declare that UV-C has similar effect to plastics as UV-A and UV-B rays usually have. Likewise, similar protection methods like painting avoids aging in UV-C radiated parts. It is important to emphasize that the influence of UV-C and water absorption is much higher to impact strength than to tensile strength. Also, as side discovery, dipped stain painting method gives us a valuable tool for diagnosing structural defects in 3D printed plastic parts. Those defects are easily detectable after destructive tensile or impact tests.

5. References

[1] Wohlers Report 2020, 3D Printing and Additive Manufacturing Global State of the Industry, ISBN 978-0-9913332-6-4

[2] Zachary W. Lipsky, Guy K. German. 2019, Ultraviolet light degrades the mechanical and structural properties of human stratum corneum. Journal of the Mechanical Behavior of Biomedical Materials, Volume 100, https://doi.org/10.1016/j.jmbbm.2019.103391

[3] Redwood B, Schöffer F, Garret B 2017, The 3D Printing Handbook: Technologies, design and applications, 3D Hubs B. V, Amsterdam

[4] ISO 15714:2019(en). Method of evaluating the UV dose to airborne microorganisms transiting in-duct ultraviolet germicidal irradiation devices

[5] Kallakas H, Poltimäe T, Süld T-M, Kers J, Krumme A 2015 The influence of accelerated weathering on the mechanical and physical properties of wood-plastic composites, Proceedings of the Estonian Academy of Sciences

[6] Gupta A 2005 Improving UV Resistance of High Strength Fibers. North Carolina State University

[7] Wang W, et al., 1998 "Surface nature of UV deterioration in properties of solid poly (ethylene terephthalate)", J. Appl. Polym. Sci. 67, 705-714

[8] Andreasena J W, Gevorgyana S A, Schlepützb C M, Krebsa F C, 2008 Applicability of X-ray reflectometry to studies of polymer solar cell degradation, Elsevier B.V.

[9] Gibson I, Rosen D W, Stucker B 2010, Additive Manufacturing Technologies: Rapid Prototyping to Direct Digital Manufacturing, Springer New York.

[10] Matsi B, Sonk K, Otto T, Roosimölder L 2009, Increasing of Rapid Prototyping Performance by 3D Printing Technologies, Journal of Machine Engineering 121-129

[11] McKeen L, W. 2019 Effect of UV Light and Weather on Plastics and Elastomers (4th Edition), Elsevier

Self-driving Shuttle Bus Use Case in City of Tallinn

K Kalda[1] R Sell[1] R-M Soe[2]
[1]Department of Mechanical and Industrial Engineering, Tallinn University of Technology, Tallinn, Estonia
[2]FinEst Twins Smart City Center of Excellence, Tallinn University of Technology, Tallinn, Estonia

E-mail: krister.kalda@taltech.ee

Abstract. Autonomous vehicles (AVs) are entering from test areas to the streets, which is one of the key components of smart cities and the future of mobility as a service (MaaS). In 2020, two AV services were operated in Tallinn. This paper focuses on one of the services, including how it was set up. The route was set up in Ülemiste City, a tech park with 10,000 people daily working in the area. It connected the offices with the airport and a shopping centre. Autonomous shuttle iseAuto was used for the service (streets with heavy traffic, including some complex crossings). Our findings associated with the Ülemiste experiment in the context of legal requirements set upon autonomous vehicles to be street legal are pointed out. Events that took place during the operation (including an accident) are addressed. Summary of the feedback from clients is presented. Further studies should focus on the extended concept of smart cities with a roadmap for the nearest future.

Keywords: Autonomous vehicle, smart city, self-driving vehicle

1. Introduction

Autonomous driving is revolutionizing transportation and will change the whole paradigm in near future. The autonomy is divided into 5 levels by SAE J3016 [1] where level 0 has no autonomy at all and level 5 means full autonomy in any conditions. Most of today's advanced and wellknown self-driving cars like Tesla autopilot, are in level 2 while low-speed autonomous vehicle (AV) shuttles are up to level 4. Last-mile (AV) shuttles have been gaining a lot of interest recently by cities to pilot new transport opportunities in the specific area or route to extend the main public transport lines. The pilot experiments allow us to introduce self-driving technology to citizens of the community and test their maturity in real traffic and urban environments. Tallinn University of Technology (TalTech) started to develop a self-driving autonomous last mile shuttle - iseAuto [2, 3] in June 2017 in cooperation with a private company Silberauto. By using open source software and a chassis of existing electrical vehicles it was feasible to achieve a fully functional AV shuttle with one year intensive development time only. The first prototype shuttle called "iseAuto" was ready a year later in 2018 for the university anniversary celebrations. The next logical steps taken were to get the vehicle into real traffic, with an ultimate goal to create a whole new autonomous traffic solution with a safer and environmentally sustainable mobility. Based on the prototype vehicle several more mature vehicles were created by the industrial partner AuVe Tech - the successor of Silberauto.

IseAuto 2020 has been developed to the level (according to both Estonian and EU rules) of being street legal. Simultaneously during the past summer months, two pilot testings in real traffic were created.
The focus was on the shuttle usage by all demographics and several groups of public transport users, whether for business or pleasure under two different kinds of density of traffic. In 2020, in Tallinn, two AV services were operated. Both projects were funded by European Commission, one called Sohjoa Baltic [4] and the other FABULOS [5]. In the Sohjoa project, which was run in a recreational park area of Kadriorg, a French Navya Evo shuttle was used, but in FABULOS, that had its Estonian leg of test area in Ülemiste City (a techpark with offices for thousands of people and Estonian international Tallinn Airport in the area) - the Estonian iseAuto. FABULOS brought together 4 partners - TalTech, Auve Tech, Fleet Complete and Modern Mobility. In Estonia, there were also some outside consortium partners like Ülemiste City, Ministry of Economic Affairs and Communications, Tallinn Transport Authority, Estonian Road Authority, Bercman Technologies, and Ericsson Eesti. In the Sohjoa project, a mobility acceptance factor was investigated [6] and specific feedback gathered. In this paper, the feedback is analysed and findings are presented.

2. Use Case in Ülemiste City

Cities around the world are experimenting with driverless autonomous vehicles, in particular AV shuttles to be prepared for future mobility. Several studies in the US [7] and Europe [8] are demonstrating that innovative cities are willing to deploy autonomous shuttles in the near future and citizens have mostly positive attitudes towards AV shuttles on the streets. Setting up the use case in Ülemiste as required by the project, several modifications and upgrades were necessary in order to get legal permission to drive on the public roads. So far iseAuto has been tested in closed areas only (e.g., on a dedicated road in the university campus), but the initial idea was to turn iseAuto also into the first street-legal autonomous vehicle in Estonia. This process generally is, and in this case was, closely observed by all different traffic authority parties - The Estonian Road Authority, the Police and Estonian Rescue Board. The legal environment is not yet mature for self-driving vehicles and differs country by country quite a lot even within Europe. Detailed analysis was conducted by the Sohjoa Baltic project which also included Baltic sea region legal environment [9]. First, they tested the hardware (like they do with every vehicle) whether the lights, windows, seats, tires etc are in adequate condition. Then test drives in designated testing spaces were conducted, where they followed how the safety person operates the shuttle in case of need and emergencies. For example, how the control of the bus from automated driving was taken over to manual control. The ability of the braking system to identify objects in different sizes while driving in automated mode was tested, and how it acts when it has problems with IT systems. Last but not least, also a driving exam was created in the area where iseAuto started to operate, to see all different situations in traffic and how it acts in reality on crossings and with other members both static and moving, including people. After tests and modifications according to the predefined rules of the authorities, the shuttles met the standards and were granted permission to enter the public road traffic as EU M1 category vehicle. In parallel, new intelligent functions like human-autonomous vehicle interaction methods [10] were experimented in TalTech campus by the Autonomous Vehicles research group.

The route was chosen due to its different kinds of crossings, turns and the density of both traffic and people to connect the offices of Ülemiste City with a large shopping mall and a hotel in the area and the Tallinn International Airport. The route and list of bus stops are shown in figure 1.

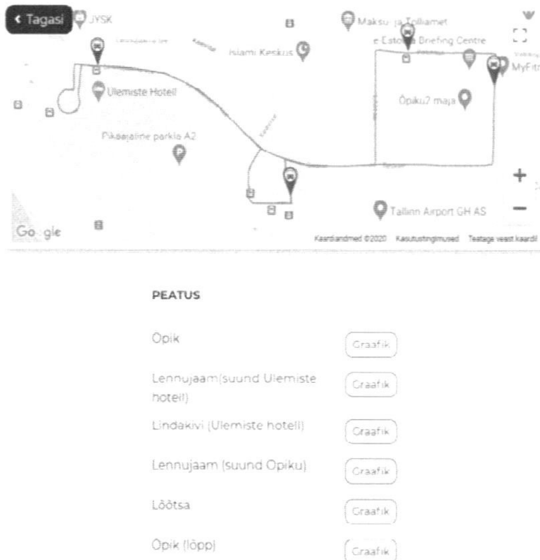

Figure 1. Internet interface for passengers about the map and stops in Ülemiste City

Three AV shuttles were dedicated to the service, with engineers and technicians working on the details daily to advance the busses and make them better for the service. Also, the operator's interface was developed on the accompanying tablet of controls connected to the controls of the shuttle. The important function under the testing was teleoperation, see remote operation workstation in figure 2. Teleoperation enables the vehicle to be taken over in case human supervision is needed. This happens usually when unexpected situations occur on the route and shuttles cannot handle them on their own.

Figure 2. Teleoperation engaged in Ülemiste remote operation station.

Shuttles were not provided any special separate lanes but were mixed with real traffic. The maximum speed allowed on the route was 30 km/h, but the speed was kept around max 20, just for the safety reasons. The drive included both left and right turns, crossing the streetcar and bus lanes prioritized for public transport, which serviced the Airport. At the same time, it was important not to forget the pedestrians, bikes and especially electrical scooters of the people working in a tech savvy environment. For passengers, it was free of charge to use the services and the operational schedule was daily between 10:00 and 16:30. All vehicles on the route and the backup bus doing the demo ride can be seen in figure 3.

Figure 3. iseAuto AV shuttle deployment in Ülemiste City

One of the main ideas was also to test smart bus stops, with real life info about the arrivals and departures, displayed all the time in the stations, while also recalculating, if there had to be longer stops on the way, or if the shuttle had to be taken off from the rotation. These stops were crucially located two in the business park, one at the public transport hub of the airport (where also city trams and busses had their airport stations, linking the whole route also to the city center) and the second between the hotel and the shopping center. Since the test pilot ran only during summer months, we could not test the operation in that drastic weather conditions like occasionally Estonia has during autumn and especially winter months, but there was heavy rain and winds in the mix of mostly nice and sunny weather. There were cars parked wrong on the shuttles predetermined way, but since it can easily pass obstacles, nothing more severe than a sudden stop happened. But as all the passengers have to wear a seatbelt and be seated, then nobody was bothered (according to our survey on board and over the internet during the service). The aim of the survey was to get to know more about what people really think and to reflect their experience. The only reasons to postpone the start of the day or sometimes stopping the service were connected to making technical updates to the shuttles.

3. Use case survey

Altogether we had 514 passengers. The operation on the public streets lasted for two months (mostly June and July of 2020). The length of the route was 2.4 km and the total mileage the shuttles gathered was 2500km.

During the pilot, we also conducted a survey. While talking with the passengers during their drive, they expressed their opinion that once using the AV, they appreciated it more than they thought they would. For the project, this is also shown by the unexpectedly high marks on the safety. While 10% (51) of all the passengers participated in the survey, some examples of the research results are presented in tables 1 and 2 and in the charts in figures 4 and 5.

IOP Conf. Series: Materials Science and Engineering 1140 (2021) 012047 doi:10.1088/1757-899X/1140/1/012047

Table 1. Pedestrian feedback about safety and attitude

Average score for overall experience	6.57 (out of 7,00 maximum)
Average score for traffic safety	5.95
Average score for personal safety	6.27
Average score for ease of use	6.45

Table 2. Pedestrian feedback about the usage intention

I would use automated shuttle bus during day-time	94.1% (out of 51 pax)
I would use automated shuttle bus during night-time	59.9%
I would never use automated shuttle bus	0.04%

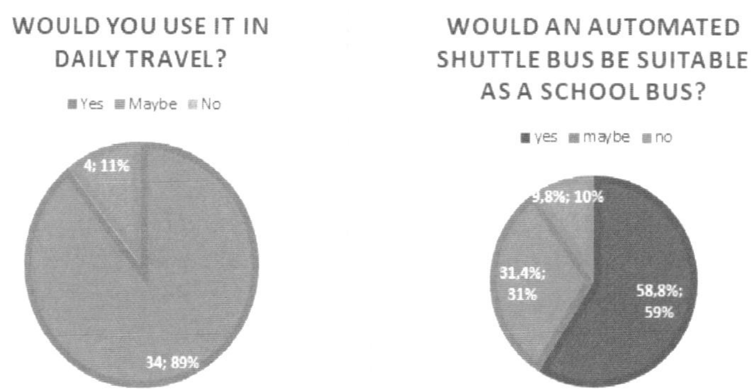

Figure 4. Two charts about passengers safety perception

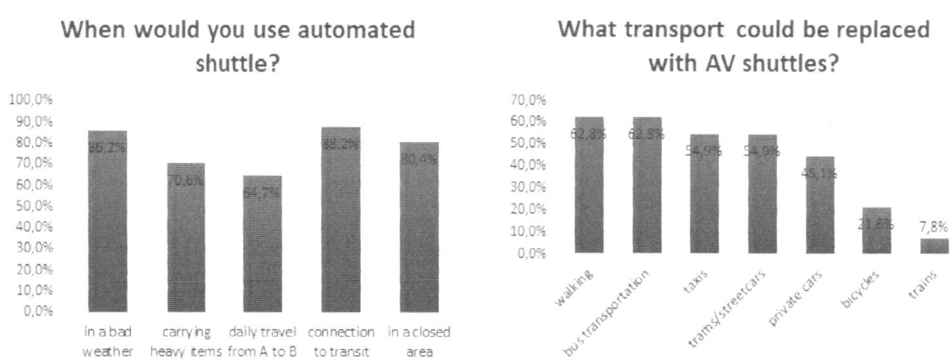

Figure 5. Survey results about passengers' estimations of AV shuttles

4. Conclusion

As the project introduced many new challenges that had not been tested before, it also gave a lot of new insight and pointed out daily what changes are needed to make shuttles safer and more convenient for both the operators and the clients. People are interested in trying autonomous vehicles, although quite many seem to be sceptical of its chances to ever fully take over the vehicles in the traffic.

Since through this use case, iseAuto has received licence to drive in real traffic, the developments towards towards becoming part of MaaS components in public city transport have taken a huge step forward. We have received valuable feedback from daily users, that confirms the need and safety expectations of AV shuttles coming into service. After the successful pilot in Tallinn, a pilot under FABULOS consortia was taken to the streets of the Greek city Lamia. The two major next steps to follow these 2020 projects are a shuttle service without any operator onboard in 2021. The aim is to create transport corridors, to service rural areas and to focus on the overall AV integration into the public transport ecosystem in 2022. The ultimate goal is to encourage more and more people to use public transport instead of contributing to ever intensifying traffic on the streets with their private cars.

Acknowledgments

The research is supported by the EC H2020 project Finest Twins (grant No. 856602). Experiments and pilots were supported by Fabulos and Sohjoa Baltic EC projects.

References

[1] Committee O-RAD (orad), On-Road Automated Driving (ORAD) committee. Taxonomy and Definitions for Terms Related to Driving Automation Systems for On-Road Motor Vehicles. DOI: 10.4271/j3016_201609.

[2] Sell R, Leier M, Rassõlkin A, et al. Self-driving car ISEAUTO for research and education. In: *2018 19th International Conference on Research and Education in Mechatronics (REM)*. 2018, pp. 111–116.

[3] Rassõlkin A, Sell R, Leier M. Development case study of the first estonian self-driving car, iseauto. *Electrical, Control and Communication Engineering* 2018; 14: 81–88.

[4] Mäkinen S, Kantala T, Haapamäki T, et al. *User experience and impact on public transport – Sohjoa Baltic. The roadmap to automated electric shuttles in public transport*. Metropolia University of Applied Sciences, 2020.

[5] FABULOS project puts robot buses to the streets. *Future Automated Bus Urban Operation Systems*, https://fabulos.eu/ (accessed 10 January 2020).

[6] Soe R-M, Müür J. Mobility Acceptance Factors of an Automated Shuttle Bus Last-Mile Service. *Sustainability* 2020; 12: 5469.

[7] Norton BE. *Autonomous Shuttles in Santa Fe Springs*. California Polytechnic State University, 2020.

[8] Hagenzieker MP, Boersma R, Velasco JPN, et al. Automated Buses in Europe: An Inventory of Pilots.

[9] Werdin S. *The Roadmap to Automated Electric Shuttles in Public Transport - The Legal Framework*. Sohjoa Baltic, 2018.

[10] Wang R, Sell R, Rassolkin A, et al. Intelligent functions development on autonomous electric vehicle platform. *Journal of Machine Engineering* 2020; 20: 114–125.

Multi-layer cyber-physical low-level control solution for mobile robots

H Pikner and K Karjust

Department of Mechanical and Industrial Engineering, Tallinn University of Technology, Ehitajate tee 5, Estonia

E-mail: heiko.pikner@taltech.ee

Abstract. Self-driving vehicles and mobile robots are used more and more in public transportation and industrial companies. Multiple experimental platforms, which can be operated in an urban or industrial environment have been developed recently. The key development for robust and safe control of the robot's operation is relying on the low-level cyber-physical system (CPS). CPS is composed of a collection of tightly integrated computational (cyber) units which are communicating with the physical world. CPS integrates computation and communication aspects with control and monitoring techniques. In this paper, the scientific goals underscore the analysis of the existing state of the art solutions to increase security, safety, and reliability of the multiple experimental platforms.

1. Introduction

Digitalization and robotization is a constant challenge for the industry to make production more effective, robust, and reliable. The fourth industrial revolution, called Industry 4.0 puts smart technology into the focus where robots, production units, and services are interconnected. Despite the interconnection of hardware, the integration of the robots with the manufacturing engineering systems should be developed and tested [1]. Thus, the standard automated guided vehicles (AGV) cannot fulfill more complex tasks and must be upgraded to more flexible mobile robots that can reroute on-demand and intercommunicate with other machines. Several types of research already investigated how to integrate autonomous vehicles into the industry 4.0 environment [2]. The modularization in mobile robotics is essential to offer flexible reconfiguration, efficient design, and reduce the development and implementation time. The proper architecture and modular concept must be considered already in the early design stage and proper methodologies are proposed for mechatronic system design. Christophe et. al. proposes OPAS [3] to synthesize conceptual design solutions in early design stages [4]. Early design and simulation toolkit for mobile robot platforms [5] relying on these methodologies helps to design modular universal mobile robots for the industrial logistic environment, inside the manufacturing area. From the development and later implementation side, the inside manufacturing area is adding additional limitation parameters for the development stage like (communication limitations, localization, shorter distances and space, limitations of safety, etc.).

The lower-level cyber-physical solution shown in figure 1 is a critical part of controlling modular mobile robots using modules that interact with the physical world. These modules are divided mainly into sensors, actuators, and computation units. Also, inside cyber-physical solutions, the modules are usually distributed, and therefore communication is an important part of the system. The separate constituent parts (sensors, actuators, software, etc.) of the cyber-physical solution collaborate to create some global behavior [6]. Nowadays it is very important to get real-time data from the production area and from the mobile robot to make real-time analysis and management tasks [7].

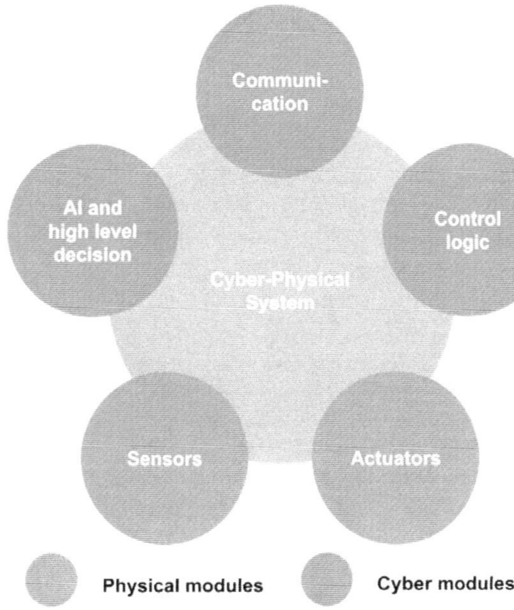

Figure 1. Cyber-Physical system

Mobile cyber-physical systems have significant computational resources to maintain localization, obstacle detection, safety functions, and path following. Computational resources can be divided into two different categories like artificial intelligence (AI) based on high-level decision-making and lower-level control logic. AI and high-level decision making based on some special computer to run robotic operating systems (ROS). In many cases, it is a regular PC. The low-level control logic is near or inside actuator or sensor modules. It handles a regulation for actuators and makes the first information processing for information received from sensors. Also, it controls and forwards information between the modules.

The communication inside the mobile cyber-physical solutions is usually based on the CAN (Controller Area Network) and the Ethernet networks [8]. CAN is an existing multi-master broadcast serial bus communication protocol for connecting electronic control units (ECU) in automotive applications. However, several vulnerabilities such as the lack of authentication and the lack of data encryption have been pointed out by several authors, which may ultimately render vehicles unsafe [9]. Security and safety-critical applications in the automotive industry require a new protocol to transfer the data at a higher speed than the existing CAN protocol allows. FlexRay is a newly introduced communication protocol for an automotive control system developed to fulfill the increasing demands for higher safety and data rate [10]. Also, multiple wireless communication mechanisms, such as WiFi, 4G, EDGE, Bluetooth used to interconnect several devices, to connect robots or connect robots to the internet. Right now, new developments are going to use and test 5G in similar conditions.

A mobile cyber-physical system may have multiple sensory input devices. The most common sensors for vision and 3D imaging systems are lidars, radars, and cameras [11]. There may be sensors for localization, for example, GPS and inertial sensors, also sensors that can detect the presence of nearby objects without any physical contact. Such sensors are ultrasound or infrared distance sensors. Different sensors may measure internal parameters. For example, battery current and motor speed. Common actuators are propulsion and steering systems for wheeled robots. There may be lifting mechanisms for cargo, robot arms, and other moving devices. Some robots have legs with electrical or hydraulic drive mechanisms. All these modules are interconnected over a communication network and controlled computational resources. In this paper, the focus is on a practical approach and implementation of a cyber-physical system on mobile robots and autonomous vehicles.

2. Overview of similar projects

Public transportation and logistics inside industrial companies include different types of robots, drones, and autonomous vehicles. Smaller robots that can be used inside the warehouses and factory floors are under the consideration. Different robots are brought out in Table 1, according to the producer, and are selected to analyze the onboard cyber-physical system development. Robots are chosen based on the sort of lifting or goods carrying mechanism, which allows them to move the payload from one location to another. These robots usually have good navigation capabilities inside the rooms, which are crucial in narrow corridors and cases where limited space is available.

Table 1. Listed robots are used in the logistic task.

Developer (Robot name)	Software	Sensors
Amazon Robotics (Drive Unit)	Proprietary agent-based software running on the robots.	Downward-facing camera, upward-facing camera, infrared sensors, and collision-detection bumpers.
Fetch Robotics (Freight)	ROS + custom navigation packages	Two Intel RealSense 3D cameras, SICK TIM lidar
Locus Robotics (LocusBot)	Proprietary	Lidar
Vecna Robotics (RC20 Conveyor)	Proprietary Pivot.al (multi-agent AI-based orchestration engine)	NA
InVia Robotics (Picker Robot)	Proprietary inVia Robot Management System (RMS)	Industrial-grade global shutter cameras
Robotnik Automation (RB-2 BASE)	ROS	2x Lidar, RGBD devices for the detection of obstacles
MiR - Mobile Industrial Robots (MiR100)	ROS	2x SICK microScan3 safety system, 2x 3D camera
HOMAG Group (TRANSBOT)	NA	2x Lidar
TalTech (BoxBot)	ROS + Autoware	4x Lidar, cameras

The robots brought out in Table 1 have quite different software, sensors, and control systems. There is a trend to use open-source solutions for different kinds of mobile robot control software. A popular choice is a Robot Operating System (ROS) with a specific software stack or add-on modules [12]. There are few such robots, for example, BoxBot, Robotnik, MiR, and Freight. Others are using proprietary software, which may be a disadvantage, because it may be difficult to add new functionality on an ongoing basis of changing tasks. High-quality sensors like 3D lidar (light detection and ranging) are not very common. Most robots use 2D lidars, cameras, and ultrasound sensors. This set and functionality of sensors may limit the robot's performance in new or more complex environments and in the manufacturing environments where the changes are very often taking place.

The long-term goal is to create a situation, where multiple robots work together for one specific task or serve the manufacturing process. For example, logistics robots serve the production stations by supplying the raw material and bringing production results into the warehouse. This kind of multi-robot environment requires an additional layer on top of the system, which controls the high-level goals and guides a single robot's individual goal. This kind of robot swarm can handle very complex tasks but need complex control algorithms. Individual robots can be also partly remotely controlled by humans [13] which makes initial implementation easier. Swarm robotics is trying to follow biological systems and theoretical developments have been conducted to apply a bio-inspired approach for robot Swarm in smart factories [14].

3. Concept of multi-layer cyber-physical architecture

Tallinn University of Technology (TalTech) developed a low-level control architecture of the CPS first for a self-driving last-mile bus called iseAuto [15, 16]. Then based on the same architecture, a new small-scale logistic robot (BoxBot) has been developed to move boxes in rather tight spaces. During the tests, it was determined that the robot is capable of transporting packages without human interference. It is not trivial to implement robot-based logistics as many specific issues need to be solved and addressed. Corridors and transportation spaces are usually designed for humans and the robot must navigate between obstacles that may change at any time. The tasks of the robot may also change.

The iseAuto has been designed to be a minibus that is going to operate primarily on the territory of the university campus, therefore the speed of the vehicle was limited to 20 km/h. The architecture of the vehicle's CPS is divided into layers as described in figure 2. The AI and high-level decision-making layer make autonomous driving decisions based on the sensor's input layer. The vehicle speed and direction commands are sent to the actuators layer that has a mission-critical functionality to take care of the vehicle's actual control. Control logic is divided into two layers – the master controller layer and the drive controller layer. The main task of the master controller is to act as a central gateway between all the nodes. The most sophisticated low-level functionality is integrated into the drive controller which controls the vehicle movement and steering. For safety reasons, a separate safety controller has been developed to stop the vehicle, when some fault is detected. The communication layer runs on separate CAN networks and Ethernet.

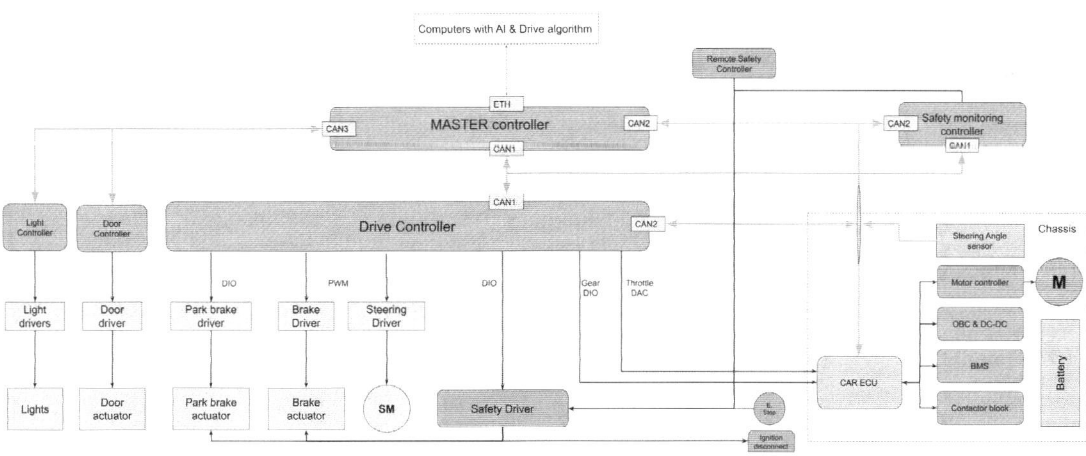

Figure 2. Low-level control solution for TalTech iseAuto

Small scale logistic robot BoxBot is developed to operate inside warehouses and to transport materials from and to the production units and empty boxes between washing stations and production units. The control architecture of the logistic robot is like iseAuto and is divided into the same layers as described in figure 3. Also, it is adapted to fit a much smaller space. The upper layer provides input to the ROS high-level control system. The AI & drive algorithm layer is based on the NVIDIA Jetson AGX Xavier developer kit. The logistic robot commands are sent to the low-level control layer that has a mission-critical functionality to take care of the robot's movement control. The central unit for this layer is the master controller. The difference is that the drive controller which is originally meant to control the iseAuto OEM platform and steering is not needed and replaced with a simpler motor controller for two-wheel hub motors. For lifting mechanism and other system dedicated controllers included.

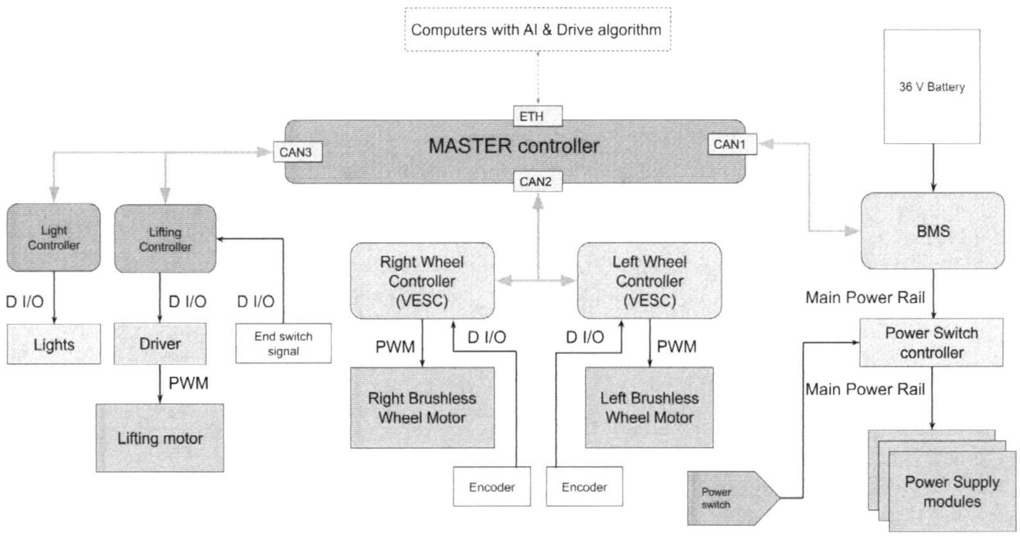

Figure 3. Low-level control solution for TalTech BoxBot

Both CPSs have three layers of control. High-level running AI and drive algorithms, mid-level master controller mediating and prioritizing the messages, and low-level control for actuators. The proposed unified concept includes an automatic connection level for all sensor and actuator modules. Plug and play modules are listed and information is sent to the upper level. The modules start automatically when they are connected to the communication network. A separate safety controller has been developed for the iseAuto platform, but it is not suitable for other applications, therefore new safety functions are built-in to the system and tightly integrated with all layers of CPS.

4. Discussion and future work

The first objective is to establish a multi-model cyber-physical architecture for a new industrial mobile robot BoxBot 2 which can be used later to build different types of smaller or bigger scale autonomous robots. The long-term goal is to create a situation where multiple robots work together for one specific task or serve the manufacturing process. More common and not platform-specific real-time tracking and monitoring systems should be further developed.

5. Conclusion

The paper analyzes the existing state of the art solutions of cyber-physical systems of mobile robots and concludes that many existing mobile robot platforms are non-modular and not compatible with extensions. Previously developed low-level control architecture of the CPS of a self-driving last-mile bus called iseAuto and mobile industrial robot BoxBot is analyzed in more detail and compared. By using the same architecture as a full-scale self-driving vehicle, a new small-scale logistic robot BoxBot 2 was developed for industrial indoor logistics. The target is to move materials in rather tight spaces in industrial areas and corridors. The concept derived from self-driving vehicles is converted to a multi-model cyber-physical architecture for an industrial autonomous mobile robot where the long-term goal is to create a situation where multiple robots work together for one specific task or serve the manufacturing process.

Acknowledgments
The research is supported by the EC H2020 project Finest Twins (grant No. 856602); the Estonian Centre of Excellence in Zero Energy and Resource Efficient Smart Buildings and Districts, ZEBE, grant TK146 funded by the European Regional Development Fund; Interreg project Inform VIR19004.

References

[1] Riives J, Karjust K, Küttner R, Lemmik R, Koov K and Lavin J 2012 *Proc. of the 8th Int. DAAAM Baltic Conf. "Industrial Engineering* Software development platform for integrated manufacturing engineering system (Tallinn: Tallinn University of Technology) pp 555–560
[2] Sell R, Rassõlkin A, Wang R and Otto T 2019 *Proc. of the Estonian Academy of Sciences 68* Integration of autonomous vehicles and Industry 4.0 pp 389–394
[3] Christophe F, Sell R, Bernard A and Coatanéa E 2009 *ASME 2009 Int. Design Engineering Technical Conf. and Computers and Information in Engineering Conf.* OPAS: Ontology Processing for Assisted Synthesis of Conceptual Design Solutions (American Society of Mechanical Engineers) pp 249–260
[4] Sell R, Coatanéa E, Christophe F 2008 *Proc. of the 6th int. conf. of DAAAM Baltic industrial engineering* Important aspects of early design in mechatronic pp 177–182
[5] Sell R and Petritsenko A 2013 *Int. Journal of Product Development 18* Early design and simulation toolkit for mobile robot platforms pp 168–192
[6] 2016 *Guide to Cyber-Physical Systems Engineering*
[7] Snatkin A, Eiskop T, Karjust K and Majak J 2015 *Proc. of the Estonian Academy of Sciences 64* Production monitoring system development and modification p 567
[8] Sawant A, Lenina S, Joshi MD 2018 *Int. Journal of Innovations & Advancement in Computer Science IJIACS* CAN, FlexRay, MOST versus ethernet for vehicular networks pp 337 – 339
[9] Buttigieg R, Farrugia M, Meli C 2017 *18th Int. Conf. on Sciences and Techniques of Automatic Control and Computer Engineering (STA)* Security issues in controller area networks in automobiles (Monastir) pp 93-98
[10] Pikner H 2020 *19th Int. Symp. 'Topical problems in the field of electrical and power engineering. Doctoral school of energy and geotechnology. III' : Tartu, Estonia, January 14-17, 2020* Overview of cyber-physical control systems for self-driving vehicles (Tallinn University of Technology) pp 105–106
[11] Arnold E, Al-Jarrah OY, Dianati M, Fallah S, Oxtoby D and Mouzakitis A 2019 *IEEE Transactions on Intelligent Transportation Systems, vol. 20, no. 10* A Survey on 3D Object Detection Methods for Autonomous Driving Applications pp 3782–3795
[12] Hellmund A, Wirges S, Taş ÖŞ, Bandera C and Salscheider NO 2016 *2016 IEEE 19th Int. Conf. on Intelligent Transportation Systems (ITSC)* Robot operating system: A modular software framework for automated driving (Rio de Janeiro) pp. 1564–1570.
[13] Sell R, Otto T 2008 *Proc. of 19th EAEEIE Annual Conf.* Remotely controlled multi robot environment pp 20–25
[14] Rohrich RF, Teixeira MAS, Piardi L and de Oliveira AS 2020 *Advances in Intelligent Systems and Computing* A Bio-Inspired Approach for Robot Swarm in Smart Factories pp 303–314
[15] Rassõlkin A, Gevorkov L, Vaimann T, Kallaste A, Sell R 2018 *Proc. of 25th Int. Workshop on Electric Drives* Calculation of the traction effort of ISEAUTO self-driving vehicle pp 1-5
[16] Sell R, Leier M, Rassõlkin A, Ernits J-P 2018 *Proc. of the 19th Int. Conf. on Research and Education in Mechatronics* Self-driving car ISEAUTO for research and education pp 111-116

Supervised learning vs. unsupervised learning: A comparison for optical inspection applications in quality control.

J Lehr[1], J Philipps[1], V Nguyen Hoang[1], D von Wrangel[1] and J Krüger[2]

[1]Fraunhofer IPK, Berlin, Germany
[2]TU Berlin, Berlin, Germany

E-mail: jan.lehr@ipk.fraunhofer.de

Abstract. For the establishment of a successful quality management system in companies, the quality control of e.g. newly produced goods or the return of old and used parts is an essential component. One solution for this is the optical inspection of the surface of objects with the help of image processing algorithms. Using the case study of printer cartridges, this paper evaluates the extent to which different methods of machine learning can contribute to a successful quality control. Established methods of supervised learning have the advantage that they are already proven in many applications and have a very high detection accuracy. However, they require a lot of labelled training data and this high effort also means high integration costs. A new approach is a data-reduced variant from unsupervised learning. Here, the algorithm is trained only with defect free objects, for example as they come to a large extent from the production. If the objects are defective, the method from the field of anomaly detection or even novelty detection detects something that is different from the learned norm. This has the advantage that not all defects have to be known beforehand. And this in turn avoids acquiring a large amount of training data for each of these defects. This paper compares the effort required to acquire training data and compares it with the detection accuracy of the different methods in order to give an assessment of the extent to which the use of unsupervised learning methods is beneficial. Newly produced and used printer cartridges are used for this purpose. Image data is acquired from 18 different printer cartridge models. Afterwards they are fully annotated (labelled). A smart separation into training, validation and test data allows the training of supervised and unsupervised methods as well as a complete evaluation regarding the effort for data acquisition, annotation and detection accuracy of the defects. Finally, an outlook for chances and risks of the respective procedures is given.

1. Introduction

In the context of ongoing globalization, companies are confronted with increasingly strong competition on the worldwide economic market. In addition to the pricing of products, the quality of these products also determines whether a company can sell its products on the world market or not. Those who cannot produce at the cheapest price must be able to provide high quality products. An optical inspection of produced goods can be essential when it comes to assuring quality requirements. But also topics such as environmental protection and sustainability are increasingly becoming the focus of attention. An efficient value chain in the remanufacturing sector can be a key factor in competitive advantage if the processes are designed to be lean and efficient. The inspection of used parts is an essential element and is nowadays largely performed by humans. An employee visually and haptically checks whether a product can be remanufactured, is recyclable or has to be thrown away. The manual inspection process, especially the optical, visual inspection, is a time-consuming and tiring process, which in addition can only provide subjective inspection results. Latest developments in the field of intelligent image processing have the potential to support the employee in this task to lean down the process. The methods from the field of supervised learning, require a large amount of training data to learn patterns to be able to carry out either the quality of a new part or the diagnosis of a used part.

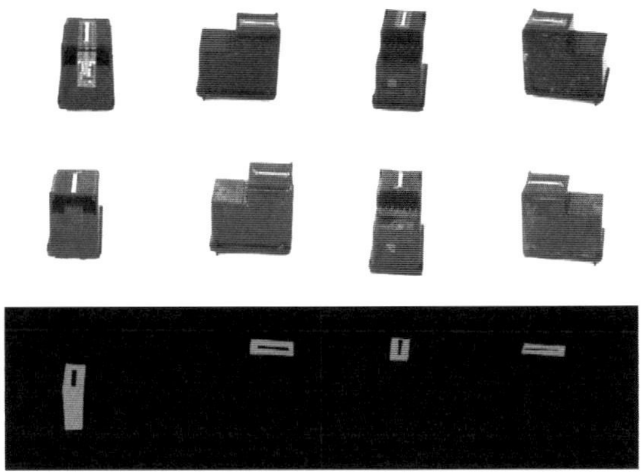

Figure 1: Example object and corresponding images from the created dataset. Top row: Defect-free printer cartridge. Middle row: Defective printer cartridge with the same serial number as in the top row. Bottom row: Pixel-wise annotation map to highlight the defects.

Figure 2: Example images for different defect classes. From left to right: Defect, dirt, sticker.

The implementation of these methods from the field of machine learning becomes more efficient if the inspection task is reformulated. Instead of finding certain defect patterns, all deviations from a previously defined standard should be found. For this purpose, only products with no defects or only defects that are acceptable are captured visually. Any deviation from the image data of the good parts is then recognized by the so-called anomaly detection. This paper examines the optical inspection of new and used printer cartridges using methods of supervised and unsupervised learning. It compares the effort of data collection and the corresponding detection accuracy for defects of different types.

2. Related works and methods used

For an automated optical inspection (AOI) this paper uses machine learning methods. Both supervised and unsupervised methods are used.

2.1. Supervised learning

Supervised learning methods require not only image data but also the corresponding labels for the application of an AOI. If a method is to detect defects on objects, corresponding image data must be acquired. The defects to be detected are annotated or labeled pixel by pixel on the image and named according to the corresponding defect category. With this training data the supervised procedure learns how the defects look like and can then recognize them in the real process. The methods presented in this work are explained below.

He et al. proposed an extension to Faster R-CNN to perform binary mask regression [2]. Mask R-CNN adopts the first and second stage of Faster R-CNN with a Region Proposal Network (RPN) followed by RoIPool, a feature extractor inspired by a combination of ResNet and a Feature Pyramid Network (FPN), which is used for parallel classification and bounding box regression. The authors added a binary mask branch to the second stage, allowing parallel computation of the bounding boxes, classifications, and binary masks for each of the regions of interest. In particular, Mask R-CNN computes K binary masks for K classes with a resolution of m x m, which matches the resolution of the RoI. The mask loss is defined as a pixel-wise sigmoid function, which is added to the cumulative class and bounding box loss.

2.2. Unsupervised learning

Unsupervised learning methods, in contrast to supervised learning, don't need any labels. They teach themselves by identifying patterns in the training data. For AOI, these methods are used in a way that only defect-free objects are used for training. It is no longer necessary to label defects pixel by pixel. In the real application, the method can then be used in such a way that it detects deviations from the learned defect-free norm, thus identifying defective objects. The methods used in this work are described in the following.

Bergmann et. al present in their work a Convolutional Autoencoder (CAE) for anomaly detection [3]. The CAE is a convolutional encoder-decoder architecture with a L2-loss function. It consists of nine encoder and nine decoder layers. After the first eight layers the image processing ends with leaky ReLUs in both encoder and decoder. During the training stage the CAE tries to reconstruct the given input image. It learns to generate images of defect-free objects. The industrial application is simulated in a test stage. Therefore, the anomalies are detected by calculating the difference between the input image with a defective object and the reconstructed image of the CAE. By using a threshold artifacts such as noise can be removed. The threshold is calculated by the largest difference between input image and reconstructed image found during the training stage.

3. Dataset

The creation of the used data set consists of two parts: The acquisition of the images and the annotation of the defect regions which are supposed to be recognized. In this work 18 different printer cartridges are analyzed, new and used. The used objects are contaminated by being either damaged, dirty or having important cartridge parts covered in another way. Accordingly, 36 different objects are considered. The data set contains 72 images for each object leading to a total amount of 2,592 images. The images are taken with the help of an automatic image acquisition setup with one camera which allows to take photos with a resolution of $2,590 \times 1,942$ Pixel in 10° steps. Every cartridge has to be placed individually on the set up. The acquistion of the whole data set took four hours, two hours for the printer cartridges without defects and two hours for the used printer cartridges with defects. This equals a time effort of seven minutes for each printer cartridge. Figure 1 shows the three parts of the dataset: Defect-free images, defective images and corresponding annotation maps.

In order to test the performance of the different methods the data is split into 75 % training and 25 % test data. After pre-processing the data by i.e. removing non-faulty images belonging to the bad ink cartridges, there are 1,839 training images and 607 test images. Each image was randomly assigned to either training or test data. The defects are categorized into three classes: Dirt, defect and sticker. The pixel-wise segmentation of the regions with defects i.e. damage, dirt or covered parts was done manually by carefully annotating the belonging pixel areas. The time required to annotate the entire data set is 57.5 hours. This results in an annotation effort of three minutes per image and 1.7 hours per cartridge.

4. Experimental setup and results

4.1. Supervised learning for defect detection

For the training of a supervised defect detection only the images of used printer cartridges are taken. This is necessary because the three defect categories dirt, defect and sticker must be trained explicitly. Since there are no defects on the images of the new printer cartridges, they are not used here for training. The testing stage include defect free and defective images. The Mask-R-CNN shows good results with an accuracy of 78 % when it comes to detecting defect-free, i.e. new printer cartridges. Only 2 out of 10 images are falsely detected as defective, although they do not have any defects. The detection rates for defects are at an average of 90 %. It is clear to see that the most prominent of the defect categories is also the best detected: Dirt. The results of the Mask-R-CNN are shown in Table 4.1.

Table 1: Results for the AOI of used printer cartridges with supervised training of Mask-R-CNN.

Prediction	Good	Dirt	Defect	Sticker
True positive	0.78	0.93	0.90	0.87
False positive	0.22	0.07	0.10	0.13

4.2. Unsupervised learning for anomaly detection

To train the AOI methods in an unsupervised manner it is necessary to held out any information about possible defects. Only the defect-free images are used to train the networks. Thus, only half of the dataset is used. In addition, no annotations are necessary. The CAE as presented in Chapter 2 is only able to detect anomalies in general. A classification of the detected anomaly is not possible. Considering the average accuracy of 72 % of defects detected, this makes a difference of 18 percentage points compared to supervised learning methods. The detection rate for defect-free images is also very low at 65 %. This is surprising, since the CAE was trained with defect-free images. However, the adjustment of the threshold based on the training data is problematic. By choosing this value, the accuracy can shift considerably. Table 4.2 shows the results in summary.

Table 2: Results for the AOI of used printer cartridges with unsupervised training of a CAE. Note: All defects are summarized, since there was no distinction made between different anomaly classes.

Prediction	Good	Dirt	Defect	Sticker
True positive	0.65		0.72	
False positive	0.35		0.28	

5. Discussion

With a maximum accuracy of 78 % for defect-free images and 90 % for defective images the supervised methods perform significantly better than the unsupervised methods. The unsupervised methods are 13 percentage points worse in terms of detection accuracy for defect-free images and 18 percentage points worse for defective images. The unsupervised methods detect one in four images incorrectly when it comes to detecting defects. When it comes to the time saving potential for the acquisition and annotation of training data, a different situation appears. The fact that the unsupervised methods do not require annotations reduces the effort for data collection by almost 97 %. Table 5 shows the time required for data acquisition and data annotation for each method. Furthermore the recognition accuracy is listed.

Table 3: Comparison of the effort of data acquisition and data annotation with the corresponding dataset split of 75 % training data and the accuracies of finding defects.

	Training data [hours]		Accuracy	
Prediction	Acquisition	Annotation	Good	Defect
Mask-R-CNN	3	43.1	**0.78**	**0.9**
CAE	**1.5**	**0**	0.65	0.72

6. Conclusion and outlook

The approaches presented in this paper can lead to a useful and helpful automatic inspection of printer cartridges. But if the effort to collect training data is to be kept to a minimum, it is necessary to accept a reduction in the detection accuracy. A solution to integrate these methods into industrial processes would be the implementation as an assistance system. This was already mentioned by the authors in an earlier publication [5]. So the question to either use supervised or unsupervised learning methods depends on the trade-off between the needed accuracy and time being invested for annotation. A system with a defect detection accuracy of 72 % can reliably inspect objects to a large extent, supporting the application of unsupervised learning methods. But there are decisions that are just above the defined thresholds. These objects could be sorted out separately to be inspected by employees with expert knowledge. Furthermore, additional image data is generated during the current process and most of it is already pre-sorted. The effort to annotate this data is far less, because the defect regions and classes are already proposed by the assistance system. This suggestion can then be accepted or rejected by the employees. If the supervised learning methods are being rejected in favor of the unsupervised learning methods the annotation time per image, which would be crucial for supervised learning, decreases from three minutes to one or two seconds.

Acknowledgement

The authors thank the german federal Ministry of Education and Research for funding the project VIADUCT.

References

[1] Deng, J. et al., 2009. Imagenet: A large-scale hierarchical image database. In 2009 IEEE conference on computer vision and pattern recognition. pp. 248–255.
[2] He, K., et al., Mask R-CNN, In 2017 IEEE International Conference on Computer Vision, pp. 2980-2988.
[3] Bergmann, P., et al., Improving Unsupervised Defect Segmentation by Applying Structural Similarity to Autoencoders. In 2019 14th International Joint Conference on Computer Vision, Imaging and Computer Graphics Theory and Applications, pp. 372–380.
[4] Bergmann, P., et al., MVTec AD — A Comprehensive Real-World Dataset for Unsupervised Anomaly Detection, In 2019 IEEE/CVF Conference on Computer Vision and Pattern Recognition, pp. 9584-9592.
[5] Lehr J., et al., Automated Optical Inspection Using Anomaly Detection and Unsupervised Defect Clustering, In 2020 25th IEEE International Conference on Emerging Technologies and Factory Automation, pp. 1235-1238.

Effects of Boundary Conditions and Operating Parameters on Temperature Distribution during the Friction Stir Welding Process

M. Boukraa[1], M. Aissani[1], N. Lebaal[2], D. Bassir[3*], A. Mataoui[4], N.Tal Ighil[1], Hao YUE[5]

[1]Research Center in Industrial Technologies CRTI, P.O. Box 64, Cheraga 16014, Algiers, Algeria

[2]Université Bourgogne Franche-Comté, UTBM, Laboratoire interdisciplinaire Carnot de Bourogne,

[3]UMR CNRS6303/ 90010 Belfort France

[4]Theoretical and applied laboratory of fluid mechanics USTHB

[5]FRDISI, ENSEM, Hassan II University, Morocco

Corresponding author: david.bassir@utbm.fr

Abstract. This work deals with a numerical simulation of the friction stir welding FSW process of alloy material AA2195-T8. A 3D transient thermal model for simulating the heat transfer phenomena in the welding phase is applied. In this model, the FSW tool is considered as a circular heat source moving in a rectangular plate having a cooling surface and subjected to non-uniform and non-homogeneous boundary conditions. The thermal problem is solved using the finite element method as part of a Lagrangian formulation. The obtained results allow us to determine the maximum value of the temperature in the Nugget zone of the welded joint. During this process, the thermal cycle and the temperature distribution were determined for different values of the welding process parameters. The obtained numerical results are in good agreement with the one available in the literature.

Keywords: Friction Stir Welding, heat transfer, finite element method, AA2195-T8.

1. Introduction

In aeronautics and aerospace domains, the use of structurally hardened aluminium alloys represents a major advantage. However, the difficulty of assembling them through fusion welding techniques (TIG, MIG-MAG...) remains a crucial problem. In 1991, a new solid-state welding technique [1] was introduced by England welding institute. This technique is called the Friction Stir Welding process (FSW). Its uses a specific rotational tool *(formed by a shoulder and a shaped probe)* to provide required weld properties [2]. The heat energy requires to join the two sides of the plates *(retreating side RS and advancing side AS respectively)* [3] is generated by the friction between the rotating tool and the mentioned surface as described in Figure 1. In the last decade, many researchers have investigated numerically and experimentally how to optimize the FSW technique. In particular, the heat transfer phenomena and materials flow [4], [5]. Underlined in different references, the perpendicular distribution of temperature is nearly isothermal under the shoulder tool. Besides, increasing the pressure and the rotation speed enhances the maximum value of welding temperature.

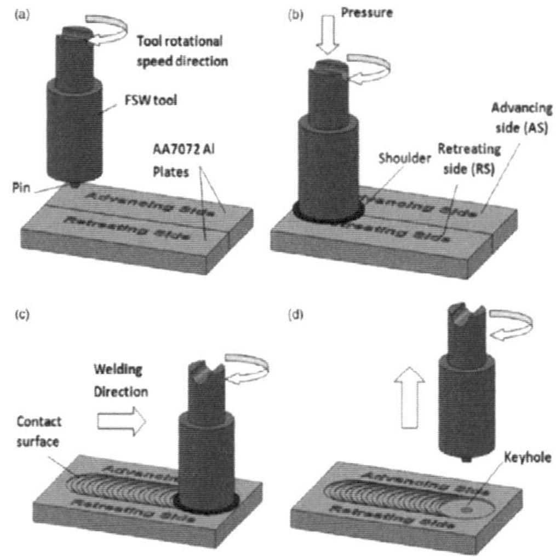

Figure 1. Presentation of the different steps of the FSW process: (a) start, (b) penetrations, (c) welding and (d) finishing [3].

It is important to mention that an inadequate heat generation in this welding process could lead to failure of the tool-probe or the workpiece shape. Thus, understanding heat flow behaviour is important to achieve good welding [1]–[3], [6]. Simar et al. [7] determined the thermal history experimentally at positions near the weld line during the FSW process. Also, they have developed a numerical model based on the finite element method (FEM), considering the effect of the welding parameters such as the welding speed, rotational speed of the tool, axial force and tool geometry. Moraitis et al. [8] have calculated using semi-analytical methods, the total energy and the heat produced by stirring the material during the FSW process. They also proposed a 3D numerical simulation to analyze the thermal histories during the welding process. The obtained results were introduced into a thermo-mechanical model to determine residual stress. Results were validated experimentally. A recent study [9], the authors analyzed experimentally and numerically the thermo-mechanical behaviour for FSW of PMMA plate. Their objective was to look for suitable conditions to attain optimal thermomechanical properties. The developed model was used to determine the thermal profile in the longitudinal and transversal directions of the welded plates, while the temperature of the surface plate was measured experimentally and compared to the simulation. The authors underlined that at higher heat inputs, with an increase of the axial force within the plastic material led to material sticking around the probe and to the formation of defects.

In this work, a numerical simulation of the FSW process is carried out, making it possible to study the thermal phenomena and heat transfer. A 3D thermal model is applied in order to consider the thermal transient in the welding tool, regarded as a mobile heat source. The effect of boundary conditions and the operating parameters (*Welding speed, rotational speed, axial force and probe radius*) on the heat transfer during the welding process were investigated.

2. Physical model and numerical simulation

The workpiece is composed of two plates to be welded by the FSW process. its dimensions are $L\times W\times H$=610×102×8.1 mm^3. The intersection of the workpiece with a tool is represented by a circular red section that moves throughout the weld joint line. Boundary conditions are also specified to updtae the mathematical model presented in the next session. Since the plates were made from the same material (*aluminium alloy*) and had similar dimensions, the symmetric condition is applied along the joint, so the calculation was performed on one side (*Figure 2*).

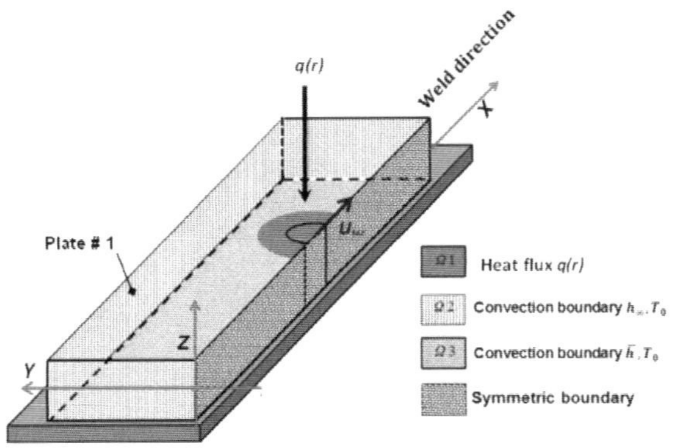

Figure 2. Applied heat transfer boundary conditions on the computational domain

Mathematical model

The modelling of the FSW heat phenomena is based on the resolution of the heat conduction equation in the workpiece with an appropriate model of the heat source. In this work, a modified heat source Q_{source} is used for the friction effects of the welding tool on the workpiece, taking into account the appropriate boundary and initial conditions [10].

Hypothesis

Some hypotheses were necessary to solve the obtained system:

- The heat transfer process was considered 3D and transient in the entire workpiece,
- The peak temperature in the workpiece during the welding always remains below the melting temperature of the studied material (T_m below 773 K),
- The heat exchange between the workpiece and the environment occurs only through a natural convection process with a heating coefficient h_∞ and radiation effect during the process is neglected,
- The heat due to the plastic deformation of the material by the FSW tool is neglected compared to the heat generated by the rotational friction,
- The force applied vertically by the welding tool generates a uniform pressure at the workpiece /shoulder interface.

Heat transfer equation

The tool moves at constant speed $\boldsymbol{u}_w$ along the joint line according to the above assumptions. The thermal energy equation in the workpiece can be written in the Lagrangian coordinates as:

$$\rho(T)c_p(T)\frac{\partial T}{\partial t} = \nabla(K.\nabla T) + S \tag{1}$$

Where

- *$\rho(T)$, $c_p(T)$ and $\boldsymbol{K}(kx, ky, kz)$ are the material density, the thermal capacity and the thermal conductivity, which depend on temperature.*
- ***S** is the volumetric heat source due to the plastic deformation of the material, it is often neglected compared to the main source of the heat created by the friction of the shoulder $\boldsymbol{q}_s$ and the probe $\boldsymbol{q}_p$ on the workpiece.*

In the FSW process, the heat is directly linked to the contact conditions of the *workpiece/tool interfaces* and the process parameters. Therefore, there are two main sources of energy produced by friction at the two interfaces: the heat $\boldsymbol{Q}_p$ created at the *workpiece/probe interface* and the heat $\boldsymbol{Q}_s$ created at the *workpiece/shoulder interface*. These frictional contacts depend on the surface state, which is defined by the friction coefficient $\boldsymbol{\mu}$. Our model is based on Chao at al. model [11] (Figure 3) as follow:

$$q(r) = \frac{3Q_{source}}{2\pi R_S^3} r \text{ for } r \leq R_S \tag{2}$$

Where

$$Q_{source} = \frac{\pi^2 \omega P}{45} \mu\left(R_S^3 + 3R_P^2 h_p\right) \tag{3}$$

$\boldsymbol{q(r)}$ *is the heat flux source,* $\boldsymbol{R_s}$ and $\boldsymbol{R_p}$ are the shoulder and probe radii, $\boldsymbol{P}$ is the pressure, $\boldsymbol{\mu}$ is the friction coefficient, $\boldsymbol{\omega}$ rotation speed and $\boldsymbol{h_p}$ is the height of the tool probe [10]

3. Numerical procedure and validation

The numerical calculation of the temperature fields during the FSW process was performed by ANSYS, which requires the resolution of the governing equations by combining the appropriate boundary conditions and a moving heat source. To perform the numerical resolution of the problem and validate the thermal model, it is important to provide experimental data *(operating parameters table 1 and observed thermal cycles observed in the following reference* [11]*)*.

Table 1. Operating parameters for welding the AA 2195-T8 alloy

Probe radius	Shoulder radius	Rotational speed	Welding speed	Axial force
5mm	12.7mm	240 tr/min	2.36 mm/sec	25 kN

The material tool is M2 steel which is widely used in industries. It has a small and distributed carbides which give the tool high wear resistance. The evolution of the temperature at *P1 (L/2, 12.7, H/4)* and *P2 (L/2, 25.4, H/4)* are described in Figure 3. It can be observed that the temperature pick decreases with the increase of distance regarding the mid plan for both experimental and numerical analysis. Based on the good agreement between the simulation and the experiment, the proposed model can be used to predict the temperature profiles and cooling rates.

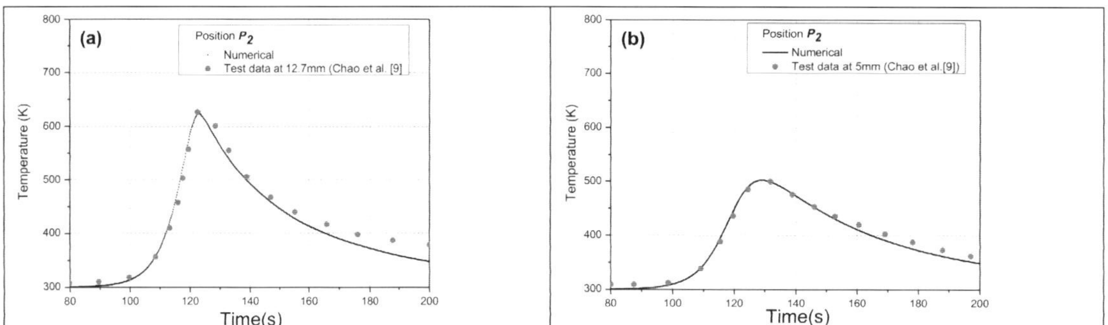

Figure 3. Comparison of the numerical thermal histories obtained by Chao et al. and the experimental results at two positions of the workpiece *(a) P (L/2, 12.7, H/4) and (b) P (L/2, 25.4, H/4)*.

The results above in Figure 3 confirm the fast temperature increase in welded plates followed by slow cooling down. They also show a small difference in the cooling phase between numerical and experimental data due to the fact that the model does not take into account the coupling with other phenomena such as the material flow, mechanical deformations and radiation of the heat with the ambient environment. It is also related to uncertainties measures (Uncertainty of sensors) and also to the gap between the real position of the sensors and its location in the FE meshing.

4. Results and discussions

The results based on the predictive numerical model fit well the available experimental measurements. Influence of the welding parameters such as the speed of advance, rotation speed and the diameter of the welding tool on thermal cycling and temperature distribution in the workpiece were investigated.

4.1. Temperature distribution

In Figure 4 the contours of the temperatures in the plane perpendicular to the welding line during the FSW process are shown. These contours were obtained at longitudinal positions X=0.305m of the workpiece, for the welding speed values *(1.5mm/s, 3mm/s and 4.5mm/s)* and rotation speed values *(300tr/min, 400tr/min and 500tr/min)*. We can notice that the temperature at a section increases rapidly when the tool passes through this point for the three cases studied. Meanwhile, the temperature decreases as the welding speed of the tool increases. Moreover, it is clearly observed that the lines of the isotherms are retracted towards of rotation axis of the welding tool when increasing the welding speed. The heat-affected zone near the welding tool is larger when the welding speed decreases because it has more time to conduct heat to the surrounding.

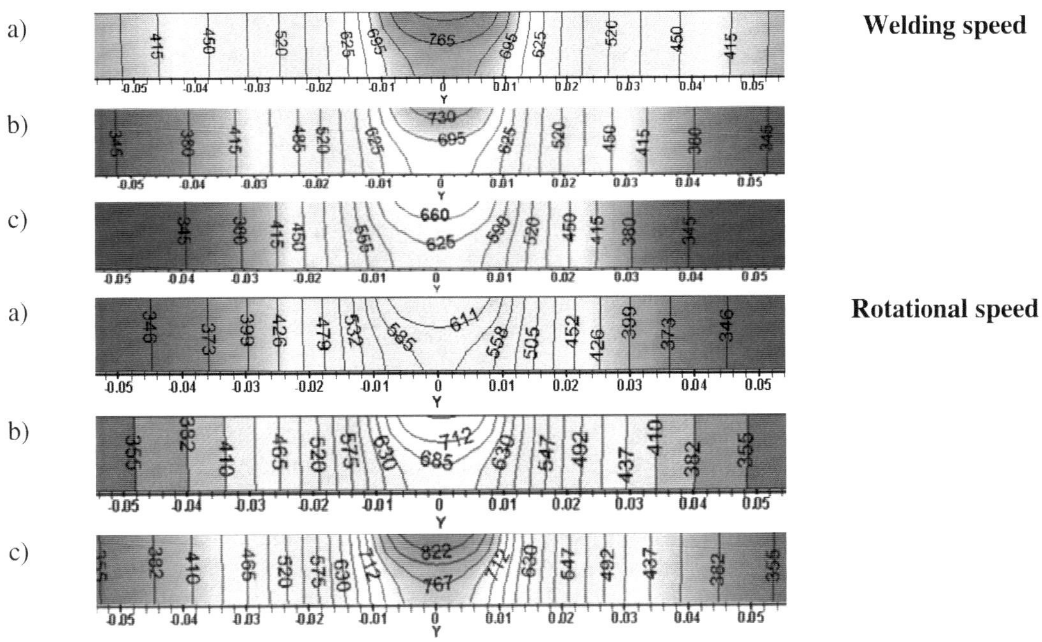

Figure 4. Effect of welding and rotational speed on the temperature field in the perpendicular plane to the welding line *a) 1.5mm/s, (b) 3.0mm/s, (c) 4.5mm/s). ((a) 300tr/min, (b) 400tr/min, (c) 500tr/min).*

Increasing the rotation speed from 300 to 500 tr/min leads to an increases of the welding heat input, the maximum value of the temperature and the overall level of the temperature. The distance between the peak temperatures of the contour 523 K (250°C) and the *shoulder/workpiece interface* increases also.

4.2. Effect of the heat transfer coefficient h_{inf} and h_{lat}

Figure 5 shows the temperature distribution in the perpendicular plane to the welding direction, for the following parameters of the welding process *(u_w=3mm/s, w=400tr/min, F=15kN and R_S=12.7mm).*

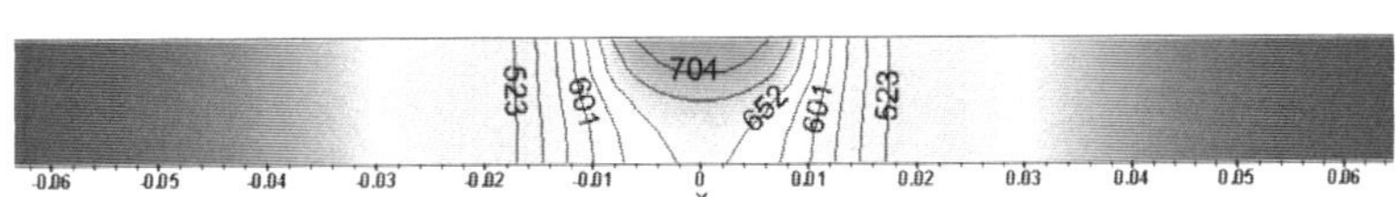

a) h_{lat} = 350, h_{inf} = 350

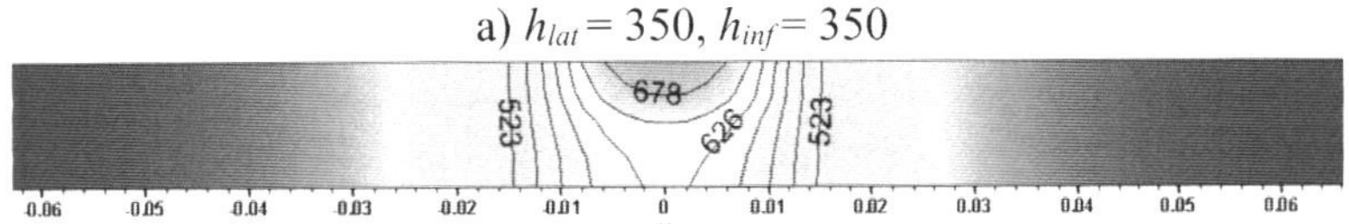

b) h_{lat} = 1000, h_{inf} = 350

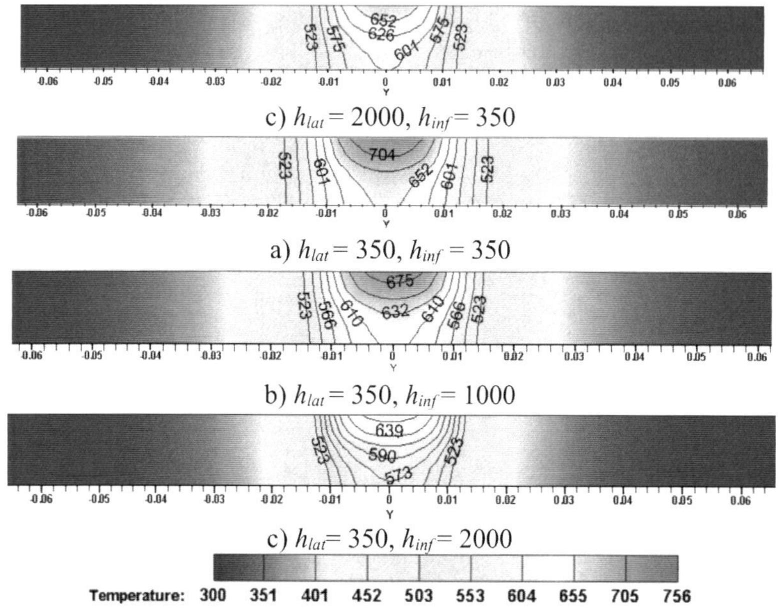

Figure 5. Effect of h_{lat} and h_{inf} on the thermal field in the perpendicular plane the welding direction at x = 305mm *(u_w =3mm/s, w = 400tr/min, F =15KN, R_S =12.7mm).*

When h_{lat} increase, the hottest zone of the workpiece is located more and more around the welding tool. The calculations with different h_{inf} values were made at the *workpiece/backing interface* and with the same operating parameters previously used for h_{lat}. We observe, the same effect for h_{lat}, when h_{inf} increased, the maximum value of the temperature at the *tool/workpiece interface* to be welded is slightly reduced as well as the hottest zone around the tool.

4.3. Thermal cycles

Figure 6a, 6b and 6c show the thermal cycles at point P1 (x =305 mm, y =12.7 mm, z =2 mm) for three welding speed (1.5, 3.0 and 4.5 mm/s), for three rotational speed (300, 400 and 500 tr/min) and for three tool shoulder diameters (10, 12.7 and 15 mm), respectively. For the three values of the welding speed (Figure 6a), very rapid variations of the temperature are observed during the passage of the tool. These variations are represented by a rapid increase in temperature (heating phase) up to a maximum value at the time of passage of the welding tool by position P1; as further than this, the temperature decreases to a uniform value of 320K (cooling phase). When the welding speed is doubled or tripled, the maximum temperature value decreases to 68.5 °C and 122.6 °C, respectively. This difference in terms of temperature is due to the rate of the amount of heat dissipated in the workpiece.

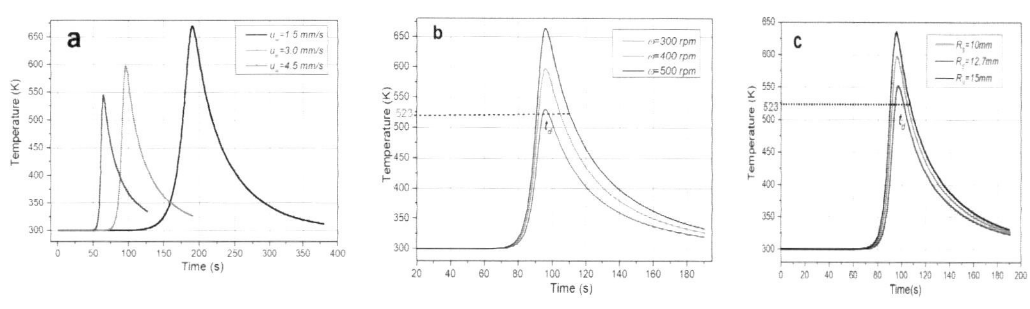

Figure 6. Effect of the welding parameters (a) welding speed, (b) rotational speed of the tool and (c) diameter of the tool shoulder on thermal cycles at point P_1 *(x = 305mm, y = 12.7mm, z = 2mm).*

The rotational speed is a significant variable that influences the thermal cycle through the FSW process. When the rotational speed increases, the temperature gradient increases to the maximum value, with a heating time almost identical (t = 96s). Also, the thermal cycle obtained for the rotational speed 300 tr/min has a higher cooling speed compared to other cases (Figure 6b). Also, a larger shoulder increases the quantity of heat created by the friction, but also results in a greater quantity of entrained material. This implies an increase in the amount of heat generated as shown in Figure 6c for three values of shoulder diameter tools 10, 12.7 and 15mm.

5. Conclusions

In this article, transient thermal model for FSW process was used with a good prediction of temperature distribution, temperature profiles, cooling rates compared to the experimental data available in the literature. The three-dimensional model can be coupled with optimization [12] method to improve the welding process and reduce the manufacturing cost. Further investigation would be performed on other aeronautical or automotive materials.

6. References

[1] R. Nandan, T. DebRoy, and H. K. D. H. Bhadeshia, "Recent advances in friction-stir welding - Process, weldment structure and properties," *Progress in Materials Science*. 2008, doi: 10.1016/j.pmatsci.2008.05.001.

[2] A. S. Hassan, T. S. Mahmoud, F. H. Mahmoud, and T. A. Khalifa, "Friction stir welding of dissimilar A319 and A356 aluminium cast alloys," *Sci. Technol. Weld. Join.*, 2010, doi: 10.1179/136217110X12720264008358.

[3] Z. Sun, C. S. Wu, and S. Kumar, "Determination of heat generation by correlating the interfacial friction stress with temperature in friction stir welding," *J. Manuf. Process.*, 2018, doi: 10.1016/j.jmapro.2018.01.010.

[4] B. T. Gibson *et al.*, "Friction stir welding: Process, automation, and control," *J. Manuf. Process.*, 2014, doi: 10.1016/j.jmapro.2013.04.002.

[5] B. S. Taysom, C. D. Sorensen, and J. D. Hedengren, "A comparison of model predictive control and PID temperature control in friction stir welding," *J. Manuf. Process.*, 2017, doi: 10.1016/j.jmapro.2017.07.015.

[6] K. Singh, G. Singh, and H. Singh, "Review on friction stir welding of magnesium alloys," *Journal of Magnesium and Alloys*. 2018, doi: 10.1016/j.jma.2018.06.001.

[7] A. Simar, J. Lecomte-Beckers, T. Pardoen, and B. De Meester, "Effect of boundary conditions and heat source distribution on temperature distribution in friction stir welding," *Sci. Technol. Weld. Join.*, 2006, doi: 10.1179/174329306X84409.

[8] G. A. Moraitis and G. N. Labeas, "Investigation of friction stir welding process with emphasis on calculation of heat generated due to material stirring," *Science and Technology of Welding and Joining*. 2010, doi: 10.1179/136217109X12537145658779.

[9] H. A. Derazkola, F. Khodabakhshi, and A. Simchi, "Friction-stir lap-joining of aluminium-magnesium/poly-methyl-methacrylate hybrid structures: thermo-mechanical modelling and experimental feasibility study," *Sci. Technol. Weld. Join.*, 2018, doi: 10.1080/13621718.2017.1323441.

[10] M. Aissani, S. Guessasma, A. Zitouni, R. Hamzaoui, D. Bassir, and Y. Benkedda, "Three-dimensional simulation of 304L steel TIG welding process: Contribution of the thermal flux," *Appl. Therm. Eng.*, 2015, doi: 10.1016/j.applthermaleng.2015.06.035.

[11] Y. J. Chao, X. Qi, and W. Tang, "Heat transfer in friction stir welding - Experimental and numerical studies," *J. Manuf. Sci. Eng. Trans. ASME*, 2003, doi: 10.1115/1.1537741.

[12] J. Majak, B. Shvartsman, M. Pohlak, K. Karjust, M. Eerme, and E. Tungel, "Solution of fractional order differential equation by the Haar Wavelet method. Numerical convergence analysis for most commonly used approach," 2016, doi: 10.1063/1.4952346.

Analysis of the reciprocal wear testing of Aluminum AA1050 processed by a novel mechanical nanostructuring technique

Babak Omranpour[1,2,3*], Lembit Kommel[1], Fjodor Sergejev[1], Julia Ivanisenko[2], Maksim Antonov[1], Marco A L Hernandez-Rodriguez[4], Edgar Garcia-Sanchez[4].

1 Department of Mechanical and Industrial Engineering, Tallinn University of Technology (TalTech), 19086 Tallinn, Estonia.
2 Institute of Nanotechnology (INT), Karlsruhe Institute of Technology (KIT), 76021 Karlsruhe, Germany.
3 Hydrogen Research Institute, University of Quebec in Trois-Rivieres, 3351 des Forges, Trois-Rivieres G9A 5H7, Canada.
4 Facultad de Ingeniería Mécanica y Eléctrica (FIME), Universidad Autónoma de Nuevo León (UANL), 66455 San Nicolás de los Garza, Mexico.

omranpou@uqtr.ca

Abstract. This research aims to investigate the impact of a novel technique in mechanical nanostructuring on the wear resistance of materials. This technique with the name of High Pressure Torsion Extrusion (HPTE) can produce bulk nanostructured materials with enhanced mechanical properties. Results of microstructural analysis and microhardness testing showed significant enhancement in materials after HPTE. Microstructural characterization by using Electron Back-Scattered Diffraction (EBSD) method illustrated the presence of Ultra-Fine Grained (UFG) materials in the specimens Analysis of the wear by implementing reciprocal wear testing revealed that the amount of displaced volume markedly decreased after processing. This change in the wear behavior can be explained by referring to the hardness increase and the reduction of plasticity in materials which confined the plastic shearing and diminished the built-up edge around the wear track.

1. Introduction

The scientific study of wear and tribology has a long history. The modern development of tribology dated back to the 15th century initiated by Leonardo Da Vinci [1]. However, understanding the mechanism of wear and selecting a strategy to reduce it or to improve the tribological properties have always been a challenge for engineers. It is believed that harder materials generally provide lesser wear and lower friction coefficient [2]. Several mechanical nanostructuring techniques have been employed to enhance the mechanical and functional properties of metals and alloys [3]. Severe Plastic Deformations (SPD) are renowned for their impact on the grain refinement in metallic materials and have become more famous during the past years because of their significant contribution to the enhancement of mechanical properties [4]. Ultra-fine grained (UFG) materials processed by SPD proved to demonstrate better mechanical and functional properties as compared to their coarse-grained counterparts [5]. In the case of tribological properties, several studies evaluated different SPD techniques including High-Pressure Torsion (HPT) [6], Equal Channel Angular Pressing (ECAP) [7], and Accumulative Roll Bonding (ARB) [8] in terms of the wear and friction coefficient. The enhancement of wear resistance in SPD induced materials is a controversial subject among scientists [9,10]; particularly in aluminum alloys, some studies showed an improvement in the wear resistance of the processed samples [7,11], while others reported a reduction in the wear resistance after processing [12,13]. This work for the first time is going to evaluate the impact of a new technique with the name of High-Pressure Torsion Extrusion (HPTE) on the wear resistance of materials; aluminum series 1050 was selected as model materials in this work.

HPTE is a combination of two processes of HPT and extrusion and is capable of producing bulk samples with UFG microstructure [14]. Figure 1 shows the die and a 3D wireframe model. The HPTE die is composed of an upper die, which is fixed, and a lower die that rotates by an electromotor. Samples are inserted through the upper die, extruded by a punch, and then, exit through the lower die. Two process parameters of HPTE, e.g. the extrusion speed (v), and rotational speed (ω) will dictate the amount of imposed strain and consequently, the level of grain refinement in this technique. Earlier studies about HPTE revealed a considerable grain refinement as well as remarkable enhancement in the mechanical and functional properties of materials after processing. Further details about the process and the die geometry can be found elsewhere [15–17]. Due to the impact of torsion in this technique, materials demonstrate a gradient distribution in grain refinement and hardness at the cross-section of the specimens, from the center to the edge; however, as a general rule, increasing the ratio of ω/v in HPTE increases the amount of strain imposed on the material, and accordingly, boosts the amount of grain refinement [16]. This research will evaluate the wear resistance and friction coefficient of aluminum after HPTE processing by implementing dry sliding wear test.

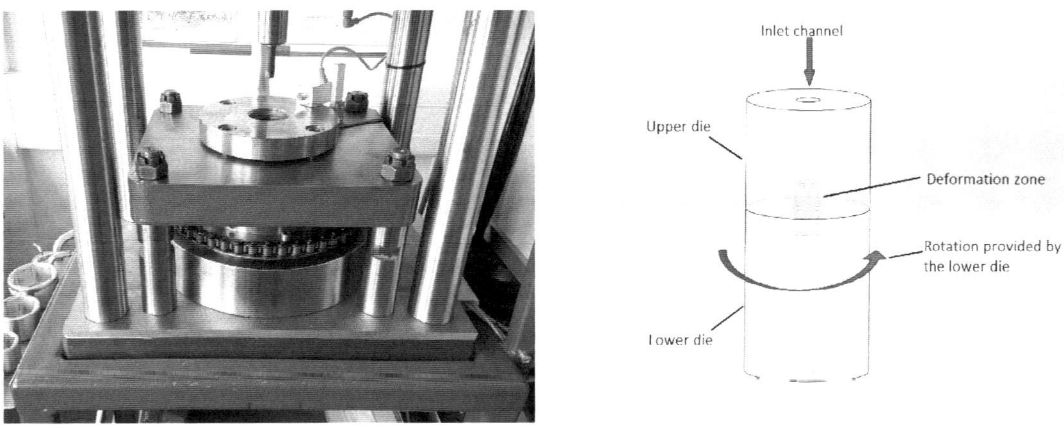

Figure 1:HPTE die; the assembled configuration (left), and 3D wireframe model (right)

2. Experimental procedure

Commercially Pure aluminium (CP-Al) with the following chemical composition was annealed at 350 °C for half an hour and then processed via HPTE: Al > 99.5 %, Fe < 0.20 %, Si < 0.15 %, Cu < 0.05, Mg < 0.02 % (wt.). Three different HPTE regimes adopting different levels of deformations were selected for evaluation: v= 7 mm/min, ω = 1 rpm (v7w1); v= 1 mm/min, ω = 1 rpm (v1w1); v= 1 mm/min, ω = 3 rpm (v1w3). One sample was kept unprocessed (only annealed) for comparative purposes. The billets after processing were cut, sectioned, and prepared for reciprocal wear testing. A "Buehler Micromet-5104" tester with a load of 0.2 Kg and a dwell time of 15 sec was implemented to apply the micro-indentation at the cross-section of the specimens on several points along the diameter to demonstrate the distribution of hardness in the specimens. The tests were carried out on three specimens in each group to ensure the consistency of the results. The reciprocal wear testing was performed by a Wazau SVT-500 tribometer on the cross-sectional surface of the samples with diameters of 11 mm in dry sliding conditions. Considering the fact that processed materials have a gradient distribution in microstructure and hardness at the cross-section from edge to center, different wear tests were performed at three different sections to study the effect of torsion-induced gradient on the wear resistance of the samples: center zone (r ≈ 0.5 mm), mid-radius zone (r ≈ 2.5 mm), and the edge zone (r

≈ 4.5 mm). Bearing steel balls made of chrome steel 100CR6 (Redhill balls ™) with a diameter of 3 mm with Vickers hardness of 63±3 HRC were used as a counterpart for wear testing of the samples. The reciprocal sliding wear tests were performed at room temperature (21 °C) and relative humidity of 50 %. Each test was repeated three times. All samples were tested under a load of 0.2 N. Material properties of the aluminum specimens and steel balls are summarized in Table 1. The wear tests were conducted at a frequency of 3 Hz and an amplitude of 1 mm with a sliding speed of 6 mm/s for a total period of 100 min. The mean and maximum values of contact pressure were ~0.16 GPa and ~0.25 GPa respectively for the Hertzian initial point contact in the present tribosysem. The volume loss in the wear tracks was measured using a confocal microscope, Brukers Contour GT-K white light optical profilometer. The microstructural analysis of the samples was performed by using FEI Quanta 3D scanning electron microscope (SEM) operating at 20 kV. The EBSD images were obtained by using the orientation imaging (OIM) software.

Table 1: Mechanical Properties of the specimens and the steel balls

	Yield strength [MPa]	Ultimate strength [MPa]	Modulus of Elasticity [GPa]	Poisson's ratio
AA1050-O workpiece	41	108	71	0.33
100Cr6 steel bearing ball	586	824	203	0.29

3. Results and discussion

3.1. Mechanical and microstructural evolution

Microstructural analysis of the specimens demonstrated a considerable refinement in the grain size after HPTE. Figure 2 shows the SEM investigation of materials. The annealed specimen is shown in Figure 2-a, and the processed specimens are represented by orientation image mapping in Figure 2- b, c, d. The EBSD images were taken from the cross-section of the HPTE specimens near the edge (where the highest amount of strain was imposed on the materials) with an approximate distance of r ≈ 4.5 mm from the center. As illustrated in this figure, the grain size in the annealed specimen with an average value of 120 µm (Figure 2-a) reduced to 1.4 µm in v7w1 (Figure 2-b), 0.9 µm in v1w1 (Figure 2-c), and 0.7 µm in v1w3 (Figure 2-d).

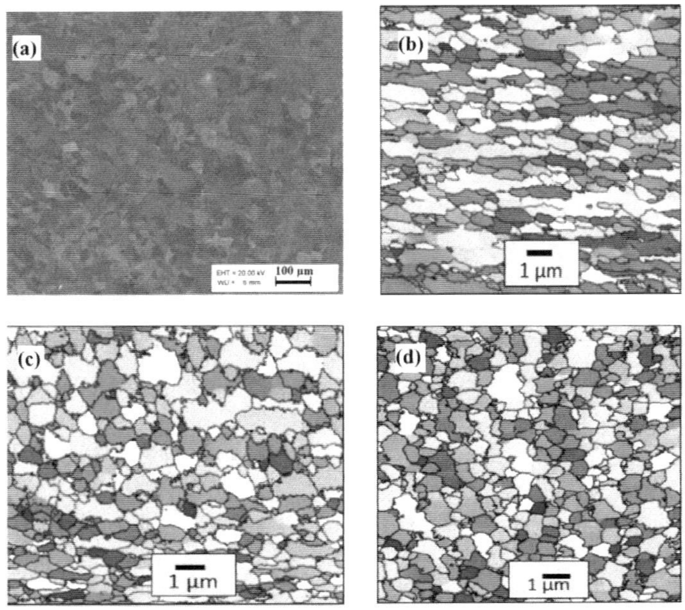

Figure 2: Microstructural evolution of aluminum by HPTE processing; (a) Initial specimen (annealed), (b) v7w1, (c) v1w1, and (d) v1w3.

Results of hardness testing are represented in Figure 3. As a result of microstructural refinement, there is a substantial improvement in the hardness [18]. All the processed samples demonstrated a V-shape distribution as a result of torsion which was applied to the material. As shown in this figure, the initial hardness with an average value of 29 HV0.2 increased to 51, 64, and 67 HV0.2, at the edge zone of v7w1, v1w1, and v1w3, respectively.

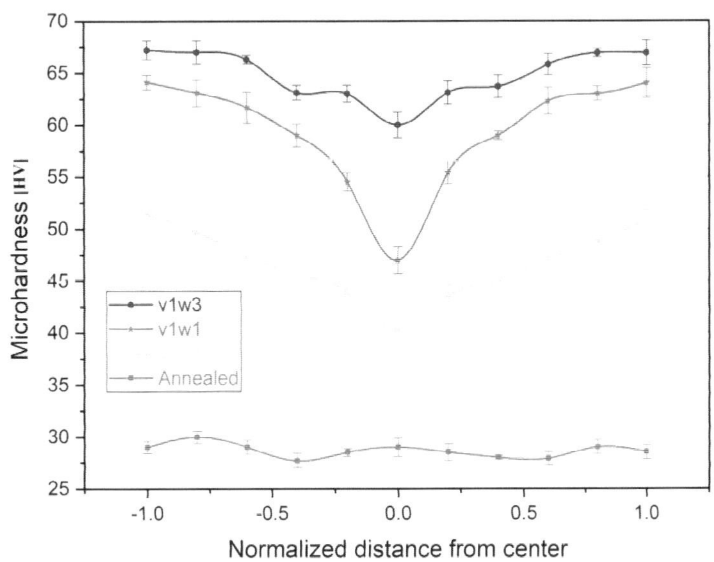

Figure 3:Hardness evolution in the cross-section of HPTE specimens

3.2. Wear test

Evaluation of the coefficient of friction (COF) in the samples showed that HPTE did not make any significant influence on the results; COF in all samples, including the annealed and processed ones, was in the order of 0.31 ± 0.03.

Analysis of the wear track using a 3D profiler is shown in Figure 4, and Table 2 collects the results of this analysis. The positive volume in this figure is the transferred material to the edge of the wear track as a result of plastic shearing during the wear test, and the negative volume is the groove that was formed during the test.

The net Missing Volume (M.V.) in Table 2 is the difference between the negative and positive volume, and the total Displaced Volume (D.V.) is the sum of absolute values of positive and negative volumes. As demonstrated in this table, the net missing volume (M.V.) did not decrease in v1w1 and v7w1 regimes; the only reduction of M.V. was found in v1w3 where the highest hardness values were achieved (see also Figure 3). However, all regimes made a noticeable reduction in the Displaced Volume (D.V.). This enhancement is due to reducing the amount of positive volume (built-up edge) forming around the wear track.

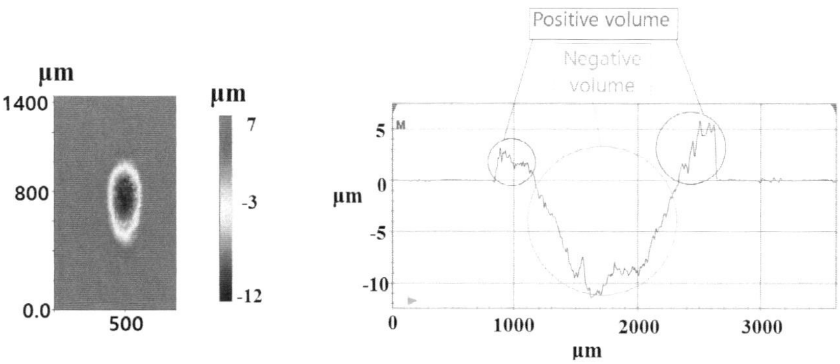

Figure 4: Illustration of the wear track in v1w1 specimen in the center zone; left: 3D-profile analysis of the wear track; right: cross-section of the wear track; the built-up edge was considered as the positive displaced volume, and the groove space as the negative volume.

Table 2: Volume loss of the wear tracks

	Net Missing Volume (M.V.) [$10^6.\mu m^3$]			Total Displaced Volume (D.V.) [$10^6.\mu m^3$]		
Track location	Center	Mid-radius	Edge	Center	Mid-radius	Edge
v1w3	1.08	1.36	1.23	1.15	1.38	1.33
v1w1	2.51	4.92	5.80	3.51	5.20	6.50
v7w1	2.85	4.50	5.95	6.35	6.93	7.75
Annealed	2.06	1.86	2.70	7.80	8.74	8.34

According to the results, the processed samples generated lower amounts of positive volume, and hence, lower D.V. This phenomenon is associated with increasing the hardness and resistance against the plastic shearing in the HPTE samples. As shown in Figure 3, the processed samples showed higher hardness values as compared to the initial one. In the case of the v1w3 regime, materials received the highest amount of strain and accordingly, represented the highest resistance against plastic shearing; therefore, the main portion of the displaced volume was the negative volume which was worn away through abrasion. On the other hand, the annealed sample with the lowest hardness value was mainly worn through the mechanism of adhesion, forming a thick layer of built-up edge (named as positive volume) around the wear track. This positive volume took up a large portion of the displaced volume in the wear. Indeed, the presence of positive volume makes a great contribution to the intensification of adhesive wear. Further investigations will be pursued by the authors to study the effect of this technique on the mechanism of the wear in materials.

4. Conclusion

- A new approach in the microstructural refinement of metals by means of mechanical nanostructuring was examined for the wear resistance and tribological properties of aluminum AA-1050.
- Analysis of the microstructure by EBSD showed remarkable grain refinement in the processed samples.

- Results of reciprocal wear testing at the cross section of the specimens showed that this technique was able to decrease the amount of displaced volume during the wear and thereby, reducing the adhesive wear. This fact corresponds to the enhancement of hardness and increasing the resistance of materials against plastic shearing.

Acknowledgment:
This work was sponsored in part by the Estonian Research Council grant (PUTJD1010) and in part by the institutional research funding IUT-1929. M. Antonov would like to acknowledge the research grant "DuraCer" MOBERA9 ETAG18012.

References
[1] Stachowiak G W and Batchelor A W 2013 *Engineering Tribology: Fourth Edition* (Elsevier Inc.)
[2] Semenov V I, Shuster L S, Huang S J, Rajendran R, Chertovskikh S V. and Shibakov V G 2016 Comparative evaluation of the tribological properties of low-and medium-carbon steels after heat treatment and severe plastic deformation *Rev. Adv. Mater. Sci.* **45** 28–35
[3] Jung A, Diebels S and Lach E 2015 Improved Mechanical Properties by Nanostructuring-Specific Considerations under Dynamic Load Conditions *Handbook of Mechanical Nanostructuring* vol 1 (Weinheim, Germany: Wiley-VCH Verlag GmbH & Co. KGaA) pp 181–209
[4] Kommel L, Shahreza B O and Mikli V 2019 Microstructure and physical-mechanical properties evolution of pure tantalum processed with hard cyclic viscoplastic deformation *Int. J. Refract. Met. Hard Mater.* **83** 104983
[5] Toth L S and Gu C 2014 Ultrafine-grain metals by severe plastic deformation *Mater. Charact.* **92** 1–14
[6] Edalati K, Ashida M, Horita Z, Matsui T and Kato H 2014 Wear resistance and tribological features of pure aluminum and Al-Al2O3 composites consolidated by high-pressure torsion *Wear* **310** 83–9
[7] Abd El Aal M I, El Mahallawy N, Shehata F A, Abd El Hameed M, Yoon E Y and Kim H S 2010 Wear properties of ECAP-processed ultrafine grained Al-Cu alloys *Mater. Sci. Eng. A* **527** 3726–32
[8] Talachi A K, Eizadjou M, Manesh H D and Janghorban K 2011 Wear characteristics of severely deformed aluminum sheets by accumulative roll bonding (ARB) process *Mater. Charact.* **62** 12–21
[9] Gao N, Wang C T, Wood R J K and Langdon T G 2012 Tribological properties of ultrafine-grained materials processed by severe plastic deformation *J. Mater. Sci.* **47** 4779–97
[10] Kommel L, Põdra P, Mikli V and Omranpour B 2021 Gradient microstructure in tantalum formed under the wear track during dry sliding friction *Wear* **466–467** 203573
[11] Zhang Z, Hosoda S, Kim I S and Watanabe Y 2006 Grain refining performance for Al and Al-Si alloy casts by addition of equal-channel angular pressed Al-5 mass% Ti alloy *Mater. Sci. Eng. A* **425** 55–63
[12] Wang C T, Gao N, Wood R J K and Langdon T G 2011 Wear behavior of an aluminum alloy processed by equal-channel angular pressing *J. Mater. Sci.* **46** 123–30
[13] Oh-Ishi K, Horita Z, Furukawa M, Nemoto M and Langdon T G 1998 Optimizing the rotation conditions for grain refinement in equal-channel angular pressing *Metall. Mater. Trans. A Phys. Metall. Mater. Sci.* **29** 2011–3
[14] Ivanisenko Y, Kulagin R, Fedorov V, Mazilkin A, Scherer T, Baretzky B and Hahn H 2016 High Pressure Torsion Extrusion as a new severe plastic deformation process *Mater. Sci. Eng. A* **664** 247–56
[15] Omranpour B, Kulagin R, Ivanisenko Y and Garcia Sanchez E 2017 Experimental and numerical analysis of HPTE on mechanical properties of materials and strain distribution *IOP Conference Series: Materials Science and Engineering* vol 194
[16] Omranpour B, Ivanisenko Y, Kulagin R, Kommel L, Garcia Sanchez E, Nugmanov D, Scherer T, Heczel A and Gubicza J 2019 Evolution of microstructure and hardness in aluminum processed by High Pressure Torsion Extrusion *Mater. Sci. Eng. A* **762** 138074
[17] Omranpour B S, Kommel L, Sanchez E G, Ivanisenko Y and Huot J 2019 Enhancement of Hydrogen Storage in Metals by Using a New Technique in Severe Plastic Deformations *Key Eng. Mater.* **799** 173–8
[18] El-Danaf E A, Soliman M S, Almajid A A and El-Rayes M M 2007 Enhancement of mechanical properties and grain size refinement of commercial purity aluminum 1050 processed by ECAP *Mater. Sci. Eng. A* **458** 226–34

The influence of high-energy ball milling and nanoadditives on the kinetics of heterogeneous reaction in Ni-Al system

Kh T Nazaretyan[1(a)], H V Kirakosyan[1], S V Aydinyan[1,3], M K Zakaryan[1,2], L S Abovyan[1], M Kulak[4] and B Khina[4]

[1]A.B. Nalbandyan Institute of Chemical Physics NAS RA, P. Sevak 5/2, 0014, Yerevan, Armenia
[2]Yerevan State University, A. Manoogian 1, 0025, Yerevan, Armenia
[3]Tallinn University of Technology, Ehitajate 5, 19086, Tallinn, Estonia

E-mail: khachik.nazaretyan@mail.ru

Abstract. Kinetic studies were performed utilizing high-speed temperature scanner in the Ni-Al system including those with and without mechanical activation (MA) of different duration in a planetary ball mill, with and without using carbon nanoadditives (NA). The temperature profiles were taken and treated at different heating rates from 100 up to 2600 °C/min considering the influence of activation duration and the role of nanoadditive on the characteristic points of thermograms. Kissinger method allowed to evaluate activation energy (E_a) for non-activated, activated (1, 2, 3, 5 min), nanoadditive (1 wt.%) containing and nanoadditive (1 wt.%) containing mechanoactivated (1, 3, 5 min) mixtures. The beneficial influence of NA on the interaction between Ni and Al in the non-activated and moderately mechanoactivated mixtures was demonstrated. The influence of MA and NA on the microstructure features and phase formation sequence at various heating rates were revealed. For all the mixtures under study, T* characteristic temperatures (the temperature, where the maximum exothermic effect was observed) were found to increase with increasing heating rates. It was unravelled that mechanical treatment leads to significant changes in the reaction kinetics and phase formation laws. Particularly, in an activated mixture, the formation of Ni_3Al is followed by NiAl intermetallic, in contrast to non-activated mixture, where the reaction proceeds only with the NiAl formation. The both MA in 1 min and addition of 1 wt.% NA decreased the activation energy of the Ni-Al reaction, exhibiting commensurate impact on the effective activation energy value of the Ni-Al system. However, ≥3 min MA in the presence of 1 wt.% NA have prohibitive effect on the reaction in the Ni-Al system.

1. Introduction

Intermetallic compounds based on Ni-Al are being recognized as potential high temperature structural materials because of their low density (5.86 g cm^{-3}), high melting point (1911 K), excellent oxidation resistance up to 1573 K, as well as good thermal conductivity [1-2]. The combination of distinctive properties has made Ni-Al intermetallic compounds potentially useful for many structural applications, such as gas turbine hardware, corrosion resistant materials, heat treatment fixtures, high temperature dies, molds and cutting tools [3-4].

Previous studies have demonstrated the fabrication of nickel aluminides by means of powder metallurgy [5-9], mechanical alloying [10-12] and self-propagating high temperature synthesis (SHS) (also known as combustion synthesis (CS)) [13-16]. The synergistic influence of the both MA and SHS of nickel aluminides was manifested to be perspective in terms of phase regulation and impressive enhancement of intrinsic properties [17]. In [18, 19] the decrease in apparent activation energy of the reaction was attributed to the change of interaction pathway due to mechanically assisted combustion synthesis of Ni-Al. The preparation of Ni-Al intermetallic compound under the activated combustion mode was successfully implemented, as described in [20]. As a promoter, negligible amount of polytetrafluoroethylene (PTFE) was added into initial Ni-Al mixture.

Therefore, it is relevant to develop novel approaches for SHS processes to provide in-situ control on the phase and structure, as well as on the properties of target materials. In this regard, preliminary MA and the introduction of nanoadditives (NA) to the initial charge become promising.

Considering high temperatures and high self-heating rates of substances in the combustion front, the in-situ exploration of interaction mechanism of the combustion process is impeded. The modeling of the process at controllable conditions (e.g. with programmed heating rates and tuning the process within the time at conditions closer to heating rates and temperatures in the combustion wave) by thermal analysis methods will tackle the obstacle of mechanism examination.

High-Speed Temperature Scanner (HSTS setup) exclusively designed at the Institute of Chemical Physics NAS RA [21-23] for the kinetic investigations of the powder mixtures under high heating rates (up to 10000 $^{o}C\ min^{-1}$) and temperature up to 1300 ^{o}C, was utilized to explore the influence of MA and NA on the structure and phase formation peculiarities in the Ni-Al system, as well as to estimate effective values of activation energies.

This approach provides enhanced opportunities to reveal the stepwise nature of reactions in the Ni-Al system. By varying heating rates of reagents, it was possible to separate the main stages and analyze intermediate quenched compounds for the efficient exploration of interaction mechanism in the systems under study. Moreover, implementation of the process with different heating rates allowed to use the model-free non-isothermal Kissinger method [24, 25] to calculate Arrhenius parameters of the process.

This study is motivated by the need to investigate the influence of MA (1-5 min) and presence of NA (carbon nanotubes) on the structure formation patterns and interaction kinetics of the Ni-Al system (with 1:1 molar ratio of metals) at non isothermal conditions using HSTS technique.

2. Experimental

Nickel (PNK-1VL7, reagents grade, particle size $\mu<10\ \mu m$), aluminium (ASD-4, reagents grade, particle size $\mu<40\ \mu m$) and carbon nanotubes (CN) were used as precursors. Carbon nanotubes were synthesized by Chemical Vapor Deposition (CVD) method on a nickel-plated surface [26]. The substrate was a copper with a galvanic nickel coating with a thickness of about 5 μm. The synthesis was carried out at atmospheric pressure. Heating was controlled by the temperature of the plate using a thermocouple. The operating temperature was 710 ^{o}C. The reactor was heated and cooled under constant nitrogen purging with a flow rate of 1.5 $l\cdot min^{-1}$ (the volume of the reaction chamber was 0.6 l). When the operating temperature was reached, vapors of the propane-butane mixture with a flow rate of 0.5 $l\ min^{-1}$ were fed into the reactor. The resulting carbon material was mechanically cleaned from the surface. The mass fraction of nickel in the CN was from 1.5 to 5 wt.%. The material contained disordered filaments and fibers with a wide variation in a diameter (from dozen nanometers to several microns) (Fig. 1). The nanoadditive was added to Ni-Al mixture after mechanoactivation (m(nanoadditive)=1 wt.%) by gently mixing in a ceramic mortar for 10 min.

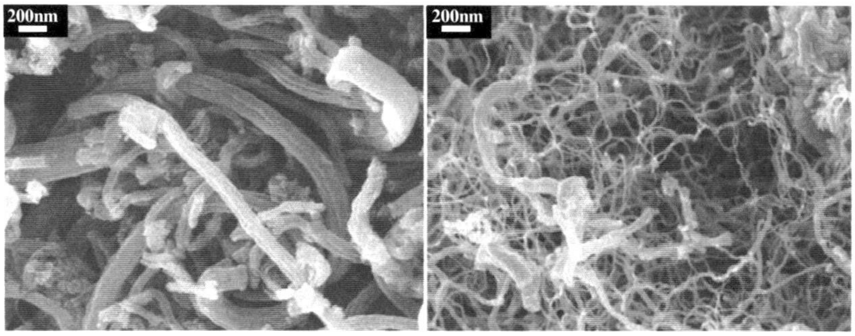

Fig. 1. SEM images of carbon nanostructured material

MA of initial mixtures was carried out from 1 to 5 min in a planetary ball mill (RETSCH PM-100) at fixed rotation speed of 500 rpm. The mass charge ratio of milling media (10 mm steel balls) and powder was 20:1.

The experiments were carried out by HSTS setup under programmed high heating rates (100-2600 °C min^{-1}) and temperature up to 1300 °C. The reactive powder mixture (50-100 mg) was placed into the metallic envelope made from thin nickel foil with 0.1 mm thickness, and a K-type thermocouple (chromel-alumel) spot-welded directly to the foil in the central area of powder location. All measurements were conducted at 0.1 MPa argon (>99.98 %) pressure. The foil heater was heated by passing electric current with desired temperature time schedule provided by PC-assisted controller. Typical temperature profile for the inert experiment provides linear temperature time history, which defines the heating rate and coincides with the reactive T profile at low-temperature region when the reaction does not occur. The reaction onset temperature (T_o), the maximum peak temperature observed during self-heating (T_{max}) and the temperature prescribed by linear heating, where the maximum exothermic effect is observed (T*), were determined from the heating thermograms (Fig. 2).

Kissinger's isoconversion method [24], based on the principle of temperature shift corresponding to the maximum advance on the DTA curve, as a function of heating rate (V_h), was applied to calculate the effective activation energy value for the process under study:

$$\ln\left(\frac{V_h}{\left(T_{max}^{DTA}\right)^2}\right) = \ln A - \frac{E}{R}\left(\frac{1}{T_{max}^{DTA}}\right) \quad (1)$$

where, V_h is the heating rate (K min^{-1}), T_{max}^{DTA} is the temperature corresponding to the maximum advance in the DTA curve (K), A is a constant, E is the effective activation energy of the process, (kJ mol^{-1}), R is the universal gas constant (8.31 J K^{-1} mol^{-1}).

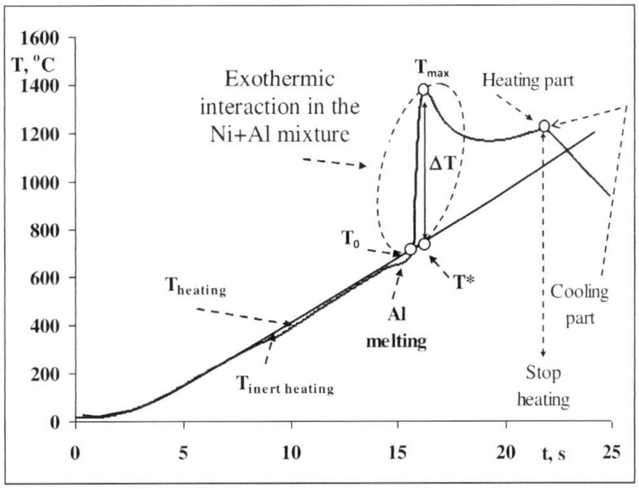

Fig. 2. Typical heating thermograms of Ni-Al mixture

The process was interrupted at different characteristic stages. Phase composition and microstructure of the cooled samples were examined. When interrupting the process, high cooling rates were achieved (up to 12000 °C min^{-1}) to exclude further interaction over the cooling stage.

Phase compositions of the intermediate and final products were analyzed by X-ray diffraction (XRD, DRON-3.0, Burevestnik, Russia) with monochromatic CuKα radiation (λ=1.5406 Å) operated at 25 kV and 10 mA. To identify the products from the XRD spectra, the data were processed using the JCPDS-ICDD database. The microstructure of powders was examined by BS-300 and LEICA440i scanning electron microscopes (SEM).

3. Results and discussion

3.1. Mechanical activation of Ni+Al powder mixture

The mechanical activation of initial Ni+Al powder mixture was implemented by dry mixing in air at different durations (from 1 to 5 min). The results of XRD examinations demonstrated no change in phase composition regardless the MA duration (Fig. 3). However, the characteristic peaks of metals were noticeably expanded with increasing the duration of MA.

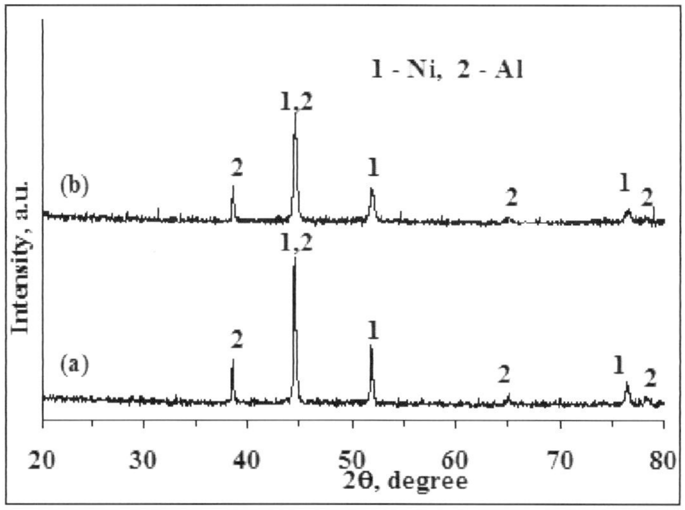

Fig. 3. XRD patterns of Ni+Al mixtures, a) t_{MA} = 0 min, b) t_{MA} = 3 min

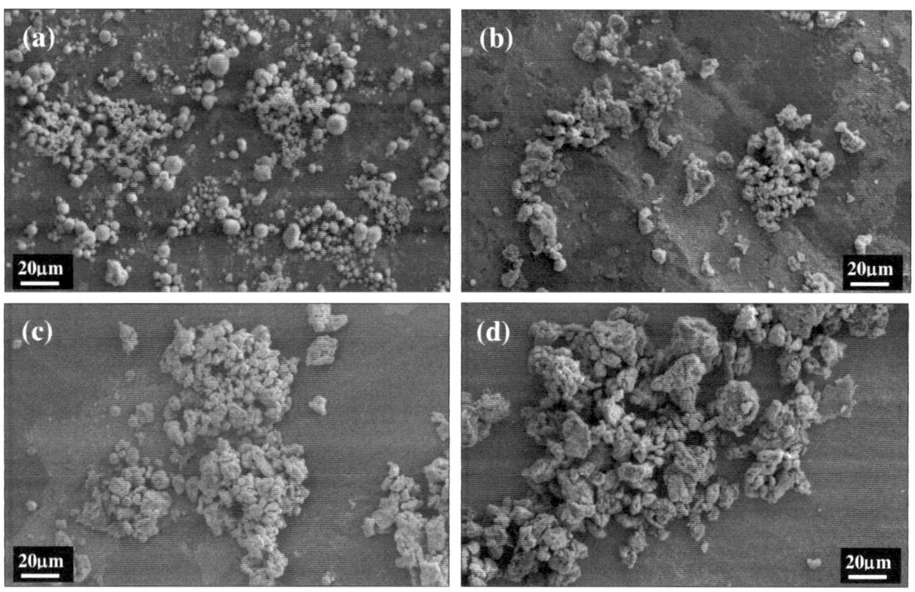

Fig. 4. SEM images of Ni+Al mixtures,
a) t_{MA}=0 min, b) t_{MA}=1 min, c) t_{MA}=3 min, d) t_{MA}=5 min

According to microstructural analysis (Fig. 4), the increase in MA duration expectedly leads to a change in the form and size of particles. In particular, homogenization was observed at 1 min MA, while agglomeration of particles was registered at 3 and 5 min activation.

In figure 5 the microstructures of the initial mixtures in the presence of nanoadditive are shown before and after mechanical activation. The non-activated powder mixture is characterized by a smaller particle size (1-10 μm) compared to 3 min mechanoactivated mixture (10-20 μm).

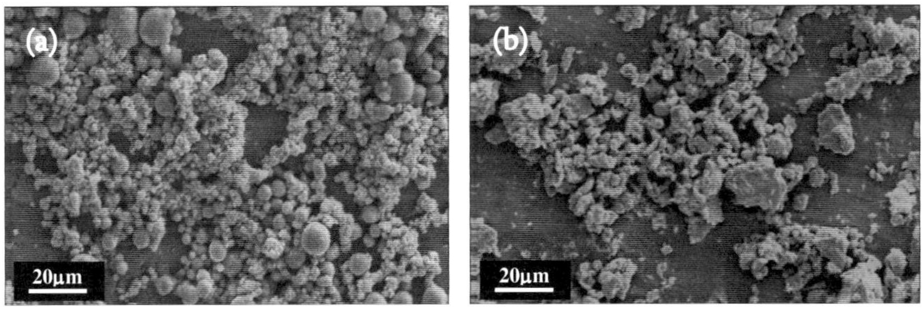

Fig 5. SEM images of (Ni+Al)+1wt%(NA) mixtures after mechanical activation, a) t_{MA}=0 min; b) t_{MA}=3 min

3.2. Influence of mechanical activation and nanoadditives on the Ni–Al system at 2600 °C min^{-1} heating rate

Mechanical activation can lead to either increase or decrease in the efficiency of heat exchange between particles, which depends on the particle shape and size, on the particle size distribution, the interparticle contact surface area and the thermal resistance of interparticle contacts. The effect of activation duration is demonstrated on the typical temperature profiles of Ni+Al mixtures at 600 and 2600 °C min^{-1} heating rates. Single stage exothermic Ni-Al interaction was observed regardless the activation duration (Fig. 6). However, in non-activated mixtures and at the short duration of MA (up to 2 min) interaction starts after the melting of aluminum (630, 670 °C), while starting from 3 min of activation the process begins before the melting point of aluminum (at 580 °C). This pattern was observed regardless the heating rates (from 100 up to 2600 °C min^{-1}). The increase in self-heating temperature (characterizing the amount of heat released) in the 1 and 2 min mechanoactivated mixtures occurred as a result of increase in the contact surface area between particles of the mixture and intensification of bulk and grain boundary diffusion. The further decrease of self-heating temperature (3 and 5 mins) is in agreement with SEM analysis and is conditioned by the reduction of the specific heat release of the activated mixture and to blockage of the contact surface of the reactants.

The addition of 1%NA didn't have significant influence on starting temperature (T_o) of the processes (Fig. 7. 0 and 0', 3 and 3', 5 and 5'), however stongly contributed to the self-heating temperature of non-activated mixtures.

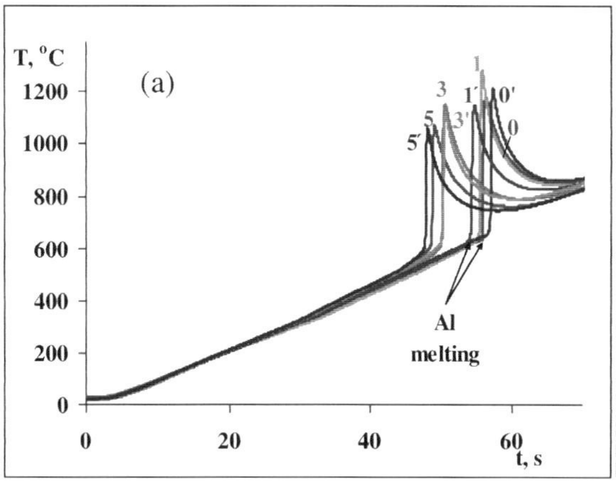

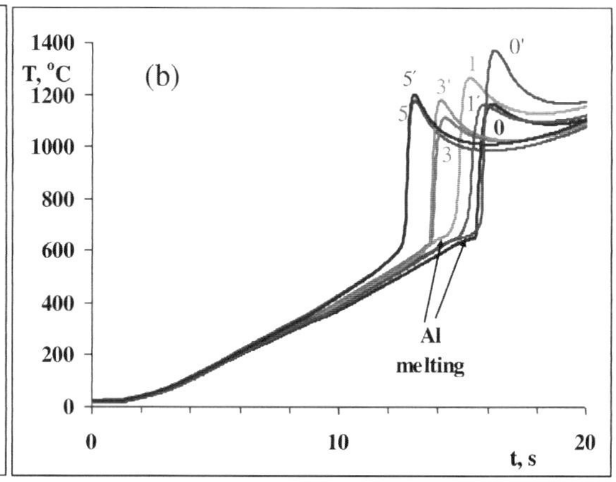

Fig. 6. The influence of MA and NA on heating curves of various Ni+Al mixtures with and without MA, a_V_h=600 °C min^{-1} , b_V_h=2600 °C min^{-1}: 0_ t_{MA}=0 min; 0'_t_{MA}=0 min+1%NA; 1 _ t_{MA}=1 min; 1'_t_{MA}=1 min+1%NA; 3 _ t_{MA}=3 min; 3'_t_{MA}=3 min+1%NA; 5 _ t_{MA}=5 min; 5'_t_{MA}=5 min+1%NA

3.3. Influence of heating rate on the activation energy and phase formation of the various Ni–Al system

In the all studied non-activated Ni+Al mixtures the reaction T* shifts to a higher temperature area from 625 to 750 °C for T* and from 1060 to 1160 °C for T_{max} with increasing the heating rate from 100 to 2600 °C min^{-1}, while the exothermic peak intensity of the reaction remains almost the same. data obtained are close to the result of DTA/TG studies at low heating rates (5-20 °C min^{-1}, 650-750 °C; 159 kJ mol^{-1}) [20].

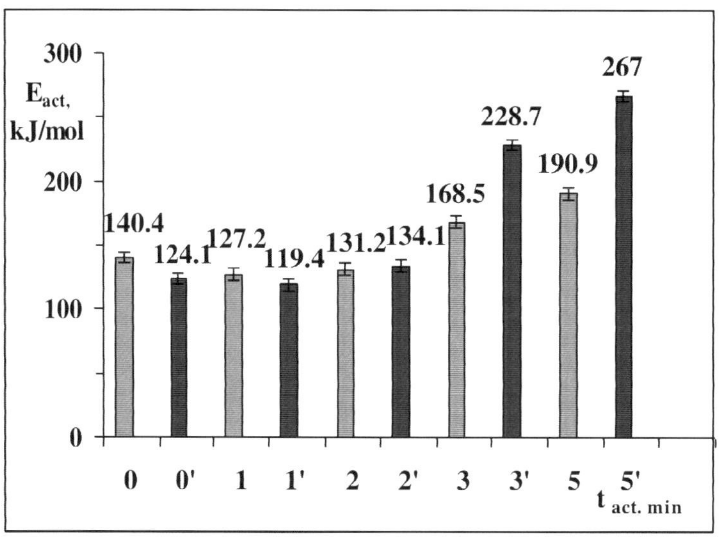

Fig. 7. The activation energy of Ni-Al mixture with and without nanoadditives dependent on the duration of mechanoactivation

This type of thermal behavior is characteristic to exothermic systems with high activation energies. In particular, according to Kissinger's method the activation energy of Ni-Al exothermic interaction was calculated to be 140.4 kJ mol^{-1} (630-750 °C, 100-2600 °C min^{-1}). The

The duration of mechanoactivation has detrimental effect on the effective activation energy value. In particular, 1 or 2 min activated mixtures demonstrated lower value of activation energy compared to that for non-activated mixtures and for the mixtures processed 3 and 5 min.

Considering the ductile nature of aluminum, it was preliminary expected that longer duration of activation (here it was only 3 min) may have reverse effect, i.e. lead to increase in particle size and decrease in the surface-to-volume ratio, or the specific surface area (SSA), which in turn affects the prevailing mechanism and hence the activation energy sensitive to particle size. Thus, effective activation energy decreases, and then increases vs activation duration (Fig. 7).

The change in interaction mechanism under the MA was disclosed by XRD examinations revealing that in the non-activated Ni-Al mixtures before the exothermic interaction, samples quenched immediately before the aluminum melting point (630 °C), contained only initial metals, whereas at 900 °C the product represented pure NiAl intermetallic compound (Fig. 8a). The mechanical activation promoted the formation of Ni_3Al intermetallic along with NiAl at 900 °C, while continuing the heating up to 1300 °C resulted in the formation of only NiAl intermetallic (Fig. 8b), which has a wide range of homogeneity according to the both parameters: composition and temperature. The formation of NiAl was predictable due to stoichiometry of the mixture (1:1), as well as by the studied temperature interval (up to 1300 °C). The formation of Ni_3Al intermediate in the mechanoactivated mixture might be explained by the lack of aluminum or diffusion barriers originated from aluminum agglomeration.

From other hand, the question about first formed phase in the Ni-Al interaction was discussed elsewhere and there is no unanimity about it, because some authors supposed $NiAl_3$, Ni_3Al, NiAl, Ni_2Al_3 as one, in addition the studied temperature interval was limited (up to 1000 °C). Enayati et al. [27] reported different behavior in Ni-Al powder mixture. They found that a Ni(Al) solid solution forms initially as an intermediate phase and then transforms to the Ni_3Al intermetallic compound on further milling. In [28] the increase of MA time had shown practical disappearance of Al with simultaneous increase of NiAl amount.

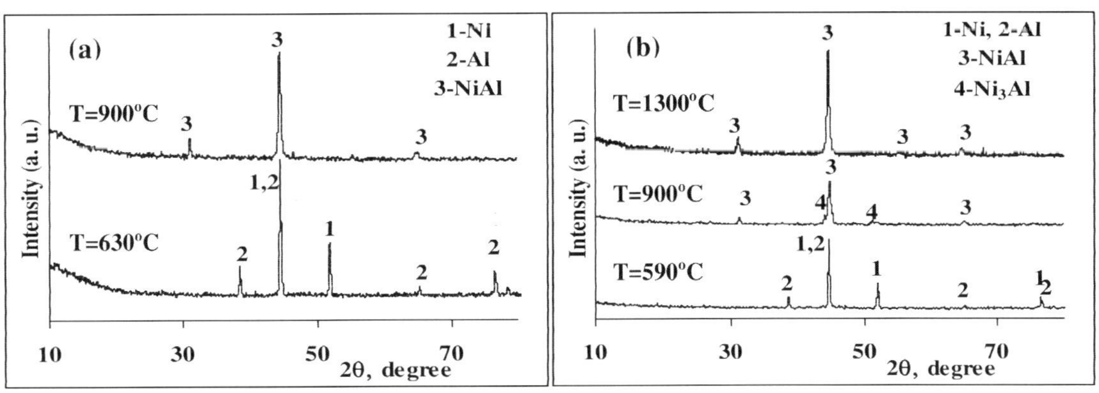

Fig. 8. XRD patterns of non-activated (a) and 3 min mechanoactivated (b) Ni+Al mixtures, 600 °C min^{-1}

The results obtained were accomplished by the microstructural examinations of the quenched samples. In the non-activated mixtures, the intermediate product didn't undergo to significant microstructural changes at 630 °C as compared to initial mixture (Fig. 9a,b). However, as the interaction process started after the melting of aluminum, particle coarsening occurred from 660 °C up to 900 °C, and at the end of the process (900 °C) agglomerates of NiAl with average particle size of 20 μm were formed (Fig. 9c). Mechanoactivation in 3 min led to an agglomeration of Ni and Al particles in the initial mixture (Fig. 9a, 10a), but noticeable change was not observed before the exothermic peak at 590 °C. As in the mechanoactivated mixtures with a duration of 3-5 min the interaction proceeded in solid-solid mechanism, before the melting of aluminum, no additional coarsening was detected, and at the end of exothermic process (900 °C) fine-grained product (up to few microns) was registered.

The microstructural differences between the non-activated and mechanoactivated mixtures are associated with (i) the size of particles in the initial mixture and (ii) onset temperature of exothermic reaction (before or after Al melting). The recent is more presiding over the interaction mechanism (s+s, s+l) and the activation energy value.

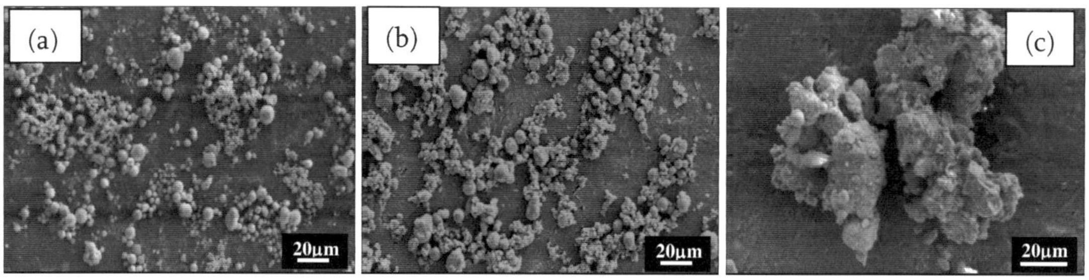

Fig 9. SEM images of quenched samples (V_h= 600 °C min^{-1}) for non-activated Ni+Al mixtures a - initial mixture, b - 630 °C c - 900 °C

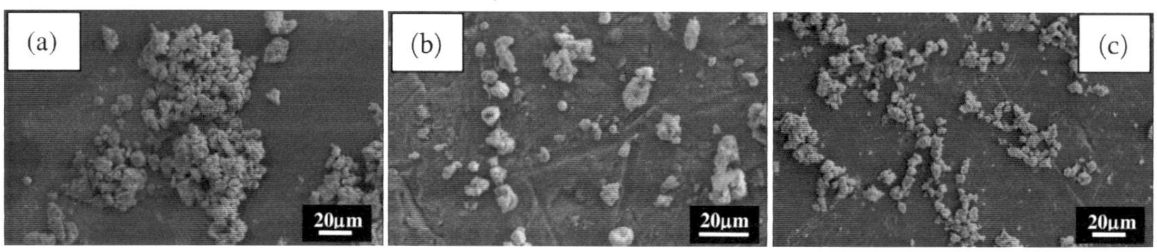

Fig. 10. SEM images of quenched samples (V_h= 600 °C min^{-1}) for 3 min mechanoactivated Ni+Al mixtures; a - initial mixture, b - 590 °C c - 900 °C

3.4. The influence of nanoadditives on the activation energy and phase formation of the non-activated and activated Ni–Al system

The addition of 1% of carbon nanotubes has stimulating influence on the non-activated Ni-Al mixture. As a result, Ni-Al interaction started earlier and exhibited increase in maximum temperature, as compared to the non-activated mixture without nanoadditives. According to the activation energy values calculated by Kissinger method, the addition of nanoadditives to a non-activated mixture significantly lowered the activation energy value from 140.7 to 127.2 kJ mol^{-1}. As it can be seen from the microstructural analysis of the samples cooled at 640 and 900 °C (V_h=600 °C min^{-1}) the introduction of nanoadditives into the system led to a significant decrease in the particle size of the product (Fig. 11) conditioned by the small grain size of the initial mixture and the homogeneous distribution of carbon nanotubes. This fact, in turn, affects the heterogeneous reaction and the crystallization of the alloy from a "short" existing melt, changing the composition and structure of the resulting material, contributing to the formation of a more homogeneous granular structure of the product (Fig. 11c).

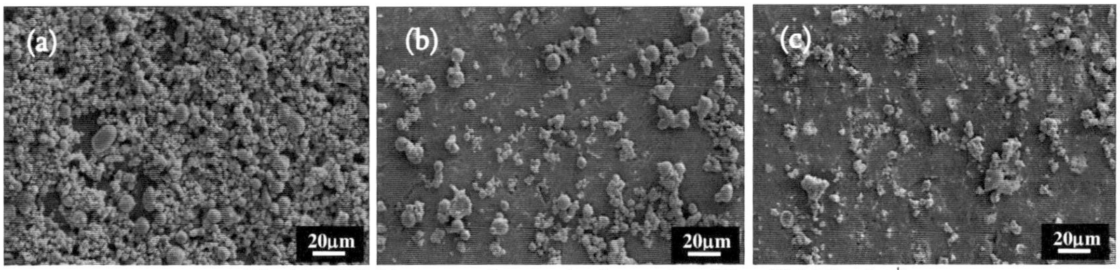

Fig.11. SEM images of quenched samples (V_h= 600 °C min^{-1}),
a - initial mixture, b - 640 °C c - 900°C

The addition of 1% nanoadditives to the mechanoactivated mixture of a duration in 1 min led to an increase in characteristic temperatures (T_o=595-680 °C, T*=600-780 °C, T_{max}=1000-1160 °C) according to heating rate (V_h=100-2600 °C min^{-1}), however as compared to the nanoadditive containing non-activated mixtures, T_{max} temperatures are lower in the studied heating rate region (Fig. 12).

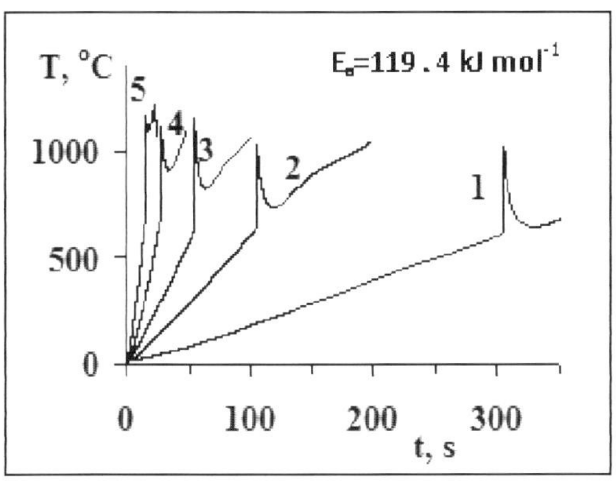

V, °C/min	T_o	T*(T_c)	T_{max}	ΔT
100	595	600	1000	400
300	630	633	1020	385
600	640	665	1140	480
1200	680	710	1110	390
2600	675	780	1160	380

Fig. 12. Heating curves of a mixture containing Ni+Al+1%NA activated for 1 min
V_h=100 (1), 300 (2), 600 (3), 1200 (4), 2600 (5) °C min^{-1}

According to XRD analysis results, regardless of the presence of nanoadditive in the non-activated mixtures (Ni+Al+1%NA and Ni+Al), one phase NiAl product was observed in the product quenched immediately after the exothermic peak. Ni_3Al was observed along with NiAl in the mechanoactivated mixtures with any duration (1, 3, 5 min), immediately after the exothermic reacton, and single-phase NiAl was formed at the end of the heating at 1200 °C (V_h=600°Cmin^{-1}).

Conclusion

The characteristic temperatures derived from the thermograms of the Ni+Al powder mixture, as well as the activation energies of the processes significantly depend on both the mechanoactivation duration and the presence of the nanoadditives. The increase in the mechanoactivation up to 2 min leads to the decrease in the activation energy (E_a) values. In the non-activated mixtures and at the short duration of MA (up to 2 min) interaction starts after the melting of aluminum (630, 670 °C), while starting from 3 min of activation the process begins before the melting point of aluminum (at 580 °C) regardless the heating rate (from 100 up to 2600 oC min-1). The both MA in 1 min and addition of 1 wt.% NA decreased the activation energy of the Ni-Al reaction, exhibiting commensurate impact on the effective activation energy value of the Ni-Al system. Mechanical treatment leads to significant changes in the reaction kinetics and phase formation laws. Particularly, in an activated mixture, the formation of Ni3Al is followed by NiAl intermetallic, in contrast to non-activated mixture, where the reaction proceeds only with the NiAl formation. In the non-activated mixtures the interaction process started after the melting of aluminum, then particle coarsening occurred from 660 °C up to 900 °C, and at the end of the process (900 °C) agglomerates of NiAl with average particle size of 20 μm were formed. As in the mechanoactivated mixtures with a duration of 3-5 min the interaction proceeded in solid-solid mechanism, before the melting of aluminum, no additional coarsening was detected, and at the end of exothermic process (900 °C) fine-grained product (up to few microns) was registered.

Acknowledgment
This work was supported by the RA MES Science Committee and Belarusian Republican Foundation for Fundamental Research the frames of the joint research project SC 18BL-011 and BRFFR T19APM-032 accordingly and Estonian Research Council grant PSG220 (S. Aydinyan).

References

[1] Czeppe T and Wierzbinski S 2000 *Int. J. Mech. Sci.* 2000 **42** pp 1499-1518

[2] Portnoy VK, Blinov AM, Tomilin IA, Kuznetsov VN and Kulik T 2002 *J. Alloys Compd.* 2002 **336** pp 196-201

[3] Kecskes LJ, Woodman RH, Barber RE, Hartwig KT and Gardiner DM, 2006 *Metall. Mater. Trans. A* **37** pp 449-57

[4] Hwang KS, Yang TH and Hu SC 2005 *Metall. Mater. Trans. A* **36** pp 2801-6

[5] Darolia R, Lewandowski JJ, Liu CT, Martin PL, Miracle DB and Nathal MV 1993 *Minerals, Metals and Materials Society* (Warrendale, PA, United States) p 916.

[6] Stoloff NS and Sikka VK 2012 *Physical Metallurgy and Processing of Intermetallic Compounds* (Springer Science & Business Media, United States) p 684

[7] Morsi K, Moussa SO and Wall JJ 2006 *J. Mater. Sci.* **41** pp 1265-8

[8] Lu L and Lai MO 1995 *Mat. Des.* **16** pp 33-9

[9] Atzmon M 1990 *Phys. Rev. Let.* **64** pp 487-90

[10] Ivanov E, Grigorieva T, Golubkova G, Boldyrev V, Fasman AB, Mikhailenko SD and Kalinina OT 1998 *Mater. Lett.* **7** pp 51-4

[11] Suryanarayana C 2001 *Prog. Mater. Sci.* **46** pp 1-84

[12] Takacs L 2002 *Prog. Mater. Sci.* **47** pp 355-414

[13] Philpot KA, Munir ZA and Holt JB 1987 *J. Mater. Sci.* **22** pp 159-69

[14] Kim JK and Chung DS 2007 *Solid State Phenom.* **124** pp 1099-102

[15] Lebrat J-P, Varma A and McGinn PJ 1994 *J. Mater. Res.* **9** pp. 1184-92

[16] Rogachev AS, Mukasyan AS and Varma A 2002 *J. Mater. Synth. Process.* **10** pp 31-6

[17] Morsi K, Shinde S and Olevsky EA 2006 *J. Mater. Sci.* **41** pp 5699-703

[18] Shteinberg AS, Lin Y-C, Son SF and Mukasyan 2010 *AS J. Phys. Chem. A* **114** pp 6111-6

[19] Kovalev DY, Kochetov NA, Ponomarev VI and Mukasyan AS 2010 *Int. J. SHS* **19** p. 120-5

[20] Baghdasaryan AM, Hobosyan MA, Khachatryan HL, Niazyan OM, Kharatyan SL, Sloyan LH and Grigoryan YG 2012 *Chem. Eng. J.* **188** pp 210-5

[21] Hobosyan MA, Kirakosyan KG, Kharatyan SL and Martirosyan KS 2015 *J. Therm. Anal. Calorim.* **119** pp 245-51

[22] Nepapushev AA, Kirakosyan KG, Moskovskikh DO, Kharatyan SL, Rogachev AS and Mukasyan AS 2015 *Int. J. SHS* **24** pp 21-8

[23] Kamboj N, Aghayan M, Rubio-Marcos F, Nazaretyan K, Rodriguez MA, Kharatyan S and Hussainova I 2008 *Ceram. Int.* **44** pp 12265-72

[24] Kissinger HE 1957 *Anal. Chem.* **29** pp 1702-6

[25] Akahira T and Sunose T 1971 *Res. Rep. Chiba Inst. Technol.* **16** pp 22-31

[26] Abuhimd HM, Kiyashko MV and Grinchuk PS 2020 *Mat. Res. Express* **7** 015060

[27] Enayati MH, Sadeghian Z, Salehi MAHDI and Saidi A 2004 *Mat. Sci. Eng. A* **375** pp 809-11

[28] Sharafutdinov MR, Korchagin MA, Shkodich NF, Tolochko BP, Tsygankov PA and Yagubova IY 2007 *Nucl. Instrum. Methods Phys. Res.* **575** pp 149-51

Electric current distribution in thin Ti-Cu base coatings

M Urbaha[1], V Vorohobovs[2], K Savkovs[3], J Fescuks[4], V.Feofanovs[5], J.Korhs[6]

1-6 Department of Transport systems and logistics, Institute of Aeronautics, Riga Technical University, Latvia

E-mail: urbaha.m@gmail.com

Abstract. The electrical resistivity of thin layer (film, coating) about 5 microns' thick has been measured by the classical four-point probes method. The layer includes titanium-copper alloy grains up to 5 microns in size. Unexpectedly we encountered a physical phenomenon that spoils the measurements: the fluctuation of measurement results, i.e. the error of measurement may exceed 1000%.. Under certain conditions such crystals may exert some long-range influence leading to tenfold fluctuations of results of measuring the electrical resistivity, despite of the fact that these crystals are evenly distributed and do not create large-size clusters It was found that the relative scatter of measurement results decreases along with the increase of the distance between the electrodes (probes) due to averaging of currents, however, the error decreasing occurs inversely proportional to the logarithm of the distance between the probes. . That is why the scatter remains significant at distances between the applied probes thousands times bigger than crystal grains. It decreases much slower than it occurs in case of one dimensional long strip where the error of measurement is simply inversely proportional to the distance between the probes. It creates huge difficulties for measurement of surface resistance. To overcome these difficulties, the method for the statistical proceeding of the non-uniform results of multiple measurements of electrical resistivity in two-dimensional systems has been proposed, which enable to extract information about average surface resistance even from such confusing set of measurements. By help of such statistic the four probe method can be potentially used in industry, for example, for controlling the quality of metal film coatings, used for heating by electrical current for de-icing of aircrafts or wind turbines.

Keywords: Two-dimensional resistivity, four-point probe method, erosion resistant coatings, anti-icing properties, electrical resistivity.

The Paper type: *Research paper*

1. Introduction

During the experimental measurement of the electrical resistivity of electro-conductive coatings made of a mixture of titanium and copper deposited on CFRP samples obtained by the method of vacuum sputtering using NNV-6.6-I1 unit within the ERDF project No 1.1.1.1/16/A/073 "High Performance Erosion Resistant Multifunctional coatings for Aircraft Composite Structures (PEROMACS)", in which a separate source for each metal was used. More information about vacuum sputtering used in our lab, can be found here [11] , [12].

IOP Conf. Series: Materials Science and Engineering 1140 (2021) 012053 doi:10.1088/1757-899X/1140/1/012053

During this study it became necessary to measure two-dimensional surface resistance of samples for quality control of obtained coatings.

1.1. Definition

The two-dimensional resistivity of a thin film is defined as the resistance R of a square, in which two opposite sides are connected to two perfectly conducting ohmmeter probes along the whole length. R is expressed in ohms. If the film structure is completely homogeneous, the resistance R does not depend on the size of the square.

Figure.1 Definition of the surface resistance.

1.2. The Four-Point Probe Method.

The four-point probe method for measuring the resistance of a homogeneous surface involves the application of four probes (electrodes) to the surface. Desired direct current I is supplied to the outer probes, while voltage U is measured between two other (inner) probes. Thus, two-dimensional resistivity can be calculated as voltage divided by current and multiplied by some coefficient that takes into account the geometry of the system.

The four-point probe method allows fully exclude the influence of contact resistance between the medium and probes. Meanwhile in more primitive two probe method the contact resistance often exerts upon the result and sometimes may even exceed the resistance of the film itself. This is a classical method: The four-point probe system is sold by various suppliers, for example, [10]. It can be successfully used for uniform homogeneous films, but, as will be shown below for heterogeneous coatings it caused huge random errors of measurement, and that is why multiple measurements and statistical proceeding is a must.

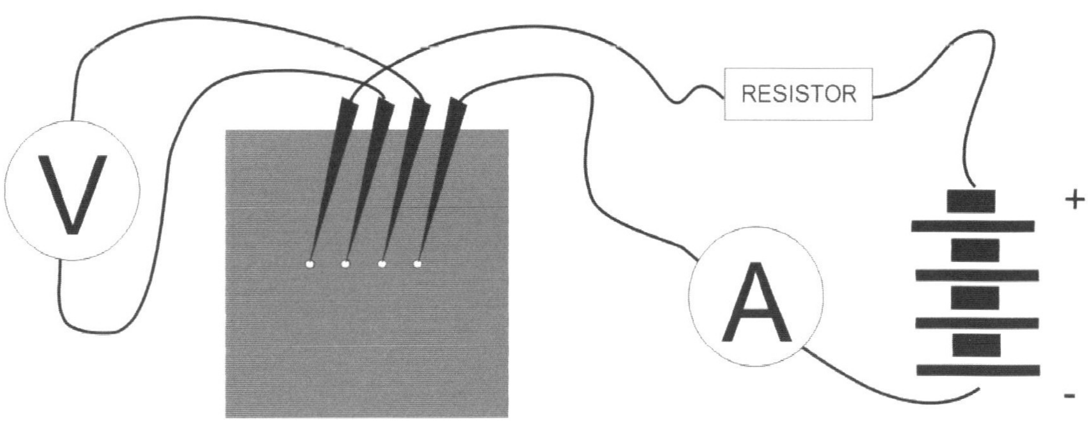

Figure 2. A plane which is practically infinite in all directions (red) and four-point probe.

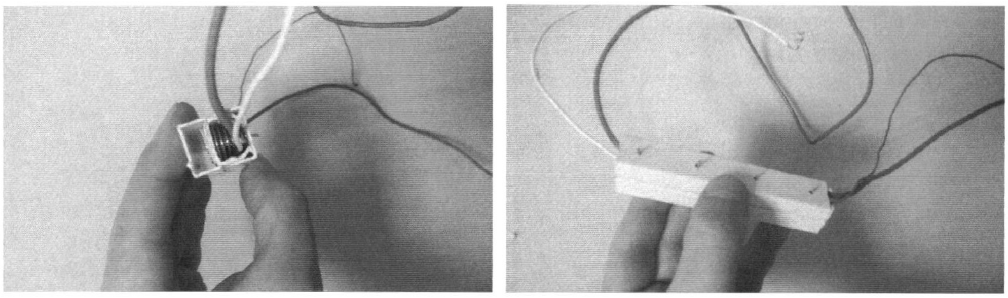

Figure 2A and 2B. One of the many probes with four spring-loaded contacts made in our laboratory.

1.3. The classical formula for an absolutely uniform coating for a plane which is relatively infinite in all directions (i.e. for essentially two-dimensional geometry)
To measure resistance in the centre of a plate, a probe with four spring contacts is used as shown at Figure 2.
Current I is supplied to two external points, while voltage V is measured between midpoints. As a source of direct current, it is possible to use a battery with a high electromotive force and a high internal resistance, which considerably exceeds the resistance of the sample being measured.. In such conditions, the additional resistor completely determines the value of current passing through the sample.
The two-dimensional resistivity of a thin film is calculated by the following formula:

$$R = \frac{V}{I} * \frac{\pi}{ln(2)} = \frac{V}{I} *4.532360142 \quad (1)$$

The derivation of this formula is available, for example, in [1].
The formula works only if the borders of a sample are far away from all probes. Practically, it is enough if the distance to the border is triple size of the distance between the probes.
Now it is difficult to determine the authorship of the method; for instance, [1] contains a reference to a work by Thomson Kevin, 1882. However, the edition [3] of 1872 of this work does not describe this method, it only sets out the ideas leading to the method, for example, the reflection method.
Work [2] describes the practical application of the given method and proposes to perform several measurements alternating the purpose of contacts (a total of 6 variants). The work states that such measurements can compensate for errors related to the inaccurate position of contacts. There is a reference to an earlier work describing the same method [4], [5]. They all present theoretical ratios that do not include distances between contacts, which as a result excludes the geometric factor (distance between the contacts) and related random error of measurement. Both these and other theoretical studies we are aware of suggest that sheet conductivity is uniform, and it does not have even micron grains being fully amorphous.

2. Problem

During our experiment, an approximately 10-micron thick coating was created as a result of simultaneous vacuum sputtering of copper and titanium from two different evaporator. When studying the cross sections under the electron microscope, it is possible to see grains that differ by the content of titanium and copper atoms which are approximately 5 microns in size. However, on a large scale, this coating is rather uniform because the grains do not create large-scale conglomerates.
Measuring electrical conductivity with the help of the four-point probe method, when the distances between the applied probes are thousands of times larger than grain sizes, it would be logical to expect that inconsiderable non-uniformities will be averaged. However, it was revealed that the value of the results of measuring the sheet resistance can be 10-fold different, and it changes dramatically at the slightest shift of the probe.
Therefore, reasonable presumption is possible: peculiarity of the nature of electrical resistance distribution leads to fact that averaging in case of two-dimensional systems occurs very slowly along with the increase of the scale (logarithmically).

IOP Conf. Series: Materials Science and Engineering 1140 (2021) 012053 doi:10.1088/1757-899X/1140/1/012053

2.1. Measuring the conductivity of coatings with a fine-grained heterogeneous structure

Unlike in the above-mentioned mathematical models, in which conductivity is presumed to be absolutely homogeneous, in reality we had to deal with coatings that have some structural non-uniformities with a typical size of up to 5 microns. Therefore, the following arguments are relevant in this case.

Assume for simplicity that the whole surface consists of areas with two conductivities. Let us denote the higher conductivity in blue and the lower conductivity in red. And let us denote the contact points in white. Intuitively you can wait that if the typical size of inhomogeneities is substantially smaller than the distance between the contacts, the measurement will give an average value between the two conductivities. But it may turn out to be a delusion because the mathematical model of the averaging process may turn out to be non-trivial. It will be discussed below.

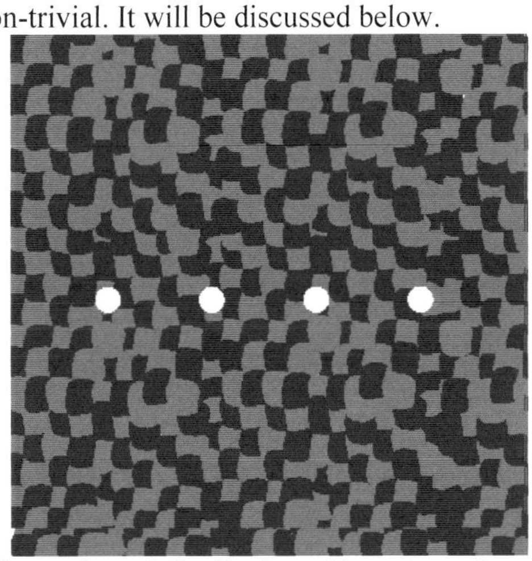

Figure 3. Just a scheme for illustration of distribution of locations of areas with different conductivity.

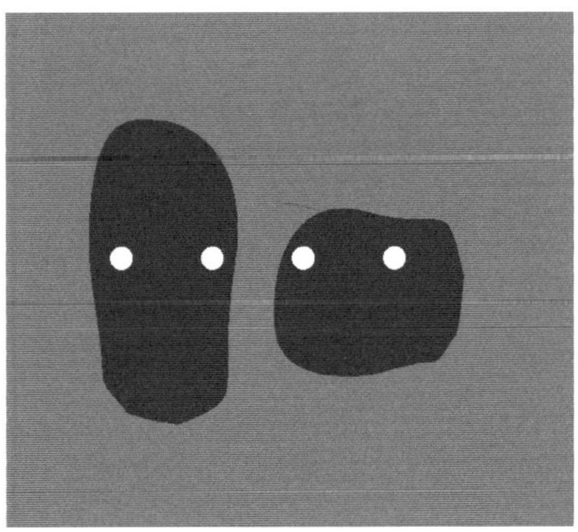

Figure 4. Distribution of areas with different conductivity, which leads to overestimating the surface resistance, because the central voltmeter is practically directly connected to battery. (It is just a scheme for illustration).

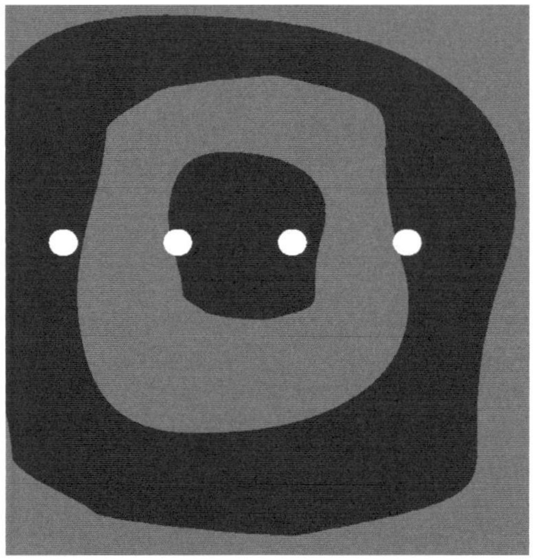

Figure 5. Distribution of areas with different conductivity, which leads to underestimating the surface resistance, because the central voltmeter is short circuited, and the battery is also short circuited. (It is just a scheme for illustration).

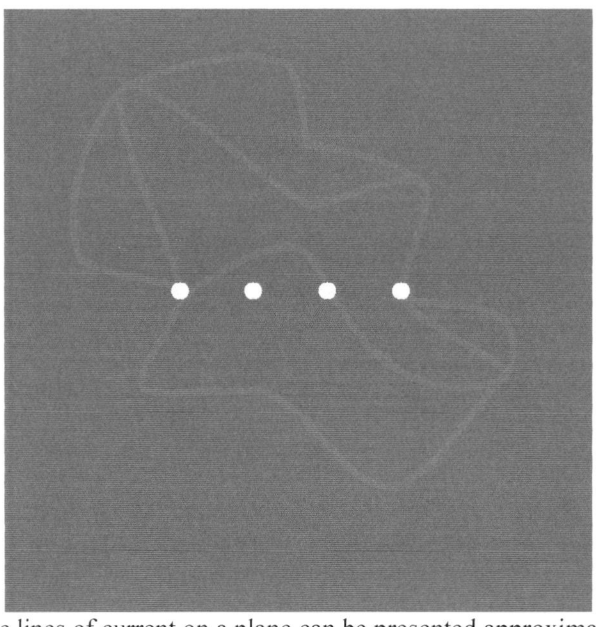

Figure 6. The lines of current on a plane can be presented approximately like this.

This is only illustrative sketch, but for real picture please look at Figure 9.

On a two-dimensional plane, the current flows unevenly, like rivers flows along curved beds on a slightly uneven land. Therefore, the result of the readings from a voltmeter connected to internal points depends on randomness: whether the probes got into a "river bed" or not.

3. Explanation why in the two-dimensional case the scatter of measurements decreases logarithmically along with the growth of a distance between the probes.

Let us consider geometry, in which one probe is positioned in the centre and is just a small circle and the second probe is a big concentric outer circle.

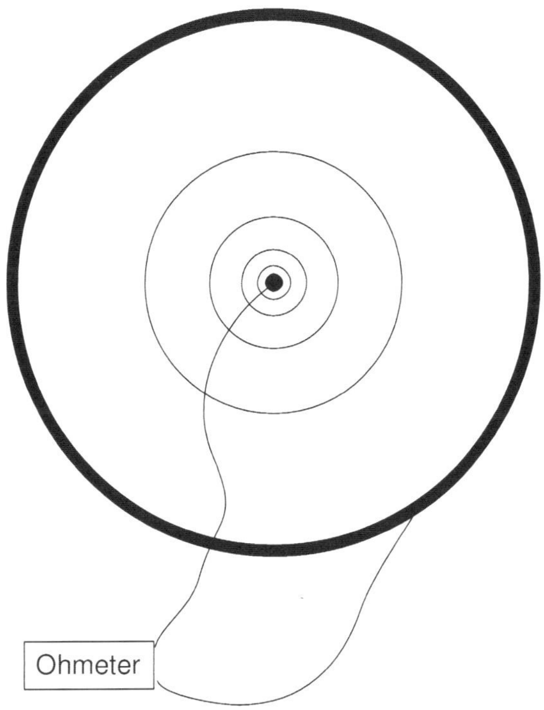

Figure 7. Illustration of imaginary experiment. One thin probe is in the centre and the second probe occupies the whole boundary of the circle.

It is possible imaginary to cut the circle into concentric rings, each of which is geometrically similar to the previous one. For instance, in the figure, every next concentric ring is two times bigger than the previous one. They are geometrically similar. It can be easily proved that all flat geometrically similar two-dimensional figures have equal electrical resistance independently on their sizes. So all rings have the equal resistance. Let say the resistance of each ring is R. As they have a serial connection, their total resistance is just a sum of all resistances R+R+R+R+R. So, adding a new ring adds just one more R. It means that the combined resistance is proportional to the number of rings. On the other hand, it should be noted that the geometry grows exponentially to the number of rings. As the logarithm is an inverse function from the exponent, the resistance is proportional to the logarithm of the ratio of small and big probe (electrode) diameters. The scatter of resistance parameters for every next ring will be smaller due to averaging on a larger area. It means that the error of measurement results is caused by non-uniformity in several first stripes, and next stripes almost do not add anything significant to it. Once having appeared, the error, which is expressed in absolute units (ohms), will not disappear along with the introduction of new bigger rings. Meanwhile, the relative error, i.e. the error divided by the total resistance, will decrease. But it will decrease slowly and inversely proportional to total resistance or other words inversely proportional to the logarithm of distance between the probes. Thus, it becomes clear that the error of measured resistance value initiates mostly in the vicinity of the small probe and does not disappear afterwards. That is why the relative error, i.e. the error related to the total resistance, will decrease creating an effect of averaging. But it will decrease inversely proportional to the logarithm of the big probe size, i.e. it is inversely proportional the logarithm of the distance between the probes.

If we measure the resistance between the two small probes located at a distance substantially exceeding their sizes, the main idea will remain the same: a basic contribution to the instability of measurements is made by the areas in the direct vicinity of the both probes. But the decrease due to averaging is very slow – approximately inversely to logarithm of distance between the probes.
Now we can understand why such a huge scatter is observed when applying the four-point probe method onto relatively infinite planes that is in two dimensional case.
But there are no such problems on practically one-dimensional thin stripes. In fact, in a one-dimensional system, the error of measured resistance values also appears in the microscopic zones around the areas where sharp probes touch the surface. However, in one-dimensional case the total resistance of the stripe grows proportional to the distance between the probes, so the relative error decreases inversely proportional to the distance between the probes. That means - not logarithmically but much faster.

4. Measurement on elongated rectangles (i.e. in almost one-dimensional geometry)

During the first series of experiments, elongated rectangular samples were used – 200 mm long and 20 mm wide; and a coating of titanium and copper with various percentage of components was deposited on the samples by using the method of vacuum sputtering.
As the samples have very elongated geometry, it is optimal to position two contacts that create the desired current, i.e. clips 1 and 4, at the ends of the samples. While voltage can be measured with the help of contacts 2 and 3, which can be easily applied at any points.

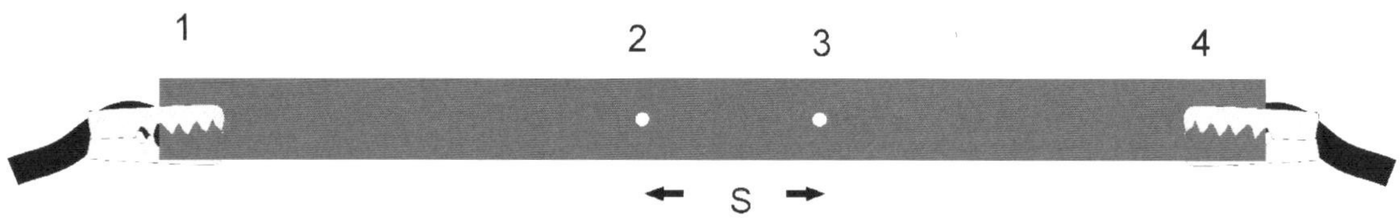

Figure 8. Measuring two-dimensional resistivity on a thin stripe.

On the elongated samples, the source of current was connected with clips to the ends of rectangle – 1, 4. The voltage was measured in the central part with two contacts – 2, 3. Thus, the sheet resistance will be as follows:

$$R = \frac{V}{I} * \frac{A}{S} \qquad (2)$$

Where V is the voltage between contacts 2 and 3, I is the current supplied to contact clips 1 and 4, A is the width of the sample, S is the distance between voltmeter contacts 2 and 3. But the length (distance between probes 1 and 4) has no influence and is not included in formulae.
Voltage is measured approximately in the middle of the plate with the help of a voltmeter.
In such utmost one-dimensional geometry, when the probes are positioned rather far from each another, large errors of measured resistance are not observed. Occurs effective uniform distribution of electrical current along the rectangle length, despite the presence of a heterogeneous structure like small grains.
The difference of results obtained on a thin stripe and on a plane, which is relatively infinite in all directions, can also be explained in a different way: On a thin stripe, current is clamped on both sides and cannot divert. And by this the random factor is eliminated.
It is possible to cut thin elongated stripes and take measurements on such one-dimensional samples and use simplest formula (2). It is the most precise way how to improve the four probe method. Obtained surface resistance can be used for reference and calibration.
But unfortunately, cutting is a destructive testing method that can be used only in a laboratory but not in the field on real aircraft and wind turbine wings.

IOP Conf. Series: Materials Science and Engineering 1140 (2021) 012053 doi:10.1088/1757-899X/1140/1/012053

5. Statistical description of the scatter of electrical resistance measurement results

The analysis of the samples allowed us to reveal large non-uniformity of results of electrical resistance measurements. Therefore, it is important to formulate a statistical processing method for such data.

5.1. Determining the mean resistance.

As prior to the experiments the nature of distribution is unknown, one of the two algorithms is suitable for determining the mean value in a natural way:

The first one: the results are put in a row in the increasing order and the mean resistance is determined as a result, in comparison to which exactly one half of the measurement results are lower and the other half are higher.

The second one: though the nature of distribution of resistance measurement results will not be Gaussian, it can be hoped that the distribution of the logarithms of measurement results will be similar to the Gaussian distribution; and in this case, as a criterion, it will be possible to use the logarithms of resistance values rather than the values themselves. Then the mean resistance *Ra* will be determined via the mean logarithm.

$$\ln(Ra) = \frac{(\ln(R1)+\ln(R2)+\cdots+\ln(Rn))}{n} \qquad (3)$$

Our practical experiment shows that these two ways of determining of mean value yield the same result, which is an evidence to the assumption that the logarithms of resistance measurement results are distributed according to the normal Gaussian law. (Though it is now too early to make a final conclusion – a more statistic is required).

5.2. Determining the mean deviation from the mean resistance.

After determining the mean value of resistance, it is possible to determine the mean deviation from the mean value and indicate some measure for measuring such deviations ⚠(R/Ra). (The triangle with the cross means average error). It can be defined as a mean deviation of logarithm of resistance from its mean value and calculated as root-mean-square deviation (RMSD) :

$$\ln ⚠\left(\frac{R}{Rs}\right) = \left(\left(\left(\ln(R1) - \ln(Ra)\right)^2 + \left(\ln(R2) - \ln(Ra)\right)^2 + \cdots + \left(\ln(Rn) - \ln(Ra)\right)^2 \right) /n \right)^{1/2} \qquad (4)$$

In order to understand by how many times the results of measurements may differ on average, it is necessary to take a natural exponent of this number:

$$⚠\left(\frac{R}{Rs}\right) = \frac{1}{2}\exp\left(\left(\left(\left(\ln(R1) - \ln(Ra)\right)^2 + \left(\ln(R2) - \ln(Ra)\right)^2 + \cdots + \left(\ln(Rn) - \ln(Ra)\right)^2 \right) /n \right) \right) \qquad (5)$$

6. Analysis of actual experimental data obtained on samples with a deposited mixture of two metals: titanium and copper

This is an example of a data set obtained as a result of measuring resistance on one and the same sample.

6.1. The measurements were performed with the help of a probe system, in which the contacts were installed at a distance of 1.2 mm from each other and positioned along one line with a total length of 3.6 mm respectively. Units of measurement – ohms. Calculated using measured voltage, current and formulae (1).

We present these numbers here specifically so that the reader can independently plot them on the axis and make sure that these results are distributed according to a statistical law that is very far from the normal Gaussian distribution. But if you plot the logarithms of these numbers on the axis, you will easily see that the distribution resembles a Gaussian one.

Table 1. Just set of results of measurements. The distance between probes 1.2 mm from each other

0.95	16.89	15.4	5.82
0.40	14.11	16.25	3.82

0.34	9.58	23.7	4.20
3.49	12.3	7.44	5.05
0.77	6.79	6.08	16.7
1.81	2.13	12.6	23.0
4.40	5.76	77.2	3.75
26.0	6.92	2.71	1.61
14.9	58.2	14.8	2.97
3.62	84.4	9.58	3.43

One can easily make sure that the scatter may be tenfold, and this is surely not the Gaussian distribution; so, in this case it is impossible to apply traditional statistical processing methods (like the calculation of the mean) directly to these numbers. It is better to apply them to the logarithms of these numbers, as proposed in formulae (3) , (4) , (5).
Thus, applying the above described statistical processing method, we will get the following result:
The mean resistance R = 6.52 ohms, the logarithm of root mean square deviation from the mean value calculated by the logarithmic formula = 1.24, which corresponds to the mean error of measurement = 247% (i.e. the scatter is on average 3.47 times upward or downward).

6.2. The following values of resistance in ohms were obtained by increasing the distance between the probes up to 30 mm and repeatedly measuring one and the same sample as before:

Table 2. Just set of results of measurements. The same sample as before
But the distance between the probes up to 30 mm

2.05	1.04	1.52
1.95	1.02	1.42
0.964	0.932	1.25
0.582	1.29	1.57
3.08	1.62	1.42
1.69	1.72	1.68
1.36	5.27	1.60
0.854	2.07	1.65

The above-mentioned statistical processing method yields: the mean resistance R = 1.44 ohms, the logarithm of root mean square deviation from the mean value calculated by the logarithmic formula = 0.41, which corresponds to the mean error of measurement = 50% (i.e. the scatter is on average 1.50 times upward or downward).
As we see, the application of a larger probe system has decreased the mean deviation of the measured resistances from the mean resistance, however this decrease is still insufficient for objective studies.

7. Control testing of four-point probe method on a glass with a good uniform high-quality amorphous coating without any grains.

Table 3. Just set of results of measurements on a glass with a good uniform high-quality coating.

67.4	77.6
68.8	73.8
73.4	70.5
88.8	75.9
73.2	77.1

The mean resistance R = 74.5 ohms, the logarithm of root mean square deviation from the mean value calculated by the logarithmic formula = 0.074, which corresponds to the mean error of measurement = 7% (i.e. the scatter is on average 1.07 times upward or downward).
As it is seen, the four-point probe method works well on uniform films. Therefore, it is widely used of uniform films, for example, in [9].

8. Infrared thermal image

To make sure that such non-uniformity does exist, the samples were photographed with an Infrared thermal imaging camera at various moments when the current heats the plate.
In the photo the formation of elongated fractal "fingers" corresponding to the places where the flowing current forms the above-mentioned "channels" is clearly seen.

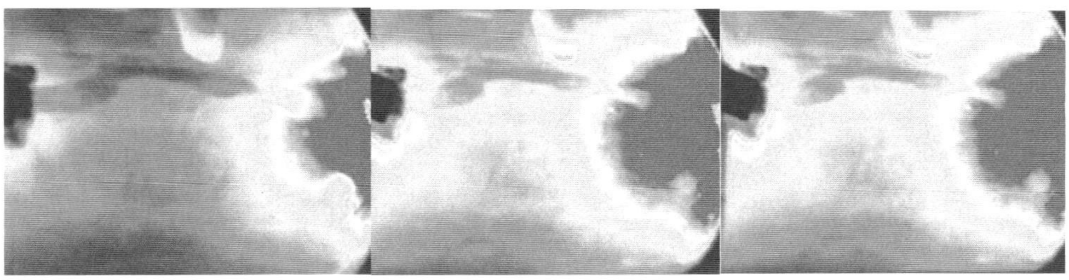

Figure 9. Infrared thermal image. Three times moments. The size of the sample is 200x150 mm.
It can be clearly seen that heat generation is not uniform, which corresponds to non-uniform current distribution. Which fully supports our original hypothesis.

9. Sample preparation

The vacuum coating method is not a topic of discussion in this article. We discuss here only measurement of electrical conductivity. Still few words should be told about the sample preparation:

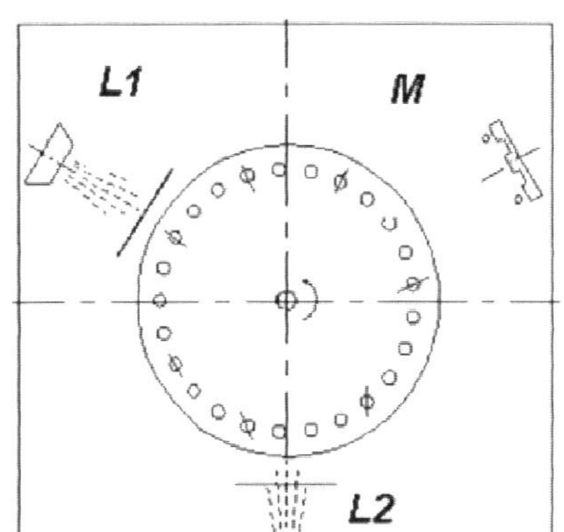

Figure 10. Sputtering chamber of NNV-6,6-I1 and the deposition scheme
Two arc sources are used: one for Titanium (L1) and the other for Cuprum (L2). The CFRP substrate is rotated on a rotating table to achieve an even coating.

The arc source has a minimum evaporator current of 40 A. During the evaporation of titanium, rather large number of droplets are formed. They are later visible as micro grains < 5 mcm.
More information about coating is available in other publications of our co-authors [11], [12], [13].

IOP Conf. Series: Materials Science and Engineering 1140 (2021) 012053 doi:10.1088/1757-899X/1140/1/012053

10. Analysis of the coating

Under the electron microscope it is possible to see numerous crystal grains up to 5 microns in size with various content of titanium and copper.

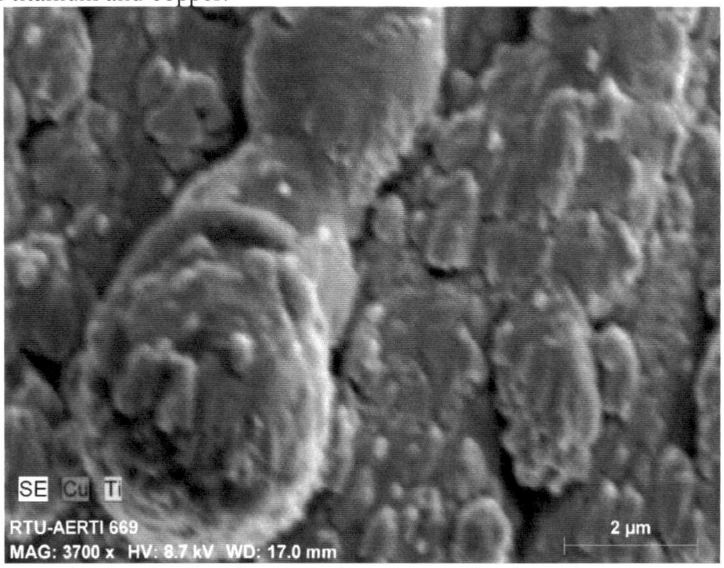

Figure 11. A photograph taken by the electron microscope. Coating microstructure. Different colours indicate the areas of copper and titanium concentration.

The resistance of titanium is thirty times higher than the resistance of copper. Thus, being included parallel with copper, it will hardly conduct current. Any measurements of resistance with the help of thin probes will lead to a huge scatter of results depending on the area that will come into contact with the sharp probe. This is what could be observed during the experiment. Such non uniformity can be explained by Stranski – Krastanov instability [6]. More information here [8]. This structure looks like solidified liquid drops. Such solidification was discussed here [7]. These references are given for further information but they are not a topic of presented article, because here only electrical resistance is discussed.

Conclusions

1. The surface films created in our laboratory are applicable for heating wind turbine or aircraft wings for preventing icing even if the film consists a mixture of very small crystals of copper and titanium. The de-icing properties of the coatings have been tested in our laboratory and it will be presented in future publications. Here we concentrate only on measurement of surface resistance, which is required to control coating quality.
2. The methods of measuring resistance with classical four-point probe method can be applicable but only after statistical proceeding of hundreds measurement results using statistical method described above. Compare: this is very different from measuring of two-dimensional resistance of homogeneous films, for which the four-point method gives the result immediately after the first measurement.
3. Alternatively producing some larger contacts covering a representative sample of all types of grains can be used. But such idea is not realistic, because it can be done only by soldering. It will not be possible quickly move such tester from one place to the other along big airplane wing.
4. Statistical distribution of electrical currents in two-dimensional system strongly depends on sizes and electrical resistivity of internal small grains. As our experiments have shown, the

currents for such system can be irregular even on large scales because the irregularity decreases approximately inversely to logarithm of scale. But to obtain precise statistical formulae, the further theoretical mathematical study of such dependence is needed. This article is the first to pose such a non-trivial complex mathematical two-dimensional problem (which could probably be solved using computer modelling).

Acknowledgement
This work has been supported by the European Regional Development Fund within the project No 1.1.1.1/16/A/073 "High Performance Erosion Resistant Multifunctional coatings for Aircraft Composite Structures (PEROMACS)"

NATIONAL DEVELOPMENT PLAN 2020

EUROPEAN UNION
European Regional Development Fund

1. INVESTING IN YOUR FUTURE

References

[1] F.M. Smits. Measurement of sheet resistivity with four point probe 1957. http://www.four-point-probes.com/smits.pdf
[2] Burlakov R.B. Kovivchak V.S. To the question of measuring the specific resistance of conducting layers by the four-probe method.
[3] Sir William Tomson Electrostatic and magnetism. London, Macmillian 1872 https://archive.org/[details/reprintofpaperso00kelv/page/n5
[4] Van der Pauw L.J. A Method of Measuring Specific Resistivity and Hall Effect of Discs of Arbitrary Shape // Philips Research Reports. 1958. Vol. 13. No. 1. P. 1-9.
[5] Van der Pauw L.J. A Method of Measuring the Resistivity and Hall Coefficient on Lamellae of Arbitrary Shape // Philips Technical Review. 1958/1959. Vol. 20. No. 8. P. 220-224.
[6] Stranski, IN; Krastanov, L. (10 February 1938). "Zur Theorie der orientierten Ausscheidung von Ionenkristallen aufeinander". Monatshefte für Chemie und verwandte Teile anderer Wissenschaften.
[7] Volmer, M .; Weber, A. (1 January 1926). "Keimbildung in übersattigten Gebilden". Zeitschrift für Physikalische Chemie.
[8] Oura K., Lifshits VG, Saranin AA et al. Introduction to surface physics / Ed. V.I.Sergienko. - Moscow: Nauka, 2006 .-- 490 p.
[9] Richard S. Waremra, Philipus Betaubun Analysis of Electrical Properties Using the four-point probe method. ICENIS 2018. https://www.e3s-conferences.org/articles/e3sconf/pdf/2018/48/e3sconf_icenis18_13019.pdf
[10] Four-Probe measurement equipment advertising: https://www.ossila.com/products/four-point-probe-system?variant=31916570945
[11] Urbaha, M., Bogdanova, S., Urbahs, A.
Evaluation of the physical and mechanical characteristics of ion-plasma decorative-protective coatings (2019) AIP Conference Proceedings, 2077.
[12] Urbaha, M., Bogdanova, S., Urbahs, A.
Multi-criteria assessment of physical properties of nanostructured decorative protective coatings (2019) Key Engineering Materials, 799 KEM, pp. 43-51.
[13] Urbahs A. Urbaha M., Savkovs K. and Bogdanova S. Wear resistant nanostructured multi-component coatings. NATO Science for Peace and Security series, Series B: Physics and Biophysics, Nanodevices and Nanomaterials for Ecological Security, Part 1, 2012, pp. 161–170.

Effect of laser remelting of plasma sprayed coating of Cr-Ni-Re

P Śliwiński[1], M S Węglowski[1], J Dworak[1], S Dymek[2], I Kalemba-Rec[2], A Wrona[3], K Kustra[3], M Lis[3]

[1]Łukasiewicz – Institute of Welding, Błogosławionego Czesława 16-18 St., 44-100 Gliwice, Poland, EU

[2]AGH, Faculty of Metals Engineering and Industrial Computer Science, Czarnowiejska 66 St., 30-059 Kraków, Poland, EU

[3]Łukasiewicz - Institute of Non-Ferrous Metals, Generała Józefa Sowińskiego 5 St., 44-121 Gliwice, Poland, EU

Abstract. Atmospheric plasma spraying (APS) consists in injecting a powder feedstock material in a plasma jet to melt and accelerate the injected particles and spray them onto a substrate. However, this mechanism of coating formation induces the presence of pores and micro-cracks. In order to eliminate those defects, laser remleting may be used to improve the properties and performances of plasma sprayed coatings. In the present paper, the Ni-Cr-Re coatings fabricated by plasma spraying on stainless steel substrate were remelted by CO_2 laser, and the effects of laser remelting were studied. The structure and chemical composition of plasma sprayed and laser remelted coatings were analyzed using scanning electron microscopy (SEM) and energy disperse spectroscopy (EDS). The results show that the laser remelted coating becomes much denser and, moreover, the chemical composition of the coatings becomes homogeneous.

Keywords: coatings, rhenium, thermal spraying, laser remelting, microstructure

1. Introduction

Rhenium has found application in industry as a refractory metal due to its unique properties. It is characterized by very high tensile and creep rupture strength over a wide temperature range and has the third-highest melting point and second-highest boiling point of any stable element [1, 2]. This element is also often used as an alloying additive. The addition of rhenium in nickel- or cobalt-based superalloys increases their creep strength [3]. Therefore it is also used for manufacturing of coatings [4] or reinforced composites [5]. To obtain coatings with the desired chemical composition the plasma spraying process can be used. This technology enables the production of layers with desired thickness even on elements with complex shape. However, a significant disadvantage of plasma sprayed layers is the lack of uniformity and compactness within their structure, caused by the presence of pores and cracks. The porosity of these layers can be reduced in the process of laser beam [6-10] or electron beam remelting.

This paper presents the effects of laser remelting processes of thermal sprayed coatings in relation to rhenium content and process parameters.

2. Experimental procedure

The materials used in this procedure were Ni20%Cr, Ni20%Cr + 10%Re, Ni20%Cr + 20%Re, Ni20%Cr + 30%Re alloys. The procedure consisted of 3 stages. The first stage was the production of powders at the Łukasiewicz - Institute of Non-Ferrous Metals. Powders were produced from commercial NiCr powders through their modification with rhenium. The modification technique consisted of a thermo-chemical treatment for producing metallic rhenium from NH_4ReO_4 (ammonium perrhenate) directly on the surface of modified powders.

During the second stage, manufactured powders with different rhenium content were plasma sprayed on an austenitic stainless steel (AISI 316Ti) substrate. The surface of the substrate was prepared by abrasive blasting using corundum abrasive. The plasma spraying parameters: current 530 A, arc voltage 690 V, shielding gas flow rate (Ar) 54 l/min, plasma gas flow rate (H_2) 9 l/min, transport gas flow rate (Ar): 5 l/min, spraying distance 140 mm, travelling speed 400 mm/s.

The last stage was laser remelting that was carried out using a laser CO_2 generator (Trumpf Lasercell 1005). 9 samples were made for each rhenium content with different process parameters (the parameters changed were the power and beam travelling speed). Argon gas (ISO 14175-I1-Ar) was used to prevent oxidation during the process. The gas flow rate was 12 l/min, the working distance of the laser head was 100 mm. Other process parameters are given in Table 1.

Table 1. Laser beam remelting parameters

Variant no	**1**	**2**	**3**	**4**	**5**	**6**	**7**	**8**	**9**
Laser beam power, W	3000	3500	3800	3000	3500	3800	2000	1500	2000
Laser beam travelling speed, mm/min	300	300	300	500	500	500	300	500	500

The microstructures of the etched samples were investigated by light microscopy (LM; Zeiss Axio Imager) and scanning electron microscopy (SEM; FEI Nova NanoSEM 450) with energy dispersive spectroscopy (EDS). The coatings thickness were measured by light microscopical examination of the cross-sections. The thickness error was about 1 μm samples. The SEM observations and EDS line analysis were carried out with an acceleration voltage of 15 kV, at working distance of 5 mm.

3. Results and discussion

The cross-section images of the coatings are shown in Figure 1. Figure 1a shows photos of plasma sprayed coatings: Ni-Cr and NiCr + Re. Those are characterized by high porosity and contain a lot of cracks and have typical plasma sprayed lamellar structure. In turn, Figure 1b illustrates the cross-section of coatings after laser remelting (for given remelting parameters: speed 300 m/min and laser beam power 3800 W). As shown in Figure 1b laser remelting effectively reduces the presence of pores and microcracks resulting in much denser coating. After laser remelting the lamellar structure of the plasma sprayed coating was replaced by fine homogenous microstructure.

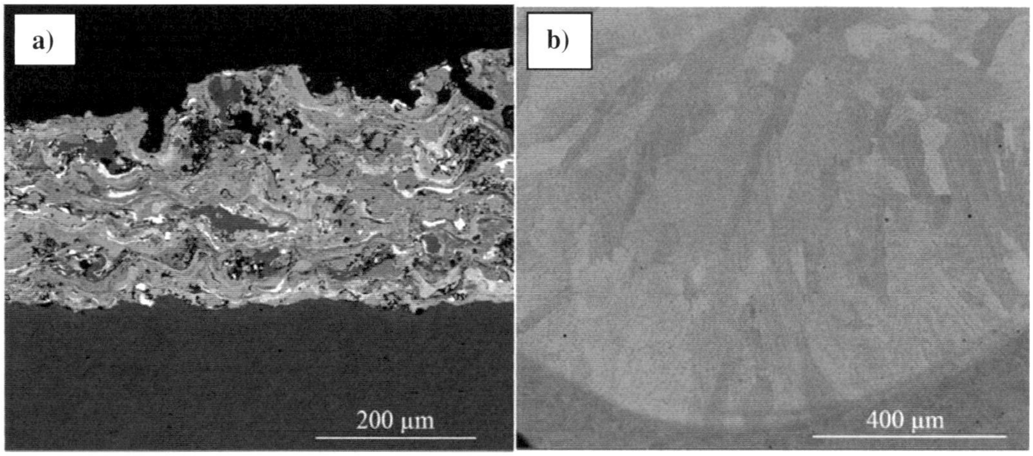

Figure 1. The cross-section images of coatings with different rhenium content: a) plasma-sprayed; b) laser remelted .

Based on SEM observation, lower adhesion Ni-Cr coating to the substrate than NiCr+Re coatings is visible. The detailed quantity analysis of the adhesion phenomena base on scratch tests will be carried. On the other hand, in the coatings with the highest rhenium content, the largest clusters of pores (5.6 %) were observed. After the laser remelting process, the coatings were characterized by good adhesion to the substrate and the absence of porosity (0.2%) and material discontinuities. For variant 8 (the lowest laser power), the Ni-Cr and Ni-Cr-20% Re coatings are discontinuous. Lack of continuity was also observed for the Ni-Cr-10% Re coating from variant 6 (highest power and speed). In the Ni-Cr-30% Re coatings remelted at a speed of 300 mm/min (for all laser power values) cracks were visible. Cracks were also seen in the variant 9 Ni-Cr-20% Re coating.

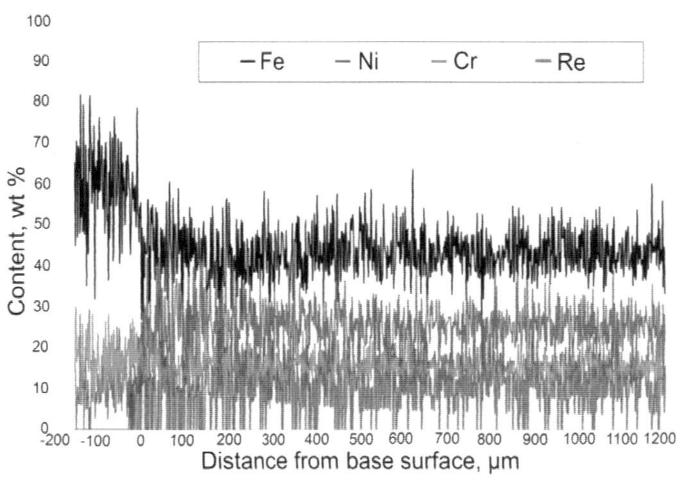

Figure 2 EDS line scan results of the plasma-sprayed Ni20%Cr + 20%Re coating after laser remelting

The micro-area chemical composition of the coatings before and after the laser remelting was analyzed using EDS. It was found that the distribution of elements before remelting is not uniform. The analysis revealed that the ratio of Ni, Cr and Re in different areas is not the same.

The result of EDS line scan of laser-remelted coating is shown in Figure 2. After laser remelting the lamellar structure becomes homogeneous in composition. The Fe (black line) content changes from about 70% in the base material surface region to about 48% in the remelted region. The content of Cr (green line) in base material is in the range up to 25% and decreases to approx. 15% (average value) in the remelted zone. Inversely, the Ni (blue line) content in the remelted region is much higher (about 26%) than that in the surface region (about 10%). Only the remelted region contains approx. 10% rhenium (green line).

All of the samples were measured before and after the process of laser beam remelting.

Table 2. Average thickness of plasma sprayed coatings

Coating material	Average thickness, µm
Ni-Cr	287 ± 21
Ni-Cr-10%Re	280 ± 20
Ni-Cr-20%Re	164 ± 21
Ni-Cr-30%Re	255 ± 30

The thickness of the plasma sprayed coatings is summarized in Table 2. The Ni-Cr coating with 20% rhenium content was characterized by the smallest thickness. This is caused by higher density of powder with Re and higher kinetic energy. For all used alloys the thickness of the coatings after laser remelting is shown in the graphs below (Figure 3).

The laser energy density (laser energy per area unit) is proportional to laser power while it is inverse to scanning velocity and beam size. Laser energy density increases as the laser scanning speed declines. From the Fig.3 it is clearly visible that increase of the laser power caused increase the layer thickness (distance between top of surface and max. penetration depth in substrate). Therefore, the layer thickness can be even twice higher after remelting. If the travelling speed increases, the penetration depth becomes lower and thus, total layer thickness decreases. However, it should be noted that when the laser scanning speed drops to some extent, the coatings will become more homogenous and compact, and this results in the relatively high thermal conductivity of the laser-treated coatings. After the laser

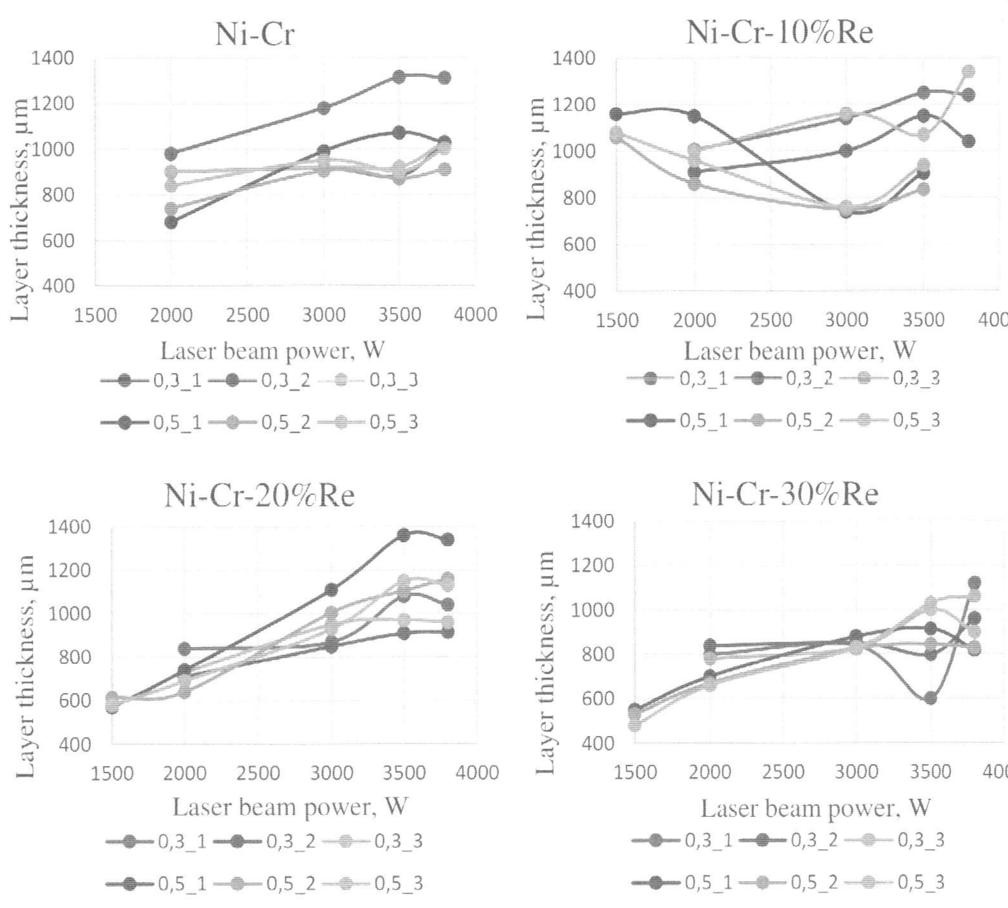

Figure 3. Graphs showing the relationship between the applied process parameters and the thickness of the resulting coatings

beam passes, the heat in the molten regions is conducted quickly to the substrate and external environment. On the other hand, at higher travelling speed the remelted coatings are cooling down faster, solidification process become promptly, thus cracks happen as a consequence of the restriction of the substrate and side coatings. The cracking tendency is more visible. Therefore, it is important and necessary to explore the proper laser parameters to obtain the relatively better performance of laser-remelted coatings.

4. Conclusion

In this paper, the effect of laser remelting on plasma-sprayed coating was studied. Ni-Cr-Re coatings were fabricated by plasma spraying and then remelted by CO_2 laser. The plasma sprayed coatings are characterised by high porosity and presence of cracks. These coatings also show the presence of a lamellar structure, thus, resulting in not-homogeneous chemical composition

Coatings with rhenium addition show better adhesion to the surface than Ni-Cr. Highest rhenium content coatings are characterized by high porosity. After the laser remelting process, the coatings present good adhesion. Laser remelting process also reduces the presence of the pores and microcracks. The structure and chemical composition becomes homogenous and uniform.

The thickness of the coatings usually increases with the power of the laser beam. Too high or too low laser beam power causes discontinuities in the coatings. The travelling speed of the laser beam significantly affects the quality. Layers produced at lower beam travelling speeds become more homogenous and compact. Coatings produced at higher speeds show cracking tendency.

5. Acknowledgements

This work was performed with funding from National Science Centre in Poland within the frame of the research grant UMO-2018/29/B/ST8/01206 entitled: “Multi scale analysis of microstructural changes in wear resistant coatings produced by plasma spraying followed by remelting with an application of concentrated energy sources in view of the improvement of their mechanical properties”.

References

[1] Campbell I E, Rosenbaum D M, Gonser B W 1959 *J. of the Less Common Metals* **1(3)** 185-191
[2] Zhang Y, Evans J R, Yang S 2011 *J. of Chemical & Engineering Data* **56(2)** 328-337
[3] Naor A, Eliaz N, et al. 2010 *The AMMTIAC Quarterly* **5(4)** 11-15
[4] Trefon-Radziejewska D, Juszczyk J, et. al. 2020 *Int. J. of Refractory Metals and Hard Materials* **87(2)** 105147.
[5] Węglowski M St, Wrona A, Richert M, Nejman I 2018 *Építőanyag – J.of Silicate Based and Composite Materials* **70(6)** 181-185
[6] Panziera R C, de Oliveira A C C, Pereira M et al. 2020 *J. Braz. Soc. Mech. Sci. Eng.* **42** 119
[7] Soveja A, Costil S, et. al. 2010 *J. of Thermal Spray Technology* **19** 439-447
[8] Bien A 2020 *Conference paper: Laser Technology VI* 137-141
[9] Afzalo M, Ajmal M, et. al. 2014 *Optics & Laser Technology* **56** 202-206
[10] Hazra S, Das J, et. al. 2015 *Materials Chemistry and Physics* **154** 22-29

Simulation based Feasibility Analysis of Autonomously Movable Robot Arm

Kristo Vaher[1], Kashif Mahmood[2], Tauno Otto[2], Jyri Riives[2]

[1]TTK University of Applied Sciences, Tallinn, Estonia.
[2]Tallinn University of Technology (TalTech), Estonia

*E-mail: vaher.kristo@gmail.com

Abstract. The use of industrial robots in production is rapidly growing. However, the vast use of industrial robots and implementation of new manufacturing technologies are mostly adopted by large industrial companies. It is due to the nature of the production volume, as robots have a lot of the same kind of work in one specific position in the production process. In smaller companies, where the robots often do not have enough workload in a single specific workplace, the process of robotization has not been so successful. SME-s need a solution where the robot can be moved from one workplace to another in order to utilize the resources like robot arm efficiently. This paper aims to analyse the feasibility for the usage of a robotic arm (a collaborative robot) to serve more than a single production cell intermittently. Production machines are located at particular distance and the movement of the robotic arm between the machines is carried out autonomously with the help of an autonomous mobile robot. Moreover, simulation and 3D visualization were used to conduct and analyse the two different scenarios. Utilization of production equipment assigns as a key performance indicator of the comparison.

1. Introduction

In manufacturing industry, the automation via robots for different applications is inevitable, the recent evolvement of the Industry 4.0 concept and new industrial communication technologies like internet of things in manufacturing led the automation to the level of autonomous.

The applications of industrial robots (industrial robots are automated, programmable and capable of movement on three or more axes) are observed mainly for repetitive and high-precision tasks or monotonous tasks demanding physical effort. The development of collaborative robots ensures the safe working condition for human workers and it allows human workers to confidently share the workspace with robots [1]. It means there is omission of fences for industrial robots and utilization of space can be increased on the factory floor. The new industrial robots can have an ability to move freely and can execute several complex activities just like humans [2]. Furthermore, the free movement of cobots (collaborative robots) in the shop floor enables better utilization of surface area in production and enhanced the usage of resources, possible reduction in costs for automated cells, timely and easier access to the process machines and lower downtime, it may encourage smaller companies to implement robotic solutions [3]. One of the possibilities to autonomously move a robotic arm among several machines is to use the Autonomous Mobile Robot (AMR) platform and mount a robot arm on the AMR. Calibration of the robot arm position relative to the workstation is solved in the authors previous article [8].

The feasibility to analyse a change or test a solution can be performed digitally by creating a similar scale virtual environment and simulate it. In the digital manufacturing context, the use of simulation tools allows the performance evaluation of manufacturing systems and production cells with the help of certain performance indicators [4, 5]. In this case, the physical resources are not needed to acquire and the changes can be validated through a digital manufacturing approach [6].

There are articles analysing the need for mobility of industrial robots [7, 8] and identified possible solutions for doing so. It was analysed that the solution for moving a robot should be manual by a human or automatically by a mobile robot [7]. The automatic positioning accuracy of the robot arm moved by the mobile robot was analysed using machine vision [8]. There have been interesting approach [9] for replacement of manual work by hybrid production using autonomous dual arm robots, enhancing operator working conditions and maintaining the same production levels. Nielsen et al showed in [10], that mobile robot arms can continuously perform meaningful industrial tasks as so-called bartender concept in Cloud Manufacturing (CMfg)

This article analysed the efficiency and cost-effectiveness of a solution where robot arm is moved by mobile robot. A 3D simulation software is used for this purpose. The aim of the simulation is to analyse the different production scenarios of an autonomously moving robot and to compare it with a situation where people do the same work.

The research was done in the framework of development of semi-industrial Industry 4.0 Lab, involving robotised production line, 3D printing stations and automatic storage system (see Fig 1).

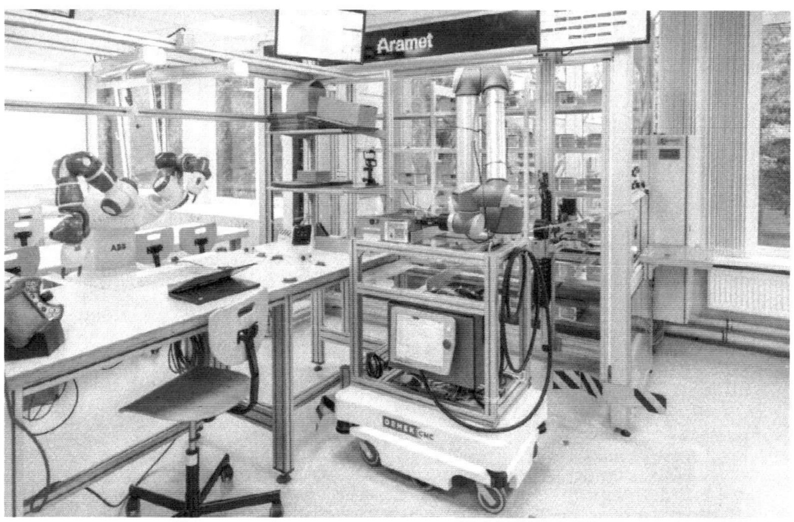

Figure 1. Semi-industrial Industry 4.0 Lab with autonomous cobot platform

2. Description of simulation

In the simulation (Fig.2), the Robot Arm (RA) is moved from Work Position 1 (WP 1) to work position 2 (WP 2) by a Mobile Robot (MR). Upon arrival at his work position, the robot arm positions it's self-relative to the workplace using machine vision. Once the position has been detected, the robot arm connects itself to the communications network using a special plug. The start signal is transmitted to the MES system, which in return sends the correct command to start the correct program inside the robot arm controller. Communication between MES and RA and MR is made via WiFi (IEEE 802.11). At the end of the work, the robot disconnects itself from the communication network, transmits a signal that the work is finished and waits for transport to the next task. The whole process takes place without human interaction.

Two different use cases were analysed. Use case A is a common way how robot arm is used by SMEs. Use case B is a new method proposed by author where robot arm is moving constantly between WPs to execute tasks and reduce downtime.

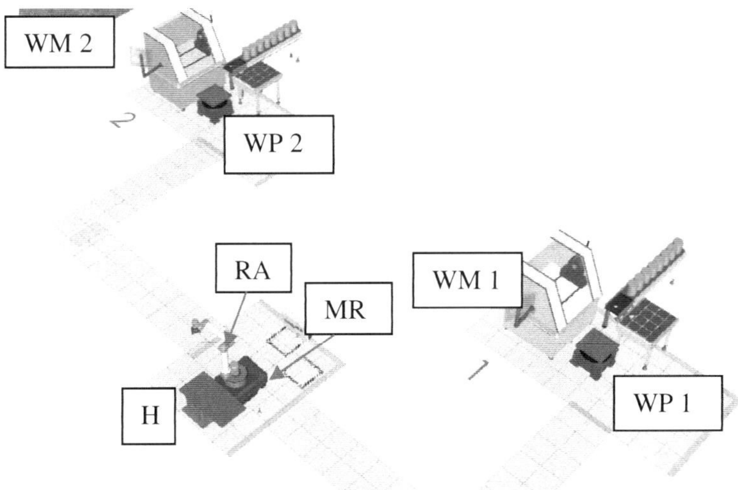

Figure 2. Robot arm (RA) moved by mobile robot (MR) between working positions (WP 1, WP 2). WM 1 and WM 2 are working machines, H – is home position for mobile robot.

2.1. Use case A

- Two workplaces, one robot arm, one mobile robot.
- Lot size 8 pieces (parts that are produced with machines, see Table 1)
- Operation time for machine 1 is 3 minutes per part and for machine 2 is 5 minutes per part.
- All parts are produced first in machine 1 (WM 1) and then goes to machine 2 (WM 2). Robot stays with machine 1 until parts are ready and then moves to machine 2 and stays there until all parts are done.
- The Mobile robot leaves after it has moved the robot arm.

2.2. Use case B

- Two workplaces, one robot arm, one mobile robot.
- Lot size 8 pieces (parts that are produced with machines)
- Operation time for machine 1 is 3 minutes per part and for machine 2 is 5 minutes per part.
- Machine 1 and 2 are working at the same time. Robot arm is moving constantly between machine 1 and machine 2.
- The Mobile robot leaves after it has moved the robot arm.

Data that is collected from use cases are working and hours of mobile robot, robot arm and working machines. In parallel, a mathematical calculation was performed with parameters given in Table 1. for the use cases and compared with the simulation results.

Table 1. Input data for mathematical analyse

Name	Value	Formula unit
Lot size	8 parts	x_1
Time to insert part to WM with RA	15 sec	x_2
Time to take out part from WM with RA	20 sec	x_3
RA positioning time	30 sec	x_4
MR movement time from home(H) to WM1	20 sec	x_5
MR movement time from home(H) to WM2	20 sec	x_6
MR movement time from home WM1 to WM2	50 sec	x_7
Working time of WM1	180 sec	x_8
Working time of WM2	300 sec	x_9

IOP Conf. Series: Materials Science and Engineering 1140 (2021) 012055 doi:10.1088/1757-899X/1140/1/012055

3. Simulation data analysis

The set of data obtained from the Visual Components analysis is presented graphically below (Figs.3-4). Comparing the WM 1 and WM 2 utilization of use case A and B, it can be seen that the utilization of use case A is very low but for use case B it is very high, close to 90%. At the same time, a large decrease in production time can be seen. Production time for user case A is 68 minutes and for use case B it is 44 minutes, i.e. the production time is reduced by approx. 35%.

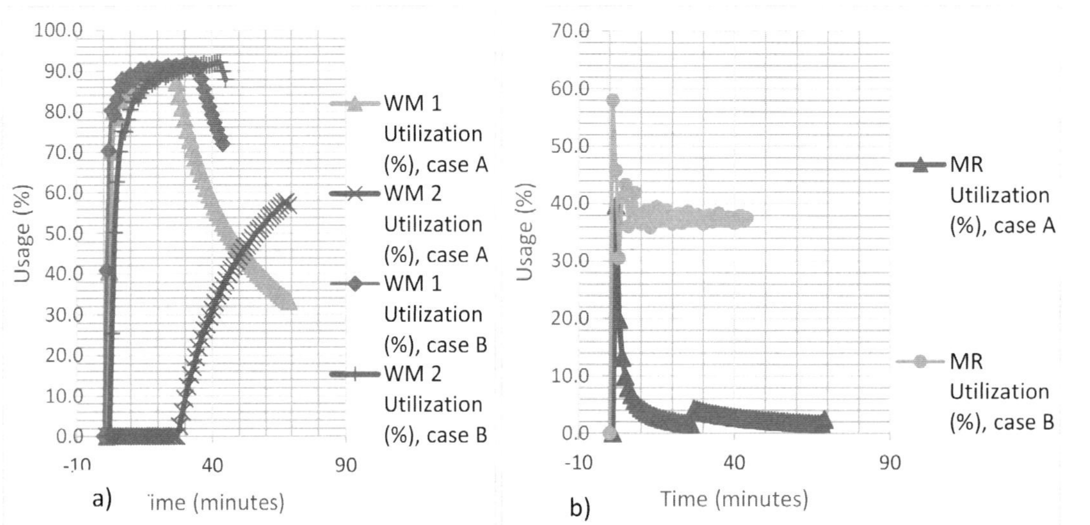

Figure 3. Work machine utilization (a) and mobile robot utilization (b)

In use case A, it can be seen that the determination of the use of a mobile robot is very low. The average utilization of a MR is 3,8%. For use case B the utilization is much higher, it is 37,5%. It is roughly 10x higher.

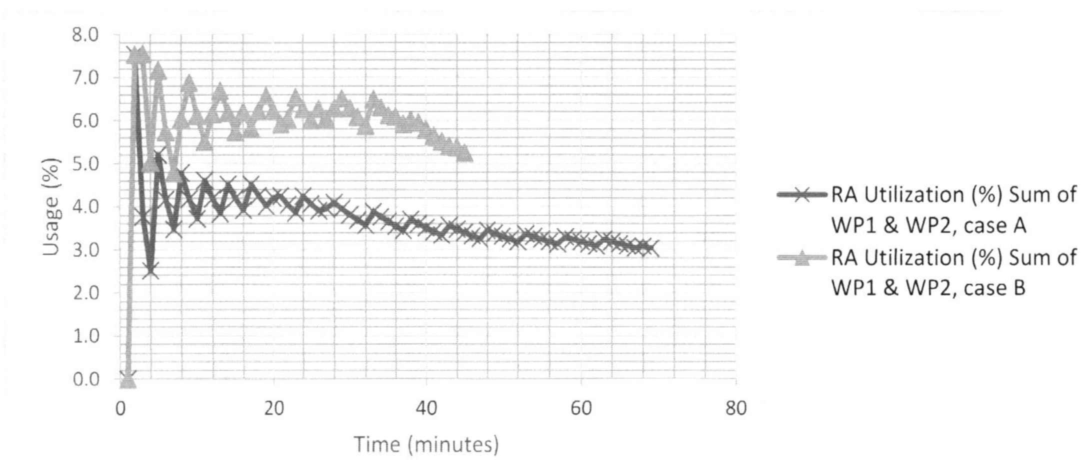

Figure 4. Robot arm utilization

Robot Arm Utilization time is given as sum of both work positions (WP 1 and WP 2). For use case A utilization is low, 3,7 %. This is due to the fact that the RA does not work during the working hours of the WM. In use case B the RA is moved between two WP constantly and utilization of RA is increasing almost 2 times. Average utilization of RA for use case B is 6%.

In addition to the Visual Components simulation, the results were also analysed mathematically. In the case of different use cases, the operating time and idle time of the devices were calculated and calculations of operating time and utilization were made on the basis of them (see Tables 2 and 3).

There are some differences between the simulation and the mathematical computer, because it is not possible to enter certain time parameters for the operation of the RA in the visualization program. A mathematical model gives more accurate results than a simulation. Input data units are in seconds. Output values are given in hours.

$$RA_{wh.A} = \frac{(x_2+x_3)\times x_1+(x_4\times 2)+(x_2+x_3)\times x_1+(x_4\times 2)}{3600} \quad (1)$$

$$RA_{wh.B} = \frac{(x_4\times x_1\times 2+x_2\times x_1+x_3\times x_1)+(x_4\times x_1\times 2+x_2\times x_1+x_3\times x_1)}{3600} \quad (2)$$

$$MR_{wh.A} = \frac{(x_5\times 3+x_7+x_6\times 3)}{3600} \quad (3)$$

$$MR_{wh.B} = \frac{(x_5+(x_5+x_7+x_6)\times(x_1-1)+(x_6+x_7+x_5)\times x_1}{3600} \quad (4)$$

$$Total\ process\ time_A = \frac{(x_8\times x_1+x_9\times x_1)+(x_5+x_4\times 2+x_7+x_4\times 2+x_6)+((x_2+x_3)\times x_1\times x_2)}{3600} \quad (5)$$

$$Total\ process\ time_B = \frac{(x_9\times x_1)+((x_4+x_2+x_3+x_4)\times x_1\times 2}{3600} \quad (6)$$

Average utilization time for RA and MR is calculated as follows:

$$RA_{ut.x} = \frac{RA_{wh.x}\times 100}{Total\ process\ time_x} \qquad MR_{ut.x} = \frac{MR_{wh.x}\times 100}{Total\ process\ time_x} \quad (7)$$

Table 2. Working and idle hours of RA and MR

Use case (x)	RA_{wh}	MR_{wh}	RA_{ph}	MR_{ph}
A	0,19	0,05	1,09	1,07
B	0,42	0,38	0,23	0,78

wh - working hours
ph - pause hours

Table 3. Utilization of RA and MR

Use case (x)	RA_{ut}	MR_{ut}	Total process time
A	14,8%	3,7%	1,28
B	38,8%	34,9%	1,09

ut - utilization, working time from total time (%)

The results of the simulation and mathematical calculation shows that the results are different. The reason is that simulation software does not have option to add additional time for RA adjustment before starting working with WM. Comparing mathematical calculation of use case B with use case A, it can be seen that the time required for production decreases by about 14% and the use time of RA and MR devices increases significantly by about 2 times.

By changing the input parameters of the process, the performance indicators also change. Extending the working time of the WM significantly reduces the total production time for use case B to compare with use case A. By increasing the distance between the working machines, i.e. by increasing the travel path of the MR, the production time as well as the utilization rate of RA and MR decreases for use case B.

4. Conclusion

The use of industrial robots in SMEs has been modest so far, one of the reasons is the lack of work for the robot in one particular location. Moving the robotic arm between working positions would significantly increase its usability and increase the rate of use of the device itself. This article compared two different ways of using a robot, the second of which (use case B) allows the robot arm to move autonomously between different workstations on the factory site with the help of a mobile robot. Autonomous movement of the robot arm between different work positions significantly increases the usability of the robot and increases production efficiency. As a result virtual production line was added to the Industry 4.0 semi-industrial lab, allowing better pre-planning and faster reconfiguration without stopping the real time production.

References

[1] Kangru T, Riives J, Mahmood K and Otto, T, 2019, Suitability analysis of using industrial robots in manufacturing, *Proceedings of the Estonian Academy of Sciences*, vol. 68 (4), pp. 383−388.

[2] Barosz P, Gołda G and Kampa A, 2020, Efficiency Analysis of Manufacturing Line with Industrial Robots and Human Operators. *Applied Sciences*, vol. 10 (8): 2862.

[3] Unger H, Markert T and Müller E, 2018, Evaluation of use cases of autonomous mobile robots in factory environments, *Procedia Manufacturing*, vol. 17, pp. 254-261.

[4] Mahmood K, Lanz M, Toivonen V and Otto T, 2018, A performance evaluation concept for production systems in an SME network, *Procedia CIRP*, vol. 72, 603 – 608.

[5] Kangru T, Riives J, Otto T, Pohlak M and Mahmood K, 2018, Intelligent Decision Making Approach for Performance Evaluation of a robot-based Manufacturing Cell. *Proceedings of the ASME 2018 International Mechanical Engineering Congress and Exposition: IMECE2018*, November 9-15, 2018. Pittsburgh, PA, USA.

[6] Lima F, de Carvalho C, Acardi M, dos Santos E, de Miranda G, Maia R and Massote A, 2019, Digital manufacturing tools in the simulation of collaborative robots: towards Industry 4.0, *Brazilian Journal of Operations & Production Management*, vol. 16 (2), pp. 261-280.

[7] Vaher K, Kangru T, Otto T, Riives J, 2019. The mobility of robotised work cells in manufacturing. *Proceedings of the 30th International DAAAM Symposium "Intelligent Manufacturing & Automation":* 10th International DAAAM Symposium "Intelligent Manufacturing & Automation", 23-26th October 2019, Zadar, Croatia, DAAAM International, Vienna, Austria.

[8] Vaher K, Otto T, Riives J 2020. Positioning error correction of autonomously movable robot arm. Journal of Machine Engineering, 20 (6), 152−160.

[9] Kousi N, Michalos G, Aivaliotis S, Makris S 2018. An outlook on future assembly systems introducing robotic mobile dual arm workers. Procedia CIRP, 72, 33-38.

[10] Nielsen I, et al 2017. A methodology for implementation of mobile robot in adaptive manufacturing environments. J Intell Manuf, 28, 1171–1188.

Chemical Bonded PA66-PTFE-Oil Composites as Novel Tribologically Effective Materials: Part 2

Thanh-Duong Nguyen[1,2,a], Lionel Simo Kamga[3], Michaela Gedan-Smolka[1], Bernd Sauer[3], Stefan Emrich[4], Michael Kopnarski[4] and Brigitte Voit[1, 2]

[1]Leibniz-Institut für Polymerforschung Dresden e. V., Hohe Str. 6, 01069 Dresden, Germany
[2]Technische Universität Dresden, Faculty of Chemistry and Food Chemistry, Dresden, Germany
[3]Inst. of Machine Elements, Gears & Transmissions (MEGT), Technische Universität Kaiserslautern, Gottlieb-Daimler-Straße 42, 67663 Kaiserslautern, Germany
[4]Institute of Surface and Thin Film Analysis (IFOS) Trippstadter Str. 120, Kaiserslautern, Germany.
[a]nguyen@ipfdd.de

Abstract. Polytetrafluoroethylene (PTFE) exhibits excellent non-stick properties and a very low coefficient of friction under tribological stress, but it is incompatible with almost all other polymers. In the first part of this study we presented the generation of the novel tribological material based on unsaturated oil, radiation-modified PTFE (MP1100) and Polyamide 66 (PA66). To get a better understanding of the chemical properties and chemical composition of the compounds, the PA66-MP1100-oil-cb (chemical bonded) compounds were examined by differential scanning calorimetry (DSC) and thermal gravimetric analysis (TGA). In this part, the mechanical properties of the compounds are compared with plain PA66 and PA66-MP1100-cb. The tribological investigation was carried out using the Block-on-Ring tribometer. It was found that the mechanical properties of PA66-MP1100-oil-cb with 20 wt.% MP1100-oil-cb only show slight differences compared to PA66, but the tribological properties of the compounds have been significantly improved through chemical coupling between the three components. Finally, the amount of the compound that was deposited on the surface of the steel disc counterpart was analyzed after the tribological testing.

1. Introduction

PTFE offers excellent tribological properties in a wide temperature range (-250 °C to +260 °C). However, PTFE exhibits poor mechanical properties and has low adhesion to metal surfaces [1]. Contrary to the chemical stability of PTFE, it is highly sensitive to high-energy radiation. Compared with PTFE, polyamide has got very good mechanical properties on one hand, but it has a high coefficient of friction against steel on the other hand. PTFE itself can be modified by using high-energy radiation to obtain persistent perfluoroalkyl (peroxy) radicals and functional groups (-COF and -COOH) [2, 3]. There are two approaches to link radiation modified PTFE to other polymers. One method is the modification via persistent peroxyalkyl radicals and the other route is using the generated functional groups (COOH and COF). The persistent perfluoroalkyl (peroxy) radicals can be used to form the covalent bond with other polymer/monomers via radical addition reaction [4]. The COOH- functional groups of radiation modified PTFE can be used to create a covalent bond with polyamide by transamidation reaction, at the same time the compatibility between radiation modified PTFE and the polyamide can be improved [5]. In this study, four olefinic oil-types were used, which contain at least one C=C double bond and different functional end groups, and as radiation modified PTFE, MP1100 was used. The olefinic oil molecules can be covalently bonded to MP1100 via radical reactions, and the oil functional end groups can generate a variation in affinity to the metal surface. The manufacture and characterization of the compounds based on MP1100, selected oil-types and PA66 were presented in the first part of this study [7]. Here, to illustrate the influence of oil on the mechanical properties of the PA66-MP1100-oil-cb compounds, the mechanical and tribological properties of PA66-MP1100-oil-cb were investigated and compared to the reference materials (PA66-MP1100-cb and origin PA66). For the mechanical testing, the compound was injection molded to multipurpose test specimens according to ISO 3167 standard, and subsequently tensile stress, strain at break, the modulus of elasticity, the tensile strength, and the Charpy notched impact strength were tested as well. For the tribological testing, the material extrudates were processed by injection molding and were manufactured in pins with a width of 10 mm and a working length of 3 mm afterwards. Finally, the tribological behavior of the compounds was investigated by block-on-ring rig test.

2. Experimental and Methods

2.1. *Materials and reactive extrusion*

The materials were processed in two steps. During the first step the four different oil types [Methyl oleate (MO), Oleyl alcohol (OA), Oleic acid (OES), and Oleyl amine (OAMIN)] were chemical bonded to radiation modified PTFE (MP1100, Chemours, US) through solid-phase reactive extrusion by using HAAKE Rheomex PTW 16/25 twin-screw extruder (Thermo Fisher Scientific, Germany). MP1100 is radiation modified PTFE emulsion polymer and was used how it was available. The ratio of MP1100/oil was 10/90 wt%. The maximum processing temperature was 170°C. After the first extrusion, the MP1100-oil-cb compound was cooled, pulverized and then used to couple with PA66 (Technyl A205F, Solvay). Here the MP1100-oil-cb and PA66 were dosed in a ratio 20/80 wt% and processed by Leistritz ZSE 27 Maxx twin-screw extruder (Leistritz Extrusionstechnik GmbH, Germany) in melt. The process temperature was 330 °C. Theoretically, the final sample contains 80 wt% PA66, 18 wt% MP1100 and 2 wt% oil

2.2. *Injection Molding*

In order to determine the mechanical properties and tribological properties of the compounds, multipurpose test specimens (according to ISO 3167 standard method) on one hand, and semi-finished products were prepared by injection molding (Allrounder 420 C, Arburg, Germany) at 350 °C on the other hand.

2.3. *Mechanical testing*

For the mechanical testing, tensile tests were carried out according to the standard testing method DIN EN ISO 527-2/1BA/1-50 (ZwickRoell 1456 universal testing machine, Germany). The notched impact bending test and the Charpy notched impact strength tests were carried out according to the DIN EN ISO 179/1EA standard using a ZwickRoell pendulum impact tester 4J.

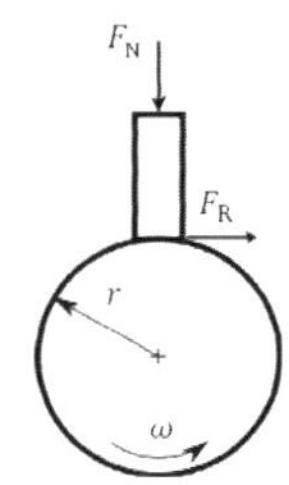

Figure 1: Scheme of tribological testing.

2.4. *Tribological testing.*

Tribological investigations of the compound materials were carried out on a block-on-ring test rig. Figure 1 shows a scheme of the test configuration. For the experiments, rectangular pins with a nominal contact surface of 10 x 3 mm² were manufactured from the respective compounds. The pins were pressed onto the rotating steel disc, whereby a new disc was used for each measurement. The steel disc made of

16MnCr5, which was hardened (60 HRC) and ground in the axial direction at a 3D arithmetical mean roughness of $Sa = 0.2$ µm, has a diameter of 80 mm and a width of 13 mm.
After the pins and the steel discs were manufactured, they were cleaned in an ultrasonic bath (using cyclohexane, isopropanol and acetone) and dried under the ambient condition in the laboratory. The tests were carried out at constant nominal pressure (p = 5 MPa) and constant speed (u= 0.3 m/s) for 5 h (approx. 6 km). The wear of compounds was determined by measuring the mass wear after the experiments.

3. Results and discussion

3.1. Results of mechanical testing.

Figure 2 shows that the compound PA66-MP1100-cb has got a significantly higher E-modulus compared to the original PA66. It is supposed that this behavior is due to the chemical bonding between the components and the very good breaking down and distribution behavior of the MP1100 particles in the PA66 matrix. Furthermore, according to the literature [6], the crystalline proportion of the compounds was increased through the reactive extrusion and injection molding process. The compounds with chemical bonded oil (PA66-MP1100-OA-cb, PA66-MP1100-MO-cb, PA66-MP1100-OES-cb, and PA66-MP1100-OAMIN-cb) generally have got lower elasticity values than original PA66 and the PA66-MP1100-cb. This is in fact due to the two steps extrusion process. In the first extrusion step, it was expected, that the bonding between oil molecules and MP1100 proceeds by radical reactions selectively. However, PA66-MP1100-OAMIN-cb compound showed the lowest modulus of elasticity in contrast to the other types of PA66-MP1100-oil-cb at the end. This is due to the fact that the NH_2-end-group of oleyl amine has reacted with the carboxylic group of the MP1100 during the first extrusion step too, resulting in a much higher coupling degree between OAMIN and MP1100, how it was detected by FT-IR-spectroscopy. During the second extrusion step of with PA66 in melt, there are only less or no -COOH functional groups of MP1100 still available for the chemical reaction that led to a very low bonding of PA66. So, the MP1100-OAMIN-cb is mainly physically distributed in the PA66 matrix.
The reactive compounding of PA66 with 20 wt% of MP1100 has reduced the tensile strength values by approximately 10% in the opposite to the plain PA66. The chemically bonded PA66-MP1100-oil-cb compounds with several oil types show only a slight additional decrease effect of the tensile strength (Figure 2) compared to PA66-MP1100-cb.

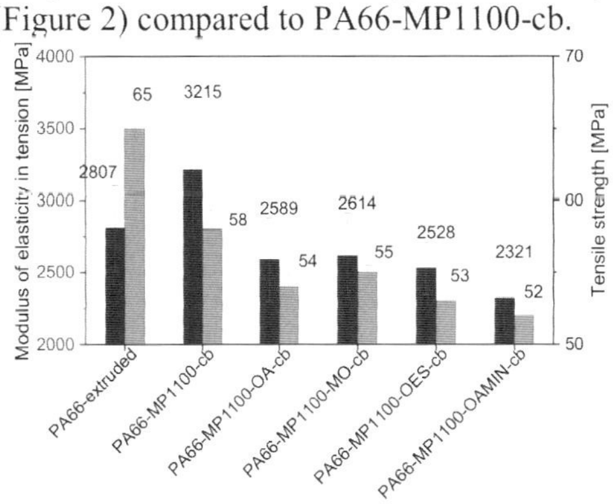

Figure 2: Modulus of elasticity in tension and tensile strength.

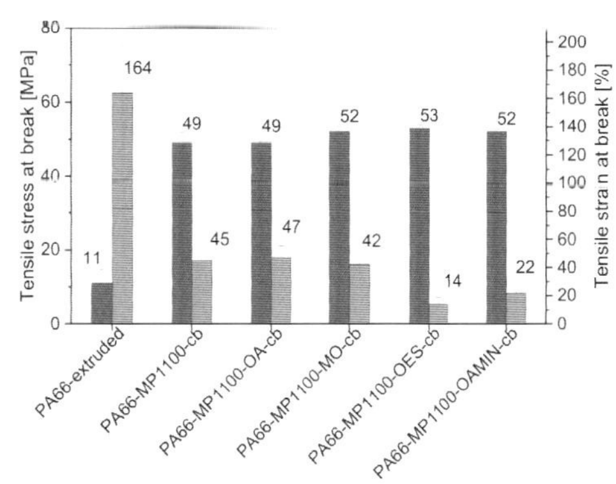

Figure 3: Tensile stress at break and tensile strain at break.

Figure 3 shows the result of tensile stress at break. It is obvious that in the case of PA66-MP1100-cb, the value is four times higher than the value of plain PA66. The samples with chemical bonded oil (PA66-MP1100-OA-cb, PA66-MP1100-MO-cb, PA66-MP1100-OES-cb and PA66-MP1100-OAMIN-cb) did not change the tensile stress at break compared to PA66-MP1100-cb. It is assumed that the oil content (2 wt%) is too low compared to PA66 and MP1100 (80 wt% and 18 wt%) to get a further significant effect. On the other hand, the values of tensile strain at break have drastically decreased (approximately 70%) when PA66 was compatibilized with MP1100 (approximately 70%). Concerning the various oil types, it was found that for PA66-MP1100-OA-cb and PA66-MP1100-MO-cb, there was almost no sign of any influence of chemical bonded oil on the tensile strain at break. Hypothetically it can be assumed that there is a transamidation reaction between excess Oleic acid and PA66, so that the probability that MP1100 reacts with PA66 could be reduced. The result is that the compatibility between MP1100 and PA66 is lower with the presence of Oleic acid, so the tensile strain at break is reduced compared to the other oil-modified compounds.

In case of Charpy notched impact strength (Figure 4), the compound PA66-MP1100-cb has the highest value, and this is approx. 40% more than the value of plain PA66. As realized before in other combinations this behavior should be the result of chemical bonding as well as the much better breaking down of MP1100 aggregates and distribution of MP1100 particles in the PA66 matrix. The Charpy notched impact strength of the other compounds in the presence of more or less bonded oil are lower than those of PA66-MP1100-cb as the oil interferes with the compatibilization between PA66 and MP1100. The Charpy notched impact strength values decrease in the following order: PA66-MP1100-cb>PA66-MP1100-OA-cb>PA66-MP1100-MO-cb>PA66-MP1100-OAMIN-cb>PA66-MP1100-OES-cb.

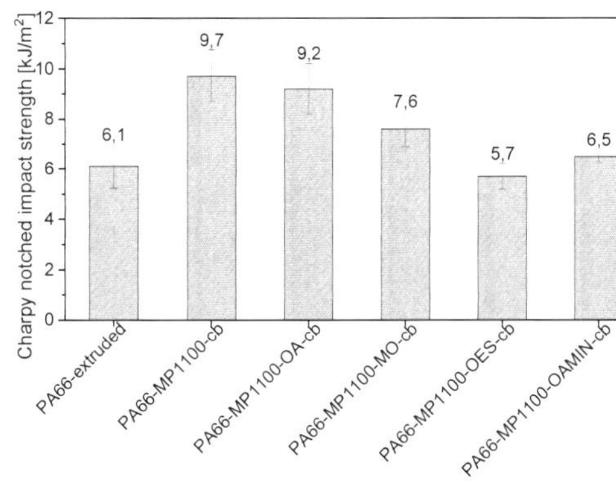

Figure 4: Charpy notched impact strength.

3.2. *Result of tribological testing*

Figure 5 shows the curve of the coefficient of friction (COF) and the disc temperature (ϑ_{Steel}) of PA66-MP1100-OES-cb in contact with the steel disc, as an example for the PA66-MP1100-oil-cb materials. The temperature was measured approximately 2 mm below the contact surface, which means that the recorded temperature is lower than the current contact temperature. For each compounds, three measurements were carried out. It can be seen that up to a sliding distance of approx. 1 km, the coefficient of friction and the temperature increase rapidly in the running-in phase. Afterwards, the coefficient of friction gradually decreases again and reaches a more or less constant value after a sliding distance of approx. 4 km. This behavior of the coefficient of friction can be attributed to the temperature dependency of the material parameters of the compounds, which influences the adhesive and the deformation parts of the friction forces.

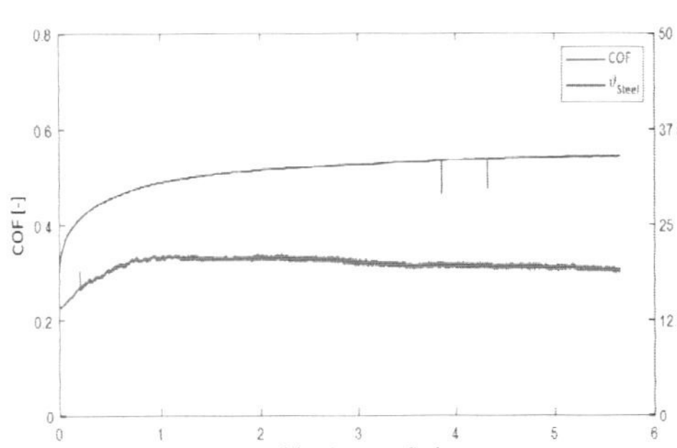

Figure 5: Characteristic behavior of the coefficient of friction and the temperature of PA66-MP1100-OES-cb

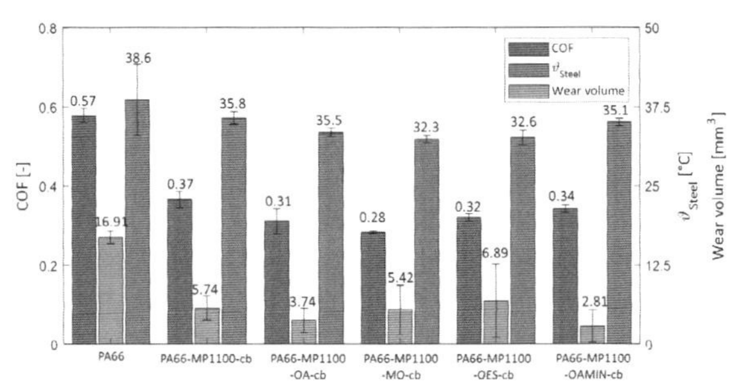

Figure 6: Influence of oil type on the friction behaviour of the PA66-MP1100-oil-cb compounds

Figure 6 shows the influence of four oil types on the friction behavior of PA66-MP1100-oil-cb and the heating of the counter body. The values of the coefficient of friction and temperature shown below correspond to the mean value of the coefficient of friction from a running distance of 5 km upwards, see Figure 5. Additionally, Figure 6 shows the volume wear of the compounds, calculated with the determined loss of mass at the end of the respective tests using the determined density of the compounds from Table 1. It can be seen that in the load range investigated, the oil type methyl oleate shows a more favorable friction behavior in contact with the steel disc.

Table 1: Density of the compounds

Compound	PA66	PA66-MP1100-cb	PA66-MP1100-OA-cb	PA66-MP1100-MO-cb	PA66-MP1100-MO-cb	PA66-MP1100-OAMIN-cb
Density (g/cm^3)	1,1355	1,2470	1,2435	1,2519	1,2472	1,2515

4. Summary and Outlook

Although only 2 wt% of oil is chemically bonded in PA66-MP1100-oil-cb, its effect is visible in their mechanical properties. The elasticity of the compound with chemically bonded oil is generally lower compared to PA66-MP1100-cb, especially for the compound with oleyl amine (reduce approx. 28%). The Charpy notched impact strength values of the compounds decreases in the following order:
PA66-MP1100-cb>PA66-MP1100-OA-cb>PA66-MP1100-MO-cb>PA66-MP1100-AMIN-cb>PA66-MP1100-OES-cb.
The friction behavior of chemically bonded PA66-MP1100-oil is generally better than for PA66-MP1100-cb, among them the compound containing methyl oleate shows the best result.
In the future works:

- We strive for single-step reactive extrusion for the producing of the compounds.
- We use other oil types containing phosphate functional groups to increase the affinity of the compound to the metal surface.
- The influence of the transferred tribo-film on the steel disc will be investigated in the steel/bronze contact.

5. Acknowledgements

The authors thank the Deutsche Forschungsgemeinschaft (DFG) for financial support within the Project SPP 2074, T. Schmidt for carrying out lots of experiments, D. Pahlitzsch for the extrusion, A. Scholze and C. Lehmann for the injection molding and H. Scheibner for the mechanical testing.

References

[1] Kaiser, W., *Kunststoffchemie für Ingenieure* (Karl Hanser Verlag GmbH & Co. KG: Germany, 2011)

[2] Lunkwitz, K.; Lappan, U.; Lehmann, D., *Modification of fluoropolymers by means of electron beam irradiation.* (Radiation Physics and Chemistry **2000),** *57* (3), 373-376.

[3] Lehmann, D. German Patent DE102014225672A (2016).

[4] Lehmann, D. German Patent DE102014225671. (2016).

[5] Pompe, G.; Häußler, L.; Adam, G.; Eichhorn, K.-J.; Janke, A.; Hupfer, B.; Lehmann, D., *Reactive polytetrafluoroethylene/polyamide 6 compounds. II. Study of the reactivity with respect to the functionality of the polytetrafluoroethylene component and analysis of the notched impact strength of the polytetrafluoroethylene/polyamide 6 compounds* (Journal of Applied Polymer Science **2005)** *98* (3), 1317-1324.

[6] Fornes, T.; Paul, D., *Crystallization behavior of nylon 6 nanocomposites.* (Polymer **2003)** *44*, 3945-3961.

[7] T-D. Nguyen, M. Gedan-Smolka, L. Simo Kamga, B. Sauer, S. Emrich, M. Kopnarski, B. Voit, *Chemical Bonded Oil-PTFE-PA66-cb Composites as Novel Tribologically Effective Materials: Part 1* (Solid State Phenomena, **2021**, ISSN 1662-9779) Vol 320, pp113-118

The effect of laser beam transmission on the thermoplastic composite joint formation

Simonas Mindaugas Jankus[1], Regita Bendikienė[1]

[1]Kaunas University of Technology, Faculty of Mechanical Engineering and Design, Department of Production Engineering, Studentų str. 56, Kaunas, Lithuania

simonas.jankus@ktu.edu, regita.bendikiene@ktu.lt

Abstract. The goal of this work was to analyse the influence of laser welding process parameters using different plastic transparency. To achieve this task the elemental composition of plastic PPAGF40 with 20 % and 30 % transparency parts were examined using EDS analysis and cross - sections were evaluated by using optical microscope. Burst pressure was performed to evaluate the strenght of welded joints produced with different laser welding process parameters. Studies showed that welded joints with 20 % of transparency possessed burn spots on the surface of samples. Cross - section analysis revealed tendency, that plastics with 30 % transparency led to higher temperatures in welding seam using the same process parameters used for 20 % transparency, which caused pores to form inside joint. Laser power of 390 W, scan - speed 1000/800 mm/s and 3000 N of clamping force used for 20 % transparency plastic parts are not suitable for 30 % transparency plastic parts. Welding process optimization achieved by using 1400/1200 mm/s scan - speed and 2700 - 3000N clamping force, which showed improved burst pressure values from 7.2 to 7.9 bar. Reduced welding cycle - time achieved from 2.2 to 2.3 s with 30 % transparency of polymer PPAG40 material.

1. Introduction

Laser welding of plastics is performed in the overlap process. Usually two plastics to be joined are used. The upper part is a laser transparent thermoplastic, selected according to the laser wavelength, which partially heats up on the passage of the laser beam [1]. For a weld seam to be produced, the second join part must absorb the laser radiation (figure 1). When lower substance absorbs energy it begins to fuse and transmits it's energy to the upper partner [2, 3]. The quality of welded parts is mainly influenced by interaction between concentrated laser radiation and plastic material. Polymer optical transparency is crutial for weld joint mechanical performance [4]. The degree of difficulty in laser-welding depends on the laser transparency of the upper and the laser absorption of the lower of the two join partners. The better the upper part transmits the laser energy and the better the housing absorbs energy from the laser beam [5, 6].

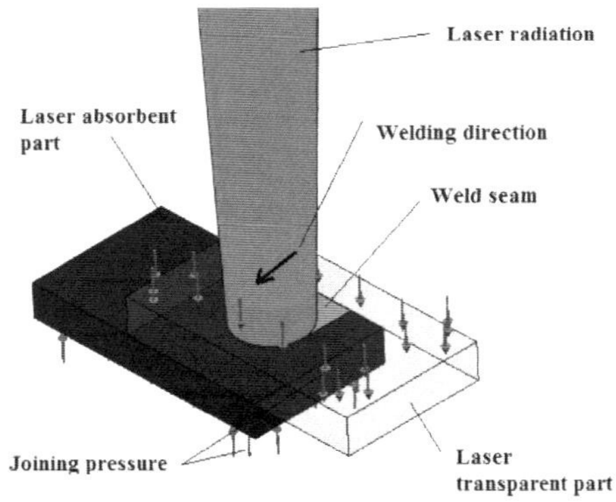

Figure 1. Basic representation of the laser transmission welding method [7].

In order to achieve absorption of laser radiation by the lower joining partner, carbon black is often used as an absorber [8]. Low upper part transparency, typically from 10 % to 20 %, is achieved by adding pigment additive black carbon 58 - 60 wt. % [9]. Studies have shown that higher transparency of upper part, obtained by reducing black carbon concentration to 50 - 55 wt. %, can led to higher temperatures and energy concentrations in the connection between cover and housing [10, 11]. This might lead to pores in the welding seam after welding process are reducing strenght of joint [12]. Therefore, it's important to understand what are the possibilities of the use of higher upper part transparency.

2. Materials and Methods

Two groups of samples were produced by using quasi-simultaneous laser welding method (LPKF InlineWeld 6600 welding equipment). For the first sample group, 20 % of transparency for the laser beam plastics were used (named sample 1). Second group of samples were welded using 30 % of transparency plastics (sample 2). Laser absorbent parts (sample 3) were used in both groups of samples. Polymer pairs sample 1 - sample 3 were welded using parameter set 1 and pairs sample 2 - sample 3 using all 5 welding process parameter sets (see Table 1). Laser source operates in cw (continuous-wave) mode, wavelengh $\lambda = 980$ nm with laser beam spot diameter in focus 2.5 mm. A total 72 mechanical parts were welded by using polyphthalamide PPAGF40 material. Dimensions of samples are $64.2 \times 118.7 \times 112.95$ mm with a welding seam thickness of 1.4 ± 0.1 mm and width of welding seam 1.5 ± 0.1 mm. For determination of burst pressure 60 parts were used. Ten pairs of sample 1 - sample 3 using parameter set 1 and 50 pairs of sample 2 - sample 3 (10 of each process parameter set). For cross - section evaluation of weld seam 12 samples were used by using digital microscope Ash Technologies INSPEX II. Burst pressure was perfomed to evaluate strengh of joints using pressure measurement equipment from ZEUS (measured in bar). Upper part transparency of 32 samples the laser radiation measured by Softing LTW-1. Welding depth distance measured by Keyence GT2-H12. EDS analysis performed for determination of elemental composition using Bruker Quantax XFlash® 6|10. SEM micro-graph obtained by Carl Zeiss EVO MA10.

Table 1. Parameter sets used for cover - housing pairs sample 1 - sample 3 and sample 2 - sample 3 laser welding tests.

Process parameter sets	Laser power [W]	Scanning speed [mm/s]	Joint cooling time [s]	Clamping force [N]	Welding depth [mm]
Set 1	390	1000/800	3.5	3000	0.3
Set 2	390	1400/1200	3.5	2700	0.3
Set 3	390	1400/1200	3.5	3000	0.3
Set 4	380	1400/1200	3.5	3000	0.3
Set 5	420	1000/800	3.5	3000	0.3

3. Results and Discussion

3.1. Plastic material PPAGF40 transparency for the laser beam measurement

For thermoplastics optical transparency for laser beam is determined by transparency measurements. Purpose of these measurements is to determine how much light in % can pass through sample 1 and sample 2 polymer PPAGF40 material. Sample 1 covers transparency for laser light 20.85 - 21.97 %, sample 2 covers transparency varying from 31.34 to 33.60 %.

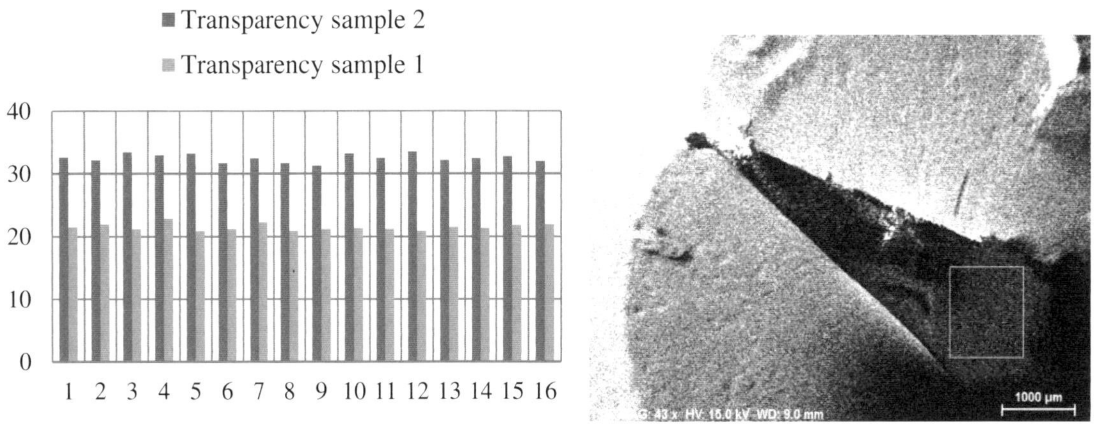

Figure 2. Sample 1 and sample 2 cover tansparency for the laser beam measurements on the left and SEM micro-graph of PPAGF40 sample on the right.

3.2. EDS chemical analysis of PPAGF40

The chemical elements of sample 1, sample 2, sample 3 were examined by usig EDS analysis. Depth of elemental measure 1 µm. Lowest amount of carbon 55.88 wt. % was found in covers sample 2 which has highest transparency of 30 %. In comparison to sample 2, sample 1 covers due to lower transparency has 56.53 wt. % of carbon. Laser light absorbing part 3 carbon concentration 58.07 wt. %.

Table 2. EDS elemental analysis of composition of the transparent covers sample 1, sample 2 and absorbing housing sample 3 [wt. %].

Chemical element	sample 1	sample 2	sample 3
C	56.53	55.88	58.07
N	19.73	22.55	23.22
O	22.53	20.29	17.46
Al	0.86	1.19	0.71
Si	0.03	0.06	0.49
Na	0.06	0.03	0.04
Ca	0.26	-	0.01

3.3. Effect of transparency for the laser beam on the plastic join formation

Cross - section studies of weld seam have been conducted to determine the structure of joint after laser beam welding. Polymer PPAGF40 pairs sample 1 - sample 3 and sample 2 - sample 3 with different cover transparency were used. First welding test was performed using 390 W laser power and 1000/800 mm/s scan - speed with sample 1 cover and sample 3 housing pair. Burn marks due to low transparency observed at the plastic surface, which is caused by 56.53 wt. % carbon concentration (figure 3 a). Higher temperature concentration due to higher transparency caused partial plastic evaporation and formation of pores inside collapsed cover and housing using sample 2 covers with process parameter set 1 and set 2 [8]. Pores covered around 50 % (figure 3 b) and 35 % of welding seam (figure 3 c).

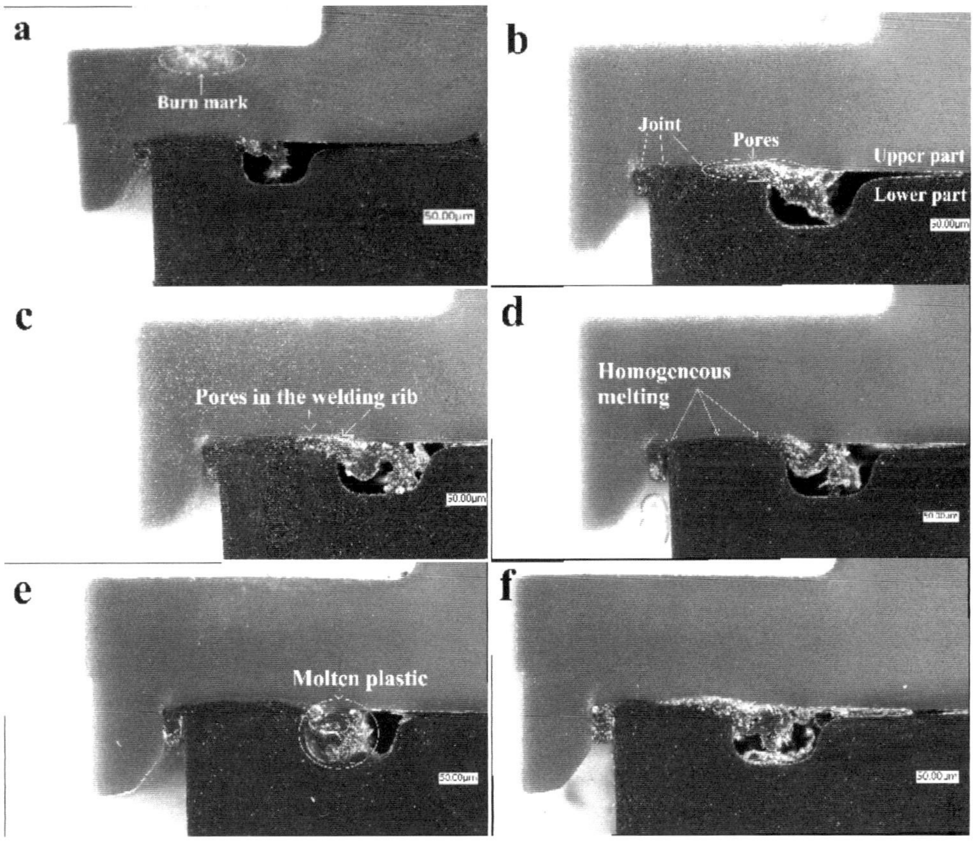

Figure 3. Weld seam cross - section view: a - sample 1 and b - sample 2 covers with parameter set 1; c - sample 2 covers with set 2 and d - sample 2 set 3; e and f - sample 2 set 4 and set 5.

No defects were observed in pair sample 2 - sample 3 cross - section area by using parameter sets 3 and 4 (see figure 3 d and e). Increased scanning speed of the laser beam in set 3 to 1400/1200 mm/s and reduced laser power by 10 W to 280 W in set 4 influenced lower energy input from radiation per time than compared to 1000/800 mm/s scan-speed 290 W laser power [11]. More homogeneous melting of the part is achieved. No burn marks, pores or vapours are generated inside welded joint after cooling process by using sample 2 covers with set 3 and set 4. Futhermore, increasing laser power up to 420 W it is observed in cross - section area that 70 - 75 % of welding seam is covered with pores (figure 3 f).

3.4. Welding process improvement by using sample 2 covers

Burst pressure test was performed for strength evaluation of polymer joints. It is observed that with welding process parameter sets 3 and 4 highest values of burst pressure were obtained from 7.2 to 7.9 bar (figure 4). Homogeneous weld seam with no defects formed (see figure 3 d and e). Lower heating time range 4.1 - 4.5 s using sample 2 covers achieved in comparison with 20 % transparency covers sample 1 from 5.4 s to 5.7 s. Joints with lowest heating time varying from 3 to 3.3 s possessed lowest burst pressure values, respectively 2.38 - 2.80 bar. Pores (figure 3 f) weakened welded plastic parts burst pressure.

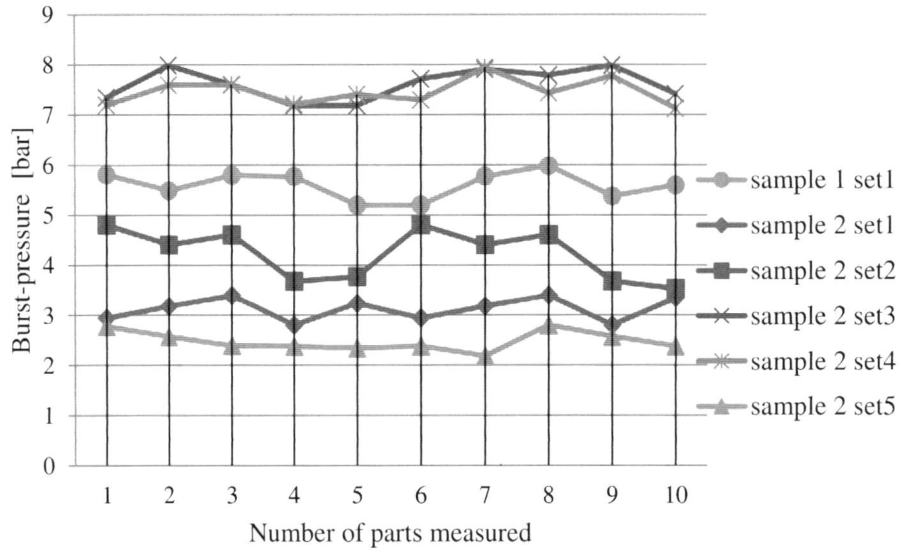

Figure 4. Polymer PPAGF40 burst pressure values using different transparency cover/housing pairss sample 1 - sample 3 and sample 2 - sample 3.

4. Summary

Transparency for the laser beam measurements performed for plastic parts. EDS elemental composition of samples used for welding trials was determined.

Increasing upper part transparency from 20 % to 30 % increases heat concentration on the welding seam locally. Welding parameters 390 W - 380 W of laser power and 1400/1200 mm/s scan - speed for 30 % transparency plastics were the most suitable. No defects were found in weld seam during cross - section analysis. Obtained burst pressure values are varying from 7.2 to 7.9 bar. Cross - section analysis revealed that using 420 W of laser power, 1000/800 mm/s scan - speed and 2700 N clamping force for 30 % transparency plastics, pores formed inside joint. As a result, burst pressure values are varying from 2.38 to 4.80 bar. Same laser welding process parameters can't be adapted for plastic parts, which have different transparency.

Welding operation time was reduced by 2.2 - 2.3 s with 30 % transparency plastics. Obtained research results of welding process optimisation are already applied in bulk production.

References

[1] Hopmann C, Bölle S, Reithmayr L 2020 Prediction of the Bond Strength of Thermoplastics Welded by Laser Transmission Welding *Advances in Polymer Processing* 247-257.

[2] Asséko A C A, B. Cosson, Deleglise M, Schmidt F, Le Maoult Y, Lafranche E 2015 Analytical and numerical modeling of light scattering in composite transmission laser welding process *International Journal of Material Forming* **8** 127–135.

[3] Acherjee B, Kuar A S, Mitra S, Misra D 2015 Empirical Modeling and Multi-Response Optimization of Laser Transmission Welding of Polycarbonate to ABS *Lasers Manuf. Mater. Process* **2** 103-123.

[4] Brosda M, Nguyen P, Olowinsky A, Gillner A 2020 Investigations on the Influence of Beam Shaping in Laser Transmission Welding of Multi-layer Polymer Films with Wavelength-Adapted Diode Laser Beam Sources *Advances in Polymer Processing* 91-100.

[5] Aden M 2016 Influence of the Laser-Beam Distribution on the Seam Dimensions for Laser-Transmission Welding: A Simulative Approach *Lasers Manuf. Mater. Process* **3** 100-110.

[6] Singh A, Pfleging W, Beiser M, Malek C K 2019 Transparent thin thermoplastic biochip by

injection-moulding and laser transmission welding *Microsyst Technol* **19** 445-453.

[7] Jankus S M, Bendikienė R 2021 The Influence of Heating Time on the Mechanical Properties of Laser Through-Transmission Welded Polyamide PPA40GF Joint *Solid State Phenomena* **320** 139-143.

[8] Acherjee B, Kuar A S, Mitra S, D Misra 2012 Effect of carbon black on temperature field and weld profile during laser transmission welding of polymers: A FEM study *Optics & Laser Technology* **44(3)** 514-521.

[9] Aden M, Mamuschkin V, Mamuschkin A 2015 Influence of carbon black and indium tin oxide absorber particles on laser transmission welding *Optics & Laser Technology* **69** 87-91.

[10] Grewell D, Rooney P, Kagan Val A 2004 Relationship between Optical Properties and Optimized Processing Parameters for through-Transmission Laser Welding of Thermoplastics *Journal of Reinforced Plastics and Composites* 23(3) 239-247.

[11] Hadriche I, Ghorbel E, Masmoudi N, Casalino G 2010 Investigation on the effects of laser power and scanning speed on polypropylene diode transmission welds *Int J Adv Manuf Technol.* **50** 217-226.

[12] Sari F, Hoffmann W M, Haberstroh E, Poprawe R 2008 Applications of laser transmission processes for the joining of plastics, silicon and glass micro parts *Microsyst. Technol.* **14** 1879-1886.

AUTHOR INDEX